ECONOMIC EFFECTS OF BIOFUEL PRODUCTION

Edited by **Marco Aurélio dos Santos Bernardes**

Economic Effects of Biofuel Production
Edited by Marco Aurélio dos Santos Bernardes

Published by InTech
Janeza Trdine 9, 51000 Rijeka, Croatia

Publishing Process Manager Petra Zobic

Technical Editor Teodora Smiljanic

Cover Designer InTech Design Team

Additional hard copies can be obtained from orders@intechopen.com

Economic Effects of Biofuel Production, Edited by Marco Aurélio dos Santos Bernardes
p. cm.
ISBN 978-953-307-178-7

Meet the editor

Dr.-Ing. Marco Aurélio dos Santos Bernardes serves as a postdoc researcher at the Centre de Recherche Public Henri Tudor in Luxembourg. His expertise is in the area of energy analysis, life cycle assessment, renewable energy and biofuels. Dr.-Ing. Bernardes has had 10 papers published in journals such as Solar Energy, International Journal of Life Cycle Assessment, ASME Heat Transfer, a book and book chapters as well. Dr.-Ing. Bernardes' areas of interest include CFD, heat transfer modelling, Solar Chimney Power Plants, thermal processes, thermodynamics. He received his Ph.D. in Mechanical Engineering at Stuttgart University in Germany and conducted a postdoctoral research at the Stellenbosch University in South Africa. He was awarded with the UNEP/SETAC Life Cycle Assessment Award for LCA Projects in Development Countries. Dr.-Ing. Bernardes served as a full professor in the Department of Mechanical Engineering at CEFET-MG in Belo Horizonte for more than 13 years.

Contents

Preface

Over the past 20 years, there has been a substantial increase in research and development in the area of biofuels. Many researchers around the world have dealt with environmental, economic, policy and technical aspects relating to these studies. In a way, this book aspires to be a comprehensive summary of current biofuel issues and thereby contribute to the understanding of this important topic. Chapters include digests on the development efforts on biofuels, their implications for the food industry, current and future biofuel crops, the successful Brazilian ethanol program, insights of the first, second, third and fourth biofuel generations, advanced biofuel production techniques, related waste treatment, emissions and environmental impacts, water consumption, produced allergens and toxins.

Relating theoretical and experimental analyses with many important applied purposes of current relevance will make this book extremely useful for researchers, scientists, engineers and graduate students, who can make use of the experimental and theoretical investigations, assessment and enhancement techniques described in this multidisciplinary field. Additionally, the biofuel policy discussion is expected to be continuing in the foreseeable future and reading of the biofuel features dealt with in this book is recommended for anyone interested in understanding this diverse and developing theme.

Dr.-Ing. Marco Aurélio dos Santos Bernardes
Researcher Energy & Environment and LCA
CRP Henri Tudor, CRTE
Luxembourg

Part 1

Crop Insurance

1

Implications for the Feed Industry

Paul A. Iji and Mohammad Reza Barekatain
University of New England
Australia

1. Introduction

The animal feed industry relies on cereal grains and pulses to supply energy and protein, respectively. Increasing amounts of both groups of ingredients, but in particular, cereal grains, are being used for the production of ethanol for biofuel. Currently, about a third of the maize crop produced in the United States is used for ethanol production and will rise to about 43 % by 2015 (van der Aar and Doppenberg, 2009). Although limited in impact, a considerable amount of oils produced from oilseeds such as canola, soybean, peanut and sunflower is being processed into biodiesel. This is causing a major strain in the supply of edible oil for feed manufacturing. An indirect effect of the increased use of maize for ethanol production is the change in land use, whereby, farmers in North America are converting land previously used for soybean production into maize production (Anon., 2011a). Although maize is the main cereal grain used by the ethanol industry, it is by no means the only grain used but plants in Canada and Europe tend to use more wheat while the two main plants currently in production in Australia and a few in the USA rely on sorghum.

Regardless of the grain type used, the ethanol industry generates a large amount of waste, principally in form of distillers' dried grains with soluble (DDGS). It is estimated that a kilogram of maize grain generates three equal parts of ethanol, DDGS and carbon dioxide (Saunders and Rosentrater, 2009). Although the two by-products, DDGS and CO_2, represent value-adding to the primary grain resource, the ethanol industry is faced with need to dispose of large volumes of DDGS. However, as more grains are used for producing ethanol, less is becoming available to the feed industry. Distillers' dried grains have become a feed resource, first explored for ruminant animal and pig feeding but lately for poultry too.

The composition of DDGS is highly variable, depending on such factors as the base grain used, the age of the manufacturing plant, the distillation process, and the preparation of the final product, especially drying and packaging (Cozannet et al., 2009; Meyer et al., 2010; Cozannet et al., 2010c). Variability in quality and nutrient composition is a major problem militating against the use of DDGS for animal feeding. Although the ethanol and feed industries both agree to the terminology of DDGS, as the residue resulting from fermentation and distillation, there are many types of DDGS.

Ethanol is also made from sugarcane and similar ingredients with readily fermentable sugars. This yields a fibrous residue known as bagasse that can be fed to ruminant animals. As with grain residue, there is great variability in the quality of bagasse according to the raw material (crop) used, age of crop and method of processing during juice extraction. Biodiesel production is the other side of the biofuel industry, resulting into the production of glycerol and press cakes as by-products.

This chapter examines the range of by-products of the biofuel industry that can be used for animal feeding; response of animals to diets containing these products; limitations to their use, and potential for increased utilization of such products.

2. By-products of the biofuel industries

Distillers' dried grains are the most important by-product of the biofuel industry for the feed industry. Distillers' grains arise from fermentation of cereal grains, followed by distillation of ethanol. The wet product could be marketed as such (wet distillers' grains, WDG) or is dried to different levels of moisture content. Some soluble material may be returned to the waste grain prior to drying, giving rise to the term, DDGS. There are plants, which use grains to produce glucose syrup rather than ethanol. The process also yields DDGS but there may also be low quality syrup, which can be mixed with the waste residue prior to drying, to form a cake.

The biodiesel industry yields only one main by-product, glycerol, which is a mixture of glycerine, impurities and any remaining alcohol plus the catalyst, usually sodium hydroxide, that is used in the process. The press cake that is derived from the oil extraction process is also different from regular meals, as has been borne out by tests conducted on pigs (McKinnon and Walker, 2009). The biofuel industry in Brazil and Australia relies more on sugarcane than on cereal grains. The fermentation of cane sugar to ethanol yields a fibrous material, bagasse, which is mainly used for feeding ruminant animals, and will be further discussed in this Chapter. The composition of key by-products of the biofuel industry that may be useful as feed is shown in Table 1.

Item (% as-fed basis)	DDGS	WDGS	HP-DDGS	Rapeseed cake
Dry matter	86.7	45	91.7	90.4
Crude protein	26.9	30.2	39.6	32.1
Ether extract	13.3	14.2	3.6	18
Starch	7.65	3.65	11.2	0
NDF	30.2	30.8	22.2	27.7
ADF	13.1	20.2	11.2	19.7
Calcium	0.01	0.03	0.02	NI
Phosphorous	1.18	0.87	0.44	NI
Sulfur	0.15	0.72	0.81	NI
Reference	(Kelzer et al., 2010)	(Kelzer et al., 2010)	(Babcock et al., 2008)	(Schöne et al., 1996)

NI: Not indicated; HP-DDGS: high-protein DDGS.

Table 1. Nutrients composition of some biofuel by-products

2.1 Cereal by-products

As has been highlighted above, DDGS are the main by-products from the fermentation of cereal grains and distillation of ethanol. In terms of yield, one kg of grain would yield about a third of its weight in ethanol, another third as DDGS and a third will be lost as carbon dioxide (Saunders and Rosentrater, 2009). Generally about 0.84 kg maize DDGS would result from the production of one litre of ethanol (Kim *et al.*, 2010b). The residue would be dried as such or some of the soluble material that is obtained is put back into it, resulting in changes in colour, moisture and nutrient composition. The drying process greatly influences the quality of the final product. Heating induces the Maillard reaction, which results in binding of sugars to lysine, with the result that the latter becomes unavailable. Several feeding trials have shown that lysine is the least available amino acid in DDGS (Jacela *et al.*, 2010b; Cozannet *et al.*, 2010; Yang *et al.*, 2010). Drying also changes the colour of DDGS, creating a range of light to dark products. Some DDGS are overheated and become dark as a result. Generally, light DDGS have a higher nutritive value than dark ones (Cozannet *et al.*, 2009; Cozannet *et al.*, 2010b) but Seabolt *et al.* (2010) have shown that pigs prefer the dark DDGS to light products and tended to lose weight when placed on diets containing light DDGS. For poultry Cozannet *et al.* (2010c) have reported a strong relationship between lightness (luminance) score and lysine:CP ratio and also lysine digestibility.

Production plants may also influence the physical quality of the DDGS. Major differences have been observed in the quality of maize DDGS from old and new plants in North America. In general, the technology involved in ethanol production continues to improve, so that the quality of DDGS is continuously improving. This will be the driver to increased utilisation of DDGS by the poultry and pig industries. In a comparative trial on DDGS from two plants, one old and the other new, Hastad *et al.* (2003) reported metabolizable energy values of 13.1 and 14.6 MJ/kg, respectively for the two products when fed to pigs. In another study involving ten plants that were less than 5 years old, Spiehs *et al.* (2002) reported wide variations in the composition of DDGS originating from the plants. The mean content of crude protein was 30.2 %, with a coefficient of variation (CV) of 6.4 %; while crude fat and crude fibre contents were (mean, CV): 10.9, 7.8 and 8.8, 8.7 %, respectively. Lysine was the most variable of the amino acids, with a CV of 17.3 % while methionine had a CV of 13.6 %. There may also be differences in composition of DDGS from different crops although Yang *et al.* (2010) reported that there were no differences in ileal digestibility of CP between DDGS from maize or wheat or combinations of these two grains when fed to pigs but values were smaller for the wheat DDGS than the other two products. Lysine and threonine were the least digestible. Ileal digesta viscosity was greater on the diet containing the mixed DDGS than single grain DDGS. Distillers' dried grains are known to compact during storage, creating a flow problem. The degree of compaction is related to moisture content after drying, handling, container filling method, filling height and container size (Clementson *et al.*, 2010; Clementson and Ileleji, 2010).

2.2 Sugarcane by-products

Bagasse is the main by-product of the biofuel industry when sugarcane or similar ingredients are used as source of carbohydrates for ethanol. It is the fibrous material that is left after the stalks are crushed and the juice is extracted (Anon., 2011b). Bagasse contains 45-55 % cellulose, making it suitable for further processing for the extraction of ethanol. It also contains 20-25 % hemicellulose, 18-24 % lignin, 1-4 % ash and about 1 % wax. In Brazil, the canes are milled rather than crushed, to maximise juice extraction and reduce the moisture content of the resulting bagasse, and to increase its value as a fuel. The juice is filtered, pasteurised and evaporated to produce a syrup. The syrup is then centrifuged to form sugar crystals and molasses, another by-product of the sugar-from-sugarcane industries. Molasses is also rich in carbohydrates and can be used as a source of energy and a sweetener in animal feeding.

2.3 Biodiesel by-products

Biodiesel production begins with the production of oil from oil-rich ingredients such as oilseeds, certain pulses and cereal grains. Oilseeds like canola, peanuts, sunflower and soybean are the ingredients, which when used will create the most strain in feed supply. The oil extraction process yields a press cake that can be used for animal feeding (McKinnon and Walker, 2009). The oil is subsequently used to produce biodiesel, yielding glycerol as the main by-product (Anon, undated). The production of biodiesel therefore creates a strain to feed supply at two points - first at the point of supply of full-fat protein meals and the second at the point following the catalytic conversion of edible oil into biodiesel and glycerol. The last process creates a shortage of oil for animal feeding, oil being essential in most non-ruminant diets as a source of energy, particle binder and sweetener.

3. Use of biofuel by-products for animal feeding

There is no doubt that the biofuel industry will create some feed shortage for the animal industry. Some of the effects are direct while there are also indirect effects. More grains are being used for ethanol production, resulting in a reduction in volumes available for animal feeding. Large areas of land that were used for soybean production in the United States are also being re-directed into maize production as maize prices rise, in response to shortage and subsidies to the biofuel industry. The requirement for EU states to increase the biofuel share in fossil fuels to at least 10 % required 19 % of the annual cereal consumption and the full output of rapeseed (Popp, 2008). To fully substitute fossil fuels, the EU would require at least twice the current output of cereals and 25 times the current rapeseed and sunflower seed production. Ileleji (2010) was of the view that the real success of the maize ethanol industry will not depend on how well maize ethanol does as a fuel but on how well its by-product, DDGS, does as a feed. The feed industry is being compelled to use the by-products arising from the biofuel industry for animal feeding. It is speculated that the proportion of these by-products in the diet will continue to rise as less conventional ingredients become available. In an assessment of the potential for use of DDGS by the Chinese livestock sector, Fabiosa *et al.* (2009) estimated that producers could make a saving of one dollar per hundredweight through the use of DDGS in diets. The report cautioned that mycotoxin contamination and nutrient profile variability must be watched. Batal (2009) has also cautioned that DDGS cannot completely replace maize in poultry diets due to their inherent limitations in terms of price, handling and logistics in the mill and nutrient variability. These limitations are further discussed below.

The increased inclusion of biofuel by-products in diets may negatively impact on animal productivity, as will be highlighted later in this chapter. However, nutritional studies are intensifying into the identification of ways by which the nutritive value of diets that are high in these by-products can be improved. Ultimately, many of the by-products are more suitable for ruminant animal feeding due to the high fibre content and nutrient deficiencies. However, producers of non-ruminant animals such as pigs and poultry are being forced to use the by-products in response to shortage and cost of grains. The use of biofuel by-products for non-ruminant animal feeding entails more diligent feed management and will be examined further.

3.1 Ruminant animal production

By their anatomical and physiological nature, ruminant animals would use by-products of the biofuel industry more efficiently than non-ruminants. Many of the by-products are fibrous and deficient in fermentable carbohydrates while being higher in non-carbohydrate constituents, particularly protein. Ruminant animals are endowed with the rumen eco-system, which aids in the digestion of fibre. Rumen microbes are also able to utilise low quality protein to generate higher quality protein and also synthesise nutrients such as vitamins from other sources. Therefore, ruminant animal production is not generally as negatively affected as non-ruminant animal production when by-products of the biofuel industry are included in the diet. The efficiency of production may, however, be compromised since DDGS, especially that made from wheat, contains significant amounts of rumen degradable protein (Belyea *et al.*, 2010). The results of studies on the use of some by-products for ruminant animal feeding are summarised in Table 2.

Species	By-product	Optimum inclusion rate or proportion (%)	Reference
Beef cattle	WDGS	40	(Larson *et al.*, 1993)
		40	(Corrigan *et al.*, 2009)
	DDGS	20	(Vander Pol *et al.*, 2006)
		20	(Huls *et al.*, 2008)
Dairy cattle	DDGS	30	(Kalscheur, 2005)
		21	(Ranathunga *et al.*, 2010)
		30	(Janicek *et al.*, 2008)
	Glycerol	430 g/day (DM basis)	(DeFrain *et al.*, 2004)
Sheep	DDGS	40	(Held, 2006)
	Rapeseed meal	30	(Mandiki *et al.*, 1999)

Table 2. Summary of selective studies using biofuel by-products in ruminant animal diets

The use of DDGS may actually be of advantage to the ruminant animal industry as the product has been shown to reduce methane emission by nearly 20 % when included in beef cattle diets (McGinn *et al.*, 2009). Distillers' dried grains have been successfully used to replace compound feed in the feeding of breeding ewes, with the material marginally improving milk yield, wool production and flock fertility (Dimova *et al.*, 2009). This is similar to the findings of Franke *et al.* (2009) who reported no effects of DDGS or rapeseed meal on milk yield of dairy cows, although milk protein content was reduced. Another study by Kleinschmit *et al.* (2006) actually reported an increase in milk yield, milk fat yield and feed efficiency on dairy cattle diets containing 20 % DDGS. The DDGS tended to increase the proportion of butyrate in the rumen although the total concentrations of volatile fatty acids were higher on the diet without DDGS.

The feed intake of lambs in response to DDGS inclusion does not appear to depend on the source grain of the DDGS, with similar effects being observed on diets containing maize, barley or wheat DDGS at 20 % of a grower diet (McKeown *et al.*, 2010). However, feed efficiency was poorer on the wheat DDGS than on the maize DDGS, due to greater production of ruminal ammonia and lower digestibility of the former. These researchers concluded that maize, wheat or triticale DDGS can replace a mixture of barley grain and canola meal at up to 20 % in diets for lambs. Mckinnon *et al.* (2008) established that wheat DDGS could replace barley grain at up to 50 % in diets for cattle.

In another study, Aldai *et al.* (2010) compared the meat quality of cattle raised on diets supplemented with barley or maize DDGS. Meat obtained from the diet containing the barley DDGS was darker at 24h and less tender than meat from animals raised on the diet containing maize DDGS.

Sugarcane bagasse is used extensively to feed cattle in many sugar-producing countries such as Brazil and Cuba (Dhore *et al.*, 2006; Murta *et al.*, 2009; Nagalakshmi and Reddy, 2010). Although such bagasse may be from the sugar industry rather than the biofuel subsidiary, the products resulting from the two industries are similar. Most of the reports point to a positive effect of the inclusion of bagasse in terms of dry matter intake and changes in rumen function, including nutrient digestibility. Cassava bagasse has been observed to alter the rumination and eating patterns of cattle when included in a silage mix at between 5 and 15 % (Silva *et al.*, 2005). The times spent ruminating and eating were reduced while inactive time increased linearly with increase in the level of the product in the diet. The efficiency of eating was not affected but the efficiency of dry matter use increased linearly with increasing levels of bagasse in the silage.

3.2 Non-ruminant animal production

Most of the by-products of the biofuel industries would naturally not be ideal ingredients for non-ruminant animals, particularly poultry, in view of their chemical composition and sometimes physical properties. However, the quality of DDGS, for example, has continued to improve with improvement in fermentation, distillation and drying processes. This will re-position the product as a useful feed for non-ruminant animals, especially poultry (Hastad *et al.*, 2003; Swiatkiewicz and Koreleski, 2008). These authors recommended that DDGS could be safely included in starter diets for broilers and turkeys at 5-8 % and 12-15 %, respectively in the grower-finisher phase. A summary of some of the studies on the use of by-products for non-ruminant animal feeding is shown in Table 3.

Ganesan *et al.* (2008) reported that more than 13 million tonnes of DDGS were produced in 2007 and the output keeps rising as new ethanol plants come into production. As has been pointed out, these products will be increasingly used for non-ruminant feeding due to feed shortages. In a trial on pigs, Avelar *et al.* (2010) reported a reduction in body weight gain when wheat DDGS was included in diets of weaned pigs at between 5 and 20 %. This contrasts with the findings of Jones *et al.* (2010) who did not observe any negative effects of maize DDGS in the diet of weaner pigs in terms of feed consumption, weight gain or feed efficiency. Sorghum DDGS, on the other hand, reduced FCE. Feoli *et al.* (2007), however, observed a reduction in weight gain as a result of a reduction in DM, protein and energy digestibility in finishing pigs on diets containing 40 % maize DDGS. Skiba *et al.* (2009) reported a similar reduction in FCE in broiler chickens with only 10 % DDGS in the diet but this was attributed mainly to feed particle selection and wastage rather than reduction in nutritive value. The relative economic efficiency of broiler chickens on 6 % DDGS was found to be approximately equal to that of chicks on a maize control diet (Shalash *et al.*, 2009).

Species	Type of by-product	Inclusion rate (%)	Reference
Swine (gestating sows)	DDGS	50	(Wilson *et al.*, 2003)
(Weanling pigs)	DDGS	30	(Burkey *et al.*, 2008)
	Glycerol	10	(Lammers *et al.*, 2007)
(Growing finishing pigs)	DDGS	30	(Xu *et al.*, 2007)
Broilers	DDGS	6-15	(Lumpkins *et al.*, 2004)
	DDGS	30	(Barekatain *et al.*, 2011)
	Glycerol	10	(Simon *et al.*, 1997)
	Rapeseed meal	10	(McNeill *et al.*, 2004)
Layers	Rapeseed meal	10	(Mawson *et al.*, 1995)
	DDGS	15	(Lumpkins *et al.*, 2005)
	DDGS	15	(Roberson *et al.*, 2005)

Table 3. Summary of selective studies using biofuel by-products in non-ruminant animal diets

In a trial to determine the optimal level of inclusion of DDGS in diets for pigs, Linneen *et al.* (2008) observed a reduction in daily gain and feed consumption as DDGS rose in the diet, particularly beyond 10 %. There was, however, a linear improvement in these measurements with the inclusion of white grease at between 0 and 6 %, along with the DDGS. The feed efficiency of the pigs was increased due to inclusion of DDGS in the diet. Min *et al.* (2008) have established that DDGS can be used at up to 30 % in broiler chicken diets without detrimental effects on feed intake but feed efficiency was reduced, a result that was linked to the reduced pellet quality. Further supplementation with glycerine, at 5 %, had no effect.

There are not many reports on the use of glycerol for non-ruminant feeding, although the product could be useful as an alternative to lipids, particularly as a source of energy. Schieck *et al.* (2010) evaluated the effect of long- or short-term feeding of glycerol to growing pigs, in diets containing 8 % of the product. Pigs that were fed on such diets for 14 weeks ate more than animals on the control diet while those that were fed only in the last 8 weeks of trial consumed similar amounts of feed to the control. Long-term feeding also resulted in higher daily gain although short-term fed pigs grew faster than the control. Hot carcass weight was greater in the long-term fed pigs than in the other groups and pork quality based on a taste panel assessment was not affected by treatment. The response to glycerol may be dependent on the quality of the product. In another study on pigs, Hanczakowska *et al.* (2010) observed a reduction in weight gain when crude glycerol was included in the diet at 10 %, but such an effect was absent when refined glycerol was used, the latter generally improved fibre digestibility. It was concluded that crude glycerol had limited value as a feed supplement for pigs. In tests with pigs, crude glycerol has been found to be adequate in digestible energy, therefore the poor response to it cannot be ascribed to poor energy supply (Lammers *et al.*, 2008).

A comparative trial has been conducted on biodiesel press cakes obtained from canola or mustard when fed to broiler chickens (Thacker and Petri, 2009). Ether extract digestibility and nitrogen retention were higher in the canola press cake group than in the canola meal or mustard press cake groups. Feed efficiency on the two press cakes was superior to that on canola meal. Mustard press cake was recommended mainly due to its lower price.

3.3 Miscellaneous animal feeding

There are not many reports on the feeding of by-products of the biofuel industry to animals other than farm animals. Bonoma *et al.* (2008) reported that weanling horses were not adversely affected by DDGS replacing up to 30 % of the concentrate portion or 15 % of diet. Juvenile hybrid tilapia have been raised on diets containing 30 % DDGS and further supplemented with fish meal, meat meal or soybean meal (Coyle *et al.*, 2004). It was found that the fish meal could be excluded from the diet without detriment to growth rate or protein efficiency ratio, thus reducing feed costs. Nagalakshmi and Reddy (2010) described a study in which sugarcane bagasse was used as sole roughage source in a complete diet that was expanded/extruded before feeding to lactating buffaloes. This was compared to a traditional diet of concentrate mixture, chopped sorghum straw and small amounts of green Napier grass. Feed intake on the bagasse-containing diet was reduced but protein digestibility was increased. There were no differences in milk yield, fat-corrected milk yield and non-fat solids. Feeding cost per unit of milk produced was substantially (>31 %) reduced with the bagasse-supplemented diet.

4. Limitations of biofuel by-products for animal feeding

The impact of biofuel by-products in animal feeding is generally more positive than negative. However, there are limitations to the use of these by-products for animal feeding, some of which would have become obvious from the discussion of animal response to their inclusion, variously described above. In this section, we will collate these constraints and discuss them in greater detail.

4.1 Availability and cost

Generally, feed cost is often regarded as the highest variable cost of production, with a strong bearing on profitability of the animal industry. Therefore, supply of any by-products seems to be a fundamental issue limiting its usage in animal diets. Currently, most biofuel plants are located in a limited number of countries around the world, including the USA and Brazil, and there is uncertainty over the number of plants being built in other parts of the world and also the extent of material being imported into developing countries, which are unable to construct local biofuel plants. In this regard, location of biofuel plant may be fundamentally crucial to the ability to supply and distribute biofuel by-products to the feed industry as transportation cost will be added to final price and hence unfavourably affect the ability of the livestock producer to use such materials. Furthermore, feed prices and their fluctuations will negatively influence the use of by-products in animal feeds, which can be noticeable more in the regions that import by-products. Besides, the demand from other markets may affect the availability of by-products for animal feeding. As a prime example, glycerol is an acceptable energy source for both ruminant and non-ruminant animals but high demand by the pharmaceutical and polymer industries as well as its use by the food industry may restrict its availability for animal feeding.

4.2 Fibre

Generally, during the fermentation process, most of the components present in cereal grains become concentrated in distillers' by-products except starch, which is converted to ethanol. Therefore, the starch content of the final by-products is much lower than that of the main grain. A large proportion of the remaining carbohydrates is regarded as fibre, also called non-starch polysaccharides (NSP). The fibre content of by-products varies according to processing methods. High-protein DDGS, for instance, are produced when the germ is removed from the main grain. Such material, therefore, contains less fibre and higher protein compared to conventional DDGS (Jacela *et al..*, 2010b). Nevertheless, neutral detergent fibre (NDF), acid detergent fibre (ADF) and total dietary fibre are approximately three times higher than those in the main grain. While in ruminant animals these fibre fractions can be readily digested due to high fibrolytic activities of rumen microbes, non-ruminants are unable to break down NSP because of the absence of such activities in their small intestine. In addition, the presence of NSP in non-ruminant gastrointestinal tract has been shown to contribute to poor performance by creating a viscous environment and subsequent impediment to nutrient digestibility. However, there are limited data on the nature of NSP in biofuel by-products. Choct and Peterson (2009) analysed 6 DDGS samples from different ethanol plants across the Midwest, USA and reported an average of 40 % total carbohydrates, most of which were identified as insoluble NSP with less than 10 % soluble NSP. Regarding the composition of NSP, they identified the main sugars as glucose and xylose followed by arabinose, galactose and mannose in descending order of concentration. This would explain the possible adverse effect of arabinoxylans and xylans in birds on DDGS-rich diets.

Although most by-products of the biofuel industry with high fibre content are regarded as desirable feed ingredients for ruminant animals, there might be a limitation to how much fibre-rich materials can be fed to lactating dairy cattle. High-fibre supplements in lactating dairy cattle diets can result to significant differences in milk composition, nutrient intake as well as nutrient apparent digestibility (Bernard and McNeil, 1991).

4.3 Bulk and storage

Basically, storage and space allocation will be primary issues for the new ingredients introduced to the feed mills as there is always a limitation for designation of space for ingredients other than the common ones being used by feed manufacturers. A wide range of factors including physical characteristics, flowability and also the time that ingredients are used will determine bin space allocation. Notably, not only space allocation for the bulky ingredients such as DDGS can be problematic due to being fairly light at around 480.6 kg/m^3 but also take considerably more time and cost to be delivered and stored, compared to the conventional ingredients and meals. Producers, therefore, have sometimes been reluctant to accept biofuel by-products, in particular, DDGS owing to inconvenience and cost of handling. There are limited reports related to bulk density and particle size distribution of DDGS. In an experiment conducted by Knott *et al.* (2003) on 16 samples from different ethanol plants in Minnesota, South Dakota to investigate average particle size and bulk density of DDGS, they reported average particle size to be 1282 microns (SD= 305, CV= 24%) and also an average of 572.4 kg/m^3 for bulk density. In the same experiment, there was moderate positive correlation between bulk density and moisture content, indicating that with a decrease in moisture content of DDGS, bulk density tends to decrease.

Generally, the ability of the solid particles and powders to flow during unloading from storage containment or transporting vehicles is defined as flowability (Babcock *et al.*, 2008). Flowability of biofuel by-products, in particular, DDGS is regarded to be an issue for transportation and storage which can be affected by a wide range of factors including cooling and drying practices, particle size and the amount of residual sugars (Shurson,

2005). It also seems that humidity greater than 60 % has an adverse effect on flowability of DDGS due to a tendency of this by-product to absorb moisture (Babcock *et al.*, 2008).
Shipment concerns arising from moisture content and bulk density can also be an issue for the other by-products such as wet dried grain with solubles (WDGS) with high moisture content (65 to 70%). It has been voiced by some feed manufacturers that bagging of WDGS is sometimes a problem as the material ends up compacting (Babcock *et al.*, 2008).

4.4 Nutrient imbalances and variability

Variations and deficiency in some nutrients are regarded to be the most important issues limiting the usage of biofuel by-products in animal feeding. Generally by-products result from several steps inherent to biofuel or ethanol production, including different temperatures and drying practices, enzymes used as well as rate of soluble incorporated, in the case of DDGS. Additionally, live yeast added during the fermentation process may interact with the factors mentioned above to influence the nutritive value of final by-products substantially. These discrepancies predominantly are reflected on protein, metabolizable energy and mineral contents of final products. Numerous studies have been conducted to investigate nutrient characteristics of DDGS. The study by Spiehs *et al.* (2002), which examined the nutrient variability of DDGS from 10 ethanol plants in Minnesota and South Dakota has been highlighted. In another study conducted by Batal and Dale (2006), TME_n evaluation of seventeen samples from different plants showed a wide range from 10.4 to 13.3 MJ/kg with a mean of 11.8 MJ/kg. Pedersen *et al.* (2007) also showed a higher average energy content (22.7 MJ/kg) of ten DDGS samples compared to corn grain. The high energy content of DDGS is associated with high level of fat in the residue (Swiatkiewicz and Koreleski, 2008).
The protein content of DDGS has been reported in several experiments to vary from 23 to 32 % (Cromwell *et al.*, 1993, Belyea *et al.*, 2004, Pedersen *et al.*, 2006, Fathi and Afifi, 2008). Several factors regulate the protein content of DDGS, as described previously, distillers' soluble and wet grain are two main components from which DDGS are formed. Substantial variations were shown by Belyea *et al.* (2004) regarding soluble composition from different batches. They postulated possible contribution of proportional rate of components to variation in protein content of DDGS. More recently, Kingsly *et al.* (2010) observed variable nutritive value of DDGS when different ratio of wet distiller grains and solubles were blended together. Additionally, yeast protein and its amino acid composition may also have an effect on protein content and composition of DDGS since yeast protein constitutes approximately 50 % of the protein in DDGS (Belyea *et al.* 2004). Furthermore, the drying process can have crucial influence not only on variability of nutrients but also on concentrations and availability of amino acids in different samples (Bandegan *et al.*, 2009, Martinez-Amezcua *et al.*, 2007). Recently, Bandegan *et al.* (2009) reported average Lys and Met concentration of 5 different samples of wheat DDGS to be 0.74 and 0.61 %, respectively. More variable data for Lys content was observed by Fastinger *et al.* (2006), ranging from 0.48 to 0.76 % in 5 DDGS samples from different sources. Amongst the most limiting amino acids, Lys and Met appeared to be the most variable in DDGS, the reason being that the heating of wet distiller's grains may adversely affect availability of heat-sensitive amino acids. Excessive heat during the drying process may accelerate reactions between reducing sugars and Lys (the Maillard reaction), which leads to unavailability of Lys in DDGS (Batal and Dale, 2006). In poultry, due to lack of enzyme to breakdown bonds between Lys and carbohydrates, this form of Lys becomes unavailable (Fastinger *et al.*, 2006).

As mentioned before, overheating during DDGS production is mainly responsible for losses in nutritional value of final by-products (Cromwell *et al.*, 1993). It has also been shown that subsequent by-products resulting from excess heat appear to be darker compared to those samples treated by less severe drying practices (Kim *et al.*, 2008). Therefore, the colour of DDGS can be predominantly used as a general guide of nutritive value and amino acid availability of DDGS samples. Cromwell *et al.* (1993) observed the lowest Lys concentration in the darkest DDGS sample, about 0.62 % and the highest amount of 0.86 % in the lightest sample. They also obtained a correlation as high as 0.67 between Hunterlab L score and Lys content for the same samples.

Nutritionists regard nutrient imbalances as the main concern when biofuel by-products are to be incorporated in animal diets. Depending on type of animals and species, different nutrients may be more important. Vander Pol *et al.* (2006) postulated that incorporation of WDGS in the diet containing more than 8 % fat for cattle may lead to a reduction in feed intake. These nutrient imbalances can be considered for a variety of nutrients such as minerals and amino acids. Batal and Dale (2003) reported a severe deficiency of sodium in an experiment conducted with laying hens using NRC values for DDGS, indicating the necessity of analysing DDGS sample prior to incorporation into diets. There is also a possibility of increase in phosphorus and sulphur in ethanol by-products which may occur with greater rate of solubles added to distillers' grain. Large amounts of sulphur (more than 0.4 %) can cause polioencephalomalacia in cattle (Babcock *et al.*, 2008).

Finally, the presence of mycotoxins in by-products of biofuel can be a potential threat for animal health and performance if the product is not monitored and becomes contaminated. Noteworthy, the process of fermentation and production of biofuels is usually incapable of destroying mycotoxins, therefore they will be reflected in by-products if the source grains were contaminated. Schaafsma *et al.* (2009) reported the mean concentration of deoxynivalenol, a toxin from *Fusarium graminearum*, in condensed distillers' solubles to be 7.11 mg/kg, which was four times higher than that in corn as main grain (1.80 mg/kg). In the same study, the concentration of deoxynivalenol was found to be 5.24 mg/kg. Nevertheless, care should be taken by regular monitoring and measurement of mycotoxins in grain and by-products to keep the concentration of mycotoxins close to the recommended level in the main grain otherwise such an anitinutitive factor will be present in the final by products, posing a risk to animal and possibly human health.

5. Scope for increased use of biofuel by-products

As the proportion of by-products produced by the biofuel industry is increasing rapidly, attempts have been made to maximise their utilisation in livestock diets. Depending upon different biofuel processes, species and the type of by-product, there has been a wide range of applications to enhance the nutritive value of by-products to facilitate their wider inclusion in animal diets, especially for swine and poultry. In this section existing applications employed to enhance the nutritional quality of by-products are discussed mainly for DDGS and WDGS, which are the most widely used by-products for non-ruminant and ruminant animals.

5.1 Further processing and management

It has been widely accepted that any alteration in the process of biofuel and in particular ethanol production will lead to changes in finished by-products (Belyea *et al.*, 2004; Gibson *et al.*, 2005). Therefore, depending upon the species that by-products are fed to a number of new technologies may be taken into consideration prior to and after the completion of by-product production. In modern bioethanol plants enzymatic milling (EM) is a new procedure in which hemicellulases and protease are used. Those enzymes facilitate separation of non-fermentable fibre and germ prior to fermentation and this leads to an increase in fat and protein content of DDGS, which makes it a more desirable ingredient for non-ruminant animals (Kim *et al.*, 2010). Further processing can be applied to the recovered germ and the pericarp and endosperm fibre to produce corn germ oil and phytosterols, respectively. In this regard, a combined separation method (Elusieve process) was developed by Srinivasan *et al.* (2005) in which sieving and elutriation were applied to separate fibre from DDGS. In the described process, an upward stream of air was created by a blower followed by sieving to separate particles based on density, size and physical form of DDGS components. Srinivasan *et al.* (2005) also evaluated the nutrient characteristics of different factions (light and heavy) obtained from the Elusieve method showing 13 to 41 % and 4 to 127 % increase in protein and fat content of the heavier fraction, respectively. Fibre content of lighter fraction of DDGS was also reduced.

As discussed previously, the presence of NSP in DDGS could impede nutrient digestibility and therefore performance in non-ruminants, in particular, broiler chickens. There are some reports on the beneficial role of extrusion in enhancement of nutrient digestibility through physical disruption of cell wall, and hence breakdown of NSP to the smaller factions (Oryschak *et al.*, 2010b, Oryschak *et al.*, 2010a, Camire, 1991). Oryschak *et al.* (2010b) found a 10-34 % increase in apparent ileal digestibility of amino acids in both extruded corn and wheat DDGS. High temperature and pressures through the extrusion process are believed to act as effective eliminators of microbial contamination and also making cell content more accessible and susceptible to enzymatic hydrolysis (Oryschak *et al.*, 2010b). Camire (1991) ascribed the improvement in protein digestibility as a result of extrusion to denaturation of protein via heat and pressure applied through extruder, which presumably increases the exposure of peptide bonds to enzymatic digestion. However, Al-Marzooqi and Wiseman (2009) recommended mild conditions of extrusion in order to avoid adverse effect of high temperature on amino acid availability.

Pelleting as a further process on DDGS may also improve flowability, a constraint that was highlighted in a previous section. It has been demonstrated that any level of agglomeration during the pelleting process can improve flowability of DDGS but this must be undertaken at the ethanol plants which would probably increase the cost of production and hence the final price of DDGS (Behnke, 2007).

The usage of WDGS for ruminants and in particular dairy cows faces some constraints largely due to high moisture, content which causes handling problems and also limits the time of storage. This can be minimized by ensiling and addition of organic acids, to prevent spoilage and improve handling as well as extended shelf life. Wet DGS has a relatively low pH, which can be an advantage for preservation, confirming that WDGS can be ensiled with other suitable companions such as soybean hulls and beet pulp that are low in protein, fat and phosphorus (Kalscheur *et al.*, 2004; Anderson *et al.*, 2009). The nutrient profile of WDGS can even be improved by ensiling as was recently demonstrated by Mjoun *et al.* (2011). They showed that a mixture of 50 % WDGS with 50 % whole plant corn can result to an optimization in nutrient and fermentation profile, therefore, maximizing aerobic stability that may ensure higher inclusion of WDGS in dairy cattle diets.

Ammoniation is a method generally used to improve the nutritive value of fibrous feed materials (Dean *et al.*, 2008). Zade *et al.* (2009) employed the technique to improve the value of rising levels of pith bagasse for beef cattle, but found no improvement in feed consumption, daily weight gain or feed efficiency although feed costs were reduced. In an earlier study, urea treatment of sugarcane bagasse was found to increase its nitrogen content and *in vitro* dry matter digestibility (Hassoun *et al.*, 1990). When demethylated, glycerine would be a useful energy source for broiler chickens but issues such as residual methanol, sodium or potassium, feed flow and handling would need to be better understood (Batal, 2009).

Meyer *et al.* (2010) have reported a study in which DDGS was fed to growing cattle in combination with rapeseed meal (RBM) and compared to soybean meal or separate feeding of RSM and DDGS. Combining the two products led to an increase in voluntary feed intake and weight gain.

5.2 Supplementation with microbial enzymes

Among the key approaches to improve nutrient utilization in animal diets, application of various exogenous enzymes has drawn enormous attention over the past decade. It has been consistently shown that exogenous enzymes may enhance animal performance and nutrient utilization by animals, in particular non-ruminants. In this regard, diets containing high concentrations of NSP have shown considerable response to exogenous NSP-degrading enzymes. In addition, the concentration of NSP in biofuel by-products is relatively high due to fermentation process, which removes most carbohydrates and almost triples the amount of NSP compared to that in the original grain. Specifically, this is a constraint to the use of this by-product by non-ruminant animals that are unable to digest NSP. Therefore, there may be a potential for the use of exogenous enzymes in the diets containing biofuel by-products.

For poultry, there are a few reports concerning the use of xylanase to break down NSP, to improve the growth performance of birds on DDGS-rich diets. Recently, Barekatain *et al.* (2011) showed significant improvement in FCE of broiler chickens fed 300 g/kg sorghum DDGS as a result of xylanase supplementation. Similarly, results of an experiment conducted by Liu *et al.* (2011) with broilers demonstrated that xylanase addition to a diet containing corn DDGS could increase hemicellulose and dry matter digestibility by 5 and 20 %, respectively. Furthermore, the combination of exogenous enzymes such as xylanase, amylase, protease and phytase could have a sub-additive effect in growth performance of birds, as has been shown by Olukosi *et al.* (2010).

Exogenous multi-enzyme complexes also appear to be beneficial for amino acid digestibility of broilers fed diets containing DDGS. Oryschak *et al.* (2010a) found between 6 and 19 % improvement in lysine, tryptophan, methionine, isoleucine, histidine and phenylalanine digestibility when a multi-enzyme complex was included in a diet with 15 % DDGS and fed to broiler chickens. The effect of enzymes on the biodiesel by-product, rapeseed press cake, is also promising. Significant improvement in feed conversion ratio (1.9 to 1.84) was found by Jozefiak *et al.* (2010) in broiler chickens on such diets.

The influence of enzymes, however, is inconsistent for other non-ruminant animals. In a series of experiments conducted by Jacela *et al.* (2010a) no effect of microbial enzymes was observed when finishing pig diets contained DDGS. It seems that variation in carbohydrate composition, difference in age and animal species as well as availability of substrate for specific enzymes to act on may be logical explanations for inconsistences in response to exogenous enzymes with by-products in the diets.

There are conflicting reports on the effectiveness of microbial enzymes in diets containing DDGS for pigs. Feoli *et al.* (2008) reported an improvement in nutrient digestibility and growth as a result of supplementation with microbial enzymes possessing β-glucanase, protease, amylase and xylanase activities. In contrast, Jones *et al.* (2008) did not observe any positive effect of microbial enzymes on diets containing sorghum DDGS when fed to pigs. Pigs on the diets with sorghum DDGS were also less efficient than those on maize DDGS (Jones *et al.*, 2008).

5.3 Probiotics

Currently, probiotics are used by the animal industries to alter the profiles of microbial populations in the digestive tract, to improve nutrient utilization and health of the animal. They are not used as such to increase the digestion of feed components although such targeted introduction of species is a possibility in ruminant animal nutrition. The relatively high level of fibre in DDGS and bagasse, for example, has been highlighted. Such material will benefit from pre-digestion with microbes and this can be achieved through ensiling, which generally enables preferred microbial species to pre-digest feed material, particularly forage. Direct fermentation of bagasse or DDGS may also be effective although such material would be more useful for ruminant animals and pigs than poultry. Ramli *et al.* (2005) have reported on such a product; bagasse subjected to solid-state fermentation with *Aspergillus sojae* prior to feeding to goats. Such a diet was found to improve the flavour, aroma and overall quality of loin meat of goats compared to animals on basal lucerne hay. A mutant strain of *Trichoderma viride* has been similarly used on sugarcane bagasse, to improve the digestion of the material through the cellulase enzymes secreted (Valino *et al.*, 2004).

6. Recommendations and conclusion

There is no doubt that by-products of the biofuel industry will continue to become increasingly important components of the animal diet. It is most likely that less and less grains will be available for animal feeding until such a time that DDGS will become the primary product utilised in animal feeding. The biodiesel industry also produces enormous amounts of glycerol, which need to be discarded and this could be useful as a source of energy provided there is improvement in technology to refine it. The volume of biodiesel press cakes relative to conventional meals will also increase as more oilseeds are used for the production of biodiesel rather than edible oil. In areas where sugarcane and similar crops are used for ethanol production, bagasse will become available and while most of it will be used for energy production, some of it can be valuable feed resource, following further processing.

The limitations that have been identified in the use of by-products of the biofuel industry for animal feeding will wane as technology becomes available for further processing or feed management. This will place the by-products in stronger position to be used at higher levels in the diet.

To initiate or improve the utilization of by-products of biofuel industry, the following recommendations are worth considering:

1. There is a need to develop rapid methods of product testing, in order to determine the nutrient composition of the products prior to feed formulation. This will reduce the effect of variability from batch to batch and between plants producing the same material.
2. More research should be conducted to determine the optimum inclusion levels of the various products. Currently very few such research have been reported and commercial use of the products is limited by lack of knowledge of how much can be included without further tests.
3. Many of the products will benefit from supplementation with other feed additives including microbial enzymes or combinations of the by-products themselves, for example, glycerine used with DDGS. This may increase their acceptability by animals and improve nutritive value.
4. A single by-product may not be useful for all classes of animals. By-products should be used for the kind of animal for which they are most suitable. While there is a strong drive for minimising feed costs, animal welfare and overall productivity should also be considered in the application of by-products of the biofuel industry.

7. References

Al-Marzooqi, W. & Wiseman, J. 2009. Effect of extrusion under controlled temperature and moisture conditions on ileal apparent amino acid and starch digestibility in peas determined with young broilers. *Animal Feed Science and Technology,* 153, 113-130.

Aldai, N., Aalhus, J. L., Dugan, M. E. R., Robertson, W. M., McAllister, T. A., Walter, L. J. & McKinnon, J. J. 2010. Comparison of wheat- versus corn-based dried distillers' grains with solubles on meat quality of feedlot cattle. *Meat Science,* 84, 569-577.

Anderson, J. L., Kalscheur, K. F., Garcia, A. D., Schingoethe, D. J. & Hippen, A. R. 2009. Ensiling characteristics of wet distillers grains mixed with soybean hulls and evaluation of the feeding value for growing Holstein heifers. *Journal of Animal Science,* 87, 2113-2123.

Anonymous (undated). Biodiesel equipment, processors, accessories and information. Downloaded from http://www.biodieselgear.com/faq.html#5.

Anonymous (2011a). Food vs fuel. Downloaded from http://www.wikipedia.com.

Anonymous (2011b). Bagasse. Downloaded from http://www.wikipedia.com.

Avelar, E., Jha, R., Beltranena, E., Cervantes, M., Morales, A. & Zijlstra, R. T. 2010. The effect of feeding wheat distillers dried grain with solubles on growth performance and nutrient digestibility in weaned pigs. *Animal Feed Science and Technology,* 160, 73-77.

Babcock, B. A., Hays, D. J. & Lawrence, J. D. 2008. Using distillers grains in the U.S and international livestock and poultry industry. First ed. Ames Iowa USA, Midwest agribusiness trade research and information center.

Bandegan, A., Guenter, W., Hoehler, D., Crow, G. H. & Nyachoti, C. M. 2009. Standardized ileal amino acid digestibility in wheat distillers dried grains with solubles for broilers. *Poultry Science,* 88, 2592-2599.

Barekatain, M., Choct, M. & Iji, P.A. 2011. Scope for the high inclusion of sorghum distillers'dried grains with solubles in broiler chicken diets. *Proceeding of Australian poultry science symposium.* University of Sydney, Australia, 20, 73.

Batal, A. & Dale, N. 2003. Mineral composition of distillers dried grains with solubles. *Journal of Applied Poultry Research,* 12, 400-403.

Batal, A. B. 2009. How much DDGS for poultry? *Feed International,* 30, 18-19.

Batal, A. B. & Dale, N. M. 2006. True metabolizable energy and amino acid digestibility of distillers dried grains with solubles. *Journal of Applied Poultry Research,* 15, 89-93.

Behnke, K. C. 2007. Feed manufacturing considerations for using DDGS in poultry and livestock diets. *Proceeding 5th Mid-Atlantic Nutrition Conference Baltimore MD March 29, 2007. Accessed 25 March 2011 avialable at* http://www.ddgs.umn.edu/articles-poultry/2007-Behnke-%20Feed%20manufacturing%20considerations%20(MANC).pdf, 1.

Belyea, R. L., Rausch, K. D., Clevenger, T. E., Singh, V., Johnston, D. B. & Tumbleson, M. E. 2010. Sources of variation in composition of DDGS. *Animal Feed Science and Technology,* 159, 122-130.

Belyea, R. L., Rausch, K. D. & Tumbleson, M. E. 2004. Composition of corn and distillers dried grains with solubles from dry grind ethanol processing. *Bioresource Technology,* 94, 293-298.

Bernard, J. K. & McNeil, W. W. 1991. Effect of high fiber energy supplements on nutrient digestibility and milk production of lactating dairy cows. *Journal of dairy science,* 74, 991-995.

Bonoma, T. A., Brogren, A. C., Kline, K. H. & Doyle, K. M. 2008. Effects of feeding distiller's dried grains with solubles on growth and feed efficiency of weanling horses. *Journal of Equine Veterinary Science,* 28, 725-727.

Burkey, T., Miller, P., Moreno, R., Shepherd, S. & Carney, E. 2008. Effects of increasing levels of distillers dried grains with solubles (DDGS) on growth performance of weanling pigs. *Journal of Animal Science,* 86, 50.

Camire, M. E. 1991. Protein functionality modification by extrusion cooking. *Journal of the American Oil Chemists' Society,* 68, 200-205.

Choct, M. & Petersen, S. T. 2009. The effect of commercial production methods on the carbohydrate and amino acid composition of corn distillers dried grains with solubles. *Proceedings of Australian Poultry Science Symposium,* University of Sydney, Australia 20, 117-120.

Clementson, C. L. & Ileleji, K. E. 2010. Variability of bulk density of distillers dried grains with solubles (DDGS) during gravity-driven discharge. *Bioresource Technology,* 101, 5459-5468.

Clementson, C. L., Ileleji, K. E. & Rosentrater, K. A. 2010. Evaluation of measurement procedures used to determine the bulk density of distillers dried grains with solubles (DDGS). *Transactions of the ASABE,* 53, 485-490.

Corrigan, M., Erickson, G., Klopfenstein, T., Luebbe, M., Vander Pol, K., Meyer, N., Buckner, C., Vanness, S. & Hanford, K. 2009. Effect of corn processing method and corn wet distillers grains plus solubles inclusion level in finishing steers. *Journal of animal science,* 87, 3351.

Coyle, S. D., Mengel, G. J., Tidwell, J. H. & Webster, C. D. 2004. Evaluation of growth, feed utilization, and economics of hybrid tilapia, Oreochromis niloticus * Oreochromis aureus, fed diets containing different protein sources in combination with distillers dried grains with solubles. *Aquaculture Research,* 35, 365-370.

Cozannet, P., Primot, Y., Gady, C., Metayer, J. P., Callu, P., Lessire, M., Skiba, F. & Noblet, J. 2010a. Composition and amino acids ileal digestibility of wheat distillers dried grains and solubles in pigs: sources of variability. *11th International Symposium on Digestive Physiology of Pigs Part 2, Costa Daurada, Catalonia, Spain.*

Cozannet, P., Primot, Y., Gady, C., Metayer, J. P., Callu, P., Lessire, M., Skiba, F. & Noblet, J. 2010b. Ileal digestibility of amino acids in wheat distillers dried grains with solubles for pigs. *Animal Feed Science and Technology,* 158, 177-186.

Cozannet, P., Primot, Y., Gady, C., Metayer, J. P., Lessire, M., Skiba, F. & Noblet, J. 2010c. Energy value of wheat distillers grains with solubles for growing pigs and adult sows. *Journal of Animal Science,* 88, 2382-2392.

Cozannet, P., Primot, Y., Metayer, J. P., Gady, C., Lessire, M., Geraert, P. A., Tutour, L. l., Skiba, F. & Noblet, J. 2009. Wheat dried distiller grains with solubles for pigs. *INRA Productions Animales,* 22, 11-16.

Cromwell, G. L., Herkelman, K. L. & Stahly, T. S. 1993. Physical, Chemical, and Nutritional Characterictics Of Distillers Dried Grains With Solubles For Chicks And Pigs. *Journal of Animal Science,* 71, 679-686.

Dean, D. B., Adesogan, A. T., Krueger, N. A., and Littell, R.C. 2008. Effects of treatment with ammonia or fibrolytic enzymes on chemical composition and ruminal degradability of hays produced from tropical grasses. *Animal Feed Science and Technology,* 145, 68-85

DeFrain, J., Hippen, A., Kalscheur, K. & Jardon, P. 2004. Feeding glycerol to transition dairy cows: effects on blood metabolites and lactation performance. *Journal of Dairy Science,* 87, 4195-4206.

Dimova, N., Ivanova, I., Mihailova, M., Todorov, N. & Naydenova, N. 2009. Wheat distiller's grains as a source of protein in dairy sheep. *Bulgarian Journal of Agricultural Science,* 15, 574-582.

Fabiosa, J. F., Hansen, J., Matthey, H., Pan, S. & Tuan, F. 2009. Assessing China's potential import demand for distillers dried grain: implications for grain trade. *CARD Staff Report Series,* 18 pp.

Fastinger, N. D., Latshaw, J. D. & Mahan, D. C. 2006. Amino acid availability and true metabolizable energy content of corn distillers dried grains with solubles in adult cecectomized roosters. *Poultry Science,* 85, 1212-1216.

Fathi, M. M. & Afifi, Y. K. 2008. Quality characteristics of distillers dried grains with solubles: Wheat and corn DDGS comparison. *Poultry Science,* 87, 26-26.

Feoli, C., Hancock, J.D., Monge, C., Gugle, T.L., Carter, S.D. and Cole, N.A. 2007. Digestible energy content of corn- vs sorghum-based dried distillers grains with solubles and their effects on growth performance and carcass characteristics in finishing pigs. *Kansas State University Swine Day. Report of Progress* 985, 131-136.

Feoli, C., Hancock, J. D., Gugle, T. L., Carter, S. D. & Cole, N. A. 2008. Effects of adding enzymes to diets with corn- and sorghum-based dried distillers grains with solubles on growth performance and nutrient digestibility in nursery and finishing pigs. *Kansas State University Swine Day 2008. Report of Progress 1001.*

Franke, K., Meyer, U. & Flachowsky, G. 2009. Distillers dried grains with solubles compared with rapeseed meal in rations of dairy cows. *Journal of Animal and Feed Sciences,* 18, 601-612.

Ganesan, V., Rosentrater, K. A. & Muthukurnarappan, K. 2008. Flowability and handling characteristics of bulk solids and powders - a review with implications for DDGS. *Biosystems Engineering,* 101, 425-435.

Gibson, M. L., Karges, K. & Giesemann, M. 2005. Overview of the ethanol industry and production of DDG/S . *Accessed on 26 May 2011, available at:* http://www.poet.com/files/division_files/13%20matt%20gibson%20-%20midwest%20poultry.pdf

Hanczakowska, E., Weglarzy, K., Szymczyk, B. & Hanczakowski, P. 2010. Effect of adding crude or refined glycerol to pig diets on fattening performance, nutrient digestibility and carcass evaluation. *Annals of Animal Science,* 10, 67-73.

Hassoun, P., Geoffroy, F., Saminadin, G., Prior, P. and Beramis, M. (1990). Studies on the ammoniation of sugar-cane bagasse by urea. Effects of moisture, urea levels, urease source and treatment periods on composition, in vitro dry matter digestibility and evolution of ureolytic bacteria. *Animal Feed Science and Technology,* 29, 113-129.

Hastad, C. W., Tokach, M. D., Nelssen, J. L., Goodband, R. D., Dritz, S. S., DeRouchey, J. M., Groesbeck, C. N., Lawrence, K. R., Lenehan, N. A. & Keegan, T. P. 2003. Use of dried distiller's grains with solubles for swine diets. *Kansas State University Swine Day 2003. Report of Progress 920.*

Held, J. 2006. Feeding soy hulls and dried distillers grain with solubles to sheep. *South Dakota university,* Available at http://pubstorage.sdstate.edu/AgBio_Publications/articles/ExEx2052.pdf.

Huls, T. J., Luebbe, M. K., Erickson, G. E. & Klopfenstein, T. J. 2008. Effect of inclusion level of modified distillers grains plus solubles in finishing steers. *Nebraska Beef Cattle Rep,* 36-38.

Ileleji, K. 2010. DDGS: the least talked about story in corn ethanol production. *Resource, Engineering & Technology for a Sustainable World,* 17, 10-13.

Jacela, J. Y., Dritz, S. S., DeRouchey, J. M., Tokach, M. D., Goodband, R. D. & Nelssen, J. L. 2010a. Effects of supplemental enzymes in diets containing distillers dried grains with solubles on finishing pig growth performance. *Professional Animal Scientist,* 26, 412-424.

Jacela, J. Y., Frobose, H. L., DeRouchey, J. M., Tokach, M. D., Dritz, S. S., Goodband, R. D. & Nelssen, J. L. 2010b. Amino acid digestibility and energy concentration of high-protein corn dried distillers grains and high-protein sorghum dried distillers grains with solubles for swine. *Journal of Animal Science,* 88, 3617-3623.

Janicek, B. N., Kononoff, P. J., Gehman, A. M. & Doane, P. H. 2008. The effect of feeding dried distillers grains plus solubles on milk production and excretion of urinary purine derivatives. *Journal of Dairy Science,* 91, 3544-3553.

Jones, C. K., Bergstrom, J. R., Tokach, M. D., DeRouchey, J. M., Goodband, R. D., Nelssen, J. L. & Dritz, S. S. 2010. Efficacy of commercial enzymes in diets containing various concentrations and sources of dried distillers grains with solubles for nursery pigs. *Journal of Animal Science,* 88, 2084-2091.

Jones, C. K., Bergstrom, J. R., Tokach, M. D., DeRouchey, J. M., Nelssen, J. L., Dritz, S. S. & Goodband, R. D. 2008. Effects of commercial enzymes in diets containing dried distillers grains with solubles for nursery pigs. *Kansas State University Swine Day 2008. Report of Progress 1001.*

Jozefiak, D., Ptak, A., Kaczmarek, S., Mackowiak, P., Sassek, M. & Slominski, B. A. 2010. Multi-carbohydrase and phytase supplementation improves growth performance and liver insulin receptor sensitivity in broiler chickens fed diets containing full-fat rapeseed. *Poultry Science,* 89, 1939-1946.

Kalscheur, K. 2005. Impact of feeding distillers grains on milk fat, protein, and yield. *Proceeding of the distillers grains technology council.* 10th Anuual Symposium, Louisville, KY.

Kalscheur, K., Garcia, A., Hippen, A. & Schingoethe, D. 2004. Fermentation characteristics of ensiled wet corn distillers grains in combination with wet beet pulp. *Journal of Dairy Science,* 87, 53.

Kelzer, J. M., Kononoff, P. J., Tedeschi, L. O., Jenkins, T. C., Karges, K. & Gibson, M. L. 2010. Evaluation of protein fractionation and ruminal and intestinal digestibility of corn milling co-products. *Journal of Dairy Science,* 93, 2803-2815.

Kim, E. J., Amezcua, C. M., Utterback, P. L. & Parsons, C. M. 2008. Phosphorus bioavailability, true metabolizable energy, and amino acid digestibilities of high protein corn distillers dried grains and dehydrated corn germ. *Poultry Science,* 87, 700-705.

Kim, E. J., Parsons, C. M., Srinivasan, R. & Singh, V. 2010a. Nutritional composition, nitrogen-corrected true metabolizable energy, and amino acid digestibilities of new corn distillers dried grains with solubles produced by new fractionation processes. *Poultry Science,* 89, 44-51.

Kim, Y. M., Hendrickson, R., Mosier, N. S., Ladisch, M. R., Bals, B., Balan, V., Dale, B. E., Dien, B. S. & Cotta, M. A. 2010b. Effect of compositional variability of distillers' grains on cellulosic ethanol production. *Bioresource Technology,* 101, 5385-5393.

Kingsly, A. R. P., Ileleji, K. E., Clementson, C. L., Garcia, A., Maier, D. E., Stroshine, R. L. & Radcliff, S. 2010. The effect of process variables during drying on the physical and chemical characteristics of corn dried distillers grains with solubles (DDGS) - plant scale experiments. *Bioresource Technology,* 101, 193-199.

Kleinschmit, D. H., Schingoethe, D. J., Kalscheur, K. F. & Hippen, A. R. 2006. Evaluation of various sources of corn dried distillers grains plus solubles for lactating dairy cattle. *Journal of Dairy Science,* 89, 4784-4794.

Knott, J., Shurson, J. & Goihl, J. 2003. Variation in particle size and bulk density of distiller's dried grains with solubles (DDGS) produced by" new generation" ethanol plants in Minnesota and South Dakota. *Report accessed at 21 Mar 2011. Available at* http://www.ddgs.umn.edu/articles-proc-storage-quality/2004-Knott-%20Variation.pdf.

Lammers, P., Honeyman, M., Bregendahl, K., Kerr, B., Weber, T., Dozier, W. & Kidd, M. 2007. Energy value of crude glycerol fed to pigs. *Animal Science Leaflet R,* 2225, http://www.ans.iastate.edu/report/air/2007pdf/R2225.pdf

Lammers, P. J., Kerr, B. J., Weber, T. E., Dozier, W. A., III, Kidd, M. T., Bregendahl, K. & Honeyman, M. S. 2008. Digestible and metabolizable energy of crude glycerol for growing pigs. *Journal of Animal Science,* 86, 602-608.

Larson, E., Stock, R., Klopfenstein, T., Sindt, M. & Huffman, R. 1993. Feeding value of wet distillers byproducts for finishing ruminants. *Journal of Animal Science,* 71, 2228.

Linneen, S. K., DeRouchey, J. M., Dritz, S. S., Goodband, R. D., Tokach, M. D. & Nelssen, J. L. 2008. Effects of dried distillers grains with solubles on growing and finishing pig performance in a commercial environment. *Journal of Animal Science,* 86, 1579-1587.

Liu, N., Ru, Y. J., Tang, D. F., Xu, T. S. & Partridge, G. G. 2011. Effects of corn distillers dried grains with solubles and xylanase on growth performance and digestibility of diet components in broilers. *Animal Feed Science and Technology,* 163, 260-266.

Lumpkins, B., Batal, A. & Dale, N. 2005. Use of distillers dried grains plus solubles in laying hen diets. *Journal of Applied Poultry Research,* 14, 25-31.

Lumpkins, B. S., Batal, A. B. & Dale, N. M. 2004. Evaluation of distillers dried grains with solubles as a feed ingredient for broilers. *Poultry Science,* 83, 1891-1896.

Mandiki, S., Bister, J., Derycke, G., Wathelet, J. P., Mabon, N., Marlier, N. & Paquay, R. 1999. Optimal level of rapeseed meal in diets of lambs. *10th Int. Rapeseed Congress, Canberra, Australia. http://www.regional.org.au/au/gcirc/1/256.htm.*

Martinez-Amezcua, C., Parsons, C. M., Singh, V., Srinivasan, R. & Murthy, G. S. 2007. Nutritional characteristics of corn distillers dried grains with solubles as affected by the amounts of grains versus solubles and different processing techniques. *Poultry Science,* 86, 2624-2630.

Mawson, R., Heaney, R., Zdunczyk, Z. & Kozlowska, H. 1995. Rapeseed meal Glucosinolates and their antinutritional effects Part 6. Taint in end products. *Food/Nahrung,* 39, 21-31.

McGinn, S. M., Chung, Y. H., Beauchemin, K. A., Iwaasa, A. D. & Grainger, C. 2009. Use of corn distillers' dried grains to reduce enteric methane loss from beef cattle. *Canadian Journal of Animal Science,* 89, 409-413.

McKeown, L. E., Chaves, A. V., Oba, M., Dugan, M. E. R., Okine, E. & McAllister, T. A. 2010. Effects of corn-, wheat- or triticale dry distillers' grains with solubles on in vitro fermentation, growth performance and carcass traits of lambs. *Canadian Journal of Animal Science,* 90, 99-108.

McKinnon, J. J. & Walker, A. M. 2008. Comparison of wheat-based dried distillers' grain with solubles to barley as an energy source for backgrounding cattle. *Canadian Journal of Animal Science,* 88, 721-724.

McKinnon, J. J. & Walker, A. M. 2009. Evaluation of feedlot backgrounding rations formulated using co-products of biofuel production. *Canadian Journal of Animal Science,* 89, 137-137.

McNeill, L., Bernard, K. & MacLeod, M. G. 2004. Food intake, growth rate, food conversion and food choice in broilers fed on diets high in rapeseed meal and pea meal, with observations on sensory evaluation of the resulting poultry meat. *British Poultry Science,* 45, 519-523.

Meyer, U., Schwabe, A., Danicke, S. & Flachowsky, G. 2010. Effects of by-products from biofuel production on the performance of growing fattening bulls. *Animal Feed Science and Technology,* 161, 132-139.

Min, Y. N., Liu, F. Z., Wang, Z., Coto, C., Cerrate, S., Costa, F. P., Yan, F. & Waldroup, P. W. 2008. Evaluation of distillers dried grains with solubles in combination with glycerin in broiler diets. *International Journal of Poultry Science,* 7, 646-654.

Mjoun, K., Kalscheur, K. F. & Garcia, A. D. 2011. Fermentation characteristics and aerobic stability of wet corn distillers grains with solubles ensiled in combination with whole plant corn. *Journal of the Science of Food and Agriculture,* 91, 1336-1340.

Murta, R.M., Chaves, M.A., Vieira e Silva, F., Buteri, C.B., Fernandes, O.W.B. & dos Santos, L.X. 2009. Carcass characteristics of feedlot sheep fed with sugarcane bagasse hydrolyzed with calcium oxide. *Ciencia Animal Brasileira* 10, 438-445.

Nagalakshmi, D. & Reddy, D. N. 2010. Effect of feeding expander extruder processed complete diet containing sugarcane bagasse on performance of Murrah buffaloes. *Animal Nutrition and Feed Technology,* 10, 1-8.

Olukosi, O. A., Cowieson, A. J. & Adeola, O. 2010. Broiler responses to supplementation of phytase and admixture of carbohydrases and protease in maize-soyabean meal diets with or without maize Distillers' Dried Grain with Solubles. *British Poultry Science,* 51, 434-43.

Oryschak, M., Korver, D., Zuidhof, M. & Beltranena, E. 2010a. Nutritive value of single-screw extruded and non extruded triticale distillers dried grains with solubles, with and without an enzyme complex, for broilers. *Poultry Science,* 89, 1411-1423.

Oryschak, M., Korver, D., Zuidhof, M., Meng, X. & Beltranena, E. 2010b. Comparative feeding value of extruded and nonextruded wheat and corn distillers dried grains with solubles for broilers. *Poultry Science,* 89, 2183-2196.

Pedersen, C., Boersma, M. G. & Stein, H. H. 2006. Digestible and metabolizable energy in ten samples of corn distillers dried grains with solubles (DDGS) fed to growing pigs. *Journal of Animal Science,* 84, 120-120.

Pedersen, C., Boersma, M. G. & Stein, H. H. 2007. Digestibility of energy and phosphorus in ten samples of distillers dried grains with solubles fed to growing pigs. *Journal of Animal Science,* 85, 1168-1176.

Popp, J. 2008. Biofuel production and its international implications. *Gazdalkodas,* 52, 13-25, 94-95.

Ramli, M. N., Higashi, M., Imura, Y., Takayama, K. & Nakanishi, Y. 2005. Growth, feed efficiency, behaviour, carcass characteristics and meat quality of goats fed fermented bagasse feed. *Asian-Australasian Journal of Animal Sciences,* 18, 1594-1599.

Ranathunga, S. D. 2008. Replacement of starch from corn with non-forage fiber from distillers grains in diets of lactating dairy cows. *Journal of Dairy Science,* 93, 1086-1097.

Roberson, K. D., Kalbfleisch, J. L., Pan, W. & Charbeneau, R. A. 2005. Effect of corn distiller's dried grains with solubles at various levels on performance of laying hens and egg yolk color. *International Journal of Poultry Science,* 4, 44-51.

Saunders, J. A. & Rosentrater, K. A. 2009. Survey of US fuel ethanol plants. *Bioresource Technology,* 100, 3277-3284.

Schaafsma, A. W., Limay-Rios, V., Paul, D. E. & Miller, J. D. 2009. Mycotoxins in fuel ethanol co-products derived from maize: a mass balance for deoxynivalenol. *Journal of the Science of Food and Agriculture,* 89, 1574-1580.

Schieck, S. J., Shurson, G. C., Kerr, B. J. & Johnston, L. J. 2010. Evaluation of glycerol, a biodiesel coproduct, in grow-finish pig diets to support growth and pork quality. *Journal of Animal Science,* 88, 3927-3935.

Schöne, F., Kirchheim, U., Schumann, W. & Lüdke, H. 1996. Apparent digestibility of high-fat rapeseed press cake in growing pigs and effects on feed intake, growth and weight of thyroid and liver. *Animal Feed Science and Technology,* 62, 97-110.

Seabolt, B. S., Heugten, E. v., Kim, S. W., Ange-van Heugten, K. D. & Roura, E. 2010. Feed preferences and performance of nursery pigs fed diets containing various inclusion amounts and qualities of distillers coproducts and flavor. *Journal of Animal Science,* 88, 3725-3738.

Shalash, S. M. M., El-Wafa, S. A., Sayed, M. A. M., El-Gabry, H. E., Ramadan, N. A. & Mohamed, M. S. 2009. Nutritive value of distillers dried grains with soluble and broiler performance at starter period. *International Journal of Poultry Science,* 8, 783-787.

Shurson, G. C. 2005. Issues and opportunities related to the production and marketing of ethanol by-products. *USDA agricultural outlook forum,* Washington, DC.

Silva, R. R., Silva, F. F., Carvalho, G. G. P., Franco, I. L., Veloso, C. M., Chaves, M. A., Bonomo, P., Prado, I. N. & Almeida, V. S. 2005. Eating behaviour of confined mixed heifers of Holandes and Zebu cross. *Archivos de Zootecnia,* 54, 75-85.

Simon, A., Schwabe, M. & Bergner, H. 1997. Glycerol supplementation in broiler rations with low crude protein content. *Archiv für Tierernährung,* 50, 271.

Skiba, F., Metayer, J. P., Clave, H. & Quentin, M. 2009. Nutritional value of a corn DDGS and effects on the growing performances of broiler chickens. *World Poultry Science Association (WPSA), Proceedings of the 8th Avian French Research Days, St Malo, France.*

Spiehs, M. J., Whitney, M. H. & Shurson, G. C. 2002. Nutrient database for distiller's dried grains with solubles produced from new ethanol plants in Minnesota and South Dakota. *Journal of Animal Science,* 80, 2639-2645.

Srinivasan, R., Moreau, R. A., Rausch, K. D., Belyea, R. L., Tumbleson, M. E. & Singh, V. 2005. Separation of fiber from distillers dried grains with solubles (DDGS) using sieving and elutriation. *Cereal Chemistry,* 82, 528-533.

Swiatkiewicz, S. & Koreleski, J. 2008. The use of distillers dried grains with solubles (DDGS) in poultry nutrition. *Worlds Poultry Science Journal,* 64, 257-265.

Thacker, P. A. & Petri, D. 2009. The effects of canola or mustard biodiesel press cake on nutrient digestibility and performance of broiler chickens. *Asian-Australasian Journal of Animal Sciences,* 22, 1531-1539.

Valino, E. C., Elias, A., Torres, V., Carrasco, T. & Albelo, N. 2004. Improvement of sugarcane bagasse composition by the strain Trichoderma viride M5-2 in a solid-state fermentation bioreactor. *Cuban Journal of Agricultural Science,* 38, 143-150.

van der Aar, P.J. and Doppenberg, J. 2009. Biofuels: Consequences for Feed Formulation *Feedinfo,* 6pp.

Vander Pol, K. J., Erickson, G. E., Klopfenstein, T. J., Greenquist, M. A. & Robb, T. 2006. Effect of dietary inclusion of wet distillers grains on feedlot performance of finishing cattle and energy value relative to corn. *Nebraska Beef Cattle Report,* 51-53.

Wilson, J., Whitney, M. H., Shurson, G. & Baidoo, S. 2003. Effects of adding distillers dried grains with solubles (DDGS) to gestation and lactation diets on reproductive performance and nutrient balance in sows. *Journal of Animal Science,* 81, 47-48.

Xu, G., Baidoo, S., Johnston, L., Cannon, J. & Shurson, G. 2007. Effects of adding increasing levels of corn dried distillers grains with solubles (DDGS) to corn-soybean meal diets on growth performance and pork quality of growing-finishing pigs. *Journal of Animal Science,* 85, 76.

Yang, Y., Kiarie, E., Slominski, B. A., Brule-Babel, A. & Nyachoti, C. M. 2010. Amino acid and fiber digestibility, intestinal bacterial profile, and enzyme activity in growing pigs fed dried distillers grains with solubles-based diets. *Journal of Animal Science,* 88, 3304-3312.

Zade, M. S., Dabiri, N., Fayazi, J. & Pour, M. B. 2009. Determination of optimum level of steam pressure hydrolyzed pith bagasse in feeding of crossbred beef. *Research Journal of Biological Sciences,* 4, 773-776.

2

The Future Crops for Biofuels

Chunlin Liu[1] and Chuanxin Sun[2*]
[1]Hunan Provincial Key Laboratory of Crop Germplasm Innovation and Utilization,
Hunan Agricultural University, Changsha,
[2]Department of Plant Biology and Forest Genetics, Uppsala BioCenter,
Swedish University of Agricultural Sciences,Uppsala,
[1]China
[2]Sweden

1. Introduction

The Sun is sending around 174 peta watts of light energy to the Earth where only about 120 peta watts can reach the Earth surface after energy losses during penetrating of light through the Earth's atmosphere. In fact, the truly used solar energy for production of biomass on land is only 74 tera watts after counting arable land and photosynthesis efficiency, which is ca. 6% (the arable land area / the Earth area) and ca. 1%, respectively. Presently, the total human consumption requires about 15 tera watts, therefore it still makes the total bioenergy production on land 5 times more than what the human being consumes. Thus, the major question now and in the near future is how to efficiently transfer bioenergy from biomass to biofuels. This chapter will describe the current production of the four forms of biomass ie sucrose, starch, ligno-cellulose and plant oil by different bioenergy crops. It will also briefly describe how to convert the biomass to biofuels by different methods. An analysis of energy conversion efficiency during converting biomass to biofuels and an economic evaluation on using the different biomass will be carried out. A suggestion and discussion will be brought up on how to change biomass from one form to another *in vivo* and how to increase biomass production in the future crops for efficient use of bioenergy and economic production of biofuels.

2. Why biofuels?

2.1 Depletion of the fossil oil and development of the oil price

Since 1870s when Julius Hock in Vienna invented the world's first oil machine, and Etienne Lenoirand and Alphonse Bear de Rochas created the first car running on petroleum based fuel (Bryan and Hellemans, 1993), the human needs of all modes of transport gradually increased, resulting in a large number of production of the modern transportation such as automobiles, ships, trains and aircrafts. These modern means of transport led to the human increased demand for oil, in consequence exacerbating the exploitation of oil. Almost everyone knows that, the worldwide surge in oil extraction will deplete the oil out of the Earth sooner or later as the Earth has a certain amount of oil stored. Figure 1 shows the last forty five years of global oil consumption and exploitation (BP, 2010). According to this trend and according to most of the statistical predictions (Duncan RC, 2000), the Earth's stored oil will run out in the coming five decades despite the projected more or less inconsistent. In Figure 1 the gap between the future demand and production and the future form of energy to supplement the gap will be the challenges placed in front of the human focus. As this gap enlarges, the attendant thought-provoking side effects increase, such as the impact on the economy. The potential side effects on the global economy can be seen in the development of crude oil price in the last thirty years with an increase of triple and even four times on crude fossil oil price (data from BP, 2010). It is estimated that the gap between the demand and production in Figure 1 can bright the world price of crude oil to sky-high in the future (see also Duncan RC, 2000). Oil consumption and depletion shake the world economy not only directly through the soaring oil prices, but also by increasing the atmospheric greenhouse effect and the indirect damage to the environment and human health to generate greater economic losses, and a threat on human survival.

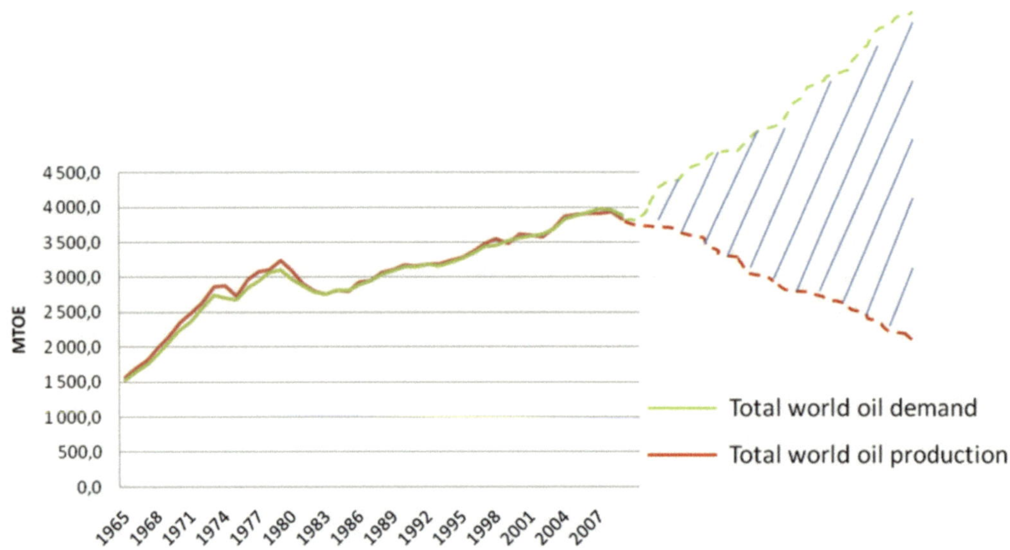

Fig. 1. The world fossil oil consumption and production from 1965 to 2009 (BP, 2010) and possible demand and production in the future. The gap with slashes indicates that additional energy supplies are needed to meet the global oil energy demand. MTOE (Million Tonnes of Oil Equivalent).

2.2 Increased concentrations of the most important greenhouse gases in the last decades

Green house effect is that the atmosphere can capture the radiation to keep temperature in different parts of the Earth relatively stable. Human activities can enhance the greenhouse effect on Earth, and may cause global warming which makes the temperature in different parts of the Earth unstable (Leaf et al, 2003). Specific meteorological data show that from 1980 to 2002, the earth's average temperature increased 0.3 degrees per year (Environment News Service, 2007). It is estimated that this upward trend will continue, by 2100 the Earth's average temperature will be 6 °C higher than the current (Barker et al, 2007). Global warming is largely caused by increased concentrations of some green-house gases.

According to the content in the atmosphere, the ranked top five gases are, water vapor, carbon dioxide, methane, nitrous oxide and ozone. They increase the impact of global warming, respectively, 60% (water vapor), 26% (carbon dioxide), 6% (methane and nitrous oxide), 8% (ozone) (Kiehl and Trenberth, 1997). The exhaust emissions from vehicles contain at least two of them, namely carbon dioxide and nitrous oxide. We first look at the most recent decades of atmospheric carbon dioxide and nitrous oxide concentration and then, briefly discuss about methane and water vapor on the impact of global warming. Concentration of carbon dioxide in the atmosphere far from the industrial revolution some 150 years ago, has been increased from 280 ppm (parts per million) to today's 380 ppm (Barker et al, 2007). Following the trend, in 2100 the concentration of carbon dioxide in the atmosphere will reach more than 550 ppm. The increases of atmospheric carbon dioxide concentration can be directly linked to human activities that have employed energy materials such as oil, coal and natural gas in the high intake of energy and release of carbon to the atmosphere. The release of carbon has significantly increased since humans used a variety of fuels after the industrial revolution, particularly in the last four decades (Yu et al, 2008). The rapid increase of the concentration of carbon dioxide in the atmosphere in the last four decades may be due to the human consumption of large quantities of fuels, especially the oil consumption. Atmospheric nitrous oxide concentration in the similar period of years (nearly 150 years), from 270 ppb (parts per billion) increased to 310 ppb (Prather and Hsu, 2010; Aneja et al, 2009). One major reason for this increase may be the use of chemical fertilizers in agriculture and the global fuel consumption, especially oil use.

The content of methane in the atmosphere 200 times smaller than that of carbon dioxide and today's concentration is about 1.8 ppm (Sass and Cicerone, 2002; Bousquet, P. et al, 2006). However, due to its greater molecular volume than carbon dioxide, easy to absorb long wavelength radiation, the molecules absorb the energy unit (relative absorption power per molecule) is much higher than carbon dioxide (Bousquet, P. et al, 2006). It creates the greenhouse effect almost one-fifth of carbon dioxide. The lead to elevated concentrations of methane in the atmosphere is mainly (2 / 3) caused by a variety of human activities. The release of methane from human activities is of the three most important sources, fuel and gas mining, rice paddies and enteric fermentation from animals (Sass and Cicerone, 2002).

Water vapor is the most important component of the greenhouse effects. However, the production of main water vapor is naturally occurring. And human exploitation and utilization of the fuel may not be immediately linked to the content of water vapor in the atmosphere. But it is worth noting that with the warming of the earth, atmospheric water vapor will be more and more, so that warming of the Earth may have a snowball effect. To reduce global water evaporation or reduce water content in the atmosphere, one of the ways may be to increase the green coverage on the Earth's land area, that is, the development of bioenergy, the two fold purposes.

2.3 Economic losses due to the climate changes

Earth's climate change or global warming has brought the threats on human health and survival. These threats are mainly for instance in the following aspects, such as melting glaciers and rising sea levels, rampant diseases, extreme weather, food crisis, the extinction of species. It is extremely difficult to use the data to accurately estimate the damage by global warming. It might be more intuitive to examine the direct economic losses in recent years and in the future, in a number of countries in the world, caused by global warming. It is estimated that over the past five years (2006-2010) the global warming has caused direct economic losses of about 350 billion USD in the world (Akomo B, 2011). Table 1 is a rough estimate, predicting that the average temperature of the Earth increased 2.5 °C, different countries in the world would meet the economic loss (loss accounts for a percentage of GDP). In 2010, for example, the world's GDP was 74,000 billion USD (World Bank, 2011), if the surge in the world average temperature was 2.5 °C, it is estimated that the global economic losses could be 74,000 x 1.9% = 1,400 billion USD.

Country	Loss (+) and Gain (-) relative to GDP (%)
India	4.9
Africa	3.9
OECD Europe	2.8
Eastern Europe	0.7
Japan	0.5
United States	0.5
China	0.2
Russia	-0.7
The world average	1.9

Table 1. An assessment of putative impacts on the world economy of possible climate changes with an increase of 2.5 °C on average temperature (data source: Unions for Jobs and the Environment, 2002).

A detailed assessment was released by Natural Resources Defense Council (Natural Resources Defense Council, 2008) on impacts of the global warming in the future years on the US economy in four sections. These four aspects are, the hurricane hit, the real estate industry, energy and flood. By 2100 (possibly an increase of 6 °C on the world average temperature, see also above), the United States in these four aspects will get the economic loss of nearly 1.9% of the GDP, similar to the average loss of the world with an increase of 2.5 °C on the world average temperature (Table 1). Some tropical or subtropical countries, such as Africa, India, Southeast Asia, South America, South Europe and other regions may get affects even more.

2.4 Biofuels are two fold, the best and maybe only resolution on the dual problems

The data from all of the above sees that the world faces two major serious energy fuel related problems. 1, Depletion of the fossil fuels is getting close. Finding of new energy fuels is imminent. 2, Air pollution and global warming caused by the consumption of fossil fuels, must be improved. New energy fuels must be beneficial to the global environment, so to ease or stop global warming. What kind of new energy fuels can simultaneously solve these two problems or meet the above two conditions? It was a period of time human beings have a great affinity for nuclear energy as it is effective, together with small occupation of land areas. But recently on March 11, 2011 the earthquake in Japan caused leakage of radioactive materials from a nuclear power plant, illustrating once again that nuclear energy may not be the best choice.

The biomass and biomass derived fuels can be the perfect solution to the two challenging problems brought to mankind by fossil fuels. Biomass of plants on Earth uses photosynthesis to transform solar energy into chemical energy stored in biomass or carbohydrates, while consuming atmospheric carbon dioxide and water on the planet. Biofuels require solar energy, atmospheric carbon dioxide and water on the planet, therefore lifelong with the Sun and the Earth. Unlike fossil fuels, they are not subject to geographical location (such as oil zone), and human factors (such as war). Their producers are the green plants that cover the land surface of the Earth. The green plants play a direct role in mitigation of the two largest greenhouse effect gases i.e. water vapor and carbon dioxide. When biofuels are consumed, they release carbon dioxide and water. This is a perfect balance for the environment on the entire planet. Use of plants (known as the crops in agriculture) to produce bioenergy and biofuels is nearly perfect for the Earth. However, most of the concerns are whether the biomass and biofuels can produce enough energy to replace fossil fuels today, that is, to fill in the gap in Figure 1.

3. Can biofuels meet the global energy requirements?

One of the most critical issues when using biofuels to replace fossil fuels is, whether bio-energy can meet the human energy consumption. One way to answer this question may be to comprehensively analyze the energy budget on the Earth and the human energy demand. In Figure 2, we do a calculation on how much energy the crops can make from the solar energy. The Sun to the Earth (using the Earth disc for the calculation) is sending the energy of about 174,000 Tera Watts (TW, 1W = J / sec) (Rhodes, 2010). After the Earth's atmosphere (the atmospheric reflection and absorption) the energy reaches the Earth surface of about 120,000 TW. In the surface of the Earth, the land accounts for about 30% and available in the land to farming (growing crops) land area for about 20%. The average crop photosynthetic efficiency (conversion of light energy to bioenergy) is about 1%. Thus, humans use crops to produce bioenergy in total: 120,000 x 30% x 20% x 1% = 72 TW. And human needs of total energy at present are about 15 TW (Rhodes, 2010). It would appear that the crop bioenergy produced is five times more than what is needed for the total energy required for human activities including food, enough to meet the needs of mankind. It must be noted that the present average efficiency of changing bioenergy into the energy of biofuels is less than 20% (Tan et al, 2010; Octave and Thomas, 2009). This means that the human production of biofuels at the current situation is difficult to meet the energy needs of mankind. Moreover, the human must reserve parts of the crops to produce food. Accelerating the increase in world population is, the demand for energy and food growing. This has given human production and use of biofuels a challenging problem at the moment. The problem is how to improve conversion efficiency from the solar energy to biofuel energy. In the long term, each step in Figure 2 should be considered as for an improvement. But for the present situation, it might be mainly to solve the bottleneck problem, i.e. converting biomass to biofuels more efficiently.

4. Targeting the bottlenecks in biofuel production

It is quite obvious from Figure 2, there are three bottlenecks in the flow chart, 1, Land limitation. As much as possible the bioenergy crops should be cultivable aquatic plants, and easy to grow on low quality of land and adverse environmental areas, such as desert, barren mountains and marsh. 2, Photosynthesis. It is extremely important to improve the photosynthetic efficiency of plants or crops. 3, Energy conversion efficiency from biomass to biofuels. The points of 1, 2 and 3 above should be all strategic bottlenecks. However, from

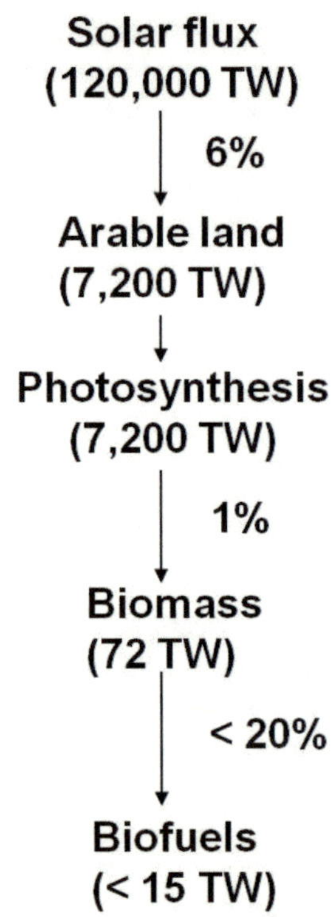

Fig. 2. A flow chart of the solar energy to the energy in biofuels based on the energy conversion efficiency of current biofuel production.

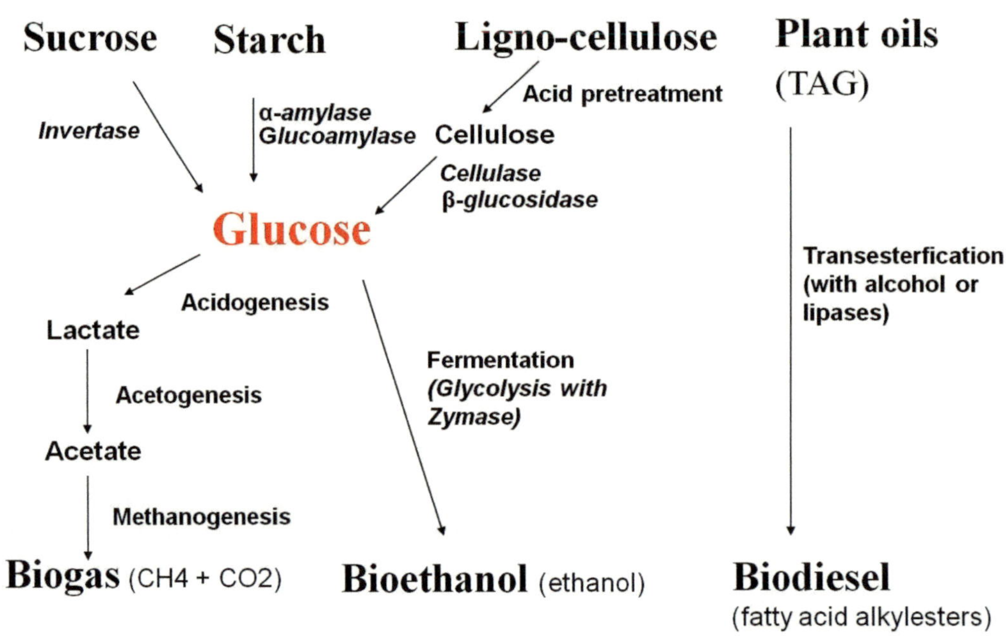

Fig. 3. A road map of biofuel production.

the current tactical point of view and existing biotechnology, the third point is an area worthy of a tremendously large effort. A basic reason is that in the coming years the production of bioenergy by crops far exceeds the total demand of mankind. How to make full and effective use of crop biological materials and turn them into biofuels, is the challenge before mankind, and it is also a very practical and realistic question. The current use of biomass to produce biofuels means that the use of sugar, starch, cellulose and plant oil to produce biogas, bioethanol and biodiesel (Figure 3; Tan et al, 2010; Octave and Thomas, 2009). Figure 3 is one of the typical presentations on production of biogas, bioethanol and biodiesel.

The chart from a technical point of view indicates that using of the sugar, starch, cellulose to produce biogas and bioethanol requires expensive enzymes. In addition, loss of energy in the fermentation process is large (Ohlrogge et al, 2009; Tan et al, 2010). In contrast, using vegetable oils to produce biodiesel is relatively simple, without the expensive enzymes and much energy loss. It might be one direction for the future development of biofuels (Ohlrogge et al, 2009).

Current crops for production of biomass and biofuels

Currently, the world's major bioenergy crops are few and listed in Table 2 according to their major production of the used biomass. Sometimes it is difficult to categorize a certain type of crop. Such as sugar cane, presently, the major biological materials being used are Syrup (Sucrose). And it can also be used as straw bagasse (cellulose) to produce bioethanol. Another example is maize, which contains the seeds for both production of a large number of endosperm starch, and the rich embryo oil. And it also contains a lot of straws (cellulose) for bioethanol production.

Biomass	Crop
Sucrose	Sugarcane, sugar beet and sweet sorghum
Starch	Maize, other cereals and cassava
Cellulose	Trees, switch grass, miscanthus and cereal straws
Plant oil	Oil palm, soy bean, rape seed, sun flower and cereals

Table 2. The currently major crops for biofuel production (data source: FAO, 2009)

4.1 Sucrose based bioethanol production

Using sugar to produce bioethanol is a rather ideal way based on the energy conversion efficiency, especially after the discovery of the bacteria *Zymonas mobilis* used for fermentation (Rogers et al, 2007). The bacteria can use both glucose and sucrose as subtracts with the energy conversion efficiency, 100% and 70%, respectively, higher than yeast. In addition, through genetic engineering, some of the bacterial strains can be improved for a higher conversion rate of sucrose. The bacterial fermentation is an ideal tool. Currently, the world's major crops for the production of sugar are sugar cane, sugar beet, and some crops, including sweet sorghum (FAO, 2009). In 2009, the world's total output of sugar from the three crops were, respectively, 1,683 M tonnes, 229 M tonnes and 0.9 M tonnes (FAO, 2009). Purely theoretical calculation according to ($H_20 + C_{12}H_{22}O_{11} \rightarrow 4C_2H_60 + 4\ CO_2$), gives that each kilogram of sucrose can yield about 538.4 grams of ethanol or 681.5 mL alcohol (using a basis of 0.790 grams of alcohol per mL). Thus, the only sugar produced from sugar cane, the total 1,683 M tonnes, can produce about 1,147 billion liters of bioethanol. According to the statistical data from the International Organization of Motor Vehicle Manufacturers (Worldometers, 2011), by 2009 around the world, there were around 600 Million passenger cars. We assume that the cars were all over the world bioethanol cars, with alcohol consumption per vehicle on average about 8.5 km / liter, the average exercise of 15,000 km per year. This is about 1,765 liters per year per vehicle bioethanol. Total worldwide cars of 600 M would consume 600 M x 1,765 L = 1,059 billion L bioethanol. That means that, by purely theoretical calculations, the only production of sugarcane bioethanol is enough for car use around the world. But, in fact, the production of sugar cane ethanol is far less from the theoretical calculation. In 2009, the total world's bioethanol production (including sugar cane), was about 76.6 billion L (RFA, 2011). One of the most important reasons is the production of ethanol from sugar cane in all aspects of technology is still not perfect (needs much improvement). In addition, sugar cane is limited to tropical and subtropical countries, notably Brazil, India, China and Thailand. Other sugar crops are also very important, such as sugar beet and sweet sorghum. Currently, the development of sweet sorghum seems to be particularly attractive. With the basic advantages of sugar cane, however, sweet sorghum has much wider planting areas on earth than sugar cane. In addition, it also has many other advantages. 1, Sorghum is a diploid plant, its genome is relatively small (approximately 730 Mbp), some variety genome have been sequenced or are being sequenced. This leads to a larger advantage of genetic breeding than sugar cane. 2, Sorghum has high resistance (biotic and abiotic). 3, And, like sugar cane, sorghum is a C4 plant with high efficiency of photosynthesis. From the current situation, sweet sorghum is a very promising sugar crop.

4.2 Starch based bioethanol production

Production of bioethanol is currently with the world's top five regions and countries, the United States, Brazil, EU, China and Canada (RFA, 2011). Among them, the United States and Brazil account for the world nearly 90% of total ethanol production. At present the main crops used in the regions are corn in the U.S.; sugar cane in Brazil; sugar beet, wheat and barley in the EU; maize and cassava in China; and corn in Canada. It is quite easy to see that in the world, currently the main crops used for starch based bioethanol production are corn, wheat, barley and cassava. Among them, corn is the main crop for production of bioethanol. Currently, the existence of important and debated issues of using starch crops to produce bioethanol is, 1, Starch crops are food crops. Using food crops to produce biofuels is the concern of most of the countries in the world. In fact, the use of food crops to produce bioethanol has directly or indirectly caused soaring food prices. 2, An economic issue with the important food crops to produce bioethanol. Corn, for example, according to a number of data produces biothanol with energy output and input ratio of about 1.3 to 1.6, much lower than that of sugar cane with 8.3 to 10.2 (Goettemoeller and Goettemoeller, 2007; Baligar et al, 2001; Gaur and Reed, 1998; Rooney, 1998; Tan et al, 2010). 3, A number of food crops such as corn, wheat and barley crops are annual crops and need a large amount of fertilizers for agricultural production, and certain amount of agricultural management of inconvenience. Large increase in fertilizers will produce greenhouse gases such as N_2O. At present, in order to solve the three problems, people are looking for a variety of ways, for example, i) Using of food crop straws to produce bioethanol, such as corn stalks, wheat and barley straws. ii) Going for some starchy crops such as cassava, which are economic, cost-effective, stress resistant, easy-managing, and none or little-fertilizing. iii) Using cellulose based bioethanol production to completely replace starch based bioethanol production, such as the use of forests, switch grass and miscanthus.

4.3 Cellulose based bioethanol production

The world's cellulose based bioethanol production is still in the very early stage, maybe in the research and development (RD) stage. The reason why this area is being prefered, may be, 1, Cellulose is the world's largest biomass. Only forests account for 80% of the biomass on earth. The use of cellulose as raw materials for biofuels may be the cheapest. 2, Most of the cellulose productions do not create any competition in land and starch production with food crops. 3, All food crops contain large amounts of straw cellulose. The use of food crop straws to produce bioethanol may add more economic value to food crops. However, the main problem of production of bioethanol with cellulose is the economic issue, in most of cases not economic at present based the current tactical situations. The main reason is that, cellulose in nature, mainly in plant cell walls with lignin and hemicellulose together. To effectively enable enzymatic degradation of cellulose, the cell wall structures should be deconstructed, the lignin has to be removed and hemicellulose should also be degraded simultaneously. To destroy the structure and to remove the lignin some harsh physical or chemical treatment processes are required (Tan et al, 2010; Octave and Thomas, 2009). In addition, fermentation of 5-carbon sugar from hemicelluloses is not effective at present (Tan et al, 2010; Octave and Thomas, 2009). These factors limit the current large-scale cellulose based bioethanol production. Prediction from the current situation indicates that in the future the main raw materials for the cellulose based bioethanol production may be, 1, Forest (the main Earth´s biomass). 2, Some fast growing plants such as switch grass, miscanthus, hybrid aspen and willow. 3, Crop residues such as corn stover; rice, wheat and barley straws and cassava stems.

4.4 Plant oil based biodiesel production

In the above, we made the assumption that if the world 600 M cars were bioethanol driven, then a total of about 1,059 billion L bioethanol per year would be of necessity. We can make a calculation on, if the similar vehicles of 600 M around the world were biodiesel cars, how much biodiesel would be required for the whole world per year. We assume that our average consumption of biodiesel per vehicle is about 17 km / L and the average requires the exercise of 15,000 kilometers per year. The total of such vehicles of 600 M worldwide would consume 530 billion L of biodiesel yearly. Currently the world every year (data from 2008) (EMO, 2011) produces about 13 billion L of biodiesel, far from the human needs. We can see from Table 3 that, in 2009 the world's total output of vegetable oil was about 136 M tonnes or 160 billion L (roughly, 1.176 L/kg). Even if all production of vegetable oil would be used for biodiesel, it would not be enough to run the total cars around the world. Among the vegetative oil production, palm oil, soy bean, rape seed and sunflower accounted for 86% of the total world´s production. Only palm oil accounted for about 35%, while palm oil is limited to the tropical cultivation. This indicates a problem that the human needs to develop oil crops, and breed new oil crops.

Oil crop	Production (M Tonnes)
Palm and kernel oil	47.1
Soy bean	36.1
Rape seed	21.2
Sun flower	13.1
World Total	136

Table 3. The major oil crops and their oil production in 2009 (data source: FAO, 2009)

5. The future crops for biofuels

5.1 What is required for the future bioenergy crops

From the analyses above and bottleneck targeting on the current situation of biofuel production, we suggest that the future crops for biofuels may need the following requirements (see also Figure 4).

1. high yielding
2. easy to grow, there is a strong resistance capacity
3. does not occupy a lot of high-quality land
4. multiple purposes
5. easy breeding in order to get more performance traits

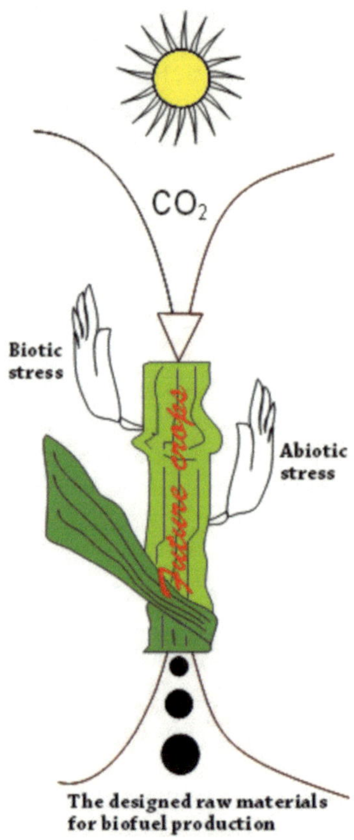

Fig. 4. The future crops with the designed raw materials for biofuel production.

1. High yielding

Many factors influence and determine the trait of high yielding. To the special nature of bioenergy crops, as a bioenergy crop its photosynthetic efficiency must be high, such as the C4 crops, sugarcane, sorghum, maize, or the crops like cassava and rice with high photosynthetic efficiency. They should be able to effectively convert the product (sucrose or carbohydrates) of photosynthesis and deposit them for production of biofuels-needed biomass, i.e. sucrose, starch, cellulose or oil. From the point of energy conversion efficiency of view, sucrose and oil based biofuel production is preferable in a short run, as the total bioenergy in the current crops exceeds the human needs.

2. Easy to grow, there is a strong resistance capacity

The future bioenergy crops should be easy to grow, including no needs for a lot of manpower, management and fertilizers. Also should have high resistance to both biotic and abiotic stresses. Switch grass, miscanthus, some cereals and cassava have shown such characters.

3. Does not occupy a lot of high-quality land

As elucidated in Figure 2, the future bioenergy crops should not occupy a lot of high-quality land. These high-quality land should be served as much as possible for the production of food crops to meet the rapidly growing human population and demands for food. This may be one of the reasons why switch grass, miscanthus, algae, trees and cassava have recently been as favorites.

4. Multiple purposes

The next generation crops should have multiple uses, especially for food and oil crops. They should produce digestible and easy-digesting straws. One can even breed a two fold crop. For example, the seeds of such crops would produce starch for food and their vegetative tissues produce vegetable oil. For multiple purpose crops, a good example is the cassava plant as reviewed recently (Westerbergh et al, 2011; Jansson et al, 2009).

5. Easy breeding in order to get more performance traits

With the development of modern biotechnology, especially in the recent systems biology and the genome sequencing of various plants, molecular breeding has been greatly facilitated. The future bioenergy crops should be very easy to change through modern breeding techniques to produce the crops needed by the human demands of the various products.

5.2 Breeding bioenergy crops in the future - some prospects

1. Starch to oil

Why starch into oil? First, we use corn as an example. Corn produces all four forms of biomass, sucrose, starch, cellulose and oil. The major production is corn starch (endosperm), cellulose (straw) and oil (embryo). From the current process data, energy conversion efficiency from corn starch, cellulose and oil into bioethanol and biodiesel are 46%, 43% and > 90%, respectively (Baligar et al, 2001; Gaur and Reed, 1998; Rooney, 1998; Tan et al, 2010). It is easy to see that if the corn could *in vivo* produce more oil, corn would no doubt increase the energy conversion efficiency. Second, in the above section we have already mentioned, there are very few world oil crops at present, namely palm oil, soy bean, rape seed and sun flower. While confined to tropical palm oil, soy bean and sun flower can occupy a very large area of high-quality land in addition to their other important uses such as soy bean as main protein recourses. Rape seed can not grow without mid-crop rotation. Thus, the development of more oil crops is imminent. Some cereals are ideal candidates, such as barley, oats. First of all, they are high yielding. They have strong resistance and they can grow in many places where the growth of rape seed is limited, such as very north of Northern Europe and North America.

1.a Converting some of the food crops to oil crops

The issue is to change storage starch into oil. With the development of system biology, a new technology known as transcription factor based technology in the field of molecular biology has just come (Century et al, 2008). The core of the technology is through up regulating or down-regulating the expression of transcription factors to enhance cell metabolism to a particular direction. The technology has been staged in animals, plants and has shown its influence and potential (Shen et al, 2010; Leaner et al, 2007; Broun, 2004). The method has become an important component in the field of molecular breeding. Currently, transcription factors controlling oil and starch synthesis have been reported (Cernac and Benning, 2004; Sun et al, 2003, 2005). We can employ transcription factor based technology to change cereal starch to oil.

1.b Production of plant oil in vegetative tissues

In some plant species, transient starch in leaves after photosynthesis during the day can be accumulated to 10% of the biomass (Stettler et al, 2009). If some of the starch could be converted into oil, those plants would produce a huge amount of vegetable oil. This goal may be through the transcription factor based technology to achieve.

2. More carbon deposition to the plant's storage organization

Some transcription factors are not only involved in the synthesis of some components, but also in source-sink communication and distribution of carbon in the source and sink tissues. An example is the transcription factor SUSIBA2 (Sun et al, 2003, 2005). We can use transcription factor based technology to increase carbon deposition to the plant storage tissue.

3. Use of plant cells as bioreactors for production of the expensive enzymes in degradation of starch and cellulose

According to the literature, some scientists have succeeded in over-production of cellulose-degrading enzyme in plants´ chloroplasts (Petersen and Bock, 2011). After harvesting, the enzymes can be used for cellulose based bioethanol production. We suggest a similar method here that one could produce starch-degrading enzymes in cellulose tissues and cellulose degrading enzymes in starchy tissues. The bioenergy plants after harvesting, smashing and mixing can accelerate the decomposition of starch and cellulose by the over-expressed enzyme mixture.

6. Acknowledgements

Funded by the following organizations and foundations:

- The Swedish Research Council for Environment, Agricultural Sciences and Spatial Planning (Formas) under the Strategic Research Area for the TCBB Program.
- The SLU Lärosätesansökan Program (TC4F) for Team 4 supported by Vinnova.
- The SLU program BarleyFunFood.
- The Swedish International Development Cooperation Agency (Sida/SAREC).
- Joint Formas-Sida funded programs for cassava and rice research, respectively, on sustainable development in developing countries.

- The Carl Trygger Foundation.
- National Natural Science Fundation of China (31071455).

We thank Xia Yan for her help in making the graphic of Figure 1.

7. References

Akomo B. (2011). http://www.worldstagegroup.com/worldstage/index.php?active=news&id=1977

Aneja V.P., Schlesinger W.H., Erisman J.W. (2009). Effects of agriculture upon the air quality and climate: research, policy, and regulations. Environ. Sci. Technol. 43, 4234-4240.

Barker, T. et al. (2007). Climate change 2007: mitigation of climate change. In: Summary for policymakers (Geneva: Intergovernmental Panel on Climate Change), pp. 1-36.

Baligar V.C., Fageria N.K., He Z.L. (2001). Nutrient use efficiency in plants. Communications in Soil Science and Plant Science and Plant Analysis 32, 921-950.

Bousquet, P. et al. (2006). Contribution of anthropogenic and natural sources to atmospheric methane variability. Nature 443, 439-443.

BP. (2010). BP statistical review of world energy. Hayward, T. ed. London.

Broun, P. (2004). Transcription factors as tools for metabolic engineering in plants. Curr. Opin. Plant Biol. 7, 202-209.

Bryan, B., Hellemans, A. (1993). The timetables of technology: A Chronology of the Most Important People and Events in the History of Technology. New York: Touchstone.

Century K., Reuber T.L., Ratcliffe, O.J. (2008). Regulating the regulators: the future prospects for transcription-factor-based agricultural biotechnology products. Plant Physiol. 147, 20-29.

Cernac A., and Benning C. (2004). *WRINKLED1* encodes an *AP2/EREB* domain protein involved in the control of storage compound biosynthesis in Arabidopsis. Plant J. 40: 575–585.

Duncan, R.C. (2000). Crude oil production and prices: A look ahead at OPEC decision-making process. PTTC workshop, Bakersfield, CA. p. 15.

Environment News Service (March 16, 2007). Cereal crops feeling the heat. Article from http://www.ens-newswire.com

EMO. (2011) http://www.emerging-markets.com

FAO. (2009). http://faostat.fao.org

Gaur S., Reed T. (1998). Thermal data from natural and synthetic fuels. New York: Marcel Dekker.

Goettemoeller J., Goettemoeller A. (2007). In: Sustainable Ethanol: Biofuels, Biorefineries, Cellulosic Biomass, Flex-Fuel Vehicles, and Sustainable Farming for Energy Independence. ISBN 978-0-9786293-0-4, Prairie Oak Publishing, Maryville, Missouri. p. 42.

Jansson C., Westerbergh A., Zhang J., Hu X. and Sun C. 2009. Cassava, a potential biofuel crop in China. Appl. Energy 86, S95-S99.

Kiehl J.T., Trenberth K.E. (1997). Earth´s annual global mean energy budget. Bulletin of the American Meteorological Society 78, 197-208.

Leaf D., Verolme H.J., Hunt W.F .Jr. (2003) Overview of regulatory/policy/economic issues related to carbon dioxide. Environ Int. 29, 303-310.

Leaner V.D., Donninger H., Birrer, M.J. (2007). Transcription factors as targets for cancer therapy: AP-1 as a potential therapeutic target. Curr. Cancer Ther. Rev. 3, 1-6.

Natural Resources Defense Council. (2008).
http://www.nrdc.org/globalwarming/cost/fcost.pdf
Octave S., Thomas D. (2009). Biorefinery: Toward an industrial metabolism. Biochimie. 91, 659-64.
Ohlrogge J., Allen D., Berguson B. DellaPenna D., Shachar-Hill Y., Stymne S. (2009). Driving on biomass. Science 324, 1019-1020.
Petersen K., Bock R. (2011). High-level expression of a suite of thermostable cell wall-degrading enzymes from the chloroplast genome. Plant Mol Biol. Epub ahead of print.
Prather M.J., Hsu J. (2010). Coupling of nitrous oxide and methane by global atmospheric chemistry. Science 330, 952-954.
RFA. (2011). http://www.ethanolrfa.org
Rhodes C.J. (2010) Solar energy: principles and possibilities. Sci Prog. 93, 37-112.
Rogers P.L., Jeon Y.J., Lee K.J., Lawford H.G. (2007). Zymomonas mobilis for fuel ethanol and higher value products. Adv Biochem Eng Biotechnol. 108, 263-288.
Rooney T. (1998). Lignocellulosic feedstock resource assessment. NREL Report SR-580-24189, p. 123.
Sass R.L., Cicerone R.J. (2002). Photosynthate allocations in rice plants: Food or atmospheric methane. Proc. Natl. Acad. Sci. USA 99, 11993-11995.
Shen B., Allen W.B., Zheng P., Li C., Glassman K., Ranch J., Nubel D., Tarczynski M.C. (2010). Expression of *ZmLEC1* and *ZmWRI1* increases seed oil production in maize. Plant Physiol. 153, 980–987.
Stettler M., Eicke S., Mettler T., Messerli G., Hörtensteiner S., Zeeman S.C. (2009). Blocking the metabolism of starch breakdown products in Arabidopsis leaves triggers chloroplast degradation. Mol Plant. 2, 1233-1246.
Sun C., Palmqvist S., Olsson H., Borén M., Ahlandsberg S., Jansson C. (2003). A novel WRKY transcription factor, SUSIBA2, participates in sugar signaling in barley by binding to the sugar-responsive elements of the iso1 promoter. Plant Cell 15, 2076-2092.
Sun C., Höglund A.-S., Olsson H., Mangelsen E., Jansson C. (2005). Antisense oligodeoxynucleotide inhibition as a potent strategy in plant biology: identification of SUSIBA as a transcriptional activator in plant sugar signaling. Plant J. 44, 128-138.
Tan T., Shang F., Zhang X. (2010) Current development of biorefinery in China. Biotechnol Adv. 28, 543-55.
Westerbergh A., Zhang J., and Sun C. (2011). Cassava - a multipurpose crop for the future. In: Cassava: Farming, Uses, and Economic Impact. ISBN 978-1-61209-655-1, Colleen M. P. Editor, Nova Science Publishers, Inc. In press.
World Bank. (2011).
http://siteresources.worldbank.org/DATASTATISTICS/Resources/GDP.pdf
Worldometers. (2011).
http://www.worldometers.info/cars
Unions for Jobs and the Environment. (2002).
http://www.angelfire.com/co4/macroeconomics302/presentation.pdf
Yu K.M., Curcic I., Gabriel J., Tsang S.C. (2008). Recent advances in CO2 capture and utilization. ChemSusChem. 1, 893-899.

3

A Comparison Between Raw Material and Technologies for a Sustainable Biodiesel Production Industry

Jorge Mario Marchetti
Av. Alem 1251. Departamento de Física. Instituto de Física del Sur (IFISUR),
Universidad Nacional del Sur – CONICET
Bahía Blanca,
Argentina

1. Introduction

Biodiesel is a renewable liquid biofuel produced from renewable lipid sources such as vegetable oils or animal fats. It results from the transesterification reaction of triglycerides with an alcohol in the presence of a catalyst; see Figure 1 (Marchetti et al., 2007, Srivastava & Prasad, 2000, Ma & Hanna, 1999).
The transesterification reaction is a series of reactions where from triglycerides diglycerides are produced, from these ones monoglycerides are produced and finally, from these monoglycerides glycerol is produced. In all these steps, fatty acid alkyl ester (biodiesel) is formed.

CH_2-COO-R_1				CH_2-OH		R_1-COO-R'
|			Catalyst	|		
CH-COO-R_2	+	3R'OH	$\leftrightarrow$	CH- OH	+	R_2-COO-R'
|				|		
CH_2-COO-R_3				CH_2- OH		R_3-COO-R'
Triglycerides		Alcohol		Glycerol		Esters

Fig. 1. Typical transesterification reaction: a triglyceride with an alcohol to produce biodiesel. (R_n are carbon chain of different length)

This reaction is most commonly catalyzed by a homogeneous base catalyst that could be sodium hydroxide among others. However, to be able to use this type of catalyst, the raw material needed is a refined vegetable oil. If more impure oil is used together with some free fatty acids and sodium hydroxide in it, the fatty acid involved in the saponification reaction will consume the catalyst. The soap produced will also be a problem for the downstreaming separation of the biodiesel and the glycerol. The saponification reaction could be seen in Figure 2 (Marchetti et al., 2007, Srivastava & Prasad, 2000, Ma & Hanna, 1999, Fukuda et al., 2001, Knothe et al., 2005, Marchetti, 2010).

R-COOH	+	NaOH	↔	H_2O	+	R-COO-Na
Fatty acid		Base		Water		Soap

Fig. 2. Saponification reaction of free fatty acid and sodium hydroxide

In order to avoid this problem, the technological solutions put forward were: the use of homogeneous acidic catalysts such as sulfuric acid (Knothe et al., 2005, Marchetti, 2010, Schuchardt et al., 1998, Noureddini & Zhu, 1997, Freedman et al., 1984, Zheng et al., 2006, Canakci & Van Gerpen, 2003a, 2003b), solid catalysts such as zeolites, solid resins (basic as well as acid) (Bournay et al., 2005, Di Serio et al., 2005, 2006, Soriano et al., 2009, Hamad et al., 2008, Suppes et al., 2004, Kulkarni et al., 2006, López et al., 2008, Cao et al, 2008), enzymatic technologies (Bajaj et al., 2010, Ranganathan et al., 2008, Antczak et al, 2009, Rodrigues et al., 2008, Dalla Rosa et al., 2008, Matassoli et al., 2008), supercritical alcohols (Demirbaş, 2002, 2003, Saka & Kusdiana, 2001, Kusdiana & Saka, 2001, 2004, Hawash et al., 2009, Gui et al., 2008, Kasim et al. 2009), membrane reactors (Dubé et al. 2007, Baroutian et al., 2011, Zhu et al., 2010, Cheng et al., 2010, Cao et al., 2008), monolithic reactors (Kolaczkowski et al., 2009, Dizge et al., 2009, Tonetto & Marchetti, 2010), etc.

All these technological solutions have many advantages over each other but also some drawbacks due to different considerations; for instance, reaction time, reaction temperature, operational cost, amount of equipment, quality of the final product, complexity of purification, and so forth (Srivastava & Prasad, 2000, Ma & Hanna, 1999,Marchetti, 2010, Schuchardt et al., 1998).

Furthermore, all these new technologies have a real advantage: they can be used for treating less pure raw materials, allowing a higher presence of free fatty acids and, in some cases, the presence of water. This is pertinent not only to the increasing Food vs. Fuel debate, but also to the one about how the vegetable oil we produce should be used. By applying these technologies, typical pollutants can be consumed, and crude oil, waste, as well as frying oil, soapstocks, and the like, can be used as raw materials for Biodiesel production. As a result, all the refined oil could be left for human consumption.

There is a need of knowing more about the different biodiesel technologies based on the type of raw materials they are able to treat. This work aims to shed light on the subject by presenting a comparison of the different qualities of vegetable oils, comparing the physical-chemical properties of these oils, and their influences over the final biofuel.

2. Global situation of the vegetable oil market

Since the beginning of the 19th century, people have used vegetable oils and animal fats based on the knowledge storage over centuries and not so much on the scientific knowledge as it are today; based on their structure, physical properties, etc. The need of understanding and knowing more about the vegetable oil arises due to the increasing value of this product as for example in its use in medicine, cosmetics as well as for fuel for lighting.

Normally, vegetable oils are obtained from different plants seeds, such as sunflower, peanut, coconut, palm, palm kernel, soybean, corn, and many other options. To chose from which of these vegetables produce the oil depends on several factors, some of them are the location of the landscape, the climate of the region, the nutrients of the soil as many other environmental as well as economic variables (it is also important which oil has a better market to be sold). Based on some types of seeds, O´Brien et al. 2000, have done a compilation showing where some of the main seeds are being cultivate around the world. An extraction from their work it could be seen in Table 1.

Seed	Amount of oil (%)	Productive areas
Canola	40-45	Canada, China, India, France, Austria, United Kingdom, Germany, Poland, Denmark, Check Republic.
Corn	3.1-5.7	USA, Mexico, Russia, Belgium, France, Italy, Germany, Spain, United Kingdom.
Cotton	18-20	China, Russia, USA, India, Pakistan, Brazil, Egypt, Turkey.
Peanut	45-50	China, India, Nigeria, USA, Senegal, South Africa, Argentina
Crocus	30-35	China, USA, Spain, Portugal
Soybean	18-20	USA, Brazil, Argentina, China, India, Paraguay, Bolivia
Sunflower	35-45	Russia, Argentina, Austria, France, Italia, Germany, Spain, United Kingdom.
Coconut	65-68	Filipinas, Indonesia, India, México Sri Lanka, Thailand, Malaysia, Vietnam, Mozambique, New Guinea, Republic of Côte d'Ivoire
Olive	15-35	Spain, Italy, Italia, Greece, Tunes, Turkey, Morocco, Portugal, Syria, Algeria, Yugoslavia, Egypt, Israel, Libya, Jordan, Lebanon, Argentina, Chile, Mexico, Peru, USA, Australia.
Palm	45-50	Malaysia, Indonesia, China, Filipinas, Pakistan, México, Bangladesh, Colombia, Nigeria, Republic of Côte d'Ivoire
Palm kernel	44-53	Malaysia, Indonesia, China, Filipinas, Pakistan, México, Bangladesh, Colombia, Nigeria, Republic of Côte d'Ivoire

Table 1. Mayor producer for several vegetable oils (O'Brien et al., 200).

In Table 1 it is presented the mayor regions where different oils are being produced as well as the percentage of oil in each seed. It could be seen that even when each seed produce different amounts of oil, there are several of them that produce an amount between 30 and 53 %.

Even more, the price of vegetable oil has become quite volatile over the last year (www.indexmundi.com (a)). In Figure 3 it can be seen the fluctuation of prices for soybean oil for the last 30 years (www.indexmundi.com (b)). In the case of soybean oil, there is a major peak in June 2008, (where the price of the soybean oil have reached values of 1414 US$)

The soybean oil situation is similar to that of the major oils such as sunflower, coconut, rapeseed, palm, palm kernel, olive, etc., being the case presented in Figure 3 just as an example. It is also important to notice that the evolution of the vegetable oil prices is link directly to the prices of petroleum oil. Figure 4 shows the evolution of the petroleum prices as well as the price of the soybean oil. When petroleum prices increases there is an increase in soybean oil; however, the increases in soybean oil are not only associated to petroleum crisis.

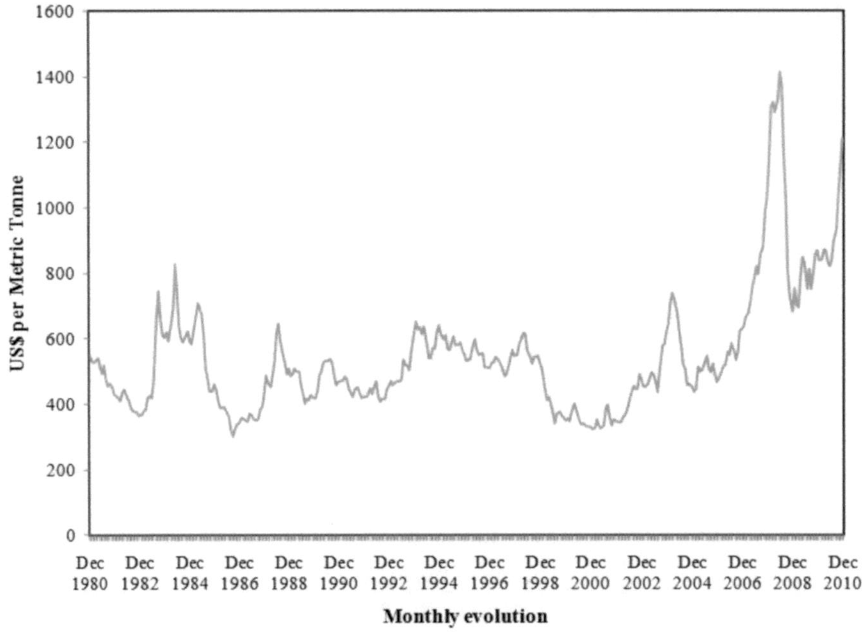

Fig. 3. Monthly evolution of the price for soybean oil over the last 30 years (www.indexmundi.com (a)).

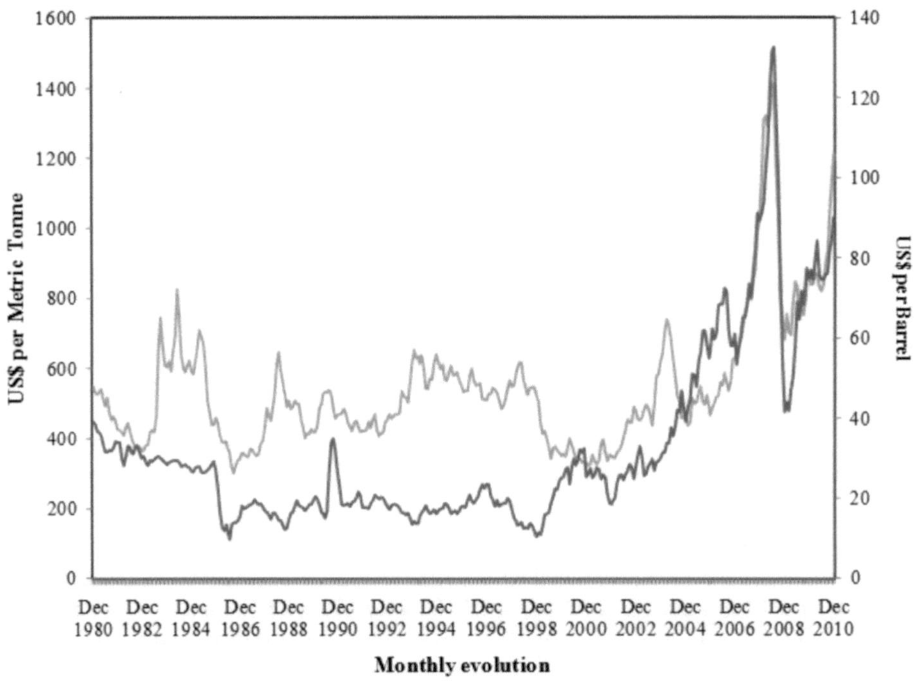

Fig. 4. Monthly evolution of the price for soybean (--) and petroleum ( ) oil over the last thirty years (www.indexmundi.com (b)).

3. Vegetable oil: types and composition

Due to the different origin of the vegetable oils, their structure and composition vary from one to another, mainly in the length of the hydrocarbon chain, as well as in the amount and location of the double bonds. O'Brien et al., (O'Brien et al., 2000) have performed a comparison with some of the main types of oil and the percentage of the different types of fatty acids involves. Table 2, extracted from reference O'Brien et al., 2000 shows these percentages.

As it could be seen from Table 2, the main composition fatty acid compositions of the vegetable oils is generally based on a C18 carbon length, with one or two double bonds, this is the case for corn, cottonseed, peanut, rapeseed, soybean and sunflower. For crambe oil, the main contribution is by C22:1. However, the second and third most abundant fatty acid is different from oil to oil. It is important to point out that, even thought there are differences in the fatty acid composition of different oils, the major participations are always from carbon chain with double bonds

Vegetable oil	Fatty acid composition % by weight								
	16:1	18:0	20:0	22:0	24:0	18:1	22:1	18:2	18:3
Corn	11.67	1.85	0.24	0.00	0.00	25.16	0.00	60.60	0.48
Cottonseed	28.33	0.89	0.00	0.00	0.00	13.27	0.00	57.51	0.00
Crambe	20.7	0.70	2.09	0.80	1.12	18.86	58.51	9.00	6.85
Peanut	11.38	2.39	1.32	2.52	1.23	48.28	0.00	31.95	0.93
Rapeseed	3.49	0.85	0.00	0.00	0.00	64.4	0.00	22.30	8.23
Soybean	11.75	3.15	0.00	0.00	0.00	23.26	0.00	55.53	6.31
Sunflower	6.08	3.26	0.00	0.00	0.00	16.93	0.00	73.73	0.00

Table 2. %wt. of the different fatty acid in the main types of oil. (O'Brien et al., 2000)

The diverse composition of the different types of oil also affects their main physical-chemical properties due to the carbon length and the double bonds. Table 3 presents some major properties for 6 typical types of vegetable oil.

It could be seen that in some cases the properties values are quite similar; that is the case of the specific gravity and the smoke point. On the other hand the flash point is quite diverse, changing from 225 °C to 327°C. Another property that has a wider range is the acid value, going from 0.6 to 6.6 accordingly.

So far, it has been presented the properties of refine oils. However, as it will be point out later in this chapter, the need for a less pure raw material, and therefore cheaper, is imperative. The main possibilities are frying, waste, cooking oils, soapstocks as well as oil from algae and non edible oils. The first four options are considered waste. On the other hand, the oil from algae is not a waste but could be considered as non edible oil.

Table 4 shows some of the main properties of refined, crude and waste oil to compare the different types of oil. There, it is clear that the sulfur content increases considerably when comparing refined oil with waste oil. Similar results could be found when comparing the acid value.

On the other hand, the viscosity values do not change that much due to the presence of impurities.

Among the non edible oils, Table 5 shows the variation of some parameters for 4 types of non edible oils.

Property	Refine [*]	Crude [+]	Waste [x]
Kinematics viscosity [mm^2/s]	30.2	36	40.2
Carbon residue [wt%.]	0.24	0.278	0.18
Cetane number	38.01	39	---
Higher heating value [MJ/kg]	39.41	39.2	24.67
Ash content [wt%]	0.012	0.01	---
Sulfur content [wt%]	0.013	0.075	5
Iodine value [centigram I/g Oil]	112.86	125	13.2
Acid value [mg KOH/g Oil]	<0.2	variable	5.96

Table 4. Properties of the mayor types of oil
* Chhetri et al., 2008, Demirbas, 2008
+ http://globalsmartinvestment.com/SoyOil.aspx
x Zhang et al., 2003, Dmytryshyn et al., 2004, Phan & Phan, 2008, Anastopoulos et al., 2009

Parameters	Non Edible Oils			
	Jatropha	Rubber	Castor	Pongamia pinnata
Viscosity	4.8	5,81	---	4.8
Specific gravity	---	0.874	0.960	---
Calorific value [MJ/kg]	39.23	36.50	39.5	---
Flash point [°C]	135	130	260	150
Cloud point [°C]	---	4	-12	---
Pour point [°C]	2	---	-32	---
Ash content [wt%]	0.012	...	0.020	0.0005
Acid value [mg KOH/g]	0.400	0.118	---	0.620

Table 5. Properties of biodiesel from different sources. Extracted from reference (Gui et al., 2008)

Gradually, a third type of oil is bearing relevance due to a number of advantages. This third type is produced from algae. Algae are microscopic organisms which could be used for the production of different types of fuel such as biodiesel, bio-hydrogen production, methane, etc. Algae have several advantages to be used for producing biofuel, some of them are: *i)* they required CO_2 to grow, *ii)* they grow in non drinkable water, not competing with drinkable sources, *iii)* algae could grow in land fields where no other vegetable oil could grow, *iv)* algae could be use as raw materials for several other chemical compounds easy to produce.
Even more, as in the case for vegetable oils, each type of algae has a percentage of oil in within. Table 6 shows some of the most common algae and the amount of oil in each of them (Chisti, 2007).

It could be seen that this percentage varies from 15 to over 75 % being the last quite suitable for producing the oil to be use in the production of biofuels.

Microalgae	Oil content (% dry wt.)
Botryococcus braunii	25-75
Chlorella sp.	28-32
Crypthecodinium cohnii	20
Cylindrotheca sp.	16-37
Dunaliella primolecta	23
Isochrysis sp.	25-33
Monallanthus salina	>20
Nannochloris sp.	20-35
Nannochloropsis sp.	31-68
Neochloris oleoabundans	35-54
Nitzschia sp.	45-47
Phaeodactylum tricornutum	20-30
Schizochytrium sp.	50-77
Tetraselmis sueica	15-23

Table 6. Oil content for several microalgae. Extracted from reference (Chisti, 2007).

Mata et al. (Mata et al., 2010) have done a research on how much biodiesel could be produce by growing different seeds in one ha. They work shows that, among those sources studied; corn has the lowest one production rate per year in 1 hectare (151 kg of biodiesel); sunflower oil is place in middle position with a production close to 1000 kilos per year per hectare; while microalgae are place quite high above. In regard to microalgae, they are classified depending on their oil content, which is divided into low, medium and high. The result is a production of 51.297, 86.515 and 121.104 kilos of biodiesel per year and per hectare respectively. This shows algae as an interesting new alternative source of vegetable oil for biodiesel.

So far it has been presented the difference in the vegetable oil accordingly to the quality of them as well as to be from different seeds, also it was presented a comparison among non edible oil and oil from algae. Thus, when comparing biodiesel from different sources, the nature of the vegetable oil is to be considered because of its effect over some of the major physical and chemical.

As it could be seen from Table 7 (Moser, 2009), when comparing the methyl esters from fatty acid with no double bonds, the increase in the chain length produce a higher viscosity. However, when double bonds are being considered, the viscosity tend to decrease as long as the amount of double bonds increases, this could be seen for C18:0 ME, C18:1 ME, C18:2 ME and 18:3 ME. When looking into the type of alcohol used, the longer the alcohol chain is the higher the viscosity of the biodiesel.

A similar scenario could be seen when comparing the cetane number, when the carbon chain or the alcohol chain increases, the cetane number increases as well, however, the presence of double bonds will decrease this property considerably.

FAEE	Melting point (°C)	$\Delta_c H$ (MJ/mol)	Kinematics viscosity (mm²/s)	Oil stability index (h)	Cetane Number	Lub (μm)
C12:0 ME	5	8.14	2.43	>40	67	416
C12:0 EE	-2		2.63	>40		
C14:0 ME	19	10.67	3.30	>40		353
C14:0 EE	12		3.52	>40		
C16:0 ME	31	10.67	4.38	>40	86	357
C16:0 EE	19		4.57		93	
C16:1 ME	-34	10.55	3.67	2.1	51	246
C16:1 EE	-37					
C18:0 ME	39	11.96	5.85	>40	101	322
C18:0 EE	32		5.92	>40	97	
C18:0 BE	28		7.59		92	
C18:1 ME	-20	11.89	4.51	2.5	59	290
C18:1 EE	-20		4.78	3.5	68	
C18:1 BE	-26		5.69		62	303
C18:2 ME	-35	11.69	3.65	1.0	38	236
C18:2 EE			4.25	1.1	40	
C18:3 ME	-52	11.51	3.14	0.2	23	183
C18:3 EE			3.42	0.2	27	

Table 7. Properties of different biodiesel. Extracted from reference (Moser, 2009).

4. Some about biodiesel production

So far, it has been considered the different raw materials and the different types of oil quality. It is important to describe some of the fundamentals of biodiesel production before reflecting on selecting the appropriate technology to use. In section 5, it will be presented different technologies and how the quality of the raw material might have an effect or might be a key factor on the decision of which production alternative should be use, and which should be avoid.

Biodiesel is a renewable and alternative liquid biofuel normally produce from vegetable oils or animal fats by the transesterification reaction (see Figure 1). Not any fuel produced from vegetable oils can be called biodiesel. Therefore, there are international standards to be reached, these are, ASTM D7467, EN 14214. The European Standard EN 14214 could be seen in Table 8 (http://www.astm.org, http://www.cen.eu)

The transesterification reaction, when carried on with a basic homogeneous catalyst, sodium or potassium hydroxide or methoxide (Marchetti et al., 2007, Srivastava & Prasad, 2000, Ma & Hanna, 1999, Fukuda et al., 2001, Knothe et al., 2005, Marchetti, 2010, Vicente et al., 2004, Meng et al., 2008, Alamu et al., 2007, Dias et al., 2008), has shown great potential producing the fuel under specification in around two hours (Srivastava & Prasad, 2000, Ma & Hanna, 1999, Fukuda et al., 2001, Knothe et al., 2005, Marchetti, 2010).

The major drawback of this technology is the need of refined oil which is on the hot spot due to the Fuel vs. Food debate. Due to the need of oil for feeding purposes and for biofuel production; alternative raw materials, pointed out in section 2, are being evaluated, tested and used. The uses of these new raw materials have generated new processes to carry on the transesterification reaction, example of this are: acid homogeneous catalyst (Marchetti et al., 2007, Srivastava & Prasad, 2000, Ma & Hanna, 1999, Fukuda et al., 2001, Knothe et al., 2005, Marchetti, 2010, Zheng et al., 2006, Canakci & Van Gerpen, 2003 a,b), solid resins (Bajaj et al., 2010, Ranganathan et al., 2008, Antczak et al, 2009, Rodrigues et al., 2008, Dalla Rosa et al., 2008, Matassoli et al., 2008),supercritical alcohols (Demirbaş, 2002, 2003, Saka & Kusdiana, 2001, Kusdiana & Saka, 2001, 2004, Hawash et al., 2009, Gui et al., 2008, Kasim et al. 2009),

Property	Units	Limits	Experiments
Esters amount	% (mol/mol)	96.5	EN 14103
Density at 15°C	kg/m^3	860-900	EN ISO 3675, EN ISO 12185
Kinematics Viscosity at 40°C	mm^2/s	3.5-	EN ISO 3104, ISO 3105
Flash Point	°C	120 min	EN ISO 3679
Sulfur content	mg/kg	10.0 max	EN ISO 20846, EN ISO 20884
Carbon residue	% (mol/mol)	0.3 max	EN ISO 10370
Cetane Number		51 min	EN ISO 5165
Sulfated ash	% (mol/mol)	0.02 max	ISO 3987
Water content	mg/kg	500 max	EN ISO 12937
Total Contamination	mg/kg	24 max	EN 12662
Copper strip corrosion (3 h, 50°C)	Degree of corrosion	1	EN ISO 2160
Oxidation stability, 110C	H	6.0 min	EN 14112
Acid number	mg KOH/g	0.50 max	EN 14104
Iodo number	$g\ I_2/100\ g$	120 max	EN 14111
Linolenic acid content	% (mol/mol)	12.0 max	EN 14103
Polyunsaturated methyl ester	% (mol/mol)	1 max	EN 14103
Methanol content	% (mol/mol)	0.2 max	EN 14110
MAG content	% (mol/mol)	0.8 max	EN 14105
DAG content	% (mol/mol)	0.2 max	EN 14105
TAG content	% (mol/mol)	0.2 max	EN 14105
Free Glycerol	% (mol/mol)	0.02max	EN 14105, EN 14106
Total Glycerol	% (mol/mol)	0.25 max	EN 14105
Group I metals	mg/kg	5.0 max	EN 14108, EN 14109
Group II metal	mg/kg	5.0 max	EN 14538
Phosphorous content	mg/kg	10.0 max	EN 14107

Table 8. EN 14214 standard for biodiesel quality control (http://www.cen.eu)

membrane reactors (Dubé et al. 2007, Baroutian et al., 2011, Zhu et al., 2010, Cheng et al., 2010, Cao et al., 2008), monolithic catalysts (Kolaczkowski et al., 2009, Dizge et al., 2009, Tonetto & Marchetti, 2010), etc.
Besides the different technologies and their applicability, we would like to introduce a few thoughts in relation to the prospective future of biodiesel production. Johnston & Holloway (Johnston & Holloway, 2010) have compared 228 countries comparing their potential production of biodiesel. According to them, the five countries with the absolute biodiesel potential are Malaysia, Indonesia, Argentina, USA and Brazil. However, when considering the high potential of production in combination with the low production cost, the top five countries are Malaysia, Indonesia, Philippines, Papua New Guinea, and Thailand. This result is relevant to draw the attention to new possible production markets.

5. A comparison of the different production technologies

As mentioned before, many works have been carried on in order to find new alternative technologies to produce biodiesel from impure raw material such as crude oil, waste or cooking oil, frying oil, soapstoacks', animal fat, etc. In a typical biodiesel production flow diagram (Figure 5), a few equipments might not be needed for all the technologies available. However, to show the most general case, we have added separation and purification processes as a rule. In some cases, purification is not required, and in some other a pre-esterification before the transesterification reaction is needed. A more complete flow

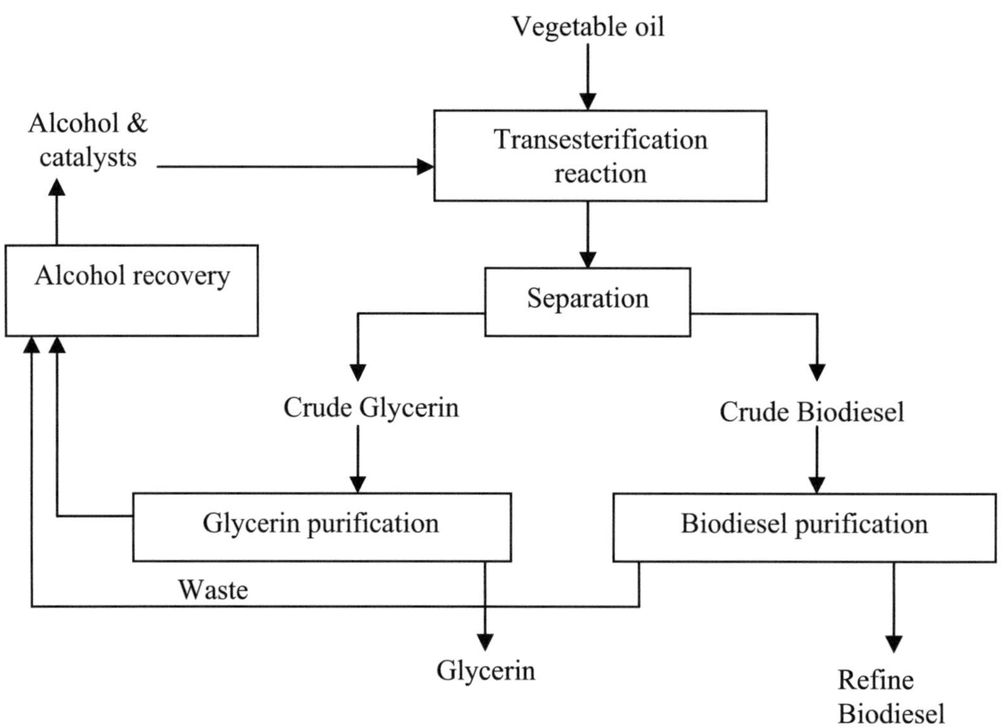

Fig. 5. Flow diagram of a conventional process (Marchetti et al., 2007).

diagram could be seen in the work done by Balat (Balat, 2011). For a more detailed process, when algae are used, it is advisable to see the work done by Lin et al. (Linn et al, 2011).

When using a homogeneous acid catalyst, like sulfuric acid, the transesterification reaction is 4000 times slower (Srivastava & Prasad, 2000, Ma & Hanna, 1999, Fukuda et al., 2001, Knothe et al., 2005, Marchetti, 2010), compared with when a base homogeneous catalyst is used. Nevertheless, the esterification reaction could also take place and be carried on without generating soaps. Therefore, this catalyst allows to treat less pure raw material with some fatty acids and/or water within it (Canakci & Van Gerpen, 2003 a,b, Marchetti & Errazu, 2008).

On the other hand, since the catalyst is in homogeneous phase, the need of neutralization, purification, and separation, are required. As a consequences, effluents as well as non desirable by-products result; this increases the amount of energy and equipments required. Thus, a bigger initial investment is needed. Finally, it is obtained a Biodiesel of good quality and a by-product, glycerin, which is of industrial grade allowing it to be used for other purposes or to be purified until pharmaceutical grade (Van Gerpen, 2005, Marchetti & Errazu, 2008).

In order to avoid some of the neutralization and purification equipments, the use of a heterogeneous catalyst appears to be a good brand new alternative. In this case, solid acid resins and basic solid resins could be used obtaining good results (Bournay et al., 2005, Di Serio et al., 2005, 2006, Soriano et al., 2009, Hamad et al., 2008, Suppes et al., 2004, Kulkarni et al., 2006, López et al., 2008, Cao et al, 2008). The advantages of using this catalyst appears in the final products obtained, more pure Biodiesel and glycerin;: and also, in the process itself, which required less equipment and ergo a lower investment. However, there are some disadvantages to take into account. In some cases, it is impossible to carry on the reaction in the presence of water as well as other impurities such as fatty acids. Some of these catalysts (solid resins) get deactivated with water; therefore, the raw material must be refined oil. In other situations, the catalyst could treat some amount of water and fatty acid, but the reaction time is so large that is not of industrial interest (Marchetti & Errazu, 2010 b).

Enzymes seem to be the best option, for being environmentally friendly. They are produced in the nature; they require not drastic conditions to work and it could treat some impurities in the raw material. It should be considered that reaction temperature could not be too high, otherwise the organisms will die. Some water is needed it to start the reaction, but too much water will deactivated the catalyst. This one will allow the transesterification as well as the esterification reaction to take place simultaneously. On the other side, the reaction time for both reactions could be considered as too long in comparison with other options and the price of this catalyst is extremely high, making it a very non competitive alternative (Antczak et al, 2009, Rodrigues et al., 2008, Dalla Rosa et al., 2008, Dizge et al., 2009). Because of this, new enzymes as well as new enzyme technologies are being developed in a daily basis aiming to reduce the price, improving the catalytic properties and find a way of reusing it over and over again so its price could be depreciated over time.

Nowadays, one of the fastest technologies available uses supercritical alcohols, either methanol or ethanol. In this case, the reaction temperature and reaction pressure provoke the alcohol to be in a supercritical state, and therefore, not catalyst is required. Thanks to this technology, a full conversion of non high quality oils could be reached in less than 5 minutes (Demirbaş, 2002, 2003, Saka & Kusdiana, 2001, Kusdiana & Saka, 2001, 2004, Hawash et al., 2009, Gui et al., 2008, Kasim et al. 2009). The absence of a catalyst allows the system to treat triglycerides, fatty acids, and water with no concerns. Even more, in order to reduce the amount of alcohol, some works have shown that is possible to reduce the alcohol molar ratio by increasing secondary supercritical fluids such as CO_2, hexane, heptanes, propane or tetrahydrofuran, being those much cheaper (Sawangkeaw et al., 2007, Tan et al., 2010, Yin et al., 2008, Han et al., 2005). In some cases, it could be found that a catalyst could be introduced into the system (Demirbaş, 2007) making possible to achieved good final conversions without compromising the down streaming separation and purification. For this process, less equipment is required and the purity of the final biodiesel and glycerin are quite good. Nevertheless, there is a need of high temperature and pressure that implies a high operational cost and makes this technology less attractive than others. But, when considering the process with different approaches it will become more economically viable (van Kasteren & Nisworo, 2007, Deshpande et al., 2010, Lim et al., 2009)

The use of membrane reactors as well as monolithic catalysts, and monolithic reactors are among the new options that being considered. Each of them has major advantages and disadvantages over the conventional process. The monolithic reactors (Kolaczkowski et al., 2009, Dizge et al., 2009, Tonetto & Marchetti, 2010), produced from powder of a basic catalyst, have as major concern the leaching of the catalyst from the heterogeneous phase to the homogeneous phase. Under this circumstance, the activity o the catalyst is lost, and also there is a need of purification and separation of the products from the catalyst. However, if the leaching problem is solved, the reaction will be more environmentally friendly and the Biodiesel produced will be of better quality. Membrane reactors have been widely used for the water gas shift reaction with great results (Dubé et al. 2007, Baroutian et al., 2011, Zhu et al., 2010, Cheng et al., 2010, Cao et al., 2008). Dubé et al. (Dubé et al., 2007) used them and succeeded in producing a final product of high quality. The catalyst employed was a homogeneous basic one; so we should consider that the separation of alcohol and other compounds, from the main flow, could be quite complicated, and the general price for the membrane is sometimes too high.

In order to compare the variables previously described, we selected those that are considered to be the most relevant and presented them in Table 9, were we show them in relation to the different technologies.

Based on Table 9, we can say that in all the cases, except from the base technology will produce ester from the presence of fatty acid. However, not all of them have the same effect when water is in the system, as it is in the case of solid resin and enzyme, where water could have a negative effect. In the case of monolithic, the leaching is due to the contact with a liquid phase, that is not necessarily water, and therefore, the effects of water itself are yet unknown.

The reaction temperature is quite low for the base, acid and enzymatic process, but could be quite high for the other three options. This is one of the main reasons for the cost of the technology to be from affordable to expensive. In the case of enzymes, even though the reaction temperature is quite low, the cost of the catalyst makes the general investment significant, and in some cases, making the technology not viable.

The purification of ester, its quality, and the quality of the produced glycerin are related to the type of technology used, especially when the catalyst is in homogeneous phase. In the last case, the catalyst needs to be separated and neutralization, separation, and purification of the products are required. For all the heterogeneous as well as the non catalytic alternatives, the need of purification is simpler and the need of equipment is also lower. In the case of the monolithic reactor, the leaching problem could make this technology less

Variable	**Base**	**Enzyme**	**Supercritical**	**Monolithic**	**Resin**	**Acid**
Temp. [°C]	60-70	30-50	200-350	50-180	60-180	50-80
Products from FFA	Soaps	Esters	Esters	Esters	Esters	Esters
*Effect of Water**	↓	↓	—	—	— ↓	—
Yield to ester	Normal	High	High	Normal	Good	Normal
Purification of glycerol	Difficult	Simple	Simple	Simple	Simple	Difficult
Reaction time+	1-2 h	8-70 h	4-10 min	6 h	variable	4-70 h
Ester purification	Difficult	Simple	Simple	Simple	Simple	Difficult
Cost	Cheapest	Expensive	Expensive	Affordable	Affordable	Cheaper
Amount of equipment	High	Low	Low	Low	Low	High

Table 9. Comparison of different technologies for Biodiesel production (Marchetti, 2010)
** in this case the pointing down arrow mean that water is a draw back while the line means that the is not effect and the system will be able to treat a raw material with some amount of water. For the Enzyme case, a down arrow has been supply, in this case is important to say that is believe that some water is require for enzyme activation; however, a lot of water will produce a deactivation of the catalyst. In the case of the resin, it could be seen a down arrow as well as a line, this is due to the fact that water has different effect over different solid catalyst. In the case of the monolithic scenario, a line has been selected due to the fact that leaching it is not causing by water per se but for a non stability of the catalyst.*
+the reaction time set in this table is what it is most likely, however, it is important to point out that other times for the same technology could be found in the open literature

suitable due to the possibility of separation and purification. So far this problem has been presented (Tonetto & Marchetti, 2010) but not much has been done in order to solve it; even though it is considered that this disadvantage might be easier to overcome.
In all the technologies studied, the yield for biodiesel is high, even thought in some cases this is much higher.

6. Conclusions

In this chapter it has been presented different types of process for biodiesel production, their advantages and disadvantages as regards the different operational variables, the different raw material qualities, and the different catalysts employed.
Due to these differences, it is not easy to select a process to use. Several operational conditions and many economic variables should be considered together with those presented in Table 9, before choosing the best alternative for each case.
About the type of vegetable oil and its quality, it is important to remember that in order not to compete with the vegetable oil for feeding purposes, the oil used is in all cases is inedible

and/or waste oil. Due to the high amount of free fatty acid that might be present on the waste oil, some technologies are more suitable than suitable than others. That is the case of supercritical alcohols over conventional process. A combination of several options produces much more reliable and environmental friendly processes. However, in those cases, the need for a much more control over other influenciable variables is stronger. When using a vegetable oil, the availability of the oil and the proximity of the plantations is also a key factor and together with a clear local policy, such as tax reduction. This will help the biodiesel companies but at the same time, it will provoke an increase in the price of the oil directly related to the need of these raw materials.
The use of waste oils should be a must. To achieve this, several options could be used based on the key factors previously explained. The use of waste oils helps to consume pollutants, does not interfere in the Fuel vs. Food debate, and demands an environmentally friendly process.

7. References

Alamu, O.J., Waheed, M.A., Jekayinfa, S.O. *Biodiesel production from Nigenerian palm kernel oil: effect of KOH concentration on yield.* Energy for Sustainable Development. XI(3), (2007), 77-82.

Anastopoulos, G., Zannikou, Y., Stournas, S., Kalligeros, S. *Transesterification of Vegetable Oils with Ethanol and Characterization of the Key Fuel Properties of Ethyl Esters.* Energies. 2. (2009), 362-376.

Antczak, M.S., Kubiak, A., Antczak, T., Bielecki, S. *Enzymatic biodiesel synthesis - key factors affecting efficiency of the process.* Renewable Energy. 34. (2009), 1185-1194.

Bajaj, A., Lohan, P., Jha, P.N., Mehrotra, R. *Biodiesel production through lipase catalyzed transesterification: a review.* Journal of Molecular Catalysis B: Enzymatic. 62. (2010), 9-14.

Balat, M. *Potential alternatives to edible oils for biodiesel production – a review of current work.* Energy Conversion & Managament. 52, (2011), 1479-1492.

Baroutian, S., Aroua, M., Raman, A.A., Sulaimna, N.M.N. *A packed bed membrane reactor for production of biodiesel using activated carbon supported catalyst.* Bioresource Technology. 102(2), (2011), 1095-1102.

Bournay, L., Casanave, D., Delfort, B., Hillion, G., Chodorge, J.A. *New heterogeneous process for biodiesel production: A way to improve the quality and the value of the crude glycerin produced by biodiesel plants.* Catalysis Today. 105, (2005), 190-192.

Canakci M, Van Gerpen J. *Biodiesel production from oils and fats with high free fatty acids.* Transactions of the ASAE. 44(6). (2003), 1429–1436.

Canakci M, Van Gerpen J. *Biodiesel production via acid catalysis.* Transactions of the ASAE. 42(5). (2003), 1203–1210.

Cao, F., Chen, Y., Zhai, F., Li, J., Wang, J., Wang, X., Wang, S., Zhu, W. *Biodiesel production from high acid value waste frying oil catalyzed by superacid heteroployacid.* Biotechnology and Bioengineering. 101(1), (2008), 93-100.

Cao, P., Dubé, M.A., Tremblay, A. *Methanol recycling in the production of biodiesel in a membrane reactor.* Fuel. 87(6), (2008), 825-833.

Cheng, L,H., yen, S.Y., Su, L.S., Chen, J. *Study on membrane reactors for biodiesel production by phase behaviors of canola oil methanolysis in batch reactors.* Bioresource Technology. 101(17), (2010), 6663-6668.

Chhetri, A.B., Watts, K.C., Islam, M.R. *Waste Cooking Oil as an Alternate Feedstock for Biodiesel Production.* Energies. 1. (2008), 3-18.

Chisti, Y. *Biodiesel from microalgae.* Biotechnology Advances. 25. (2007), 294-306.

Dalai, A,K. *Applications of vegetable oil derived esters as a diesel additive.* Energeia. 15(6), (2004), 1-6.

Dalla Rosa, C., Morandim, M.B., Ninow, J.L., Oliveira, D., Trichel, H., Vladimir Oliveira, J. *Lipase-catalyzed production of fatty acid ethyl esters from soybean oil in compressed propane.* The Journal of Supercritical Fluids. 47, (2008), 49-53.

Demirbaş, A. *Biodiesel from sunflower oil in supercritical methanol with calcium oxide.* Energy Conversion & Management. 48, (2007), 937-941.

Demirbaş, A. *Biodiesel fuels from vegetable oils via catalytic and non catalytic supercritical alcohol transesterification and other methods: a survey.* Energy Conversion and Management. 44, (2003), 2093-2109.

Demirbaş, A. *Biodiesel fuels from vegetable oils via transesterification in supercritical methanol.* Energy Conversion and Management. 43, (2002), 2349-2356.

Demirbas, A. *Biodiesel. A Realistic Fuel Alternative for Diesel Engines.* Editorial Springer 2008.

Demirbas, A. *Production of Biodiesel from Algae Oils.* Energy Sources, Part A: Recovery, Utilizations and Environmental Effects. 31:2. (2009), 163-168.

Deshpande, A., Anitescu, G., Rice, P.A., Tavlarides, L.L. *Supercritical biodiesel production and power cogeneration: technical and economic feasibilities.* Bioresource Technology. 101, (2010), 1834-1843.

Di Serio, M. Ledda, M., Cozzolino, M., Minutillo, G., Tesser, R., Santacesaria, E. *Transesterification of Soybean Oil to Biodiesel Using Heterogeneous Basic Catalysts.* Ind. Eng. Chem. Res. 45, (2006,) 3009-3014.

Di Serio, M., Tesser, R., Dimiccoli, M., Cammarota, F., Nastasi, M., Santacesaria, E. *Synthesis of biodiesel via homogeneous Lewis acid catalyst.* Journal of Molecular Catalysis A: Chemical. 239, (2005) 111-115.

Dias, J.M., Alvim-Ferraz, M.C.M, Almeida, M.F. *Comparison of the performance of different homogeneous alkali catalysts during transesterification of waste and virgin oils and evaluation of biodiesel quality.* Fuel. 87, (2008), 3572-3578.

Dizge, N., Aydiner, C., Imer, D.Y., Bayramoglu, M., Tanriseven, A., Keskinler, B. *Biodiesel production from sunflower, soybean and waste cooking oils by transesterification using lipase immobilized onto a novel microporous polymer.* Bioresource Technology. 100, (2009), 1983-1991.

Dmytryshyn, S.L., Dalai, A.K., Chaudhari, S.T., Mishra, H.K., Reaney, M.J. *Synthesis and Characterization of Vegetable Oil Derived Esters: Evaluation for their Diesel Additive Properties.* Bioresource Technology. 92. (2004), 55.64.

Dubé, M.A., Tremblay, A.Y., Liu, J. *Biodiesel production using a membrane reactor.* Bioresource Technology. 98, (2007), 639–647.

Freedman, B., Pryde, E.H., Mounts, T.L. *Variables affecting the yields of fatty esters from transesterified vegetable oils.* Journal of American Oil Chemists Society. 61(10). (1984), 1638-1643.

Fukuda. H., Kondo. A., Noda, H. *Biodiesel fuel production by transesterification of oils.* Journal of Bioscience & Bioengineering. 92(5). (2001), 405–416.

Fuse, T., Kusu, F., Takamura, K. *Determination of acid values of fats and oils by flow injection analysis with electrochemical detection.* Journal of Pharmaceutical and Biomedical Analysis. 15, (1997), 1515-1519.

Gui, M.M., Lee, K.T., Bhatia, S. *Feasibility of Edible oil vs. non-edible oil vs. waste edible oil as biodiesel feedstocks*. Energy. 33. (2008), 1646-1653.

Gui, M.M., Lee, K.T., Bhatia, S. *Supercritical ethanol technology for the production of biodiesel: process optimization studies.* The Journal of Supercritical Fluids. Doi:10.1016/j.supflu.2008.12.014.

Hamad, B., Lopes de Souza, R.O., Sapaly, G., Carneiro Rocha, M.G., Pries de Oliveira, P.G., Gonzalez, W.A., Andrade Sales, E., Essayem, N. *Transesterification of rapeseed oil with ethanol over heterogeneous heteropolyacids.* Catalysis Communications. 10, (2008), 92-97.

Han, H., Cao, W., Zhang, J. *Preparation of biodiesel from soybean oil using supercritical methanol and CO_2 as co-solvent.* Process Biochemistry. 40, (2005), 3148-3151.

Hawash, S., Kamal, N., Zaher, F., Kenawi, O., El Diwani, G. *Biodiesel fuel from Jatropha oil via non-catalytic supercritical methanol transesterification.* Fuel. 88, (2009), 579-582.

http://globalsmartinvestment.com/SoyOil.aspx

http://hypertextbook.com/facts/2000/IngaDorfman.shtml

http://journeytoforever.org/biodiesel_SVO-palm.html

http://journeytoforever.org/biodiesel_yield2.html

http://journeytoforever.org/biofuel_library/fatsoils/fatsoils2.html

http://vegburner.co.uk/oils.htm

http://www.astm.org/Standards/D6751.htm

http://www.canolacouncil.org/uploads/Chemical1-6.pdf

http://www.cen.eu/cen/pages/default.aspx

http://www.diracdelta.co.uk/science/source/f/l/flash%20point/source.html

http://www.elsbett.com/engl/oil.htm

http://www.indexmundi.com/commodities/?commodity=crude-oil&months=360

http://www.indexmundi.com/commodities/?commodity=soybean-oil&months=360

http://www.indiamart.com/company/2466820/products.html

http://www.sunflowernsa.com/oil/product-specifications/nusun-refined/

Johnston, M., Holloway, T. *A global comparison of national biodiesel production potentials.* Environmental Science & Technology. 41(23), (2007), 7967-7973.

Kasim, N.S., Tsai, T.H., Gunawan, S., Ju, Y.H. *Biodiesel production from rice bran oil and supercritical methanol.* Bioresource Technology. 100, (2009), 2399-2403.

Knothe, G., Van Gerpen, J., Krahl, J. (Eds.), *The biodiesel handbook*, AOCS Press, Champaign, Illinois, 2005, 302 pp.

Kolaczkowski, S.T., Asli, U.A., Davidson, M.G. *A new heterogeneous ZnL_2 catalyst on a structured support for biodiesel production.* Catalysis Today. 147(1), (2009), S220-S224.

Kulkarni, M.G., Gopinath, R., Meher, L.C., Dalai, A.K. *Solid acid catalyzed biodiesel production by simultaneous esterification and transesterification*. Green Chemistry. 8, 1056-1062, 2006.

Kusdiana, D., Saka, S. *Effects of water on biodiesel fuel production by supercritical methanol treatment.* Bioresource Technology. 91, (2004), 289-295.

Kusdiana, D., Saka, S. *Kinetics of transesterification in rapeseed oil to biodiesel fuel as treated in supercritical methanol.* Fuel. 80, (2001), 693-698.

Lim, Y., Lee, H.S., Lee, Y.W., Han, C. *Design and economic analysis of the process for biodiesel fuel production from transesterificated rapeseed oil using supercritical methanol.* Industrial & Engineering Chemistry Research. 48, (2009), 5370-7378.

Lin, L., Cunshan, Z., Vittayapadung, S., Xiangqian, S., Mingdong, D. *Opportunities and challenges for biodiesel fuel.* Applied Energy. 88, (2011), 1020-1031.

López, D.E., Goodwin Jr., J.G., Bruce, D.A., Furuta, S. *Esterification and transesterification using modified zirconia catalysts.* Applied Catalysis A: General. 339, (2008), 76-83.

Ma F. Hanna MA.. *Biodiesel production: A review.* Bioresource Technology. 70. (1999), 1–15.

Marchetti, J.M. *Biodiesel Production Technologies.* Nova Publishers. (2010). New York, New York.

Marchetti, J.M., Errazu, A.F. *Biodiesel production from acid oils and ethanol using a solid basic resin as catalyst.* Biomass & Bioenergy. 34(3), (2010), 272-277.

Marchetti, J.M., Errazu, A.F. *Esterification of free fatty acids using sulfuric acid as catalyst in the presence of triglycerides.* Biomass & Bioenergy. 32, (2008), 892-895.

Marchetti, J.M., Miguel, V.U., Errazu, A.F. *Techno-economic study of different alternatives for biodiesel production.* Fuel Processing Technology. 89, (2008), 740-748.

Marchetti. J.M.. Miguel. V.U.. Errazu. A.E. *Possible methods for biodiesel production.* Renewable and Sustainable Energy Reviews. 11. (2007), 1300-1311.

Mata, T.M., Martins, A.A., Caetano, N.S. *Microalgae for biodiesel production and other applications: A review.* Renewable and Sustainable Energy Reviews. 14. (2010), 217-232.

Matassoli, A.L.F., Corrêa, I.N.S, Portillo, M.F., Veloso, C.O., Langone, M.A.P. *Enzymatic synthesis of biodiesel via alcoholysis of palm oil.* Applied Biochemistry & Biotechnology. 155, (2008), 44-52.

Meng, X., Chen, G., Wang, Y. *Biodiesel production from waste cooking oil via alkali catalyst and its engine test.* Fuel Processing Technology. 89, (2008), 851-857.

Moser, B.R. *Biodiesel production, properties and feedstocks.* In Vitro Cellular and Developmental Biology - Plants. 45. (2009), 229-266.

Noureddini, H., Zhu, D. *Kinetics of transesterification of soybean oil.* Journal of American Oil Chemists Society. 74(11). (1997), 1457-1463.

O'Brien, R., Farr, W., Wan, P. Editors. *Introduction to Fats and Oils Technology*: Second Edition, 2000. Editorial AOCS.

Phan, A.N., Phan, T.M. *Biodiesel Production from Waste Cooking Oils.* Fuel, 87. (2008), 3490-3496.

Purdy, R.H. *High oleic sunflower: Physical and chemical characteristics.* Journal of American Chemists Society. 63(8), (1986), 1062-1066.

Ranganathan, S.V., Narasimhan, S.L., Muthukumar, K. *An overview of enzymatic production of biodiesel.* Bioresource Technology. 99. (2008), 3975-3981.

Rodrigues, R.C., Volpato, G., Wada, K., Ayub, M.A.Z. *Enzymatic synthesis of biodiesel from transesterification reactions of vegetble oils and short chaing alcohols.* Journal of American oil Chemists Society. 85, (2008) 925-930.

Saka, S., Kusdiana, D. *Biodiesel fuel from rapeseed oil as prepared in supercritical methanol.* Fuel. 80, (2001), 225-231.

Sawangkeaw, R., Bunyakiat, K., Ngamprasertsith, S. *Effect of co-solvents on production of biodiesel via transesterification in supercritical methanol.* Green Chemistry. 9, (2007), 679-685.

Schuchardt, U., Sercheli, R., Vargas, R.M. *Transesterification of vegetable oils: a review.* JBCS, 9(1). (1998), 199–210.

Soriano Jr, N.U., Venditti, R., Argyropoulos, D.S. *Biodiesel synthesis via homogeneous Lewis acid-catalyzed transesterification.* Fuel. 88, (2009) 560-565.

Srivastava A. Prasad R.. *Triglycerides-based diesel fuels.* Renewable Sustainable Energy Reviews. 4. (2000), 111–133.

Suppes, G.J., Dasari, M.A., Doskocil, E.J., Mankidy, P.J., Goff, M.J. *Transesterification of soybean oil with zeolite and metal catalysts.* Applied Catalysis A: General 257, (2004), 213-223.

Tan, K.T., Lee, K.T., Mohamed, A.R. *Effects of free fatty acids, water content and co-solvent on biodiesel production by supercritical methanol reaction.* The Journal of Supercritical Fluids. 53, (2010), 88-91.

Tonetto, G.M., Marchetti, J.M. *Transesterification of soybean oil over Me/*Al_2O_3 *(Me = Na, Ba, Ca, and K) catalysts and monolith K/*Al_2O_3*-cordierite.* Topics in Catalysis. 53(11-12), (2010), 755-762.

Van Gerpen, J. *Biodiesel processing and production.* Fuel Processing Technology. 86, (2005), 1097-1107.

van Kasteren, J.M.N, Nisworo, A.P. *A process model to estimate the cost of industrial scale biodiesel production from waste cooking oil by supercritical transesterification.* Resources, Conservation and Recycling. 50, (2007), 442-458.

Vicente, G., Martínes, M., Aracil, J. *Integrated biodiesel production: a comparison of different homogeneous catalysts systems.* Bioresource Technology. 92, (2004), 297-305.

Yin, J.Z., Xiao, M., Song, J.B. *Biodiesel from soybean oil in supercritical methanol with co-solvent.* Energy Conversion & Management. 49, (2008), 908-912.

Zhang, Y., Dubé, M.A., McLean, D.D., Kates, M. *Biodiesel production from waste cooking oil: 2 Economic assessment and sensitivity analysis.* Bioresource Technology. 90. (2003), 229-240.

Zheng, S., Kates, M., Dubé, M.A., McLean, D.D. *Acid-catalyzed production of biodiesel form waste frying oil.* Biomass & Bioenergy. 30(3). (2006), 267–272.

Zhu, M., He, B., Shi, W., Feng, Y., Ding, J., Li, J., Zeng, F. *Preparation and characterization of PSSA/PVA catalytic membrane for biodiesel production.* Fuel. 89(9), (2010), 2299-2304.

4

How to Balance Biofuel and Food Production for Optimal Global Health and Nutrition - The Food Crop-Feed Crop-Fuel Crop Trilemma

Leif Hambræus
Unit for Public Health Nutrition, Department of Bioscience and Nutrition, NOVUM, Karolinska institutet, Stockholm, Sweden

"Food is basically a net product of an ecosystem, however simplified. Food production starts with a natural material, however modified later. Injections of energy (and even brains) will carry us only so far. If the population cannot adjust its wants to the world in which it lives, there is little hope of solving the food problem for mankind. In that case the food shortage will solve our population problem." (Steinhardt & Steinhardt, 1974)

1. Introduction

The energy crisis 1973/74 opened our eyes for the close linkage between energy resources and agricultural production. In an analysis of the energy flow in the US Food system, Steinhardt and Steinhardt (1974) found the figures regarding the changes in energy efficiency in food production with time alarming and called for a more holistic view of the energy efficiency in agricultural production. They pointed out that the energy cost to produce animal protein foods such as milk, egg and especially meat is far more than to produce plant foods. To feed the world population with a US type of food system should have required 80 per cent of the world's annual energy expenditure in 1974. They concluded that the most effective way to reduce the large energy requirements of food processing would be a change in eating habits towards less highly processed foods.

Now 40 years later, the linkage between energy resources and food production still exists, but today also from another perspective. It is no longer only the role of agricultural production as energy consumer and the waste of primary resources as result of increased consumption of animal protein that causes a problem. The situation is further stressed by the concomitant increase in the demand of food as result of a growing population. In addition, increased energy needs, especially for transport, makes production of biofuels an interesting alternative for the agricultural sector.

The increased interest to develop biofuel production not only started a conflict of interest regarding land use. It also accentuated a discussion of the impact of indirect land use changes (ILUC) on the environment including increased risks for pollution and climate changes as result of greenhouse gas (GHG) emission (Harvey & Pilgrim, 2011). However, little attention has still been devoted to discuss to what extent changes in dietary habits in order to obtain a better public health situation is in harmony with an increased biofuel production and ILUC. Another perspective is how food and nutrition policy could be involved in the layout for the future direction of the development of biofuel technology.

During the last years a number of reviews have presented an overall picture of the recent development of biofuel technology and its economic, social and environmental impacts. (e.g. Howarth & Bringezu, 2009; Lawrence et al., 2011). Although some of them have linked the increased production of biofuels to a negative effect on food availability and the global food crisis (Fischer, 2009), this is still a matter of controversy.

2. Aims of the chapter

To discuss the changes in global population, socio-economy and lifestyle, food availability and biofuel production, and its impact on primary resources and public health;

To discuss the potentials of better dietary habits to reduce the ongoing increase in indirect consumption of primary resources;

To discuss the potential impact of first and secondary generation of biofuel technology on nutritional intake, public health and food security;

To stimulate an interdisciplinary discussion on the optimal combination of developing a sustainable food production system for optimal health with a sustainable production of renewable energy for the society.

3. Is there food for all?

More than 200 years ago Malthus (1798) presented the problem of imbalance between the population growth vs. growth in agriculture production. However, the balance between food production and population increase (fig 1) illustrates to a certain degree the "hen and egg" question. An increased food production is a prerequisite for population growth, while in a rapidly growing population food shortage leads to malnutrition and death and a "self regulation" of population growth. Nevertheless, all since then, the focus has been on food production as such and to what extent there are enough resources on the globe to feed a rapidly growing world population.

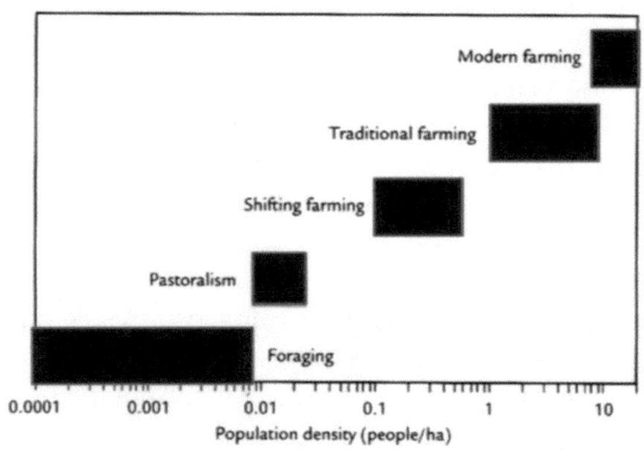

Fig. 1. Carrying capacity of different food systems vs population density (adapted from Smil, 2000)

In the 1960's, i.e. before the energy crisis, Georg Borgström (1965) revived this problem and commented that the situation was aggravated by the high consumption of animal protein, which seriously affected the food balance. The increased use of cereals for feed in animal husbandry resulted in the fact that only 10% of the cereals were consumed directly as food in the human diet in the US population versus 90 % in the Indian population. In addition, life style changes almost doubled the energy consumption per person in the US, which resulted in a three to four-fold higher consumption of primary calories in the food production line in affluent societies compared to that in low income countries. Borgström thus started the discussion whether we can afford to have such a luxury consumption pattern when resources are limited. This was followed up several decades later by Popkin and collaborators (Caballero & Popkin, 2002) when they analyzed the dietary changes during the last century as a result of the socio-economic development and introduced the *nutrition transition concept.*

The multifactorial causes for the imbalance between food production and population growth is consequently not only due to total food production and/or population growth, but also to dietary habits and life style. To solve the population problem with family planning without taking into consideration the role of public health activities to reduce infant mortality is another oversimplification of the problem.

4. The food crop – cash crop dilemma has developed into a food crop – feed crop – fuel crop trilemma

The competition between biofuel and food production represents a revival on an old conflict, i.e. the balance between *food crop* and *cash crop*. Agricultural production initially had the goal to produce food for consumption, and subsistence farming is still the basis for rural development in many low income countries (LIC). However, increased expansion of the monetary system has also increased the interest for producing cash crops for export, e.g. production of coffee and tea, sugar cane, palm oil and tobacco, fish farming of salmon and shell fish. The increased consumption of animal food in addition to the new interest for biofuel has revisited this food crop - cash crop dilemma into a *food crop – feed crop –fuel crop trilemma.*

In addition a new concept has entered the market the last years: *land grabbing* which is a hot political topic of today. Farmland has become a commodity of great interest on the world market. Countries with good economy, which are net food importers but foresee potential problems of food availability in the future, have started to outsource food production. In December 4, 2010 an investor conference was held in Riyadh for high-level delegates from African countries (Against the Grain, 2010). Thus investors from Saudi Arabia are making deals with a number of African countries to control several hundred thousand hectares productive farmland for rice production. Likewise Indian investors have been offered 1.8 million hectares farmland in Ethiopia (IANS, 2011) and China's most powerful agribusiness firms have started to acquire farmland areas in Argentina (Against the Grain, 2011). This has aggravated the competition for land use not only within countries but also between countries. It could be one way to transfer technology between developing nations and give a stimulus to country's food security. But the losers could as well be small family subsistence farmers who cannot compete when prices increase for farmland. The end result may be increased problems of malnutrition among the poor in low-income countries and less food security. What will happen if/when the big agrobusiness firms change their focus from food production to biofuel production?

5. The global food policy meetings

In connection to the energy crises in the beginning of 1970´s, the UN system recognized the global crisis in food production and food availability and the first World Food Conference was arranged by FAO and WHO in Rome 1974 with participants from most UN member states. The starting point was initially to ask for greater engagement in increasing the food production in low income countries (LIC). However, as the increase in these countries was of about the same magnitude as in affluent societies, representatives from LIC raised the question of inefficient use of primary resources in the diet of affluent societies and called for changes in dietary habits in the affluent societies. It was stated that food and nutrition policy was needed in all countries and that dietary habits in the affluent societies should not be exported to low-income countries. Eighteen years later it was followed up by a second meeting called International Conference on Nutrition in Rome 1992 also arranged by FAO and WHO in collaboration. Now the major topic was to get all countries to formulate a food and nutrition policy. This time was the first when also non-governmental organizations (NGO) as well as academics representing the food and nutrition sector were invited to participate in a UN conference for the discussions. Four years later policymakers participated in the World Food Summit in Rome (1996) and it was then stated that food availability is a human right and not only a question of nutritional requirements.

6. The concepts of security, safety and sustainability.

Interestingly the productions of food as well as of biofuels share the same basic problems to be solved: Security, Safety and Sustainability.

6.1 Food security

To have food for tomorrow is the dominant problem for mankind since beginning of history. No political system survives if food security cannot be guaranteed. More people have been killed by malnutrition and its consequences than by bullets throughout the years. Interestingly, the role of food security has increased in the modern society when political conflicts more and more involve the civil population. The responsibility for our politicians to guarantee food availability for its population has increased rather than decreased in the modern society. Increases in food prices and reduced food availability are still the major causes for rebellious movements to bring about the fall of dictatorships. Those societies who have no possibility to guarantee food sustainability for their population and need support from outside are not politically free in the world today. Consequently food power represents a biological and economic weapon in international as well as national conflicts.

6.2 Energy security

This leads to similar concerns. Energy crisis since the 1970´s has clearly illustrated how the development in all societies results in an increased demand for energy. During the 20th century this could be solved by increased production and exploration of fossil energy resources. However we also learned that political power is related to energy security and prompted interest to develop alternatives to fossil fuels in most countries. This in addition to the growing awareness that the fossil resources are limited increased the political interest for biofuel production immensely during the last few decades.

6.3 Food safety

This represents a completely different problem. Sometimes the concept of food safety is misinterpreted as equivalent to food security. However, food safety is not the question to have food for tomorrow; it represents the problem of environmental effects on the food quality. This has so far essentially been discussed in affluent societies. When resources are scarce, there is a tendency to underestimate the potential risks of neglecting the safety problems in the food production sector. The focus of interest is to get optimal yields in food production, often based on increased use of pesticides and biocides. Thus the food safety problem is now an increasing challenge in public health perspectives also in LIC.

6.4 Energy safety

This is related to the climate changes as result of increase in the transport sector, no matter if this is using fossil energy or biofuels. The environmental disturbances as result of greenhouse gas (GHG) emissions is related to increased agricultural productivity and use of land areas, including conversion of forest lands and pastures for increased food and feed crops as well as for biofuels. This calls for political actions including planning of indirect land use changes (ILUC) as well as adequate monitoring tools and sustainability certification (Scarlat & Dallemand, 2011).

6.5 Sustainability

This is a basic and common problem for the food and the energy sectors. The European Union has established some mandatory criteria in their renewable energy directives for biomass, biofuels and bioliquids (European Committee Directive, 2009). The impact on biodiversity, water resources and quality as well as on soil quality should also be evaluated. However there are no criteria for social sustainability, but it is said that the European commission will monitor biofuel consumption and impact on land use, commodity prices and food security.

7. Changes in the global public health panorama

Health statistics seem to be the only objective way to illustrate shortcomings by any political system to solve the problems related to food availability and socio-economic development. Infant and child mortality together with maternal mortality are *the* indicators of an insufficient socio-economic and public health policy. They are also difficult to hide for any type of political system. On the other hand the increased life-span, observed in all types of communities, illustrates not only the effect of decreased infant mortality. It also is a result of the benefits of a functioning health care system. Excellent illustrations of the dynamics in health statistics in relation to socio-economic development on the globe are presented by Hans Rosling and his collaborators on the homepage Gapminder world, which is continuously updated (www.gapminder.org/world).

Three problems dominate the changes in public health during the last 100 years in a global perspective: (i) the (im)balance between population and food availability as result of the population increase; (ii) the changes in the age distribution of the population, and (iii) the socio-economic development leading to changes in life style including changes in dietary habits (described as nutrition transition) as well as increased energy consumption.

Of special nutrition concern are the changes in dietary habits and its impact on public health. The prevalence of nutritional deficiencies has decreased and especially resulted in a drastic reduction in infant and child mortality. However, the diet in affluent societies has caused deleterious long-term effects on public health secondary to increased prevalence of obesity. This has resulted in chronic diseases, e.g. cardiovascular disease, diabetes, cancer, osteoporosis, leading to rapidly increasing costs for health care.

7.1 Population changes

The changes in global population during the last century are not only characterized by an increased number, i.e. a *quantitative* problem. It is also characterized by an increase in mean life span and a concomitant, although slow, decrease in fertility and a profound change in age distribution, i.e. a *qualitative* change. As illustrated in WHO international health statistics (www.gapminder.org/world) it took 40 years in Sweden for life expectancy at birth to increase from 44 to 61 years between 1880 to 1920 in a population of about 4.5 millions. Interestingly it also took 40 years to reach the same increase in life expectancy (from 42 to 62 years) in India with a population of more than 1 billion between 1960 and 2000! Corresponding changes have occurred in a number of LIC as well as in China, sometimes even in shorter time. Thus all countries can be classified as developing societies, and the only difference between affluent and low-income countries with respect to the time it takes to increase life expectancy at birth, is timing. In 1900 only countries in northern Europe and the US showed a life expectancy at birth above 50 years, 100 years later only a few countries in the world showed life expectancy less than 50 years. These changes with time are also excellently illustrated on the Gapminder world website
(www.gapminder.org/world).

7.2 From a population pyramid to a population hexagon

The increased life span has also had an effect on the classical so called population "age pyramid". A high fertility leading to high percentage of children under 5 years of age, has changed into a population "age hexagon" with lower percentage of infants and young children and an increased percentage of elderly. This results in a change in the public health perspectives on the link between diet and health.

7.3 Malnutrition and public health

With a low life span, e.g. below 50 years, public health was dominated by *malnutrition minus* problems, i.e. energy, protein and vitamin and mineral deficiencies leading to a high infant mortality. An increased life span, e.g. to 75-80 years, in combination with changes in dietary habits, the nutrition transition, as a result of the socio-economic development, has resulted in *malnutrition plus* problems, with obesity, cardiovascular diseases as well as diabetes, osteoporosis and cancer dominating the public health panorama.

If malnutrition minus represents a short-term effect of nutritional problems, malnutrition plus represents long-term effects. This is illustrated when the causes of death in low income countries are compared to those in affluent societies (table 1). In low income countries where high infant mortality exists, the dominating causes of death are infection diseases and maternal mortality. In affluent societies cardiovascular diseases, diabetes and cancer dominate. Interestingly, although nutritional problems are etiological factors for acute as well as chronic diseases they are rarely referred to in the death certificates. This may be one reason why the interest to cope with nutritional deficiencies and disorders in preventive health care is limited in most public health policy programs. Scientific resources for

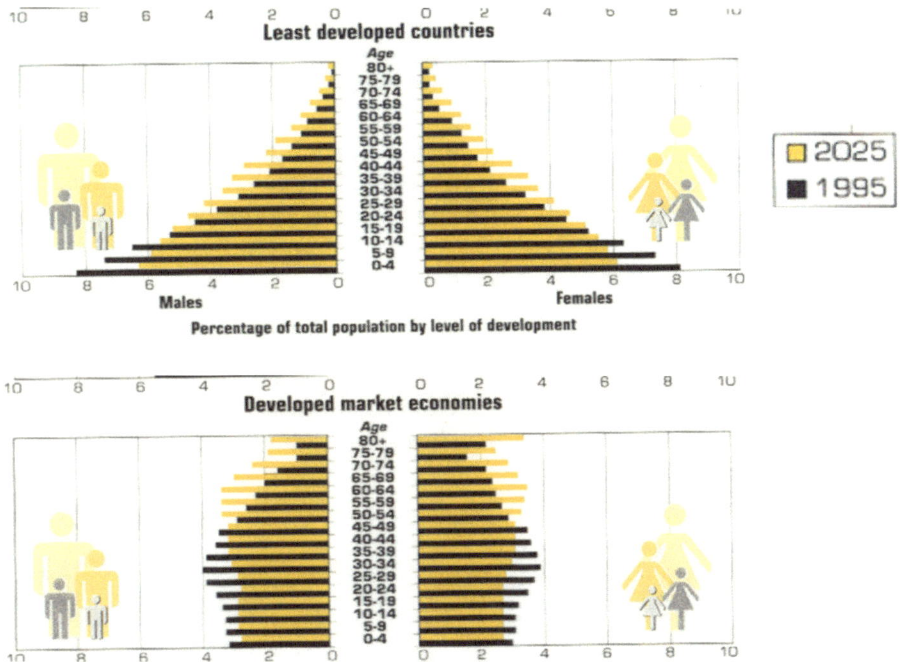

Fig. 2. Age distribution in developed and less developed communities 1995 and 2025 (Source WHO)

treatment of infectious diseases (in low income countries) and cardiovascular diseases, diabetes and cancer (in affluent societies) dominate in relation to preventive health care initiatives for better dietary habits and physical activity programs.

Cause	*Low-income countries*	*Industrialized Countries**
Infectious diseases and parasitic infestations	43	1.2
Cardio-vascular diseases	24.5	45.6
Maternal mortality	10.6	1
Cancer	9.5	21
Respiratory diseases	4.8	8.1
Other causes	7.7	23.1

*Nb Tobacco responsible directly or indirectly to 1/3 of all deaths
(Source WHO Health Statistics)

Table 1. Causes of death in the world (the values refer to per cent of deaths in 1996)

Somewhat provocative it can be said that the loss of manpower due to high infant mortality as a result of *malnutrition minus* is compensated for by high fertility and does not lead to increased public health costs per se. *Malnutrition plus* on the other hand represents a long-term effect which leads to an economic burden on the public health sector due to non-communicable chronic diseases in an increasing elderly population. What is most expensive for the society: High infant mortality or increased public health costs for adults and elderly?

Why is it so that it is easier to convince politicians about the need to develop a policy against malnutrition minus than to counteract the increase of malnutrition plus and obesity in all societies, e.g. in LIC as well as in affluent societies?

7.4 What is the solution: Family planning, food production or public health measures?

It is often argued that family planning will solve the problem. Figure 3 show the relation between fertility (number of children per mother), child mortality (per 1000 births), and life expectancy (years at birth) in 1950 and 2007 respectively. The figures are based on data from some selected industrial and low income countries on the various continents (Africa: Egypt, Kenya, Nigeria, Sudan, Uganda; Asia: Bangladesh, China, India, Indonesia, Japan, Pakistan, Thailand; Europe: Denmark, France, Hungary, Italy, Poland, Russia, Sweden, United Kingdom; Latin America: Brazil, Mexico; Middle America: Costa Rica, Cuba; North America: USA). The figures illustrate the close relation between fertility and child mortality and life expectancy indicating that public health and nutrition programs resulting in lower infant mortality and better health are the best ways to approach the population increase problem. That there is a close relation between child mortality and life expectance at birth is self-explanatory.

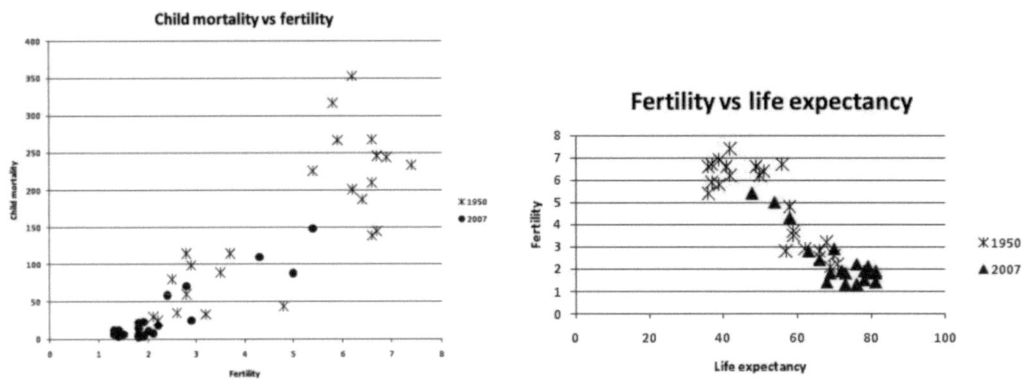

Fig. 3. Fertility versus child mortality and life expectance, respectively in 1950 and 2007. (Source: www.gapminder.org/world)

Figure 4 is based on data from UNFPA (United Nation Population Fund) and illustrates that there has been a reduction in fertility during the last 60 years. Despite this the world population is still presumed to increase although at a lower rate. (The grey shaded area refers to expected changes).

Obviously we must be aware of the fact that there is a lag period before lower fertility will have an impact on population size. Fertility lower than 2 children per woman must however result in a negative trend in world population in the long perspective. However, it is quite obvious from the diagrams that there are reasons to believe that family programs which are not based on actions to reduce infant mortality may be of limited value.

8. Nutrition transition

With the aid of national food balance sheet data from 85 countries Périssé and collaborators already in 1969 made an attempt to relate the general trends of consumption patterns as a

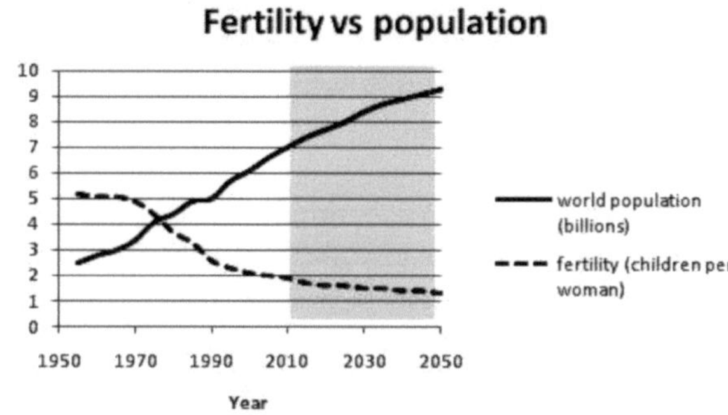

Fig. 4. Trends in fertility and global population during 100 years since 1950 (Source: www.gapminder.org/world)

function of income (figure 5) (Périssé et al. 1969). They could show that although the protein energy percent (E%) was almost the same in low-income countries and affluent societies, the amount of fat and refined sugar constituted much higher E% in the diet of affluent societies leading to increased energy density and reduced nutrient density.

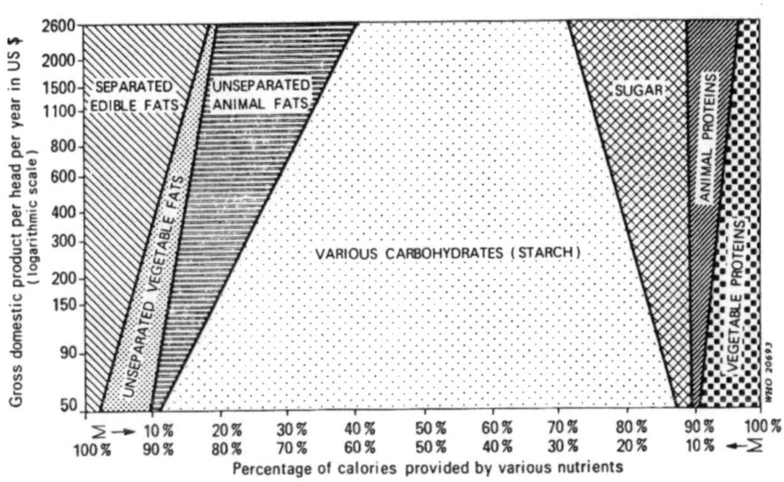

Fig. 5. Absolute and relative intake of energy constituents in relation to socioeconomic factors (Modified from Périssé et al, 1969)

Although 40 years has passed since then, the same difference in energy percentage distribution between macronutrients remains between rich and poor in affluent societies as well as in low income-countries (LIC) and seems to be related to the change in life style in all communities. Figure 5 illustrates that various carbohydrate (starch) sources (cereals, tubers) represent the major energy source (about 70%) amongst the poor, especially in LIC, while the fat consumption is low (less than 10%). In the affluent societies fat constitutes 35-40% of the energy intake, while the intake of complex carbohydrate only represents about 30-40%. Refined sugar may constitute up to 17%. Refined sugar together with separated edible fat are often characterized as "empty calories" as they do not contain any essential nutrients although they constitute about one third of the energy intake in the diet in affluent societies.

Figure 5 also illustrates the interesting fact, which is often not fully realized, that the energy in the diet derived from protein in LIC represents 10-15 E% and is similar to that in high-income groups. This leads to the following two very essential conclusions: 1) the high protein intake in high-income groups is not due to a higher protein concentration in the food *per se* but to the fact that the total energy consumption is higher; 2) The high prevalence of protein deficiency in LIC is essentially due to a too low total intake of food (i.e. energy), and not primarily due to a low protein content in the food *per se* as often presumed. The latter misconception had for a long time lead to action programs for protein and amino acid enrichment of food. However, the worst off are the poor in affluent societies whose diet is characterized by a high relative amount of empty calories as influenced by the diet in affluent societies, in addition to lack of food. Today priority is given to high yields in agricultural production rather than especially protein rich varieties for human food.

In the affluent societies of today it took thousands of years to develop from hunting and gathering to agricultural production and then another two centuries to become industrialized. Today the same transition in the LIC occurs within a few decades. The socio-economic development has resulted in changes in dietary habits, which is characterized as "nutrition transition" (Caballero & Popkin, 2002). From the nutrition point of view this transition includes not only a better availability of various food items and changes in the dietary pattern, but also in life style, e.g. reduced physical activity, which have further stressed the global food resources. Some of these changes are positive, i.e. a more differentiated food pattern. The reduced physical activity as a result of increased mechanization has reduced total daily energy turnover remarkably, e.g. in Sweden a 30% reduction of total energy turnover as observed from 1930 to 1990 (table 2). The reduced physical exercise in addition to the negative changes in dietary habits including a higher energy density, i.e. increased intake of "hidden fat" and refined sugars, has however increased the risk for the development of obesity if total energy intake is unchanged.

Year/Commodity	*1876-85*	*1886-95*	*1896-1905*	*1906-15*	*1920-29*	*1930-39*	*1950-59*	*1997*
Energy (kcal)	2164	2331	2708	2954	3036	3114	2825	2116
Protein (g)	58	63	74	81	82	87	73	82
Fat (g)	44	49	66	78	94	111	120	82
Protein (E%)	11	11	11	11	11	12	11	16
Fat (E%)	19	20	23	25	29	33	39	34
Carbohydrate (E%)	70	69	66	64	60	55	50	47

Table 2. Intake of macronutrients in Sweden 1876 – 1997. (Source: Swedish Food statistics)

8.1 Various patterns of nutrition transition

Popkin and his collaborators (Caballero & Popkin, 2002) described five patterns developed during the nutrition transition (table 3). The majority of the affluent societies still belong to the pattern 4 while the public health try to stimulate the dietary changes to be more optimal for health as in pattern 5. The question is now to what extent the dietary changes in the LIC can develop directly to pattern 5.

Transition type Behaviour profile	*Pattern 1 Collecting food*	*Pattern 2 Famine*	*Pattern 3 Receding famine*	*Pattern 4 Food surplus society*	*Pattern 5 Healthy diet concept*
Diet	Varied diet (plants, wild animals)	Cereals dominate	Less starchy staples, more fruit, vegetables, animal protein	More empty calories (fat, refined sugar); processed food introduced	Less fat, sugar and processed food; more starchy staples, fruit and vegetables
Nutritional status	Few nutritional deficiencies	Malnutrition minus; vulnerable groups affected	MCH nutrition problems; some deficiencies decrease	Malnutrition plus, Obesity	Reduced obesity improved health
Economy	Hunter-gatherers; subsistence farming	Agriculture, animal husbandry; monoculture; Subsistence farming	Crop rotation, fertilizers and agricultural mechanization; industrial revolution;	Less heavy work; sedentary life; mechanization; increased service sector	Sedentary work dominates; leisure time and leisure exercise increased
Demographic profile	Low fertility; high mortality, low life expectancy; Rural population	High fertility, low life expectancy, high infant and maternal mortality; mostly rural population; small cities	Fertility static; slow mortality decline; Mostly rural population; urbanization begins	Life expectancy increase; rapid fertility decline; increased elderly proportion; urbanization	Life expectancy high (70-80 yrs); increasing proportion of elderly; Less disability; urbanization of rural areas
Morbidity	Infectious diseases; no epidemics	Epidemics; endemic diseases; deficiency diseases	Infection and parasitic diseases; Epidemics; Tb, smallpox, polio; weaning diseases	Chronic diseases related to diet and pollution (CVD, osteoporosis, cancer); infectious diseases declines	Preventive and therapeutic health promotion;
Food processing	none	Food storage begins	Food storage processes; cooking technology	Food transforming technologies	Food and food constituent substitute technologies

(modified from Caballero & Popkin, 2002)

Table 3. The Nutrition Transition pattern characteristics

8.2 Life style changes

Life style changes are characterized not only by changes in dietary pattern, but also by increased mechanization in industry and agriculture and an enormous expansion of the transport system. Mechanization has lead to less heavy work load in agriculture, forestry and industry which has been beneficial for the public health of the population. Interestingly the reduced energy intake as result of less heavy work during working hours is to some extent compensated for by more activities during leisure time. That is one reason why in the international energy recommendations during the last decades the concept of life style was introduced (FAO/WHO/UNU Expert Consultation, 2001). Earlier focus on a reference man or woman and on the energy cost of occupational work has been substituted for by estimates for individuals instead of groups and estimates of total energy turnover now include both occupational work and discretionary physical activities during leisure time. The total daily energy turnover is expressed as a multiple of the basal metabolic turnover per day (which is related to sex, age and body weight) and called physical activity level (PAL). The PAL value is then accepted as indicator of the life style. This is a valuable help to evaluate the optimal energy requirements in man in developing as well as developed communities as it is adapted to differences in life style.

The life style changes have been based on an increased use of essentially fossil energy resources, e.g. coal and oil. There is reason to believe that the same changes in socio-economy standard and lifestyle, which has characterized the affluent societies of today, will occur in other developing societies. Why should not the population in China and India have the same right to have cars and refrigerators as those in the affluent societies when the socio-economic situation gets better? However, not only are conventional energy resources limited, they also lead to environmental problems. There is consequently an urgent need to find sustainable and renewable energy sources which are in balance with a sustainable food production.

9. Nutritional priorities

Under normal conditions body gives priority to cover its *energy needs*. This can be covered from any of the energy-yielding macronutrients (carbohydrate, fat, protein and alcohol). Thus energy need defines the amount of food needed, i.e. a *quantitative* aspect of the dietary intake. When the energy need is not covered from the diet, energy is mobilized from energy-yielding substances stored in the body (glycogen in liver and muscle, fat in subcutaneous tissue and adipose tissue). Adipose tissue is the dominating energy store, but interestingly body protein also constitutes a potential and substantial fraction (about 20% of the total energy store in the body). Nevertheless the body uses carbohydrate as energy source in the first hand to cover acute energy needs. Since blood glucose represents a minor energy resource and the store of liver glycogen, which is available for energy turnover in the body limited, endogenous carbohydrate is produced from catabolism of muscle protein in skeletal muscle during acute energy deficit before energy is mobilized from the fat stores. Protein will be used as energy source, even if protein needs are not met via the diet in an acute energy deficit. This means that energy turnover and protein turnover are closely related.

The *nutrient needs* refer to the specific need of certain nutrients, i.e. protein, trace elements and vitamin, and water. This leads to a *qualitative* aspect of the dietary intake. The requirement of essential nutrients is essentially related to the active cell compartment, the fat free mass. Thus nutrient requirement is related to the body composition which varies with age, sex and body size, as well as to growth and maturation of tissues and organs. The need of specific nutrients is, however, with few exceptions, not related to energy turnover and physical exercise.

9.1 Protein has a two-fold role

Protein need is due to (i) a *specific nutritional role* as source of essential amino acids for protein synthesis i.e. building up, repairing and maintaining tissues; as well as to (ii) *a non-specific role* as an energy-yielding nutrient. If energy needs are not met, protein is used as energy source. Likewise, if more protein is consumed than specifically needed for protein turnover, which is the case in most affluent societies, the surplus is converted through gluconeogenesis and stored as energy, since protein cannot be stored in the body.

9.2 Is there a protein problem and need for increased consumption of animal products?

During the decades after the Second World War, the international society got more and more aware of the magnitude of nutritional problems in the developing world. International agencies concerned with nutrition - WHO, FAO and UNICEF - were beginning to realize the maldistribution of protein rich food of animal origin and the frequency with which young children in low income countries failed to receive weaning foods of adequate protein value to supplement breast milk. The discussion of global nutrition problems was very much focused on the protein gap and in 1955 WHO initiated the Protein Advisory Group (PAG) to form a prestigious scientific advisory body of the UN system on protein-calorie malnutrition (PAG, 1975). In 1971 it prepared a background document and advice for preparation of the "Strategy Statement on Action to Avert the Protein Crisis" as a UN report. In its early years, PAG was focused on potential sources of relatively low-cost protein, e.g. oil seed. Later other more futuristic protein sources were identified and discussed, e.g. leaf protein concentrates, single cell, fish protein concentrates. As long as energy costs for transport and production were more or less marginal, the focus on the global nutrition problems was on the quantitative and qualitative protein intake. However two years later, as a result of the energy crisis during 1973/74, the scenery had completely changed. One now realized that the focus on protein was too narrow in a global perspective and the interest to produce protein concentrates and protein-rich food products almost collapsed. The energy efficiency in protein production was very low especially for various forms of protein concentrates, e.g. single cell protein, leaf protein and fish protein concentrates. Rapidly increasing energy costs, and the awareness that the major problem was not protein intake, but a too low energy intake and lack of food changed the scenery. A provocative article "The protein fiasco" by Mclaren (1974) started a discussion about a complete re-evaluation of the situation and the unnecessary focus on protein concentrates. As long as the energy needs are not met, the addition of protein rich foods represent a waste of resources since protein is then used as energy source. The name of the advisory group was later changed to "Protein Energy Advisory Group".

9.3 Energy crops for human consumption

As stated earlier the body gives priority to cover its energy needs. It is consequently essential to find optimal resources for high energy yields in food production. The global food supply and reserve is usually expressed as total cereal stores (wheat) on the world market, as per cent of annual consumption or as number of days to feed the world population. The importance of roots and tubers for food security and source of income for poor farmers is overlooked. The great importance of introducing potato and cassava for eradication of malnutrition and increased food availability for the poor was illustrated during the 17th century. In 1985, Horton and Fano discussed the edible energy produced by major crops and showed the potentials of tubers and roots. Their role in the 21st century was commented in a report by Scott and collaborators in 2000 (Scott et al, 2000).

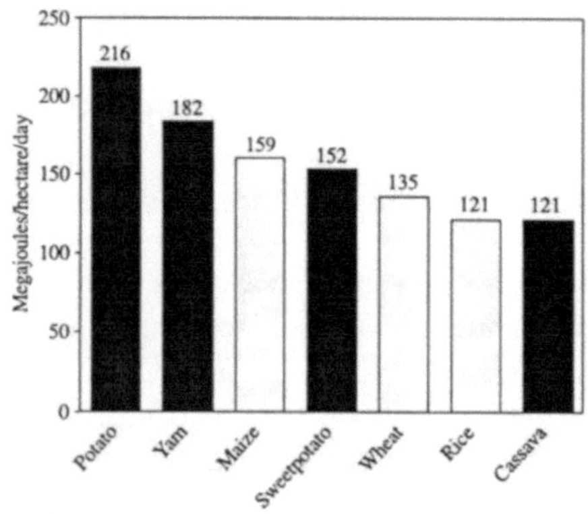

Fig. 6. Edible energy produced by major roots and tubers and cereals (Source: Horton & Fano, 1985)

Even energy yielding crops contain protein, which is often forgotten. In the comment on nutrient transition it was said that the protein energy per cent in the diets is remarkably unchanged. The explanation to this is that protein energy percent is 15% in wheat and maize and only slightly below 10% in rice and potato, as indicated in table 4.

Commodity	*Energy (edible) (kcal per 100 g)*	*Protein energy %*
Cassava	160	3.5
Potato	86	8
Sweet potato	86	7
Yams	118	5
Maize	86	15
Rice	358	7
Wheat	340	15
Meat	201	39
Milk (cow)	61	21
(human)	70	5

Table 4. Energy content and protein energy per cent in some staple foods.

Of special interest is to note the low protein content in human milk. This is partly compensated for by the high quality of milk proteins. The low growth rate in combination with a reduced tolerance for protein as result of immaturity of liver and kidneys in the human newborn has resulted in this adaptation of human milk composition. Such an adaptation is also illustrated by the large differences in milk composition in various mammalian species (Hambraeus, 1979)

The relevant question for a nutritionist in the 21st century is to what extent increased focus on using crops with high energy yields for biofuel production may represent a potential threat for the global food market not only for energy but also protein intake. But it is also essential to argue against the belief that protein requirement is not met by conventional diets and that there is a need for increased meat consumption, protein rich food and/or protein supplements in the affluent society.

10. Is there an optimal diet?

As nutritionist you are often asked to give advice of the optimal diet. This is an almost impossible question to answer, as the human being is a very adaptable object which is not always realized. However, it is impressive to what extent man can adapt to the environmental situation. When we are discussing the concept of an optimal diet the premises must be defined and analyzed. A systems analytical approach must be established where nutritionists, agronomists and those engaged in technical change and innovation within the bioenergy sphere all cooperate in the future direction of the policy for future food and energy production.

When discussing optimal dietary habits in various zones of the globe it is often forgotten that the present staple diets are mainly the result of adaptation to production resources and not necessarily an optimal diet from public health perspectives. This is probably the main reason why milk and dairy products are staple items in the food basket in populations in temperate zones with only one crop per year; ruminants can be fed on meager fields they can also deliver essential nutrients during the winter season. The production resources may also explain why potato so rapidly developed into a staple food in Europe and cassava in the tropical zones during the 17th century. Potato and cassava were both imported from the Alto Plana region in Latin America during the 17th century and represent "import" products. As they represent two crops with high edible energy yield per acre, they have probably saved more lives in poor families in times of famine than most other food items.

In his classical metabolic study, Hindhede already in 1913 (cited by Salaman, 1970) showed that a person could maintain body weight and nitrogen balance with no nutritional deficiencies for more than 100 days when he consumed a diet containing 3000 calories per day based on 2-4 kg potato plus margarine. Thus the content of protein and essential nutrients in a tuber such as potato was enough to cover the nutritional needs. This explains why the introduction of potato with such a high energy yield per area could reduce malnutrition among the poor farmers in Ireland during the 16th century. But the risk to be dependent on a monoculture was also illustrated by the serious famine in Ireland as a result of the potato crop failure during the middle of 19th century (Salaman, 1970).

On the other hand fruits and vegetables are less common in the temperate zones but dominant in the tropical regions. To be a vegan in northern Scandinavia necessitates good transport facilities and socioeconomic welfare as vegetables and fruits must be imported during the winter season often to high prices.

To argue that Paleolithic diet and or the Mediterranean diet is the optimal diet for mankind is an oversimplification of the problem. If high life expectancy is used as indicator of optimal solution of nutrition, the diets of population in Iceland, Japan and Sweden could be indicators. However although they represent populations with the highest lifespan in the world today, their diets are very different from each other. On the other hand socioeconomic factors and prosperity is not a guarantee for optimal health, as indicated by the lower life

expectancy in Denmark when compared to two other Scandinavian countries, Norway and Sweden with similar population and socioeconomic situation.
When discussing an optimal future development of biofuel production as well as of food production, the first step must be to define in what perspective they should be optimal:

- Use of primary resources (energy, water, labor, land)
- Socio-economic development (employment, national economy)
- International policy (world trade, global food distribution)
- International solidarity (vulnerable groups in the society, next generation)
- Energy and food sustainability (national as well as international)
- Optimal public health (minimal sickness benefit costs, malnutrition minus and plus)

10.1 Case study

This case study was presented in a seminar for the Agricultural Marketing Board in Sweden in order to illustrate the effect of the nutrition transition in an affluent society such as Sweden on cereal consumption (Abrahamsson, 1979).
Starting from the situation in 1972/73 in Sweden with a production of 1.5 mill tons food cereals (for direct consumption) and 3.6 mill tons feed cereals (for indirect consumption via production of animal products) the potential effect of a reduced production of feed cereals with one mill ton which was substituted for by food cereals, was studied. The reduced consumption of animal products as result of decreased feed production was calculated to be obtained by a 50% reduction in the production of pork, broiler and egg and no intensive beef production. The milk production was left intact (and the related beef production from dairy plants) in order to keep the nutrient intake in children and teenagers more or less intact. In order to keep energy intake the same, the consumption of cereals, potato, roots and legumes in the diet was increased. This led to the following change in food consumption (table 5).

Nutrient	*Decreased consumption of* *Pork (14 kg)* *Broiler (6 kg)* *Egg (6 kg)* *Beef (3 kg)*	*Increased consumption of* *Potato (5 kg)* *Roots (2 kg)* *Legumes (2.5 kg)* *Cereals (12 kg)*	*Net effect*
Energy (kcal)	-54000	+54000	0
Protein (g)	-3400	+1700	-1700
Fat (g)	-4000	+200	-3800
Carbohydrate (g)	-6	+11000	+11000
Calcium (mg)	-5000	+5300	+300
Iron (mg)	-520	+770	+250

Table 5. Calculated effect of decreased feed cereal production on intake of some macronutrients

Although there was increased direct cereal consumption, the net effect was a surplus of 900,000 tons food cereals which could be available for cereal export, or the use of a corresponding production area for other purposes, e.g. cash crops and biofuel.
The obtained change in nutrient intake per capita and day would essentially be a reduced protein intake from 78 to 73 g per capita and day, and a reduction of the animal protein energy percent in the diet from 65 to 56 %. This is however well above the minimal requirement, both with respect to the amount of protein needed and the protein quality.

To what extent could such dietary habits be realistic? The answer is indirectly given in table 6 which shows the food consumption pattern in Sweden in 1939 (before the 2nd world war), 1943 (during the 2nd world war when food production was adapted to self sufficiency with limited resources) and 1982 (when Sweden was a typical affluent society with "modern" dietary habits).

	Consumption g/person/day			
Food item	*1939*	*1943*	*1982*	*Proposed diet*
Milk	530	700	500	660
Cheese	15	10	40	25
Beef & Pork	125	100	165	60
Egg	25	15	35	20
Cereals (wheat&rye)	265	265	175	265
Potato	320	430	230	400
Sugar	130	90	110	100
Butter/Margarine	50	45	65	60
Fish	50	50	75	n.a.

Table 6. Intake of various food items throughout years in relation to proposed modified diet. (The figures refer to g per person and day.)

The table illustrates that the increased direct cereal and potato consumption was well in line with what was consumed during wartime when resources were limited. It also shows the drastic increase in the indirect consumption of cereals as result of increased animal protein consumption during the post war period in an affluent society such as Sweden as result of the nutrition transition.

The question remains: will such changes in dietary habits be accepted by a population in affluent societies? As a matter of fact there were no nutritional deficiencies affecting the public health situation in Sweden in 1943 and less stress on the public health budget due to non-communicable diseases. Why are we still accepting continued increased meat consumption as an unavoidable necessity in affluent societies and on the same time argue that the LIC population should not be allowed to increase their consumption of meat and animal protein products?

Interestingly in 2006 similar discussions were presented by Aiking and collaborators, who calculated that if consumers in developed countries would reduce their protein intake about 30 % and replace meat by vegetable protein, about 80 % of agricultural land used for feed crops could be set free for other productions (Aiking, 2011).

11. Nutritional perspectives on the future of biofuel production

11.1 Development of biofuel technology

The first generation of biofuels comprised production of ethanol from plants containing sugar and/or starch (sugar cane, sugar beet, cereals, cassava) and biodiesel from oilseed crops (rapeseed, sunflower, soybean, palm oil). These could directly compete with the global food supply.

Today the dominant part of liquid biofuel production is based on crops that can also be used for food, e.g. sugar cane, corn and rapeseed. The challenge for expansion of biofuel production is consequently representing a challenge for land area use. This is illustrated by the fact that in 2007 about a quarter of the national corn harvest in the US was used to produce ethanol. The impact of biofuels/bioenergy production on indirect land use change (ILUC) is still controversial. There is a potential for increasing the yields and/or using marginal areas for energy crops. However, probably also energy crops need a certain soil quality and can not only be using marginal land. Thus there is a risk that degraded lands and natural landscapes are restored for biofuel production (Timulsina & Shrestha, 2010).

So far bioethanol and biodiesel produced from a limited number of crops dominate the renewable biofuels for the transport sector. Sugarcane from Brazil and corn/maize from USA are dominating crops for bioethanol. Rapeseed is the dominant crop for biodiesel although oil palm in South Asia and soybean in US and Brazil and sunflower in East Europe are also coming in the focus of increased interest. There are also interesting regional differences in biofuel production. Diesel is essentially the biofuel produced in Europe, while ethanol dominates in Brazil and the US. Another interesting difference is that Brazil, by far the largest producer of ethanol, is producing not only for domestic consumption but is also the world's leading exporter of biofuels. In contrast the biofuel production in the US and Europe is essentially targeted for domestic consumption.

Global production of fuel ethanol has more than doubled from 2004 to 2009. The total production of biodiesel is much smaller but is growing more rapidly than that of ethanol, the annual growth rate being around 50% during the same period.

The second generation of biofuels is based on technologies that convert lignocellulosic biomass (e.g. agricultural and forest residues) or advanced feedstock, e.g. jatropha, microalgae. This production from non-food crops would be much less deleterious for the global food sustainability. On the contrary it could help to get better economy in the agricultural sector, which can then produce both food and fuel together, and increase food availability.

We are still in the beginning of developing biofuel technologies. Priority is presently given to find alternatives for power trains in vehicles and reduce the need for oil import in the transport sector. There are also alternatives based on production of biofuel in other sectors than the agricultural one. Anything from conventional internal combustion energy vehicles using alternative biofuel, i.e. ethanol, methanol, biodiesel, to hybrid and battery powered electric vehicles and fuel cell vehicles are of interest. Of special interest is the potential for energy production from waste products in our consumer society. Furthermore waste products from sewage treatment in industries and the forestry industry may have a great potential, in addition to waste products from the agriculture sector, e.g. rapeseed cakes and straw. From the nutritional point of view these alternatives are of great interest as it means less conflict between production of biofuels and food crops regarding the use of agricultural land.

11.2 A challenge for nutritional science

Changes in life style as a result of socioeconomic development has lead to a nutrition transition which so far has included a deterioration of dietary habits in relation not only to public health but also to the optimal use of primary resources. The concept of optimal intake must be better defined and related to individual needs as well as to food availability and security. This must also take into consideration the optimal use of agricultural resources and land to be balanced in relation to land use for biofuel production. This necessitates a realistic and scientifically sound base for dietary intake recommendations.

Increased production of biofuel may be deleterious for food production if the wrong crops are chosen. But it could also be one of the ways to increase economic power in the agricultural production and thereby help to increase food security. Optimal use of production area also calls for optimal selection of crops under the prevailing circumstances and to increase the efficiency in food production, transport, food industry and food handling. This calls for balance between food production, socioeconomic development and public health perspectives on food and nutrition policy in all countries. This necessitates scientific knowledge and skills in all disciplines and capability for cooperation and an interdisciplinary approach.

11.3 Human capital is the bearing and not the wearing sector in the society! Political power is depending on food policy and energy policy in harmony!

It is high time to reevaluate the role of human capital! Why is it so that the public health sector and educational sector are looked upon as a tearing sector of the society that costs money, and the industrial and economic sector as a wearing sector giving financial support to the society? How can a production function without healthy workers and the society develop without increased knowledge and literacy? How can a society be developed without a democratic balance between producers and consumers? How should a modern society exist without healthy workers and consumers?

The global economic turmoil during the last year has clearly demonstrated the incompetence of the economists and policy makers to handle the situation. In many respects the latest economic crisis was an artificial crisis started by unethical gamblers on the market. It was not a result of a sudden and unforeseen crisis in the availability of primary resources for feeding a population. It was a luxury crisis calling for a reduction of our overconsumption of cars and luxury products, leading to unemployment, malnutrition and public health problems. But the price was not paid by the affluent societies or the rich. Those who had to pay for the mismanagement were the poor people in LIC, who faced increased food prices, with malnutrition as a secondary result of economic mismanagement.

A political system that cannot guarantee food security for its population will never survive, as we have learnt from history and also are reminded of in today's political conflicts. After all it is food and water that represent our basic needs. We do not drink oil neither do we eat dollar or euro notes. Political power must be focused on the human capital and not only on energy or economic power. The best indicators of a healthy population are low infant and child mortality as well as low maternal mortality and a long life span. This is obtained by food availability and sustainability in combination with public health measures for the population. Food power as well as energy power should give priority to meet these goals and not be used as political weapons. New technology to help us reach these goals is expected to be in focus for our scientists. Today it is a challenge to cooperate for developing a biofuel production in balance with food production and optimal use of primary resources and fertile land. If so, the balance between food production and population size will be solved and democracy develop. Otherwise political power will still result in conflicts for control of the world's primary resources and uneven distribution persist between individuals and societies.

12. Conclusion

Today it seems as the dominating global interest is devoted to the energy situation and oil prices and its impact on the global financial situation. This might be the major reason for the increased interest to find alternatives to oil as energy source. It is also an illustration that the debate is still dominated by people from the industrialized world. The real priority for survival and optimal health for mankind from a global perspective is however still food and water. Energy comes much later on the priority list of factors for the survival of mankind. Energy might be a helper but not even the most essential component in the combat against world hunger.

Water and energy resources are prerequisites for food security, but are unfortunately not equally allocated on the globe. During the last decades more interest has been devoted to the global water resources and the impact of human activities on the global water system (Hoekstra & Hung, 2005). The impact of the cost of water on food production and food availability thus represents an additional challenge for the nutritionist, but now also for those engaged in production of biofuels. It has been estimated that food production is a dominant "consumer" of 70% of freshwater and 20% energy resources, respectively (Aiking et al, 2006).

The new interest in production of biofuels is a challenge for all of us. In the enthusiasm to test new approaches to solve global problems, there is a great risk that we do not analyse the long term perspectives and its indirect effects on other markets. The interest in biofuel production is a welcome complement for the development of the agricultural sector, n.b. in case production of biofuel can be based on surplus and waste products, also from the food production sector. A sustainable food production also calls for sustainable energy resources.

The role of the nutritionists is to define the optimal diet for a population based on the specific conditions for food production in various regions and the need to guarantee optimal health for all. The dietary recommendations must be based on objective scientific data and optimal use of available resources, and not on subjective feelings and political fashion. It is obvious that the diet which has developed in the affluent societies of today is by no means optimal, no matter whether we analyze it from the preventive health perspectives, or from the optimal use of primary resources. This calls for interdisciplinary discussions between those engaged in nutritional science, global public health problems and agricultural production, and those engaged to find new means for a sustainable and renewable energy production. Selection of crops and byproducts for renewable energy must be in harmony with the work for optimal diets for optimal health of the global population and vice versa.

The innovative capacity of those involved in developing sustainable energy systems should be combined with the scientific knowledge within the field of nutrition, public health and agronomy in the combat against malnutrition and poverty, as well as against non communicable diseases and waste of global resources. The global resources for food, water and energy in relation to population demands, will dominate national and international policy once and for ever and unfortunately probably be a cause for political unrest also in the future.

13. References

Abrahamsson, L. (1977). Producer- consumer –nutrionist perspectives on analysis of alternatives for dietary consumption in Sweden. Report presented at a meeting of the National Preparedness Commission of the Agricultural Marketing Board in Sweden, March 31, 1977. (In Swedish).

Against the Grain (29 November 2010). Saudi investors poised to take control of rice production in Senegal & Mali? http://www.grain.org/articles

Against the Grain (25 January 2011) New agricultural agreement in Argentina: A land grabber's "instruction manual" http://www.grain.org/articles

Aiking, H. (2011) Future protein supply. *Trends in Food Science & Technology*, vol 22, pp 112-120.

Aiking, H., De Boer, J & Vereijken, J.M. (2006) Sustainable protein production and consumption: Pigs or peas? In: *Environment & policy.* Vol 45. Dordrecht, Springer, The Netherlands.

Borgström, G. (1965) The Hungry Planet. The modern world at the edge of famine. *Macmillan*, New York, 487 p

Caballero, B. & Popkin, B.M. (2002) The Nutrition transition: Diet and Disease in the Developing World. *Academic Press*, London 2002;

European Committee. (2009). *Directive 2009/28/EC of the European Parliament and of the Council of 23 April 2009 on the promotion of the use of energy from renewable sources and amending and subsequently repealing Directives* 2001/77/EC and 2003/39/EC.

FAO/WHO/UNU Expert Consultation (2001) Human Energy requirements. *Food and Nutrition Technical Report Series.* FAO No 1, Rome. Pp 1-96.

Fischer, G. (2009) World food and agriculture to 2030/50: How do climate, change and bioenergy alter the long-term outlook for food, agriculture and resource availability? Expert meeting on How to feed the World in 2050 (12-13 October 2009) FAO, Rome, Italy FAO. *http://www.fao.org/wsfs/forum2050/wsfs-backgrund-documents/wsfs-expert-papers/en.*

Hambraeus, L. (1979) Composition of human milk, In Visser, HKA (ed) Nutrition and metabolism of the fetus and newborn infant. Fifth Nutritia Symposia, Rotterdam 1979. *Martinus Nijhoff Publ.* The Hague. pp 253-262.

Harvey, M. & Pilgrim, S. (2011). The new competition for land: Food, energy, and climate change. *Food Policy* volume 36, pp 540-551.

Hoekstra, A.Y. & Hung, P.O. (2004). Globalisation of water resources: international virtual water flows in relation to crop trade. *Global Environmental Change* volume 15, pp 45-56

Horton, D. & Fano, R.H. (1985) *Potato Atlas. Lima, Peru. International Potato Center (CIP) 1985*

Howarth, R.W. & Bringezu, S. (Eds.). (2009) *Biofuels: Environmental consequencies and interactions with changing lands use.* Proceedings of the Scientific Committee On Problems in the Environment (SCOPE), international biofuels project rapid assessment. 22-25 September 2008. Gummersbach, ISBN:1441488294. Germany.

Indo-Asian News Service (IANS) (13 February 2011) African farmland to Indian firms no cause for worry http://framlandgrab.org/post/view/181676

Lawrence, D., Beddington, J., Godfray, C., Crute, I., Haddad,L., Muir, J., Pretty, J., Robinson, S. & Toulmin, C. (eds) (2011) The challenge of global food sustainability. *Food Policy* volume 38 (supplement 1), pp S1-S114.

Malthus, T. (1798) *An Essay on the Principle of Population* (Penguin edition) London 1982.

Mclaren, D.S. (1974). The great protein fiasco. *Lancet* volume 304, pp 93-96.

Protein-Calorie Advisory Group (1975) The PAG compendium. *Worldmark Press Ltd*, New York. Nine volumes.

Périssé, J., Sizaret, F. & Francois, P. (1969). The effect of income on the structure of the diet. *FAO Nutrition Newsletter*, volume 7, pp 1-9.

Salaman, R. N. (1970) The history and social influence of the potato. *Cambridge university Press.* 1949, reprinted 1970)

Scarlat, N. & Dallemand, J-F. (2011) Recent developments of biofuels/bioenergy sustainability certification: A global overview. *Energy Policy* doi:10.1016/j.enpol.2010.12.039

Scott, G.J., Rosegrant, M.W. & Ringler, C. (2000) Roots and tubers for the 21st century. Trends, Projections and Policy options. Food Agriculture and the Environment. Discussion paper 31. *International Food Policy Research Institute* Washington. May 2000 pp 1-64.

Smil, V. (2000) Feeding the world: a challenge for the twenty-first century. *MIT Press,* Cambridge, Mass. USA 2000

Steinhardt, C.S. & Steinhardt, C.E. (1974) Energy use in the US Food system. *Science,* volume 184, pp 307-316

Timilsina, G.R. & Shrestha, A. (2010) How much hope should we have for biofuels? *Energy* doi:10.1016/j.energy.2010.08.023

The following websites are recommended for those who are interested to be updated on the ongoing debate on food policy and its relation to health statistics and socioeconomic development.

www.gapminder.org/world;

www.grain-org ;

Part 2

Agricultural Policy

5

The Post 2008 Food Before Fuel Crisis: Theory, Literature, and Policies

Cheng Qiu, Gregory Colson and Michael Wetzstein
Department of Agricultural and Applied Economics, University of Georgia
USA

1. Introduction

As early as 1983, research began to appear indicating the potential for biofuel production to emerge as a disruptive force in US and world food sectors (Barnard, 1983). Of particular concern in early and present research is that increased use of agricultural outputs for energy, as opposed to foodstuffs, could ultimately lead to a net welfare loss where the benefits of biofuels are outweighed by the negative consequences linked to reduced food availability. This dilemma emerges due to the direct competition between biofuel and food production for the same renewable and nonrenewable resources critical for their sustainability (Rajagopal and Zilberman, 2007 and von Urff, 2007). In 2007-2008, global food prices experienced a significant upward spike resulting in political and economic instability, conflict, and hardships in both the developed and developing world. Figure 1 illustrates the United Nations FAO monthly food price index and the cereals price index since 2000. As indicated in the figure, in 2006 food prices started to rise with the most rapid increases occurring in 2007 through the middle of 2008 when an equally rapid price decline occurred. Relative to the general food price index, the increase in cereal prices was more pronounced.

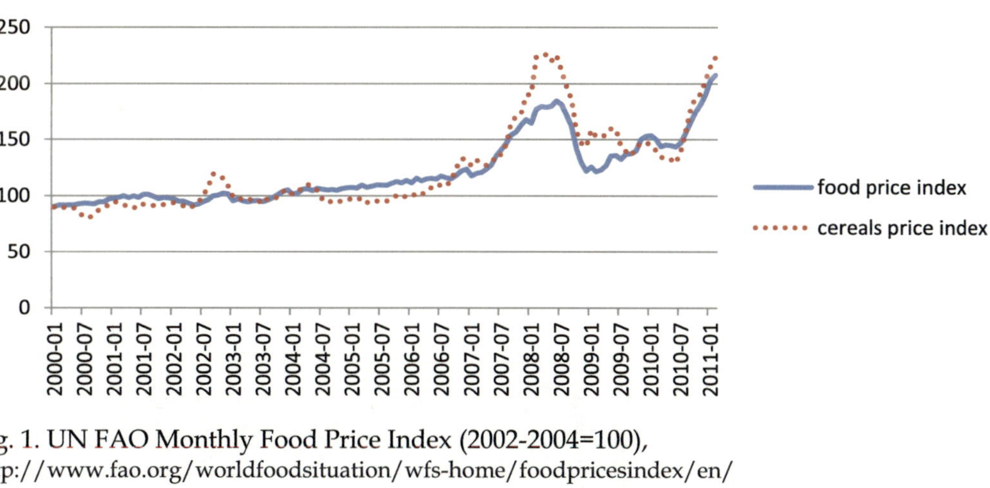

Fig. 1. UN FAO Monthly Food Price Index (2002-2004=100), http://www.fao.org/worldfoodsituation/wfs-home/foodpricesindex/en/

The effects of the spike in food prices were particularly acute in parts of Africa, Asia, the Middle East, and South America where significant portions of household budgets are spent on food (e.g., 50-70% of typical household budgets in Africa are spent on food, Diao et al., 2008). This resulted in not only a worsening of poverty statistics, but also led to aggressive national protectionist food policies, civil unrest, malnutrition, and deaths. In general, populations most vulnerable to significant rises in food prices are those in countries that suffer food deficits and import oil. These two features are directly correlated with a country's income status, with the majority of the 82 low-income countries having food deficits and being net oil importers (Senauer, 2008; Runge and Senauer, 2007). With assumption of biofuels produced mainly with corn, causing food price inflation, countries where corn is the major food grain will generally experience larger increases in food costs, while countries with rice as the major food will experience less of an increase. Countries where wheat and/or sorghum are the major food grains fall in between. Consequently, the highest percentage cost increases are observed in Sub-Saharan Africa and Latin America and the lowest percentage cost increases are in Southeast Asia (Elobeid and Hart, 2007).

A widely considered view both in policy circles and the domain of public perception is that the dominant underlying driver of the 2007-2008 price spike was increased use of crops for the production of biofuels (Diao et al., 2008; Abbott et. al, 2008). This shift from fossil fuels to biofuels, which has in large part been fostered through national agriculture and energy policies motivated by increased oil price volatility, energy security ambitions, and environmental concerns, is particularly prominent among many Kyoto Protocol signatory countries (Balcombe and Rapsomanikis, 2008). In effect, the emergence of a significant biofuel market has given producers a choice of supplying food or fuel depending on their relative net returns (Brown, 1980; Zhang et. al, 2010). However, the rapidly growing market for biofuels has given rise to the perception that rapid biofuel expansion generates upward pressure on global food prices, exacerbating global hunger problems (Runge and Senauer, 2007). Figure 2 illustrates this rapid biofuel growth for U.S. ethanol production. Some

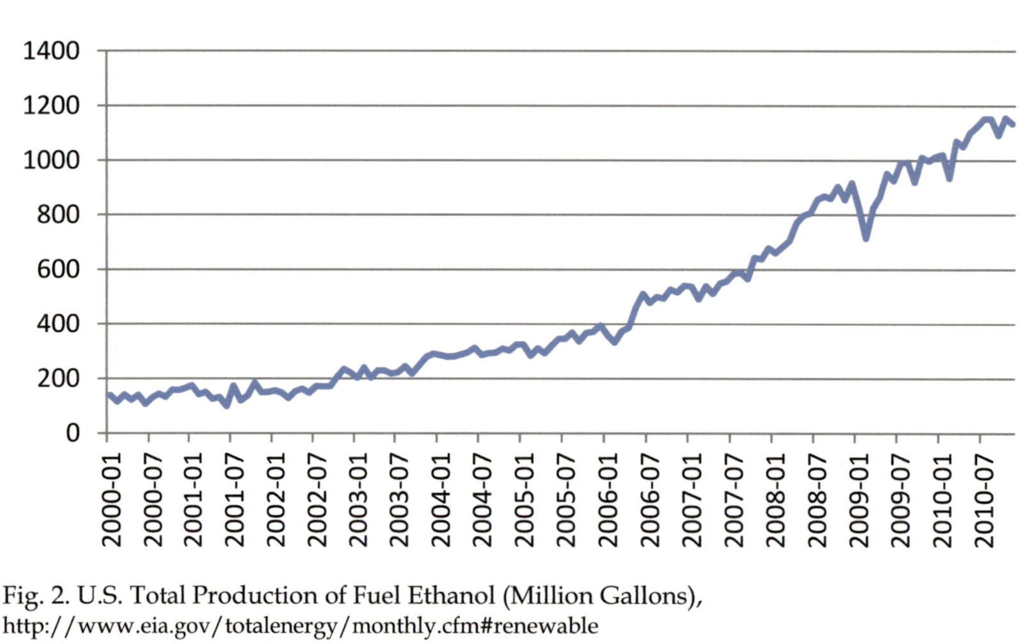

Fig. 2. U.S. Total Production of Fuel Ethanol (Million Gallons), http://www.eia.gov/totalenergy/monthly.cfm#renewable

estimates have even placed the number of malnourished people globally at 1.2 billion, twice the number without any effects on the food supply due to biofuels (Runge and Senauer, 2007). These concerns have given rise in some policy circles of calls for agricultural and energy policies be reprioritized where food takes precedence before fuel (in short food before fuel).

In contrast to this perception, evidence is provided countering the hypothesis that the 2007-2008 food price spike was the result of shifts in crop usage from food to fuel. Instead evidence is presented supporting the hypothesis that the food crisis was the result of a shift in global policies toward relying primarily on markets to provide adequate agricultural commodities in periods of supply shortfalls and demand increases. Given this evidence and underlying supporting economic theory, policies capable of adverting future food crises are presented.

This hypothesis addressing the root of the global food crisis is first framed in the context of the historical underpinnings of the 2007-2008 food price spike and the prevailing economic view at that time supporting policies contributing to the spike. The literature warning of the potential for biofuels to disrupt global agricultural commodity prices is then presented in an economic theory context. One of the key predictions of economic theory is that global competitive agricultural commodities markets will respond to commodity price shocks, restoring prices to their long-run trends. However, due to inherent frictions in the market, costly or irreversible decisions, and uncertainty, there is a lag time in such response, thus yielding potential short-run volatility in food prices.

2. Theory

Surges and downturns of ethanol and food prices are not isolated incidents, but economic consequences (Gohin and Chantretnd, 2010; Von Braun et al., 2008; Mcphail and Babcock, 2008; Chen el al., 2010; Balcombe and Rapsomanikis, 2008). Kappel et al. (2010) argue that fundamental market forces of demand and supply were the main drivers of the 2007-2008 food price spike. In a supply and demand model, economic theory suggests agriculture will respond to a commodity price increase from a biofuel or other demand shock. As illustrated in Figure 3, a demand shock will shift the demand curve outward from Q_D to Q_D'. This results in a short-run increase in the agricultural commodity price, from p_e to p_e', leading to existing firms earning short-run pure profits (total revenue above total costs). The magnitude of this increase in price depends on how responsive supply, in the short run, is to the demand shift (represented as an increase in supply from Qe to Q_S. However, in the long-run, existing firms will expand production and new firms will enter yielding a further increase in supply. Assuming no cost adjustments, this increase in supply will restore the market price to the long-run equilibrium price p_e. Furthermore, given the relative unresponsiveness of demand and supply for staple food commodities, small shifts in demand leads to a significant movement in prices.

Abbott, et al. (2008) identified three major agricultural demand shifters causing the 2007-2008 food price spike: increased food demand, low value of the dollar, and a new linkage of energy and agricultural markets. These demand shifters drove up the prices of agricultural commodities in 2007 and 2008. In 2009, high market prices spurred increased crop-production shifting supply outward and the global economic downturn at the end of 2008, sharply decreased demand and as a result led to lower agricultural commodity prices. Figure 4 illustrates this agricultural commodity price volatility for the U.S. corn market.

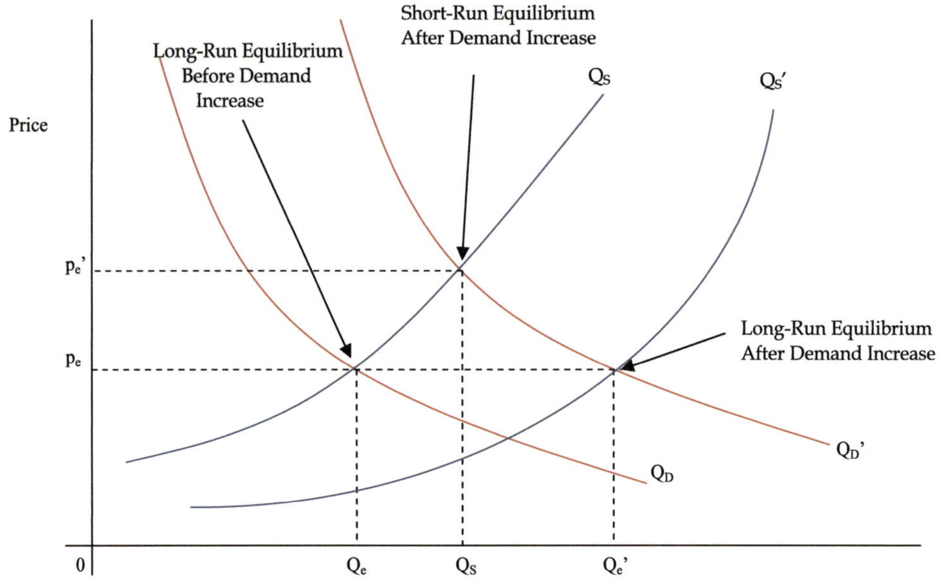

Fig. 3. Supply and Demand Short- and Long-run Shifts

U.S. corn prices rapidly increased in 2007-2008, but with a downturn in economic activity (the Great Recession), price precipitously declined. With a resurgence of current economic activity corn prices specifically rebounded along with agricultural prices in general. As indicated in Figure 2, U.S. ethanol production continued to increase during the economic downturn as corn prices fell. The high correlation of biofuel production with agricultural commodity prices during the 2007-2008 food price spike did not continue through the Great Recession.

Generally the responses to the demand shifters are rapid, while supply-utilization adjustments are slower. A shift in demand will elicit an immediate price increase response. While the supply response will take a number of months as agriculture gears up to increased production. With this supply and demand model, the issue is how rapid is this supply response and what is its magnitude. If supply is able to rapidly respond to a demand shift, then there is no food before fuel issue. If not, then there is a cause for concern.

The underlying driver of the 2007-2008 food price spike was the lack of sufficient food stocks to rapidly buffer the price spike and avoid a food before fuel issue. In the late 20th century, many economists and government policymakers assumed open markets were more efficient in stabilizing agricultural commodity prices then maintaining commodity buffer stocks. One example of this view is an article by Jha and Srinivasan (2001) where they conclude that by liberalizing trade, agricultural commodity stocks are no longer required to stabilize prices. With free trade, when a region experiences a shortfall in grains, it can supplement supply by importing from a grain surplus region. This theory works well when there are ample supplies of grains. However, when there is a global grain shortage, without food buffers a food price spike can occur as was experienced in 2007-2008 food price spike.

The global agricultural system has historically responded to changing patterns of demand (Prabhu et al., 2008). The issues are: are there sufficient agricultural endowments for a supply response to a demand shift, such as a biofuel shock, and if so, how rapid is this response.

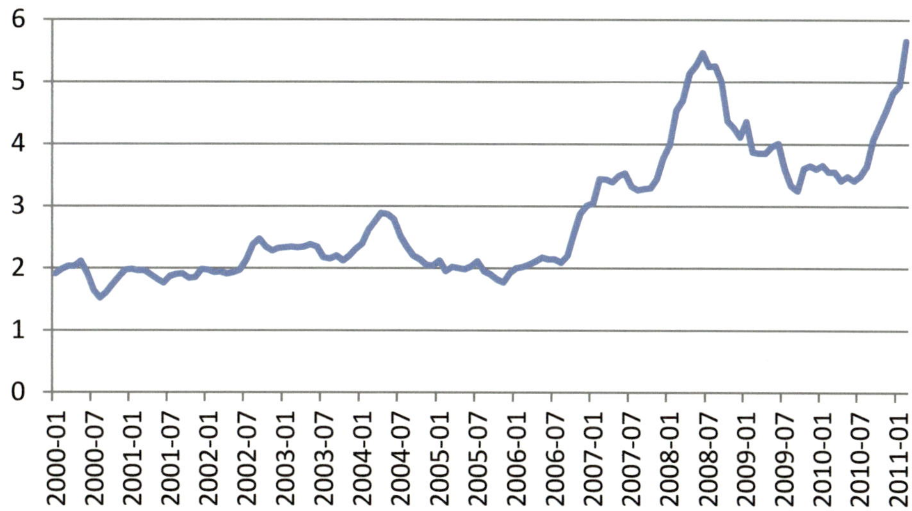

Fig. 4. U.S. Corn Price (dollar per bushel), http://quickstats.nass.usda.gov/#36836568-52F8-393F-9658-05B4E5C1DFB2

Chen et al. (2010) suggests that increasing derived demand for corn, from biofuel production, has led to acreage declines and associated price increases of other crops (wheat and rice). They see a short-run constraint on agricultural endowments, leading to commodity price increases. However, in the long run, the potential for increasing agricultural production is high. Within the U.S. there is about 35 million acres of idle cropland representing approximately 10% of current cropland in use, along with about 75 million acres of cropland in pasture (Marlow et al., 2004). Africa's abundant arable land and labor offer the potential for it to be a major exporter food (Juma, 2010). Global agriculture in general and U.S. agriculture in particular appear capable of adjusting without major difficulties to even high levels of biofuel production (Webb, 1981; Kerckow, 2007). This ability of agriculture to supply growing demand is supported by Licker et al., 2010 who indicate approximately 50% more corn, 40% more rice, 20% more soybeans, and 60% more wheat could be produced if the top 95% of the crops' harvested areas met their current climatic potential.

In 1979, Vincent et al., (1979) indicated the days of cheap corn are not over. Prices may be more stable as corn production expands to meet ethanol requirements and second generation ethanol, increased buffer stocks, and new technologies emerge (Vincent et al., 1979). This prediction of stable agricultural commodity prices would still hold if supply responses are rapid enough to mitigate demand shocks or global buffer stocks are expanded.

In a game theory context, Su (2010) illustrates how rational expectations will lead to consumers stockpiling commodities when prices are low. This type of rational expectations theory can be directly applied to governments where it would be feasible for them to stockpile agricultural commodities in times of relatively low prices to blunt possible future price spikes. Maintaining a buffer stock of agricultural commodities will provide a rapid supply response to blunt a demand shock and avoid a short-run food before fuel issue. If the world economy recovers from the economic slowdown without food production growing sufficiently to replenish stocks, food prices and hunger may rise again (Kappel et al., 2010). Currently in 2011 food prices are rising which is one underlying cause of the recent uprisings in North Africa and Middle East.

3. Methodologies

With this underlying theory of global competitive agricultural markets as a foundation, the two main methods, computable general equilibrium (CGE) and time series models, for food before fuel analysis are investigated. The advantages and disadvantages of these models are outlined in Table 1.

Computable General Equilibrium (CGE) Models
Advantages
Limited data requirements
Disadvantages
Not based on estimated time trends and price volatility
Rely on exogenously determined elasticities among food and fuel variables
Unless expressly modeled, challenging to distinguish short- and long-run impacts
Time Series Models
Advantages
Efficient in illustrating the dynamics and measuring the interaction among prices
Considers both the short- and long-run impacts
Disadvantages
Spurious results are possible for non-stationary data

Table 1. Methodologies Addressing the Food before Fuel Issue

3.1 Computable general equilibrium models (CGE)

CGE models are widely employed in addressing the food before fuel issues, although with different modeling strategies and focuses (Elobeid and Tokgoz, 2007; Ignaciuk and Dellink, 2006; Arndt et al. 2008; Rosegrant el al., 2008; Tyner and Taheripour, 2008; Yang et al., 2008; Saunders et al., 2009; Gohin and Chantret, 2010; Mcphail and Babcock, 2008; Vincent et al., 1979; Hanson et al., 1993; Saunders et al., 2009). Their advantage is a historical data set containing prices and quantities is not required. Only estimates on the elasticities (responsiveness of one variable to a change in another variable) are required. These estimates could be derived empirically, theoretically, or expert opinion. However, a shortcoming of CGE models is their failure to precisely illustrate the time trends and price volatility, and they are not directly applied to the estimation at a particular point in time (Ignaciuk and Dellink, 2006). An exception is Gohin and Chantret (2010) who model the long-run relationship between food and energy prices and examine an array of energy and agricultural commodities with a wider set of macroeconomic factors. Furthermore, CGE models rely on exogenously determined elasticities among energy and agricultural commodity variables. This leads to a predetermined relation between food and fuel which makes it challenging to distinguish the short- and long-run impacts. If these elasticities are not supported by theory and empirical evidence, the conclusions they derive concerning the linkages among food, fuel, and other variables including global economic activity are questionable.

3.2 Time series models

An alternative avenue of research attempts to determine linkages between food and fuel using time-series models estimated with historical data (Imai et al., 2008; Baek and Koo, 2009; Zhang et al., 2010; Saghaian, 2010; Esmaeili and Shockoohi, 2011). Time-series models, such as autoregressive distributed lag (ADL) models, are widely used for empirical analysis of food before fuel (Bentzen and Engsted, 2001; Dimitropoulos et al., 2005; Hunt et al., 2005; Baek and Koo, 2009; Chen et al., 2010). Such models are efficient techniques for illustrating dynamics and measuring the interaction among prices in a time series context, as well as considering both short- and long-run effects (Chen et al, 2010). For example, with a structural break considered, Baek and Koo (2009) used an ADL model to investigate the short-run and long-run impacts of market factors such as energy prices on U.S. food prices. Chen et al. (2010) built a model where the price of grain is established as a function of its own price and other current and lagged variables such as the prices of oil, soybeans, and wheat.

However, the validity of the ADL approach is questionable on unit roots grounds (Bentzen and Engsted, 2001). ADL is an efficient approach when time-series data are stationary, but for non-stationary data it could yield spurious results unless all the variables are cointegrated . Thus, cointegration tests and vector error correction models (VECM) are suggested as more appropriate techniques to capture possible non-stationary characteristics (Bentzen and Engsted, 2001). These methods are generally augmented with supplementary analysis including Granger casualty tests, pairwise correlation matrix analysis, scree tests, and proportion of variance methods.

4. Supply

With energy as a key input into producing agricultural commodities, as prices of energy rise the potential exists for food price inflation. Table 2 outlines the impacts energy has on the supply of agricultural commodities.

4.1 Energy input effects on agricultural commodity prices

Conforming to economic theory, prevailing empirical literature indicates that agricultural prices, which are a function of production costs, have a positive relationship with energy prices. The impact these higher energy prices have on agricultural production costs, short-run price volatility, and long-run price trends are investigated in terms of the underlying chapter hypothesis.

Previous spikes in food prices are usually considered as supply driven, and volatility of food prices were considered as a consequence of supply shocks (e.g. weather, pests, and diseases) (Mcphail and Babcock, 2008). Under this scenario, research on how the energy sector influences the agricultural sector considered energy as an agricultural production cost.

Supply	
	Although fuel is a key input in agricultural production, caution is required in concluding fuel prices directly cause agricultural commodity prices.
	In the long run, the potential exists for supplying biomass to meet the growing demand for biofuels.
	Increased biofuel production may impose adverse effect on environmental resources.
Demand	
	Past research establishing a direct link between food and fuel prices are not consistent with recent trends.
	The major weakness, in past research, is not differentiating short- and long-run impacts and not considering macroeconomic linkages.
	Current research trends indicate, in the short run, there is probably some causation between food and fuel, but no long-run relation exists.
	Macroeconomic activity possibly is the underlying cause of both food and fuel price instability.

Table 2. Supply and Demand Effects on Food and Fuel Markets

This increased energy cost is reflected directly in fuel costs associated with field operations, transportation, and processing and indirectly in increased cost of factors with energy as a major component (e.g., fertilizer and pesticides) (Musser et al., 2006). By substituting other inputs (e.g. reduced tillage technology, improved drying and irrigation systems, and efficient application and timing of fertilizers) the effects of higher energy costs can be mitigated (Musser et al., 2006; Von Braun et al., 2008).

Baffes (2007) indicated that the pass-through of oil price changes to fertilizer and agricultural commodities was high relative to other inputs, thus relatively high oil prices will be passed-through leading to high agricultural commodity prices. However, with lags in cost adjustments, these energy cost-push effects on agricultural commodity prices might not exist in the short-run (Gohin and Chantret, 2010; Von Braun et al., 2008).

The magnitude of these energy cost-push effects are subject to energy use relative to other inputs (Muhammad and Kebede, 2009). For energy-intensive agricultural commodities, with other factors fixed, an increase of energy prices would shift the supply curve of agricultural commodities to the left, which subsequently increases agricultural commodity prices (Chen et al., 2010). However, for labor-intensive agricultural commodities an increase in energy prices might yield insignificant impacts on agricultural commodity prices. Thus, although considered as a key production input for agricultural commodities, care is required in concluding that higher energy prices directly imply higher agricultural commodity prices, especially in the short-run. Gohin and Chantret's (2010) results indicate other factors (biofuels, trade restrictions, speculative demands, climatic events, higher demands, and lower stocks) besides oil prices affecting the cost of agricultural production may better explain agricultural commodity prices.

4.2 Supply potential of bioenergy

Perlack et al. (2005) determined within the U.S. forestland and agricultural land, the two largest potential biomass sources, there exists over 1.3 billion dry tons per year of biomass potential. This is enough to produce biofuels meeting over one-third of the current demand for transportation fuels. The United States can produce nearly one billion dry tons of biomass annually and still continue to meet food, feed, and export demands. This biomass resource potential can be produced with relatively modest changes in land use. In contrast, Reilly and Paltsev (2007) estimate that large increases in domestic biofuel production would result in the U.S. becoming a net importer of food as opposed to an importer of oil.

Within China, current biofuel development paths could pose significant impacts on China's food supply and trade, as well as the environment. Yang et al. (2009) conducted a study on the land and water requirements for biofuel in China, and found that 3.5-4% of the total corn production was used for ethanol production. They predicted that by 2020, 5%-10% of the cultivated land in China will be used for ethanol-production crops, and that biofuel development will have significant impacts on China's food supply. Food and bioenergy demands can be satisfied at the same time without rising agricultural commodity prices, but significant research and development efforts in agronomy, technology, and markets will be required to ensure efficient, sustainable land use (Rosegrant et al., 2008; Yang et al., 2008).

Natural endowment redistribution is another consequence of the food vs. fuel competition. Increased biofuel production imposes adverse effects on land and water recourses (Rosegrant et al., 2008). With the expansion of biofuels, more natural ecosystems are switched to agricultural use, releasing CO_2 originally stored in ecosystems into the atmosphere (Chakravorty et al., 2009; Fargione et al., 2008). Searchinger et al. (2008) estimated that greenhouse gas emission would double over 30 years and last for 167 years due to conversion from natural habitat to cropland caused by increased of biofuel production.

5. Demand

Although supply is considered to play a significant role in the long-run relationship between energy and agricultural commodities, the role of demand should not be ignored or underestimated (Gohin and Chantret, 2010). The 2007-2008 food price spike focused research on investigating the demand side. The expanding biofuel market has provided producers a choice of supplying food or fuel depending on their relative net returns. The issue is: can agriculture respond to the growing demand for food and fuel in a time frame sufficiently rapid to avoid commodity price inflation. The literature investigating the food versus fuel demand linkage is mixed. Research has either assumed or empirically derived a direct link between biofuels and food prices, where increased crop demand for biofuel production is limiting its supply for food and thus driving up the food prices. Along with the supply effects on food and fuel markets, Table 2 also lists the demand effects of expanding biofuels on food.

5.1 Previous research

Past research concluded, of the factors causing rising food prices (increased biofuel production, weak dollar, and increased food production cost due to higher energy prices), the most important is the large increase in biofuel production in the U.S. and the EU (Martin, 2008; Mitchell, 2008; OECD-FAO, 2007). Without these increases, global wheat and corn stocks would not have declined appreciably and price increases would have been moderate. Since the Energy Act of 2005, a stronger relationship between corn and biofuel (ethanol) has emerged (Muhammad and Kebed, 2009). Although still questionable, biofuel is considered a key transmitter of energy prices to the agricultural prices (Arndt et al., 2008;

Chakravorty et al., 2009; Chen et al., 2011; Elobeid and Hart, 2007; Hochman et al. 2010; Ignaciuk et al., 2006; Ignacuik and Dellink, 2006; Runge and Senauer, 2007; Lazear, 2008; Mitchell, 2008; Muhammad and Kebed, 2009; Rajagopal, 2009; Sexton et al., 2009; Taheripour and Tyner, 2008; Yahaya, 2006).

Recently, corn price volatility has contributed to the integration between the energy market and the agricultural commodity market (Mcphail and Babcock, 2008). However, this direct linkage between food and fuel prices are not consistent with recent trends and fail to illustrate the connection among food and fuel prices (Chen et al., 2010). The strong positive correlation between U.S. ethanol production and agricultural commodity prices during the 2007-2008 price spike, quickly reversed to a negative correlation in the years following the spike (see Figures 2 and 4). U.S. ethanol production continued to rise with commodity prices falling. A major weakness of these studies is not differentiating between the short- versus long-run food before fuel impacts. Gohin and Chantret (2010) attribute these inconsistencies to the omission in previous studies of macroeconomic linkages. Macroeconomic activity is hypothesized to be the underlying driver of both food and fuel prices.

In sum, Kilian (2009) discusses the importance of differentiating impacts (shocks) between demand and supply, given each of them is associated with different magnitudes, patterns, and persistence. But one of the main shortcomings for most papers is a failure to distinguish the source (demand or supply) and the magnitude of energy price influences on agricultural commodities (Chen et al., 2010). Of the studies which indicate a direct link between biofuels and agricultural commodity prices, they either employed models with a pre-built-in exogenous link between fuel and food, which is characteristic of CGE models or just assumed there is a relationship.

5.2 Current research trends

Other literature indicates more complex linkages with possible differing short- and long-run relations (Balcombe and Rapsomanikis, 2008 ; Diao et al., 2008; Daschle, 2007; Kerckow, 2007; Perlack et al., 2005; Prabhu et al., 2008; Webb, 1981; Senauer, 2008; and Zhang et al., 2010). This research indicates, in the short run, there probably is some causation between ethanol and agricultural commodity prices (Senauer, 2008; Zhang et al., 2009; Zhang et al., 2010). However, results indicate no long-run relationship. In support of these results, Esmaeili and Shokoohi (2011) indicate only a possible indirect relation between oil and agricultural commodity prices. Economic theory suggests global competitive markets will restore prices to their long-run equilibrium trends after any agricultural price shocks due to increased biofuel demand or other shocks (Figure 3) (Zhang et al., 2009; Zhang et al., 2010). As an example, using a world-market economic model, the rapid growth in biofuels will trigger a sharp rise in crop production at the expense of pasturelands and forests (Hertel et al., 2010). Further, Balcombe and Rapsomanikis (2008) found oil prices determine the long run equilibrium of both sugar and ethanol prices in Brazil. Sugar prices Granger-caused ethanol prices, but not the other way around. In the long run, farm prices (the prices of grains, dairy products, meats, and other farm produced commodities) and wages drive food prices. Claims that food prices are most strongly affected by energy price changes are not supported. Reducing energy prices will not reduce food prices (Lambert and Miljkovic, 2010). Furthermore, second and third generation biofuels have the potential to shift biomass production onto marginal croplands, reducing biofuel's food-price impacts.

5.2.1 Macroeconomic activity

This market response was a determinant in recent agricultural commodity price volatility: rising in 2007-2008, declining in 2009-2010, and then rising again in late 2010. Price volatility is also due to the heating up and cooling off of macroeconomic activity. Such activity is possibly the underlying cause of both food and fuel price instability (Kilian, 2009). Initial research in this direction, Balcombe and Rapsomanikis (2008) extend the supply-demand framework, which focuses only on biofuel and agricultural markets, by considering oil prices along with ethanol and sugar prices. Gohin and Chantret (2010) compared the relationship between the macro-linkages of the energy sector with the food sector, but do not consider biofuels. Additional research in this vein by Harri et al. (2009), Harrison (2009), Hayes et al. (2009), Sheng-Tung et al. (2010), and Yang et al. (2008) suggests a link between oil prices and agricultural commodity prices. Saghaian (2010) indicates that although there is a strong correlation among oil and commodity prices, the evidence for a causal link from oil to commodity prices is mixed. Considering five variables (oil, ethanol, corn, soybeans, and wheat prices) there are no causal links between the energy and agricultural sectors. However, the results of Granger causality tests indicate crude oil prices Granger cause corn, soybeans, and wheat prices.

When considering these global macro-linkages, international trade patterns and balances come into play. Hanson et al. (1993) have demonstrated that with fixed exchange rates and exogenous oil prices, U.S. agricultural commodity prices slightly declined with a doubling of crude oil prices; while with a fixed trade balance, farm prices increased. Saghaian (2010) also concludes that exchange rates are correlated with energy and agricultural markets, and attributes the correlation to oil prices denominated in U.S. dollars. A rise in oil prices increases the supply of U.S. dollars, which depreciates the dollar along with an increase in grain exports and higher food prices (Saghaian, 2010; Abbott et al., 2008).

Different baskets of agricultural commodities might lead to different conclusions on the relationship between the food and fuel prices. Imai et al. (2008) suggest the persistent impacts of a price change of oil on food might differ among countries and foods, and might be affected by the type of data used. For example, in China, their results indicate oil prices yield significantly positive effects on wheat and fruit prices, while imposing no effects on the price of rice and vegetables. In contrast, oil prices have positive effects on the India's price of wheat, rice, and fruit and vegetables.

5.3 Public policies

Public policies might be another important channel through which macroeconomic linkages of energy and food markets is built, especially in recent years. Those policies (including subsidies and mandates) are playing a more significant role in the interaction between food and energy prices, especially in developed countries such as the U.S. and EU (Von Braun and Torero, 2009; Gohin and Chanret, 2010; Balcombe and Rapsonmanikis, 2008; Vincent et al., 1979; Hanson et al., 1993). U.S. ethanol demand is mainly driven by government support, thus shocks to ethanol demand are considered as policy driven more than market driven (McPhail and Babcock, 2010). Senauer (2007) estimated that the U.S. $0.51per gallon tax credit has distorted the food vs. fuel competition, making corn valued more as a fuel than a food input. Balcombe and Rapsonmanikis (2008) using Brazil as an example, found the growth of Brazil's ethanol market has been realized not only by the supply-demand linkage between the ethanol-sugarcane market, but also by various other factors including government policies, technical changes, and the manufacturing of flex-fuels vehicles. Chen et al. (2010) indicate that production subsidies which encourage biofuel crops might result in significant impacts to the environment and the economy. They state that not only high oil prices but also government subsidies would result in a higher derived demand of corn-based ethanol, as well as price increases in various agricultural commodities.

5.4 Modeling shortcomings

Specific channels of food and fuel interaction are not clearly defined or quantified. With current empirical methodologies and data, it is challenging to distinguish simultaneous supply-demand linkages and isolate impacts from macroeconomic variables. Insufficient theoretical understanding and observations among energy and agricultural commodity prices might generate misleading causal conclusions (Saghaian, 2010). As an example, without understanding the market channels linking agricultural commodity markets with energy markets, exogenous model elasticity assumptions may be invalid. Those shortcomings led to the post 2007-2008 forecasts of relatively high agricultural commodity prices when commodity prices actually declined (Figure 4). Theoretically understanding the simultaneous supply-demand linkage and isolating the impacts from macro effects may yield improved parameter estimates (Saghaian, 2010). Structural vector autoregressive models, such as Kilian (2009) and Mcphail (2010), may offer improved estimation techniques for investigating the co-movements of food and fuel variables. With endogeneity allowed, these techniques provide for the decomposition of demand and supply impacts.

Previous research generally specified linear models leading to pairwise linear correlations. As stated by Balcombe and Rapsomanikis (2008), oil, sugar, and ethanol markets could be treated as a nexus or perceived as separate when prices move within certain thresholds. Once prices fall outside a threshold, substitution effects between oil and ethanol would induce the transmission of price from market to market, introducing nonlinear behavior. Such threshold effects could be better captured by nonlinear models. Examples of nonlinear models are Balcombe and Rapsomanikis' (2008) use of Bayesian Monte Carlo Markov chains and Azar's (2003) use of a bottom up approach to investigate the competition between biomass and food. Alternatively, Baek and Koo (2009) and Chen et al., (2010) introduced structural breaks to divide the time-series data to capture the short-run and long-run impacts of energy prices and exchange rates on the food prices.

In summary, the literature solely investigating biofuel and food prices or the literature exogenously assuming a link exists suggest that indeed there is a direct and significant relationship between food and fuel. However, when considering more complex connections in terms of short- versus long-linkages and macroeconomic impacts such a direct relationship is questionable. Demand shocks, including sharp fluctuations in biofuel prices and macroeconomic shocks, and supply shocks in agricultural production probably do cause short-run agricultural commodity price inflations but not in the long-run. The underlying driver of both energy and agricultural prices is macroeconomic activity.

6. Policy

In this section, policy implications are addressed surrounding the hypothesis that the 2007-2008 food price spike was caused by the shift in global policies toward relying primarily on markets to provide adequate agricultural commodities in periods of sharp increases in food demand. This hypothesis and accompanying support from economic theory suggest in the long-run markets will adjust to changes in crop usage, hence government policies such as food subsidies, price controls, and export restrictions are not warranted. However, in the short-run, due to inherent volatility throughout the food and biofuel production chains, tailored government policies are necessary to avoid future price spikes. As a reference for the discussion on both efficient and inefficient policies directed toward the food before fuel issue, a listing of policy prescriptions is provided in Table 3.

Short-Run Policies

- Economically Efficient
  - Completing negotiations on reducing agricultural trade restrictions
  - Global food-price monitoring
  - Precautionary agricultural commodity buffer stocks
  - Emergency response and humanitarian assistance programs
  - Educate consumers to expect greater food price volatility
- Inefficient
  - Government incentives and regulations favorable to biomass production
  - Policies directed toward maintaining fallow acreage

Long-Run Policies

- Economically Efficient
  - Allow free markets to adjust to changes in crop usages
  - Constant infusion of public sponsored research and outreach
  - Shift to sustainable perennial crops arresting topsoil erosion
  - Improving energy efficiency
  - Subsidize public transport
  - Diversify food and fuel imports
- Inefficient
  - Food and biofuel subsidies
  - Price controls
  - Export and import restrictions

Table 3. Policy Prescriptions

6.1 Long Run
6.1.1 Supply
6.1.1.1 Free Competitive markets

As indicated by economic theory and supported by empirical research, global competitive markets will lead to long-run stable agricultural commodity markets (Webb, 1981; Kerckow, 2007). U.S. farmers and technology will more than keep pace with demand not only for food but also for fuel (Daschle, 2007). Productivity gains for corn averaged nearly 3% per year, and the annual U.S. corn crop increased from 7 billion bushels in 1980 to nearly 12 billion bushels in 2006. However, competitive markets require a constant infusion of pubic sponsored research and outreach to maintain current productivity growth (Arndt, 2008; Christiaensen, 2009; Hochman, et al., 2008; Johnson, 2009; Prabhu et al., 2008; Rosegrant et al., 2008; Sexton, et al., 2009; Yang et al., 2008). Low levels of agricultural productivity in Africa are a major constraint to both poverty reduction and long-term economic growth

(Diao et al., 2008). Productivity gains in Africa are possible by increasing smallholder access to a modern package of inputs and management - improved seed, modern fertilizers and pesticides, and irrigation–along with enhanced integrated regional markets–low transportation costs, information systems, financial services, grades and standards, farmer and trader organizations, and commodity exchange systems (Diao et al., 2008; Kerckow, 2007; Prabhu et al., 2008). A shift to biofuels from mainly perennial, lignocellulosic plants and low input crops will contribute to a sustainable utilization of lower quality soils with limited water supply including degraded areas (Kerckow, 2007). However, there is concern that widespread planting of energy crops will accelerate the deterioration of the world's cropland base (Brown, 1980). In conjunction with advancing technology gains, efforts should be directed toward arresting topsoil erosion losses.

Providing more support to agencies such as the Consultative Group on International Agricultural Research (CGIAR) would be an important avenue toward stable food prices (Prabhu et al., 2008). In real 2008 dollars, U.S. investment in agricultural development abroad fell to $60 million in 2006, down from an average of $400 million a year in the 1980s. In developed countries, public investment in research, which had grown annually by more than 2% in the 1980s, shrank by 0.5% annually between 1991 and 2000. Global official aid to developing countries for agricultural research fell by 64% between 1980 and 2003. The decline was most marked in poor countries, especially in Africa. This reduction in investment is directly associated with reduced growth in agricultural productivity (Runge & Runge, 2010). A reason for this decline in public investment is that agricultural technology is difficult to ascribe to specific actions by a government and is unlikely to address the immediate impacts of food and energy price volatility (Arndt, 2008).

6.1.1.2 Inefficient market controls

The empirical relationship between biofuel and agricultural commodity prices suggests policies should be directed toward mitigating the short-run impacts on food prices. Effective adjustments require they send efficient market price signals. Imposing inflexible food subsidies or price controls distort market prices resulting in market inefficiencies leading to more volatile food prices and reduced security of the world's food supply (Collins and Duffield, 2005; Elam, 2008; Senauer, 2008). Food subsidies benefit consumers in the short-run, but at the expense of future investments due to the financial requirement for subsidization. Subsidies are not well targeted, are expensive, and exacerbate the burden of macroeconomic adjustment (Arndt, 2008). Price controls send negative price signals to producers that blunt the incentives for increasing supply (Johnson, 2009). More flexible policies should be designed that are responsive to agricultural and energy market realities (Elam, 2008). All such policy responses should reflect not just changes in world prices but also local price effects (Dewbre et al., 2008).

6.1.2 Demand

On the energy side of the equation, reducing the acceleration of global energy consumption and improving energy efficiency will lead toward sustainable energy and agricultural markets (Kerckow, 2007). U.S. and EU government policies providing incentives for biofuel production should be reconsidered in light of their impact on short-run food prices (Chen et al. 2011). As an example, increasing the U.S. Corporate Average Fuel Economy (CAFÉ) standard would cost approximately a third as much as it costs to subsidize ethanol (Doering, 2006). Alternatively, removing tariffs on ethanol imports in the U.S. and EU would allow more efficient producers, such as Brazil and other developing countries, including many African countries to produce ethanol profitable for export to meet the mandates in the U.S. and EU (Arndt, 2008; Kerckow, 2007; Mitchell, 2008). Devadoss and Kuffel (2010) determine the current U.S. $0.57 per gallon import tariff on ethanol should be a $0.09 subsidy if the U.S. is interested in efficiently achieving the policy goals of reducing reliance on imported petroleum and reducing greenhouse gas emissions. An energy policy that more strongly emphasizes energy conservation is required (Elam, 2008). An example is subsidized public transport, but public transport passengers are typically not among the most vulnerable groups to high food prices, and such public subsidies are expensive and difficult to administer (Arndt, 2008).

U.S. government incentives and regulations favorable to biomass production, rather than investing in basic research and development for conservation and renewable sources of energy, enhance the profitability of biofuels over food (Runge and Senauer, 2007). Under current U.S. government incentives and regulations, the food vs. fuel choice is tilted toward fuel (Reilly and Paltsey, 2007).

6.2 Short run

6.2.1 Trade liberalization

For food importing countries, relying on agricultural productivity gains from other countries is a passive and risky policy. Instead they should consider watching their importing countries for possible major supply changes due to biofuel production or other factors and consider diversifying their agricultural imports (Brown, 1980). Food importing as well as exporting countries should work toward completing the Doha Round of World Trade Organization (WTO) negotiations leading toward more efficient agricultural free trade with regulations on food export restrictions (Christiaensen, 2009; Johnson, 2009; Von Braun et al., 2008).

Trade liberalization is much easier to administer than a subsidy and is consistent with a fundamental open economy policy. Non-price distorting policies include expanding social protection programs but such programs come with considerable cost or require a fundamental redistribution of income from the wealthy to the poor (Christiaensen, 2009; Prabhu et al., 2008; Yang et al., 2008). In the short-run, suspending ethanol blending mandates, subsidies, and ethanol import tariffs would cause a market response and lower agricultural commodity prices (Prabhu et al., 2008).

6.2.2 Global food monitoring with buffer stocks

As far back as the 1980s it was suggested to establish a global food-price monitoring system that is sensitive to short-run price volatility from biofuel impacts or other market shocks (Brown, 1980). If such a monitoring system was in place prior to the 2007-2008 food price spike, the spike may have been avoided. However, instead policies were adopted that directly reduce supply by holding some acreage fallow as a way of reducing the cost of managing agricultural surpluses. The United States still has millions of acres enrolled in such programs. Those policies must be reconsidered in a world in which inventories have dwindled and critical food shortages can emerge and go unmet, as they did in the 2007-2008 food price spike (Johnson, 2009).

In conjunction with monitoring, global agricultural commodity stocks should be maintained to buffer short-run price spikes (Christiaensen, 2009). The dismantling of public food reserves led to the 2007-2008 food price spikes (McMichael, 2009). As in the past, if government and private grain dealers had large inventories, the 2007-2008 food price spike would not have occurred. Food vs. fuel would have not been an issue. Recently, these precautionary inventories were allowed to shrink with the idea countries suffering crop failures could always import the food they required (Jha and Srinivasan, 2001). However, with no food in reserve, the global spike in food and biofuel demand resulted in a short-run rise in food prices when agricultural trade could not satisfy this world demand (Myers and Kent, 2003). World organizations including the International Monetary Fund and the World Bank have responded with policies and programs which commit funds for both immediate food aid and long-run increases in agricultural productivity (Singh, 2009).

Markets will adjust to shocks, but in cases of global supply shortfalls, such adjustments come at a high price of social discord and stress. The recent uprising in North Africa and the Middle East is predicated on high food price inflation. The aim is to avoid or at least buffer future price spikes by governments focusing on the public good to reinsure the global food supply (Christiaensen, 2009). An example where grain stocks were used to mitigate price increases is China's use of grain stocks to moderate the domestic price rise during the 2007-2008 food price spike (Yang et al., 2008).

However, in cases of localized food shortages or an unavoidable global price spike, expanded emergency response and humanitarian assistance programs are required to assist food-insecure people along with strengthened food-import financing. A closer look at the efficiency of current U.S. food aid programs also reveals many avenues for improved efficiency. The U.S. has been slow to change its food aid policies. As just one example, the U.S. currently requires a minimum share of its food aid be shipped on U.S.-flag vessels. This requirement costs U.S. taxpayers $140 million in 2006, which is roughly equal to the cost of non-emergency food aid to Africa (Bageant et al., 2010).

6.2.3 Food vs. agricultural commodities

The distinction between high world prices for agricultural commodities and the consumer costs of food is an important one. In developed countries consumers generally do not buy raw agricultural commodities at international prices. In many cases the proportion of agricultural commodity cost in their food is relatively small compared with the processing costs. In contrast, for consumers in many developing countries, the proportion of agricultural commodity to food costs can be large. Agricultural commodity price inflation will thus have a disproportionate effect on developed relative to developing countries. The degree to which the price of traded agricultural commodities and the price of food are related depends on factors that dampen price transmission. In the search for appropriate policy response, it is important to measure consumer effects correctly and to apportion properly the causes of current high food prices (Dewbre et al., 2008).

A final public action is to educate consumers to expect greater food price volatility, so they can adjust and plan (Yang et al., 2008). Without agricultural commodity supply buffers, food and agricultural commodity prices, particularly in the developing world, will continue to be volatile.

7. Summary and conclusions

The chapter lays out evidence in support of the hypothesis that the 2007-2008 food price spike was not only caused by growing demand for biofuels but also by more complicated macroeconomic factors, such as public policies. Literature is presented in a supply and demand framework. On the supply side, how energy inputs are affecting the agricultural sector in terms of production costs are reviewed. Conforming to economic theory, results indicate agricultural commodity prices are driven by production costs with higher prices of energy inputs implying higher agricultural production costs. However, care is required in concluding that higher energy prices directly imply higher agricultural commodity prices, especially in a short-run. Other factors (biofuels, trade restrictions, speculative demands, climatic events, higher demands, and lower stocks) besides oil prices affecting the cost of agricultural production may better explain agricultural commodity prices.

Within the supply-demand framework, two main methods (CGE and econometric approaches) are employed for food before fuel analysis. CGE models are widely adopted with a consideration of macro-linkages. However, they rely on exogenously determined elasticities among fuel and agricultural commodity variables. If these elasticities are not supported by theory and empirical evidence, the conclusions derive concerning the linkages among food, fuel, and other variables including global economic activity may be questionable.

In contrast, econometric approaches attempt to determine these linkages with Granger casualty tests, pairwise correlation matrixes, cointegration tests, and VECMs. Results suggest considering both the short-run price volatility of commodities as well as the long-run commodity price trends.

Implications from this literature review suggest a possible modification in the CGE models and other numerical models which may assume a direct long-run link between fuel prices and agricultural commodity prices. The resulting forecasts of high agricultural commodity prices precipitating from high fuel prices may be misleading. Based on time series results, a reshaping of these models may be in order. Yet the results have implications far beyond suggesting modifications in economic modeling. In the short run, it is important to ensure food availability to all, but most importantly to the global poor. Spikes in agricultural commodity prices, whether caused by biofuels, climate, or just human mistakes, cause irreparable harm to the global poor. Policies, including agricultural commodity buffers, designed to blunt these short-run price spikes should be reconsidered as a tool to reduce food volatility (Zhang et al., 2010).

8. References

Abbott, P.C.; Hurt, W.E. & Tyner, C. (2008). What's Driving Food Prices? In: *Farm Foundation*, 07.2008, Available from: http://www.farmfoundation.org

Arndt, C.; Benfica, R.; Maximiano, N.; Nucifora, A. &Thurlow, J. (2008). Higher fuel and food prices: impacts and responses for Mozambique. *Agricultural Economics*, VOL.39, No.1, (November 2008), pp.497–511,ISSN 01695150

Baffes, J. (2007). Oil Spills on Other Commodities. *Resources Policy*, VOL.32, No.3, (September 2007), pp.126–134, ISSN 0301-4207

Baek, J. & Koo, W.W. (2009). Assessing the Exchange Rate Sensitivity of U.S. Bilateral Agricultural Trade. *Canadian Journal of Agricultural Economics*, VOL.57, No.2, (June 2009), pp.187–203, ISSN 1744-7976

Bageant, E.R.; Barrett, C.B. & Lentz, E.C. (2010). Food Aid and Agricultural Cargo Preference. *Applied Economic Perspectives and Policy*, VOL.32, No.4, (Winter 2010), pp. 624-641, ISSN 2040-5804

Balcombe, K. & Rapsomanikis, G. (2008). Bayesian estimation and selection of nonlinear vector error correction models: the case of the sugar–ethanol–oil nexus in Brazil. *American Journal of Agricultural Economics,* Vol.90, No.3, (August 2008), pp. 658–668, ISSN 0002-9092

Barnard, J. (1983). Gasohol/Ethanol: A Review of National and Regional Policy and Feasibility Issues. *Regional Science Perspectives,* Vol.13, No.2, (August 1983), pp. 3-14, ISSN 0097-1197

Benten, J. & Engsted, T. (2001). A revival of the autoregressive distributed lag model in estimating energy demand relationships. *Energy Economics*, Vol.26, No.1, (January 2010), pp. 45–55, ISSN 0360-5442

Brown, L.R. (1980). Food or fuel: new competition for the world's cropland, In: *Worldwatch Paper*, 03.1980, Available from: Worldwatch Institute, Washington, DC 20036, USA

Chakravorty, U.; Hubert, M. & Nosthakken, L. (2009). Fuel versus Food. *Annual Review of Resource Economics*, Vol.1, No.1, (September 2009), pp. 645-663, ISSN 1941-1340

Chen, S.T.; Kuo, H.I. & Chen, C.C. (2010). Modeling the relationship between the oil price and global food prices. *Applied Energy*, Vol.87, No.8, (August 2010), pp. 2517-2525, ISSN 0306-2619

Chen, X.; Huang, H.; Khanna, M. & Onal, H. (2011). Meeting the Mandate for Biofuels: Implications for Land Use, Food and Fuel Prices. In: *National Bureau of Economic Research*, Available from http://ageconsearch.umn.edu/bitstream/61629/2/AAEA%202010_Xiaoguang%20Chen.pdf

Christiaensen, L. (2009). Revisiting the Global Food Architecture: Lessons from the 2008 Food Crisis. *Review of Business and Economics*, Vol.54, No.3, (n.d. 2009), pp. 345-361, ISSN 2155-7950

Collins, K. & Duffield, J. (2005). Energy and Agriculture at the Crossroads of a New Future, In: *Agriculture as a Producer and Consumer of Energy*, J. Outlaw, K. Collins, & J. Duffield, (Ed.) , pp.1-29, ISBN 0-85199-018-5, CABI Publishing, Cambridge, MA, USA

Daschle, T. (2007). Food for fuel? Myth versus reality. *Foreign Affairs*, Vol.86, No.5, (September/October 2007), pp. 157–160, ISSN 0015-7120

Devadoss, S. & Kuffel, M. (2010). Is the U.S. Import Tariff on Brazilian Ethanol Justifiable? *Journal of Agricultural and Resource Economics*, Vol.35, No.3, (December 2010), pp. 476-488, ISSN 10685502

Dewbre, J. ; Giner, C.; Thompson, W. & Lampe, M.V. (2008). High food commodity prices: Will they stay? Who will pay? *Agricultural Economics*, Vol.39, No.s1, (November 2008), pp. 393–403, ISSN 0165-1587

Diao, X.; Headey, D. & Johnson, M. (2008). Toward a green revolution in Africa: what would it achieve, and what would it require? *Agricultural Economics*, Vol.39, No.s1, (November 2008), pp. 539–550, ISSN 0165-1587

Dimitropoulos, J.; Hunt, L.C. &Judge, G. (2005). Estimating underlying energy demand trends using UK annual data. *Applied Economics Letters*, Vol.12, No.4, (March 2005), pp. 239–244, ISSN 1350-4851

Doering, O. (2005). Energy Systems Integration: Fitting Biomass Energy from Agriculture into U.S. Energy Systems. In : *Agriculture as a Producer and Consumer of Energy*. J. Outlaw, K. Collins, and J. Duffield, eds. Cambridge, pp. 112-130, CABI Publishing, ISBN 9780851990187, Cambridge, MA, USA

Elam, T. (2009). Food or Fuel? Choices and Conflicts. *Choice*, Vol.23, No.3, (n.d. 2009), pp. 12–15, ISSN 0009-4978

Elobeid, A. & Hart, C. (2007). Ethanol Expansion in the Food versus Fuel Debate: How Will Developing Countries Fare. *Journal of Agricultural & Food Industrial Organization*, Vol.5, No.2, (December 2007), pp.1–21, ISSN 1542-0485

Elobeid A.; Tokgoz, S.; Hayes, D.J.; Babcock, B.A.& Hart, C.E. (2007). The long-run impact of corn-based ethanol on the grain, oilseed, and livestock sectors with implications for biotech crops. *AgBioForum*, Vol.10, No.1, (n.d. 2007), pp.11–18, ISSN 1522936X

Elobeid A. & Tokgoz S. (2008). Removing distortions in the US ethanol market: what does it imply for the United States and Brazil? *American Journal of Agricultural Economics*, Vol.90, No.4, (November 2008), pp. 918–932, ISSN 0002-9092

Esmaeili, A. & Shokoohi, Z. (2011). Assessing the effect of oil price on world food prices: Application of principal component analysis. *Energy Policy*, Vol.39, No.2, (February 2011), pp.1022–1025, ISSN 0301-4215

Fargione, J.; Hill, J.; Tilman, D.; Polasky, S. & Hawthorne,P. (2008). Land Clearing and Biofuel Carbon Debt. *Science*, Vol.319, No.5867, (February 2008), pp. 1235–1238, ISSN 0036-8075

Gohin, A. & Chantret, F. (2010). The Long-run Impact of Energy Prices on World Agricultural Markets: The role of Macro-economic Linkages. *Energy Policy*, Vol.38, No.1, (January 2010), pp. 333–339, ISSN 0301-4215

Hanson, K.; Robinson, S. & Schluter, G. (1993). Sectoral effects of a world oil price shock: economywide linkages to the agricultural sector. *Journal of Agricultural and Resource Economics*, Vol.18, No.1, (July 1993), pp. 96–116, ISSN 10685502

Harri, A.; Nalley, L. & Hudson, D. (2009). The Relationship Between Oil, Exchange Rates, and Commodity Prices. *Journal of Agricultural and Applied Economics*, Vol.41, No.2, (August 2009), pp. 501–510, ISSN 1074-0708

Harrison, W. (2009). The Food versus Fuel Debate: Implications for Consumers. *Journal of Agricultural and Applied Economics*, Vol.41, No.2, (August 2009), pp. 493–500, ISSN 1074-0708

Hayes, D.; Babcock, B.; Fabiosa, J. ; Tokgoz, S. ; Elobeid, A.; Yu, T.; Dong, F. ; Hart, C.; Chavez, E.; Pan, S.; Carriquiry, M.& Dumortier, J. (2009). Biofuels: Potential Production Capacity, Effects on Grain and Livestock Sectors, and Implications for Food Prices and Consumers. *Journal of Agricultural and Applied Economics*, Vol.41, No.2, (August 2009), pp. 465-491, ISSN 1074-0708

Hertel, T.; Tyner, W. & Birur, D. (2010). The Global Impacts of Biofuel Mandates. *Energy Journal*, Vol.31, No.1, (n.d. 2010), pp. 75-100, ISSN 0195-6574

Hochman, G.; Rajagopal, D. & Zilberman, D. (2010). Are Biofuels the Culprit? OPEC, Food, and Fuel. *American Economic Review*, Vol.100, No.2, (May 2010), pp. 183-187, ISSN 0002-8282

Hunt, L.C.; Judge, G.& Ninomiya, Y. (2005). Primary energy demand in Japan: an empirical analysis of long-term trends and future CO2 emissions. *Energy Policy*, Vol.33, No.11, (July 2005), pp. 1409–1024, ISSN 0301-4215

Ignaciuk,A.M. & Dellink, R.B. (2006). Biomass and multi-product crops for agricultural and energy production-an AGE analysis. *Energy Economic*, Vol.28, No.3, (May 2006), pp. 308–325, ISSN 0140-9883

Ignaciuk, A.M.; Vöhringer, F.; Ruijs, A. & Van Ierland, E.C. (2006). Competition between biomass and food production in the presence of energy polices: a partial equilibrium analysis. *Energy Policy*, Vol.34, No.10, (July 2006), pp. 1127-1138, ISSN 0301-4215

Imai, K.; Raghav, G. & Ganesh, T. (2008). Food and oil prices, In: *Economics discussion paper series, School of Social Sciences, The University of Manchester, Manchester* , n.d. 2008, Available from: http://www.socialsciences.manchester.ac.uk/disciplines/economics/research/discussionpapers/pdf/EDP-0801.pdf

Jha, S. & Srinivasan, P.V. (2001). Food Inventory Policies under Liberalized Trade. *International Journal of Production Economics*, Vol.71, No.1-3, (May 2001), pp. 21-29, ISSN 0925-5273

Juma, C. (2010). *The New Harvest : Agricultural Innovation in Africa*. Oxford University Press, ISBN 978-0-19-978319-9, Oxford, UK

Johnson, K. (2009). Policy Lessons Drawn from the recent Food and Fuel Price Inflation. *CESifo Forum*, Vol.10, No.1, (April 2009), pp. 15-20, ISSN 1615-245X

Kappel, R.; Pfeiffer, R. & Werner. J. (2010). What Became of the Food Price Crisis in 2008? *Aussenwirtschaft*, Vol.65, No.1, (n.d. 2010), pp. 21-47, ISSN 0004-8216

Kerckow, B. (2007). Competition between agricultural and renewable energy production. *Quarterly Journal of International Agriculture*, Vol.46, No.4, (n.d. 2007), pp. 333–347, ISSN 0049-8599

Kilian, L. (2009). Not All Oil Price Shocks Are Alike: Disentangling Demand and Supply Shocks in the Crude Oil Market. *American Economic Review*, Vol.99, No.3, (June 2009), pp. 1053-1069, ISSN 0002-8282

Lambert, D. & Miljkovic, D. (2010). The Sources of Variability in U.S. Food Prices. *Journal of Policy Modeling*, Vol.32, No.2, (March 2010), pp. 210-222, ISSN 0161-8938

Lazear, E. (May 14th, 2008). Responding to the global food crisis, In: *Council of Economic Advisers Testimony to the Senate Foreign Relations Committee*, 04.2009, Available from: http://georgewbush-whitehouse.archives.gov/cea/lazear20080514.html

Licker, R.; M. Johnston; J. Foley; C. Barford; C. Kucharik; C. Monfreda & N. Ramankutty (2010). Mind the Gap: How do climate and Agricultural Management Explain the "Yield Gap" of Croplands Around the World? *Global Ecology and Biogeography*, Vol. 19, (2010), pp. 769-782, ISSN 1466-8238

Marlow, V. ; Krupa, K.S.& Lubowski , R.N. (2004). Estimating U.S. Cropland Area. *Amber Waves*, Vol.2, No.5, (November 2004), pp. 47-47, ISSN 1545-8741

Martin, A. (2008). Fuel Choices, Food Crises and Finger-Pointing, In: *The New York Times*, 15.04.2008, Available from: http://www.nytimes.com/2008/04/15/business/worldbusiness/15food.html

McMichael, P. (2009). A food regime analysis of the world food crisis. *Agriculture Human Values*, Vol.26, No.4, (December 2009), pp. 281-295, ISSN 0889-048X

McPhail, L.L. & Babcock, B.A. (2008). Ethanol, Mandates, and Drought: Insights from a Stochastic Equilibrium Model of the U.S. Corn Market, In: *Center for Agricultural and Rural Development (CARD) at Iowa State University in its series Center for Agricultural and Rural Development (CARD) Publications*, 03.2008, Available from: http://www.card.iastate.edu/publications/synopsis.aspx?id=1071

McPhail, L.L.& Babcock, B.A. (2010). Assessing the impact of U.S. ethanol market shocks on global crude oil and U.S. gasoline: a structural VAR approach. *Proceedings of Agricultural & Applied Economics Association 2010 AAEA, CAES, & WAEA Joint Annual Meeting*, Denver, Colorado, USA, July 25-27, 2010

Mitchell, D. (2008). A note on rising food prices, In : *Policy Research Working Paper* , The World Bank, 01.2008, Available from: http://www.wds.worldbank.org/servlet/WDSContentServer/WDSP/IB/2008/07/28/000020439_20080728103002/Rendered/PDF/WP4682.pdf

Muhammad, A.& Kebede, E. (2009). The emergence of an agro-energy sector: is agriculture importing instability from the oil sector? *Choice*, Vol.24, No.1, (n.d. 2009), pp. 12-15, ISSN 0009-4978

Musser, M.; Lambert, D. & Daberkow, S. (2006). Factors Influencing Direct and Indirect Energy Use in U.S. Corn Production. *Proceedings of the American Agricultural Economics Association annual meeting*, Long Beach, CA, USA, July 23-26, 2006

Myers, N. & Kent.,J. (2003). New Consumers: The influence of affluence on the environment. *Proceedings of the National Academy of Sciences of the USA*, Vol.100, No.8, (n.d. 2003), pp. 4963-4968, ISSN 0027-8424

OECD-FAO. (2007). OECD-FAO Agricultural Outlook 2007–2016, In: *Organization for Economic Co-operation and Development (OECD)*, Available from: http://www.agri-outlook.org/dataoecd/6/10/38893266.pdf

OECD-FAO. (2008). Agricultural Outlook 2008–2017, In: *Organization for Economic Co-operation and Development (OECD)*, Available from: http://www.fao.org/es/esc/common/ecg/550/en/AgOut2017E.pdf

Perlack, R.;Wright,L.;Turhollow,A.; Graham,R.; Stokes,B. & Erbach,D. (2005). Biomass as feed stock for a bioenergy and bioproducts industry: the technical feasibility of a billion-ton annual supply, In: *USDA, USDOE*, 04. 2005, Available from: http:// www.osti.gov/bridgeS

Prabhu, P.; Raney, T. & Wiebe, K. (2008). Biofuels and food security: missing the point. *Review of Agricultural Economics*, Vol.30, No.3, (September 2008), pp. 506–516, ISSN 1058-7195

Rajagopal, D.; Sexton, S.; Roland-Hoist, D. & Zilberman, D. (2007). Challenge of Biofuel: Filling the Tank Without Emptying the Stomach. *Environmental Research Letters*, Vol.2, No.4, (n.d. 2007), pp. 1–9, ISSN 1758-6429

Rajagopal, D.; Sexton, S.; Hochman, G.; Roland-Holst, D. & Zilberman, D. (2009) Model Estimates Food-versus-Biofuel Trade-Off. *California Agriculture,* Vol.63, No.4, (October-December 2009), pp. 199-201, ISSN 0008-0845

Rajagopal, D.& Zilberman, D. (2007). Review of Environmental, Economic and Policy Aspects of Biofuels, In: *The World Bank, Policy Research Working Series*, 01.09.2007, Available from:
http://papers.ssrn.com/sol3/papers.cfm?abstract_id=1012473

Reilly, J. & Paltsev, S. (2007). Biomass Energy and Competition for Land, In: *MIT Joint Program on the Science and Policy of Global Change*, 04.2007, Available from: http://web.mit.edu/globalchange/www/MITJPSPGC_Rpt145.pdf

Rosegrant, M.W. (2008). Biofuels and Grain Prices: Impacts and Policy Responses, In: *International Food Policy Research Institute, Washington, DC,* 05.2008. Available from:
http://beta.irri.org/solutions/images/publications/papers/ifpri_biofuels_grain_prices.pdf

Rosegrant, M.; Zhu, T.; Msangi, S. & Sulser,T. (2008). Global scenarios for biofuels: impacts and implications. *Review of Agricultural Economics*, Vol.30, No.3, (September 2008), pp. 495–505, ISSN 1058-7195

Runge, C.F. & Runge, C.P. (2010). Against the Grain: Why Failing to Complete the Green Revolution Could Bring the Next Famine. *Foreign Affairs,* Vol.89, No.1, (January/February 2010), pp. 8-14, ISSN 0015-7120

Runge, C.F. & Senauer, B. (2007). How biofuels could starve the poor. *Foreign Affairs*, Vol.86, No.3, (May/June 2007), pp. 41-53, ISSN 0015-7120

Saghaian, S.H. (2010). The Impact of the Oil Sector on Commodity Prices: Correlation or Causation? *Journal of Agricultural and Applied Economics,* Vol.42, No.3, (August 2010), pp. 477-485, ISSN 1074-0708

Saunders, C.; Kaye-Blake W.; Marshall, L.; Greenhalgh, S. & De Aragao Pereira,M. (2009). Impacts of a United States' biofuel policy on New Zealand's agricultural sector. *Energy Policy*, Vol.37, No.9, (September 2009), pp. 3448–3454, ISSN 0301-4215

Searchinger, T.; Heimlich, R.; Houghton, R.A.; Dong, F.; Elobeid, A.; Fabiosa, J. ; Tokgoz, S., Hayes, D. & Yu, T.H. (2008). Use of U.S. Cropland for Biofuels Increases Greenhouse Gases through Emissions from Land Use Changes. *Science*, Vol.319, No.5867, (February 2008), pp. 1238–1240, ISSN 0036-8075

Senauer, B. (2008). Food market effects of a global shift toward bioenergy. *American Journal of Agricultural Economics*, Vol.90, No.5, (December 2008), pp. 1226–1232, ISSN 0002-9092

Sexton,S.; Zilberman,D.; Rajagopal, D.& Hochman, G. (2009). The Role of Biotechnology in a Sustainable Biofuel Future. *AgBioForum*, Vol.12, No.1, (n.d. 2009), pp. 130-140, ISSN 1522936X

Sheng-Tung, C. & Chen, C. (2010). Modeling the Relationship Between the Oil Price and global Food Prices. *Applied Energy*, Vol.87, No.8, (August 2010), pp. 2517-2525, ISSN 0306-2619

Singh,S. (2009). Global Food Crisis: Magnitude, Causes and Policy Measures. *International Journal of Social Economics*, Vol.36, No.1, (January 2009), pp. 23-36, ISSN 0306-8293

Su, X. (2010). Intertemporal Pricing and Consumer Stockpiling. *Operations Research*, Vol.58, No.4-part-2, (July 2010), pp. 1133-1147, ISSN 0030-364X

Taheripour, F. & Tyner, W. (2008). Ethanol Policy Analysis: What Have We Learned So Far? *Choice*, Vol.23, No.3, (n.d. 2008), pp. 6-11, ISSN 0009-4978

Tyner, W. & Taheripour, F. (2008). Policy options for integrated energy and agricultural markets. *Review of Agricultural Economics*, Vol.30, No.3, (September 2008), pp. 287–396, ISSN 1058-7195

Urbanchuk, J.M. (2007). The relative impact of corn and energy prices in the grocery Aisle, In: *LECG LLC* , 04.11.2008, Available from: http://www.lecg.com

Vincent, D.P.; Dixon, P.B.; Parmentier, B.R. & Sams, D.C. (1979). The short-term effect of domestic oil price increases on the Australian economy with special reference to the agricultural sector. *The Australian Journal of Agricultural Economics*, Vol.23, No.2, (August 1979), pp. 79–101, ISSN 1364-985X

Von Braun, J.; Ahmad, A.; Okyere, K.A.; Fan, S.; Gulati, A.; Hoddinott, J.; Pandya-Lorch,R.; Rosegrant,M.W.; Ruel, M.; Torero,M.; Van Rheenen,T. & Von Grebmer, K. (2008). High Food Prices:The What, Who, and How of Proposed Policy Actions, In: *International Food Policy Research Institute, Washington, DC*, 05.2008, Available from: http://www.ifpri.org/sites/default/files/publications/foodpricespolicyaction.pdf

Von Braun, J.& Maximo, T. (2009). Implementing physical and virtual food reserves to protect the poor and prevent market failure, In: *International Food Policy Research Institute, Washington, DC*, 02.2009, Available from: http://www.ifpri.org/sites/default/files/publications/bp010.pdf

Von Urff, W. (2007). Biofules–A New Chance for Agriculture or a Threat to Food Security. *Quarterly Journal of International Agriculture*, Vol.46, No.2, (n.d. 2007), pp. 99-104, ISSN 0049-8599

Webb, S. (1981). The impact of increased alcohol production on agriculture: a simulation study. *American Journal of Agricultural Economics*, Vol.63, No.3, (n.d. 1981), pp. 532–537, ISSN 0002-9092

Yahaya, J. (2006). Impacts of Biodiesel Development on the Pam Oil Industry. *Malaysian Journal of Economic Studies*, Vol.43, No.1&2, (June/December 2006), pp. 113-140, ISSN 1511-4554

Yang, J.; Wiu, H.Q.; Huang, J. & Rozelle, S. (2008). Fighting global food price rises in the developing world: the response of China and its effect on domestic and world markets. *Agricultural Economics*, Vol.39, No.s1, (December 2009), pp. 453–464, ISSN 0169-5150

Yang, H.; Zhou. Y.& Liu, J. (2009). Land and water requirements of biofuel and implications for food supply and the environment in China. *Energy Policy*, Vol.37, No.5, (May 2009), pp. 1876–1885, ISSN 0301-4215

Zhang, Z.; Lohr, L.; Escalante, C. & Wetzstein, M. (2009). Ethanol, Corn, and Soybean Price Relations in a Volatile Vehicle-Fuels Market. *Energies*, Vol.2, No.2, (June 2009), pp. 320-339, ISSN 1996-1073

Zhang, Z.; Lohr, L.; Escalante, C. & Wetzstein, M. (2010). Food versus fuel: What do prices tell us? *Energy Policy*, Vol.38, No.1, (January 2010), pp. 445-451, ISSN 0301-4215

6

Can Biofuels be an Engine for Growth in Small Developing Economies – The Case of Paraguay

Anil Hira[1] and Plinio Torres Garcete[2]
[1]*Simon Fraser University*
[2]*Universidad Americana*
[1]*Canada*
[2]*Paraguay*

1. Introduction

This analysis examines the feasibility of setting up a biofuels sector in Paraguay. As a small agriculturally-based country, Paraguay can serve as an interesting test case for the feasibility of creating ethanol in other small developing countries, such as those in Central America. Our analysis is inspired by previous work on the remarkable success of Brazil's ethanol industry (Hira, 2009). More than 80% of Brazil's cars can run on a spectrum of gas and alcohol blends based on local sugarcane supply and processing. Ethanol reduces dependency on politically volatile and expensive petroleum imports. The International Energy Administration, along with other recognized international energy institutions, states that while demand will continue to increase over the long-term, prospects for supply meeting that demand "are extremely uncertain (IEA, 2008, 3)." A viable biofuels sector could help Paraguay to improve energy security and reliability, spur economic growth and reduce external dependence, improve employment and rural development, possibly create a new export industry, and reduce greenhouse gas emissions. We focus here on sugarcane ethanol. Sugarcane ethanol is the most attractive option, the only feedstock currently providing an economically feasible substitute at an estimated oil price of $70/barrel, and is more environmentally friendly than other feedstocks as waste is burned for electricity cogeneration. Sugarcane ethanol also produces less carbon emissions than petroleum. As a recent report states, there is no explicit policy at present for biofuels in Paraguay (IICA, 2007, 54). This report is based on secondary analysis and field research conducted during July 2009. We note here the severe constraints on primary data, requiring a more qualitatively-oriented approach.
We organize our analysis around an examination of the following factors: **agricultural, economic, and governance**. The problem of sustainability must be addressed by any plan along with economic/financial feasibility, however, we believe that this question is adequately answered in a number of analyses that demonstrate that sugarcane ethanol, if conducted with safeguards, is a net reducer of carbon emissions (Goldemberg, 2007).

2. The Paraguayan context: pressures for economic change

2.1 Political context

Paraguay's geographic isolation and lack of mineral wealth left it in the periphery of Spanish colonialism. Paraguay also has a history of military dictatorship, with some exceptional periods of stability around central leaders marred by long periods of instability. Modern Paraguayan history begins with the ascension of Alfredo Stroessner through military coup in 1954. Stroessner ruled the country until his overthrow by coup in 1989. While dictatorial, he also presided over rapid economic growth in the 1970s led by increases in agricultural production and the construction of the large binational dam projects, Itaipú and Yacyretá. Stroessner used the Colorado Party and the military as the key institutions for Paraguayan political power. Economic decline during the 1980s, including the end of Itaipú construction; drops in agricultural export prices; and ongoing concerns about corruption fed growing political opposition led by the Church and eroded support for Stroessner. There was also concern about the development of large agro-businesses and the purchase of large swathes of land by foreign, mainly Brazilian, interests (Lewis, 1991, 264).

These issues, Stroessner's age (72), and uncertainty about his successor, precipitated his top assistant, General Andrés Rodríguez, to overthrow him and begin the present phase of democratic rule. After several elections and other sidesteps, Paraguay's nascent democracy seems to be maturing at a rapid pace. The current President, Lugo, is a social reformer who is the first opposition (to the Colorados) candidate to peacefully win election in Paraguay's history. Lugo's coalition government is a fragile one, reflecting difficulties in developing a strong economic vision and policy for the country during a period of transition. For democracy to survive, economic conditions have to be ripe for the inclusion of more of the population in economic activities. Lugo's government has been chastened by widespread land protests, reflecting the tense situation of inequality in the country.

2.2 Economy

Economic growth in recent years, however, has tailed off considerably, including negative growth rates in 2001-2003. The economy has become more liberalized under the new governments, with openness (% of exports and imports/GDP) increasing from averaging 27% in 1961-70 to 82% from 1991-2000! Curiously, the government share of the GDP has also slightly increased, moving from 12 to 17% over the same period (Hira calcs from PWT). The heavy reliance on the public sector, going back to the Stroessner days, is accompanied by a heavy external debt. External debt as a percentage of gross national income increased from an average of 26% in 1960-70 to 38% in 2001-7 (Hira calcs from WDI). The current economic situation shows other causes for alarm. Unemployment has shot up, from an average of 4.1% in 1971-80 to 8.7% between 2001-7. The new democratic governments have not been able to attract significant amounts of foreign direct investment, the expected engine of growth after liberalization, which increased only slightly from an average of 1% in 1960-70 to 3% from 2001-7 (Hira calcs from WDI).

The problem is particularly pressing for the long-term. While Paraguay has a population of just 6.2 million, the growth rate by decade has remained high since the end of World War II, reaching 28% from 1991-2000 (Hira calcs from PWT). This means considerable pressure for employment is occurring. Indeed, Paraguay has one of the youngest populations in the region, with 26% of the population under the age of 10, leading to an incipient bulge in the work force (IDB 2008, 5).

Paraguay has a major blessing in terms of the steady revenue and cheap electricity provided to it through the Itaipú and Yacryretá binational dams. Yet, there are serious impediments to economic growth. There is a low level of overall competitiveness, related to corruption, lack of finance, weak export diversification, income distribution concentration, and a low level of human capital. Infrastructure, including roads, are poor. Only 35% of the population has more than a primary education (IDB 2008, 7), though university enrolment more than doubled from 1990 (25,989) to 2000 (59,836). Yet funds for research and development and systems such as accreditation and evaluating are sorely lacking (IDB 2008, 16-18).

According to the Vice President's Office (2009), poverty increased in Paraguay from 30% of the population in 1995 to 35.6% in 2007. Extreme poverty increased from 6.8 to 27.4% during the same period. These are indicators that macroeconomic stability and agricultural production improvements have not reached significant levels of the population.

These facts add up to a desperate need for new sources of income and employment. Yet, it is not realistic to think that new manufacturing industries are likely to arise given the limited population centers and the isolation of the country, as well as the inability to compete with lower wage production in Asia. Also, the idea of developing advanced services, financial or technological, does not seem plausible in the short- to medium-term given major problems with educational systems and political and institutional weaknesses. While such industries should be encouraged, they are unlikely to create the numbers of jobs at lower levels of education needed by the growing workforce. This leaves agriculture, Paraguay's natural comparative advantage.

3. Agriculture in Paraguay

The Paraguayan economy is organized primarily around agricultural production. The agricultural sector accounts for 30% of GDP, 45% of jobs, and 80% of exports. Therefore world prices on principal goods such as soy and cotton have great impact upon Paraguay's well-being (IDB, 2008, 6). Colonization of the Eastern frontier during the 1970s led to major expansion of export crops. Cotton increased from 1.1% of total exports in 1960 to 44% in 1985; soy went from negligible to over 16% in 1981. During the same period, more traditional products cattle and quebracho (source of timber and tannin) declined (IDB, 2004).

Forty three percent of the population lives in rural areas. Ninety-nine percent of the population lives in the Eastern half of the country. According to a USAID report, in this Eastern portion, there are 16 million hectares, of which 9 (56%) are dedicated to agriculture and livestock. Agriculture occupies 2.8 million hectares, including 1.9 for mechanized agriculture and 0.9 that is based on family plots. Livestock occupies 5.2 million hectares and fallow land another 1 million. In the Eastern region, 84.2% of all farms and ranches have less than 20 hectares, with family incomes of less than $250, indicating extreme poverty. Those with between 20 and 200 hectares cover another 13.9% of all farms and ranches, and have annual incomes of less than $2,500/year, indicating economic efficiency. As a result of this inequitable distribution, Paraguay has experienced peasant movements throughout its history to redistribute land. These are a reflection of the fact that, particularly after the end of dam construction, there are few other industries to generate employment for campesinos (USAID, 2004). Property rights are poorly defined and there are few families with clear land title (IDB, 2008, 22).

While food exports have been increasing, there are signs of a continuing long-term migration to cities, which goes hand-in-hand with the aforementioned issues around agricultural concentration and unemployment, as noted in Fig. 1.

Food exports and rural population, 1960-2007

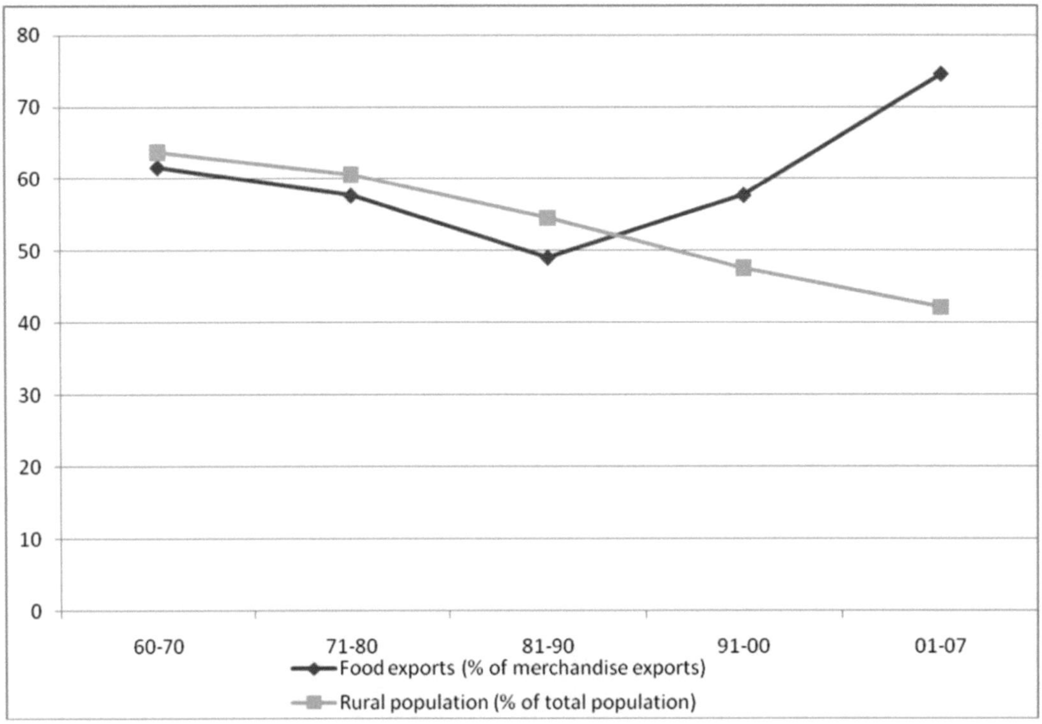

Fig. 1. Source: WDI

If agricultural productivity increases were accompanying the large scale urban migration, one could presume net positive benefits. Yet, while exports are increasing, as seen in Fig. 2 there are indicators that productivity appears to be fairly stagnant, indicating absolute production is increasing, but general productivity is not. It is likely that agricultural production is increasing through agglomeration and increasing use of land (rather than productivity).

4. Why ethanol might make sense for Paraguay

The International Energy Association/OECD (IEA, 2007a) projects that world energy demand will grow by 50% between 2005 and 2030. Developing countries, primarily China and India, will account for 74% of that increase. Fossil fuels- oil, coal, and natural gas- currently provide more than 80% of the world's energy supply (Henimo and Junginger 2009). Like most analysts, the IEA is skeptical that increases in demand can be met by traditional fossil fuel sources alone. Moreover, with increasing concerns over climate change, the most likely short-term alternative fuel, coal, is viewed with trepidation. Furthermore, the greatest increase is likely to be in transportation, where coal cannot be used easily as a fuel, at least with current technologies. These concerns have sparked a surge in interest in renewable fuels. As of 2005, combustible renewables and waste made up just 10% of the world's energy supply, with hydro accounting for another 2.2% (IEA 2007b).

Indicators of agricultural productivity, 1960-2007

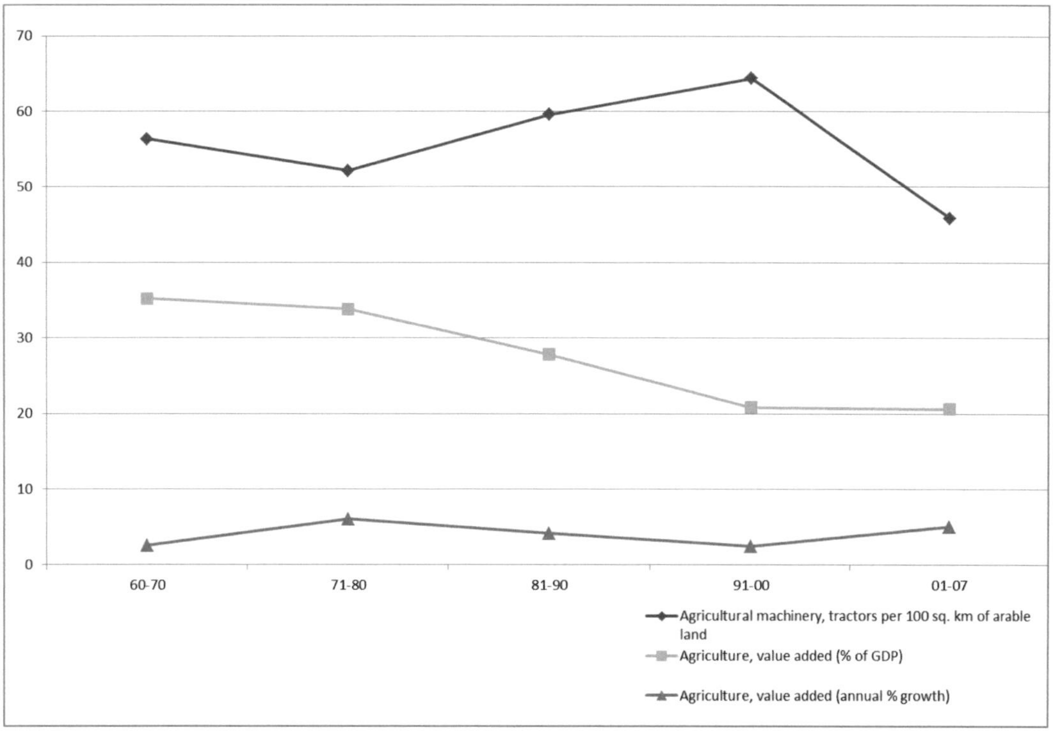

Fig. 2. Source: WDI

4.1 Sugarcane ethanol can serve as a petroleum substitute

Ethanol is a highly controversial subject, with sceptical media such as Time magazine (March 27th, 2008) calling it "The Clean Energy Scam," meaning that the subsidies given to ethanol were not worth the energy it produced. A flurry of negative press, particularly around the 2008-9 period, linked biofuels subsidies to then skyrocketing world food prices and created a food for fuels debate that continues on even as world food prices have abated. Increases in food prices are more closely tied to impressive economic growth in China and India during the last decade, which increased demand and changed diets, as well as increases in petroleum prices, which once receding in 2009 have brought down food prices with them (Hira, forthcoming a). Therefore, the choice of feedstocks, and the economic trade-offs each entails, is a key part of assessing the viability of the biofuel in question.

While we can not provide an exhaustive background, the feasibility of this study depends foremost on discussing 2 aspects of this debate. First, we need to examine whether the net benefits of ethanol allow it to be a feasible substitute for petroleum, and secondly we need to see whether sugarcane is the appropriate feedstock for making ethanol.

As with other markets, there is no "definitive" way to identify the breakeven point for sugarcane ethanol vs. petroleum. Markets are intertwined, so that when the price of petroleum goes up, it will necessarily increase the costs for producing sugarcane. Similarly, the demand for sugar in food will evidently compete with its use for ethanol. Moreover, every situation will lead to different costs for production of both ethanol and linked and substitute products- it will depend on the land, the inputs, the macroeconomic conditions, the wage rates, even the weather. The opportunity cost of producing ethanol is estimated at 1.67 X the price of producing sugar (BNDES and CGEE, 2008, 52). However, these rough calculations must be taken with caution given the major distortions to the sugar market such as quotas in the US market (Hira,forthcoming b).

However, current global production does not reflect cost efficiency; sugarcane is by far the cheapest source of biofuel. The differences in feedstock efficiency are estimated to be quite substantial. Rapeseed produces 100 gallons of biofuel per acre; corn 400; and sugarcane 660 (Coyle 2008). Another estimate puts the costs at: $835/ton of fuel in Europe based on vegetable oils; $546/ton in the US based on corn; and just $387/ton in Brazil based on sugarcane (Prakash 2007). The IEA (2007) estimates that sugarcane ethanol costs the equivalent of $.25-.35/liter of gasoline equivalent (lge), and so is competitive with gasoline at $40 per barrel oil prices. Corn in the US and sugar beets in the EU are estimated to cost $.60-.80/liter, or 3X as much. Moreover, the same report notes that sugarcane fuel leads to up to a 90% reduction in CO2 emissions compared with gasoline while corn ethanol leads to just a 15-25% reduction. In sum, there is a consensus that corn as a feedstock is considerably less efficient than other feedstocks such as palm oil and sugar (Delucchi 2005; De Oliveira 2006; Farrell 2006; Larson 2005). Some studies estimate that corn uses 29% more fossil fuels than the energy it produces, and that vast expansions in cropland and/or productivity would be needed in order to satisfy just a portion of US vehicle needs for fuel (Lokey 2007).

Brazil's example leads the world in terms of continuing reduction of production costs of ethanol (van den Wall Bake, Junginger, Faaij, Poot, Walter, 2009; Goldemberg, Coelho, Nastari, and Lucon, 2004); of providing *flexibility* in markets to adjust to price swings of sugar and petroleum; and of reducing labour and environmental costs by phasing out manual harvesting (Hira and de Oliveira, 2009). Fully 80% of cars in Brazil are now flex fuel, able to run on a spectrum of alcohol and gas blends. Brazil's ethanol industry has created 1 million jobs, and significantly reduced reliance on petroleum imports. Moreover, Brazil has perfected cogeneration using bagasse (cane waste) so that mills are now exporting power to the grid (Mathews, 2006). Brazilian sugar mills' surplus energy generation is expected to continue to accelerate as generation efficiency improves. The cogeneration aspect differentiates sugarcane efficiency from sugar beets and other feedstocks as nothing is wasted (BNDES and CGEE, 2008, 77, 85; Worldwatch Institute, 2007, 166-67). Recent studies also support the claim that ethanol creates lower emissions than gasoline, with significant reductions in CO and hydrocarbons, but slightly higher emissions of aldehydes (BNDES and CGEE, 2008, 47-8).

A number of studies put the breakeven point for Brazilian sugarcane ethanol at US $70/barrel. A 2006 study of cost comparisons suggested that Brazilian ethanol was being produced at $0.42-45/litre and was therefore competitive with the then world price of gasoline at $0.55/litre. It cautioned though that transportation costs would add significantly in the case of exports (S&T, 2006, 21). A comprehensive life cycle analysis of sugarcane bioethanol vs. gasoline concluded that, at crude oil prices in 2005 (about $50/barrel), it was at the breakeven point (Luo, van der Voet, Huppes, 2009). Prices have declined from a high in 2008 of around $91 to around $70/barrel now. We should also keep in mind that the petroleum industry has been receiving billions of dollars in subsidies including largely publicly funded transportation infrastructure over the last century.

4.2 Paraguayan dependence on oil imports

There are several reasons why expanding ethanol production might make sense in Paraguay. The first is that Paraguay is 100% dependent upon imports of foreign oil. World energy consumption is projected to increase by 50% from 2005 to 2030, fuelled by economic growth in China and India, particularly as auto purchases proliferate. Oil prices are expected to continue on a high trajectory over the period. Meanwhile, there are no signs that major new finds will allow supply to keep pace (EIA, 2008). The International Energy Association agrees with these assessments, and notes that oil supplies might peak sometime after 2030 (IEA, 2008, 91).

4.3 Potential for sugarcane production

Paraguay has the possibility of significantly increasing its production of sugarcane; climactic and soil conditions are propitious. The FAO land database suggests that just 10% of arable land is in use in Paraguay. Paraguay in the 1980s expanded the use of ethanol, but this only affected 1/6 of the 40,000 vehicles there (Hanratty and Meditz,1988, 147). Such a move would allow Paraguay to move away from its traditional concentration in meat and its new and growing dependency on soy exports, along with a high degree of vulnerability to world price swings in those products. Moreover, unlike soy, which has been linked to large scale plantations leading to deforestation, sugarcane could be developed from transforming currently under-utilized smallholder subsistence plots.

Paraguay top exports 2006

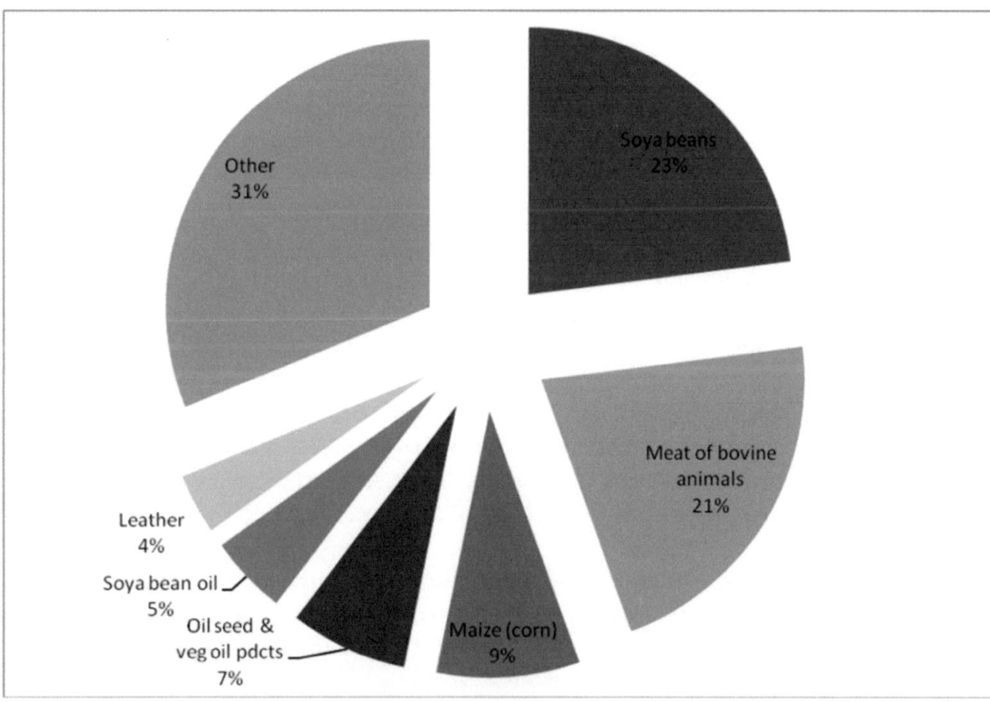

Fig. 3. Source: CEPAL Statistical Yearbook 2008

As seen in Fig. 3, besides cattle raising, which is also linked to deforestation and requires large economies of scale, maize, wheat, cassava, cotton, hen eggs, and sesame seeds are, in order of size, the other major crops grown in Paraguay. While these certainly could be useful as sources of food for smallholders, they also offer more limited value added linkages and must compete with a considerably larger pool of worldwide competitors, including heavily subsidized agricultural farmers in the North. According to FAOstat, the average farm size in Paraguay's oriental region is only 38 ha. The FAO's report on Paraguay reinforces the aforementioned USDA statistics, "Farms between 1 and 20 hectares own less than 15 percent of the livestock and represent more than 70 percent of allproducers/producer families." The initial reaction might be to push food security for small farmers, however, the World Resources Institute's country report on Paraguay estimates only 5% of children in Paraguay are underweight, as opposed to 27% in the world. Therefore, raising incomes, not starvation, is the pressing issue. Sugarcane, in its dual use, provides more income security than other food crops, should Paraguay be able to develop an ethanol system.

Other non-traditional crops, such as soy and stevia, currently have limited industrial linkages. A domestic processing industry would help to create middle class employment for Paraguayans. Stevia also faces keen competition from large agribusinesses in lower cost production areas. For example, Chinese stevia sells for \$14 to \$33/kg while Brazilian and Paraguayan stevia sells for \$140 to \$150/kg (Paraguay Vende, 2004, 39).

5. Sugarcane in Paraguay

5.1 Agricultural requirements for sugarcane

Sugarcane grows best in tropical or subtropical climates, where the average temperature ranges from 26-32 degrees Celsius. Sugarcane has a long growing season of 10-12 months. Sugarcane requires large amounts of water, therefore it must be done in areas where there are high rainfall and natural irrigation. Growing and processing 1 kilo of sugar requires 1500 to 2000 litres of water (Friends of the Earth, 2008), or 75-120 cm of rainfall per year. Long days of sunshine are essential for growth, and fairly dry and clear cool nights for ripening (FICCI, 2009).

As with all agriculture, extension and financing are needed to help with capital equipment, and treatments to prevent weeds, insects, and diseases. Also, research and development monies are needed to help with soil fertility and to develop appropriate varieties for local areas.

Harvesting of sugarcane has traditionally been done with heavy manual labour, however, in more recent years, mechanical harvesting is spreading. Manual harvesting involves separating out and burning the tops (leaves) from the sugarcane stalks. Sugarcane shoots then arise from the remaining stems. Harvesting of the same stalks can continue from 2 to 10 times before replanting.

As with any large monocrop cultivation, sugarcane cultivation in Latin America has a history of linkages to labour exploitation and environmental degradation. The large plantations create concentrations of income. There is also the danger of deforestation if sugarcane leads to destruction of rainforests. Working conditions in Brazilian sugarcane have historically been very poor. There are slave-like conditions in some camps. Sugarcane is harvested seasonally, leading to intermittent work and earnings. The burning of sugarcane creates hazards for the workers and is the most labour-intensive part. Also, the development of new croplands has displaced former cattle areas that then move to the

Amazon, leading to deforestation (Friends of the Earth, 2008). However, Brazil has improved conditions. There is a strong adoption of mechanical harvesting, particularly in São Paulo state.

5.2 Current sugarcane production in Paraguay

Sugarcane has been cultivated since 1549 in Paraguay. However, the disruption of the 2 major wars meant that local output could not meet domestic demand until mid-20th century. For most of its history, sugarcane has been grown on small scale plots with basic technologies (Hanratty and Meditz, 1988, 119-120).

Paraguay is the world's largest exporter of organic sugar. The crop for 2008/9 covers 100,000 hectares and official estimates put the potential at 450,000 tons. Sugarcane is produced in 14/17 departments, but most production is concentrated in the central part of the eastern region (USDA 2008). While organic sugar is another promising market, it does not offer the fuel savings or value added advantages that an ethanol industry does, but again flexibility to adjust to market signals is the key.

Though absolute numbers are relatively small, there are positive signs that sugarcane production can increase; indeed there is a strongly positive trajectory to production over the last 4 decades as shown in Fig. 4.

Paraguay Sugarcane Harvest: Annual Average, 1961-2007

Fig. 4. Source: CEPAL Statistical Yearbook 2008

Interviews with a structured sample of agricultural producers, government officials, and ethanol distillers in Paraguay are uniformly positive that the amount of land for sugarcane

can be increased, though it might mean changing the crop use pattern of some land, primarily small subsistence-based farming. Because sugar is seasonal, an appropriate crop rotation would be needed. Some ethanol production is based on small family plots, thus family members and not an itinerant labour force are responsible for harvesting. Around 29,000 farmers, many small (< 10 ha), are involved, but larger producers are predominantly responsible for most of the sugar produced. About 35% of the sugarcane cultivation is mechanized, with most machinery owned by the farmers themselves (IICA, 2007, 5-6).
According to extensive field research by Benitez, smallholders in Paraguay are currently producing around 50-60 tons/hectare, while Brazilian and Argentine competitors are closer to 120. Part of the problem is the need to introduce new, more productive varieties of sugarcane; he suggests that the Tucuman or RB varieties would vastly increase yields. Lack of equipment (and capital for it) and research and extension to improve local practices are other problems. These point to a general problem of lack of human capital. Yields appear to vary considerably across the country.
According to the IICA, the estimated cost of sugarcane in Paraguay is $13.7/ton (IICA, 2007, 17). This would put Paraguay behind Brazil, but in favourable position in terms of world markets, as demonstrated in Table 1.

Sugarcane production costs internationally, 2006-2007

Country	$/ton
Brazil	10.45
Paraguay	13.7
Caribbean	15
China	23
India	15
Mexico	29
USA	29

Table 1. Source: Hira forthcoming b.

5.3 Conclusion of agricultural analysis

Paraguay faces a series of dilemmas in regard to ethanol expansion that other small countries will face as well. It lacks the financing and extension services of Brazil, reducing its competitive potential. Smallholders in particular struggle with lack of access to finance and insecure property rights. It could support further production through expanding sugarcane into larger fields that would allow for economies of scale and possible approaching of competitive pricing with Brazil. It could mechanize the harvest to avoid replicating the problematic conditions of Brazil's northeastern sugarcane producers. However, this would mean yet another concentration of agricultural production, following the example of soy and cattle, with limited social benefits and possible the loss of the few smallholder sugar farmers.
Based on our own survey and the suggestions of interviewees, there are 2 other possibilities worth considering. The first is one already in motion, and is a response to changes in the European Union towards requiring sustainable certification for all ethanol imports. Paraguay seems well poised to develop its production around this requirement and thus

differentiate its product from Brazil's. A second possibility would be to try to develop the sector around cooperatives that would grow sugarcane separately and share ownership of a micro-distillery. There is some discussion, yet to be confirmed, of new technology being developed in Brazil, that would allow such small scale distillation to be competitive. If so, it could be an ideal way to develop a large part of the new sector in a sustainable way. We discuss further the challenges for a co-op-based system in the governance section below.

6. Industry requirements for sugarcane ethanol

Paraguay is still a predominantly agricultural country, though there has been some growth of maquila assembly, mainly near Ciudad del Este. Unfortunately, the key advantage of extremely cheap electricity so far has not been realized due to a lack of transmission lines from the Itaipu dam to Paraguay. In 2009, an historic agreement was reached to increase the amount of revenues received from Itaipu, to allow for the possibility of selling Paraguay's share to the private Brazilian market (rather than the previous case where it had to sell to Eletrobrás, the Brazilian state owned company), and a possibility to sell to third parties, to be further negotiated. The new revenues could be an important source of reinvestment into infrastructure and industrial promotion for targeted sectors, including ethanol.

Ethanol is produced from molasses derived from sugarcane. Yeasts are added to induce fermentation, producing alcohol. After distillation, further dehydration (removing water) takes place through chemical means. One hectare of sugarcane yields approximately 4000 liters of ethanol per year. Some alcohol is shipped directly for export in dehydrated form, as it may cut costs for shipping (eg by reducing the possibility of contamination) and may face different import barriers.

The dehydrated ethanol is denatured (purified) and moved to gasoline wholesalers/distributors, where it is blended. The blended gasoline is then shipped to retail outlets. In Paraguay, most ethanol is sent by tanker trucks from distributors to retail outlets. While pipeline transportation is possible, ethanol is corrosive and takes on water, requiring high quality and frequent inspection of materials (BNDES and CGEE 2008, 60).

We can describe the overall process of ethanol refining as occurring in two overall stages, agricultural and industrial, with finance, labour and land costs and quality, quality and costs of capital equipment, quality and costs of sugarcane and petroleum and their substitutes; costs of imports (both inputs and petroleum) and transportation costs pervading all of them.

In Paraguay, Petropar, the state oil company, by regulation, has a dominant position in wholesale distribution, covering 80% of gasoline in the country. Petrobrás, the Brazilian

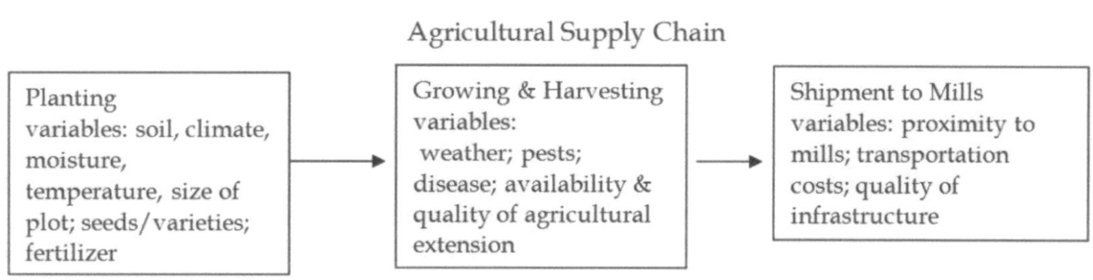

Fig. 5.

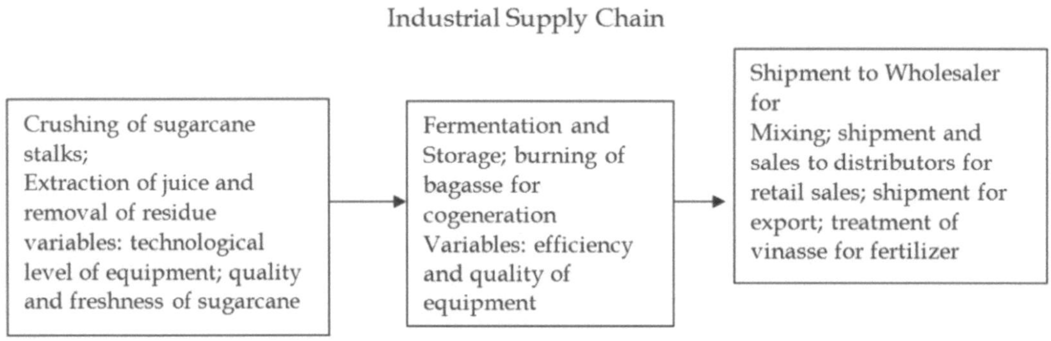

Fig. 6.

giant with state ties, is an exception in distributing its own fuel to its many retail stations, but it does not carry ethanol yet in Paraguay.

6.1 Transport requirements

Transportation infrastructure is vital yet underdeveloped in Paraguay. Eighty five percent of all domestic and international goods is transported by roads. Agro-industry and rural development are thus intimately linked with and constrained by transport costs. Import and export costs tend to be the highest in the region. While Paraguay has 0.77 km of paved roads/1000 inhabitants, Argentina has 3, and Uruguay 4. As a result of these problems and its landlocked status, Paraguay's transportation costs for trade are 43% higher than the average cost for South American economies. Road maintenance is poor, with 30% reported in bad condition in 2006, and several communities lacking all weather access. There are problems with transportation and planning in government, including inadequate information systems and monitoring (World Bank, 2006, 2-4; CEPAL, 2003).

Another key challenge is the fact that 70.4% of all cars are diesel, while only 21.3% run on gasoline (Roundtable 2008, no.1). Ethanol under current technology mixes only with gasoline. Therefore, Paraguay will have to completely revamp its vehicle fleet towards flex fuel vehicles in order to develop a viable domestic market. The Government is well aware of this challenge and has begun to change incentives through tax and tariff regulations to promote a shift in the vehicle structure.

6.2 Current state of Paraguayan ethanol industry

Of an estimated 2,280,000 tons of sugarcane produced in 2006, the Centro Azucarero Paraguayo estimates that 1.4 million were processed into sugar, 760,000 into alcohol, and 120,000 into honey, as demonstrated in Fig. 7 below (IICA, 2007, 8).

Sources of data on biofuels are quite limited. The Biofuels Roundtable, organized by REDIEX of the Ministry of Industry and Commerce, has produced the only reports known to us with data. Their first report from 2009 notes that current capacity in Paraguay is at 236 m (million) litres of ethanol per year, well beyond their estimates for demand at 95 m litres. All plants are producing under capacity, with utilization rates ranging from 20-85%. They note major limitations in regard to the availability of sugarcane, something we also noted in field research. Three companies, Petropar, INPASA (Industria Paraguaya de Alcoholes S.A.), and AZPA (Azucarera Paraguaya S.A), produced 60% of the ethanol in the country in 2009. Benitez (2007) notes that Paraguayan distillation is inefficient because of backwards technology, so that transfer of Brazilian technology is necessary. An example of this became clear during author site visits to the distillation plants of INPASA, which was using state of the art Brazilian equipment, while that of Petropar seemed considerably less efficient.

Uses of sugarcane in Paraguay, 2006

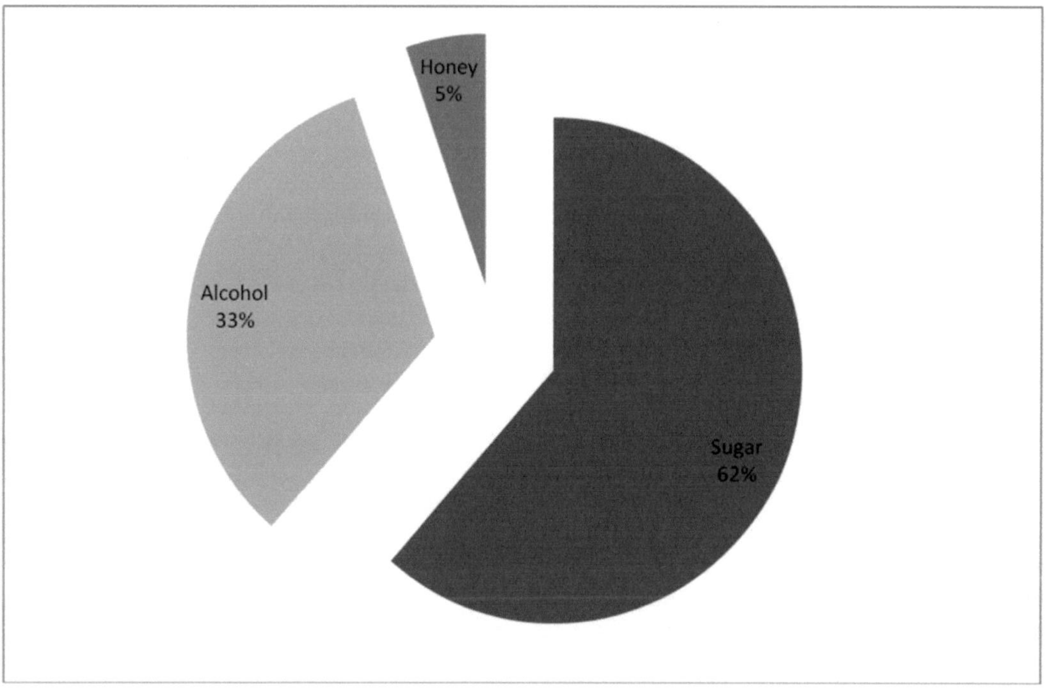

Fig. 7. Source: IICA 2007, 8

The changes in governance framework for ethanol (see below) appear to be paying off, as 2008 saw an increased production to 90 m litres of ethanol, almost all from sugarcane. Production for 2009 is forecast at 120 m litres, with projections continuing to increase afterwards. There are 8 sugar mills in Paraguay, two of which have anhydrous ethanol distilleries, and two that produce hydrated ethanol. There is also one plant that can produce ethanol from corn, manioc and sorghum (INPASA). The largest producer with estimates ranging from 26-40% is Petropar, the national oil company. Official estimates suggest that Paraguay could export up to 300 m litres of ethanol by 2013. However, currently (2009) there are neither exports nor imports (USDA, 2008 and IICA, 2007, 12). Current cost estimates for dehydrated alcohol are $.46/litre and for hydrated .41/litre (IICA, 2007, 17), however costs do seem to vary considerably from plant to plant and evidently by feedstock and other input costs.

One vital fact is to reinforce is the current cost differences with Brazil. Table 2 summarizes the differences as estimated by an ethanol engineer in Paraguay.

Item	Paraguay	Brazil	Differential
Sugarcane	45 tons/hectare	100 tons/hectare	122%
	$20/ton	$16/ton	25%
costs of harvesting	$5-6/ton	$3-4/ton	67%
time to factory after milling	120-170 hours	48-72 hours	150%
transport costs	$6-8/ton	$3-4/ton	100%
refined sugar	$430/ton	$275/ton	56%

Table 2. Source: ethanol engineer in Paraguay (anonymized)

Our source (anonymized engineer) tells us that the end result is that distilled ethanol is 20% less efficient in Paraguay than Brazil, due to lower sugar content, higher rates of infection, and outdated machinery. Paraguay still relies on small producers who have not achieved minimum economies of scale, likely at 50 hectares. However, the solution is not as simple as reproducing the Brazilian model as we discuss below.

6.3 Conclusion

Field research revealed a universal problem of lack of access to finance in Paraguay that undoubtedly impedes all sectors in the economy, and will be particularly problematic for paying for the up front costs of new equipment.

On the other hand, the development of a strong industrial base for ethanol in Paraguay seems less problematic than for the development of a competitive agricultural one. Brazilian state of the art technology is readily available, and through Petropar and Paraguay's strong tradition based on dam construction there is a ready supply of engineers. As we have seen, the transportation sector is problematic in terms of the poor state of secondary roads, needed for gathering agricultural materials and getting them efficiently to the plant. It is not only a question of costs in terms of time lost, but also that fresh agricultural goods yield higher levels of ethanol. The lack of organization and public support for the agricultural part of the chain have led to some distillation plants beginning to initiate their own efforts at agricultural production, which could be an efficient solution, but one perhaps with considerably less social benefits.

One dilemma in developing the industrial chain relates to the role of Petropar. Petropar was created in 1985 as an autonomous company, with operations beginning the following year. In Brazil, the dominant state oil company Petrobrás was key to stabilizing the ethanol market through its decided policy to create a large market for ethanol, including cross-subsidization during price swings, and its ability to vertically capture the production chain, ensuring retail availability (Hira and de Oliveira, 2009). In Paraguay, the industrial structure of the energy sector is considerably less propitious. Indeed, a 2006 World Bank report found that there was no real energy policy in Paraguay; that Petropar lacked any objective financial transparency or oversight; that Petropar was handicapped by state policies to subsidize diesel, which made it cheaper than gas; and there generally was no oversight into quality standards in gasoline. It suggested ending price fixing and subsidies, setting up an independent regulator, and allowing Petropar to compete as a separate entity with other private companies. The suggested reforms have been discussed but the legislation has not yet passed.

While Petropar as the key wholesaler is the main blender for ethanol, not all gas is mixed (despite regulations noted below). There are a wider variety of companies involved in the sector, including some fairly small independent stations, that are quite reluctant to take on the costs of new infrastructure that would afford a wide variety of blends (%s of alcohol with gas). Having a variety of blends was essential for instilling confidence in Brazilian consumers that they would have the flexibility to deal with price drops in petroleum. To reduce costs, it would make sense to simplify the number of blends from potentially 5 to possibly 3 for most stations. Petropar could then have an advantage in offering a wider variety of blends at its stations.

However, Petropar is currently in heavy external debt related to an agreement with Pedevesa, the Venezuelan state oil company, regarding contracts that did not reflect the drop in oil prices over the last year. Petropar's red ink led to the sale of Paraguay's only refinery and a continuing debt. We think that Petropar should be relieved of its wholesale distribution position, and be allowed to compete head on with other oil companies. The government could follow the example of Petrobras and give Petropar the lead in oil exploration in the country, as well as autonomy in its accounts, but in return it would require the need to move to transparency and competition at all levels. Petropar could thus begin to capture market share at the retail level, where it is quite limited, improving its options.

7. Governance requirements for an ethanol industry

7.1 General context

There are a limited number of studies of Paraguayan governance. One of the more recent is an Inter-American Development Bank-sponsored report by Molinas et. al. The report takes a comprehensive view of the Paraguayan policymaking process. It finds a strong level of "rigidity" in the process that is effective in delivering "particularistic" benefits, but not redistributive or regulatory reform. The system is a result of the combination of heavy repression and patronage under a one party system during the *stronato* (Stroessner's rule), marking a hyperexecutive system (one dominated by the executive branch). The Paraguayan case is considered a "triumvirate" of the military, the Colorado Party, and the government. During this period, the high levels of the bureaucracy were filled on the basis of loyalty, meaning that there was little turnover in terms of personnel or ideas. Public employment is the principle instrument of cooptation; public sector workers earn an estimated 17X higher wage than private sector counterparts (vs. an average of 4% for the rest of Latin America) (Molinas et. al., 2006, 10 & 43). Nonetheless, economic growth during the late 1960-70s was high, based on the development of agro-exports and dam construction. The government followed a deliberate long-term strategy of developing agricultural lands on the Eastern border with Brazil in order to reduce dependence on Argentina (Molinas et. al., 2006, 12).

There are major institutional challenges for the Paraguayan public sector, according to an article by Nickson and Lambert (2002). They note that Paraguay is ranked as one of the most corrupt countries by Transparency International. Moreover, there is a low tax base, meaning a paucity of resources, a lack of a merit-based civil service, a lack of evaluation-oriented units, and a high level of politicization. They cite the general ineffectiveness of the neoliberal reforms after 1989 focused on civil service reform, privatization and decentralization. A World Bank report (2005, ix.) notes similar problems, citing in particular the human resourcing system for the civil service; a lack of inter-institutional coordination, and of participation by citizens; poorly qualified public employees with little in service training; inconsistency in salary levels according to classification; high rotation of senior and middle management; clientelism, including promotion based solely on political and personal loyalty; lack of transparency in recruitment; and no performance evaluation measures. While a new law attempting to address some of these problems was instituted in 2001, many of the new features were suspended "within weeks" (World Bank, 2005, 170).

Nickson and Lambert suggest serious problems with the process of privatization. They suggest that buyers were closely tied to the elite and contracts were corrupt and undervalued. The state alcohol plant, APAL, was converted into a company, Cañas Paraguayas (CAPASA), in 1993, and privatized in 1995. The key purchasing company, Tekojaja SA, only paid 1/5 of the agreed price for majority share ownership, leading to the state retaking control in 1997. The company was operating at a loss at least through the time of the writing of that article in 2002. In mid-2001, the government's state reform secretariat (Secretaría Nacional de Reforma del Estado, SNRE), asked for the company, then with $1.5 million in debts, to be declared bankrupt.

The decentralization law sought to pass more control over finances to provincial and local authorities, in line with the new public management ideas in vogue in the 1990s. While a new law granting shares of royalties from the two binational dams was passed in 1998, the actual transfer of resources has not been properly implemented, and only a fraction of promised resources have been transferred. In fact, with Itaipu and Yacyreta producing a large share of total national revenues, the lack of transparency concerning both revenues and expenditures is a major obstacle to any kind of accountable budget. Such revenues could be a key source of resources for industrial policy, however at this time it is impossible to say with any clarity how revenues are being used. Besides this black hole, there is another problematic aspect to government budgeting. The controlaría nacional (national accounting office) is manned by the opposite party of that in power. While attractive at first glance, it would be more logical and consistent to have an autonomous body à la the Central Bank to produce audits and accountability in government accounts.

7.2 Current policies towards ethanol

The government has taken the logical first steps to support the industry. Paraguay has had formal provisions for blending since Decree 2162 of March 1999 that set up a 7% ethanol blend. In Oct. 2005, the Paraguayan Congress passed Law 2748 for Biofuels Promotion. The law declares biofuels to be a national interest. It sets blending for ethanol at a range of 20-24% gasoline. Biofuel use is mandatory as long as there is sufficient local supply. It encourages the use of different local feedstocks. It provides tax benefits, especially concerning investment. The Ministry of Industry controls investment and determines production levels, and that of Agriculture and Livestock certifies feedstock. In May 2008, the government passed Decree 12240 reducing the VAT on biodiesel and ethanol to 2% and eliminating import duties on flex fuel and E85 new and used cars. Ethanol imports are banned, though a special exception was given to Brazil in 2008 for 6 million litres (USDA, 2008; Souto, 2008, 46).

The government has lofty goals in regards to future exports. The Roundtable for Biofuels' National Plan for Biofuels envisions significant increases in both ethanol and biodiesel production for replacing 50% of local demand for petroleum and exports. It includes goals for employment, savings of import expenses, attraction of new investment, and environmental improvement.

There is also discussion of government support through its own procurement policies. Paraguay will need regulation that deals explicitly with ethanol in terms of fuel standards in order to prepare for possible exports. It will have to consider whether it wanted price supports, particularly in terms of stabilizing the price. It should also consider labour legislation. Lastly it needs explicit regulation in regard to investment in the sector. A neutral regulator will have to be set up to govern the sector. While there seems to be a general awareness of these regulatory requirements, legislation and the needed resources seem to be slow in coming, reflecting the fragility of the governing coalition. Moreover, field research interviewees raised a number of doubts about the extent of de facto enforcement of existing provisions. These reflect back on even more serious challenges beyond the general ones of corruption, lack of resources, and low levels of technical expertise that impede governance in all areas. It was difficult to tell through field research the exact number of stations that carry ethanol; we heard different answers to the question from different experts. We did on visual observation notice availability in some stations in Asunción.[1]

7.3 Lack of long-term vision and planning

Field research interviewees repeatedly noted the lack of a shared long-term vision for the country, though some individual offices were attempting to develop one. This is a problematic aspect, as the development of any new sector requires not only coordination but determination among multiple actors to get through the high costs and learning curve of setting up a new industry. Some interviewees suggested a vision that takes advantage of Paraguay's location to create infrastructure so that it could be used as a shipping point. They cite plans for developing railway lines to Brazilian ports and better roads and ports, as well as possibly gas pipelines from Bolivia to the markets of Argentina and Brazil. During field research there was also the exciting prospect discussed of selling surplus electricity to Chile via Argentina. This would also require major investment in transmission lines as well as clarification of the dam agreements.

The Brazilian case signals the need to have clear, long-term goal posts for the program during which progress on reducing costs can be made. The goals would be set up by the stakeholders. Brazil also set up key support policies along the following lines: Petrobrás, the oil company, guaranteed a minimal purchase each year, low interest loans were given to help set up ethanol distilleries through the Bank of Brazil, final prices were subsidised and smoothed out through the tax system on fuels and vehicles, to ensure competitiveness with gasoline, and production quotas for sugar production were set. One of the main stumbling blocks experienced in Brazil was that sugar production was initially inadequate to meet rising demand, so lag times should be a major consideration (Hira and de Oliveira, 2009).

---

[1] While biodiesel was not the subject of this study, we did find it curious that the government has promoted a processing plant using animal fat. This is not only out of step with international research on the efficiency of different feedstocks but also inherently limited in terms of total supply. Moreover, there is concern by those in the industry that animal fat leaves damaging residues in engines at temperatures below 15° C.

7.4 Coordination breakdowns

Field research revealed a gap in inter-institutional coordination among different government ministries and the private sector. There seems to be a lack of recognition of the inter-sectoral nature of ethanol, spanning across agriculture, industry, energy, and transportation. One source for hope is the recent development of the Mesa Rodonda de Biocombustibles (Biofuels Roundtable) that brings together private partners and the Ministry of Industry and Commerce.

Within the agricultural sector itself, there are severe governance problems. A 2004 USAID report notes: a total lack of clear medium- and long-term policies towards land distribution; a lack of coordination between the Min. of Agriculture and Livestock (MAG) and the Rural Development and Land Institute (INDERT), two actors key to the sector; the fact that census and survey data are extremely dated, with the last National Census taking place in 1991; and a lack of agreement between soy producers and the government in regard to the level of taxation for that product. Besides MAG and INDERT, the National Survey Service (SNC), the General Public Registry Directorate (DGRP) and the municipal governments are the other key public sector stakeholders. However, interviewees noted that MAG was largely out of the loop in terms of current developments on ethanol. The Ministry of Industry and Commerce (MIC) seems to be taking the lead without working with MAG. This means a series of potential problems with improving sugarcane yields and quality for industry growth.

Based on the example of Brazil, a multiple stakeholder approach would be needed, with partners from the private, non-profit, and public sectors all working together. The private sector partners would include farmers, agricultural supply companies, transport companies, and others who are part of the production process. The non-profit partners would include industry associations and non-governmental associations that represent environmental and labour rights interests, as well as civic associations where the industry takes hold. The public sector would require agricultural, treasury, transportation, and pro-export components. Coordination of these ministries is key. Finally, there would be a need for a research and development component to improve production costs, introduce innovation, and monitor the program according to local conditions. In the Brazilian case, public sector partners were brought together under Proálcool, a national program with explicit goals of reducing petroleum dependency through substitution for ethanol.

7.5 The Brazilian road not taken

Perhaps the more sensible direction from the point of view of feasibility as alluded to above would be at least initially to encourage Brazilian ethanol companies to invest in Paraguay for export production. This would require a stable regulatory climate and encouragement of investment. It would give time for technology transfer to take place and a ready export market while Paraguay developed the agricultural, industry, and policy requirements for domestic consumption and the knowhow and financing for exporting. A variant on this option, not mutually exclusive, would be to woo other investors, such as Japanese companies, or large multinationals such as Louis Dreyfus, that are active in Brazil. This would alleviate Brazilian concerns about quality control and facilitate customs processing.

There are signs of Brazilian interest. At the end of 2006, MERCOSUR set up a Special Working Group on Biofuels. In 2007, they identified an action plan, including evaluation of different feedstocks and production capacity; identification of research entities for partnerships, analysis of the infrastructure and distribution systems for fuels, and tools to promote investment in biofuels. In 2008, the MERCOSUR countries signed an MOU (Memorandum of Understanding) to develop a program of cooperation for biofuels. In 2007, the Presidents of Brazil and Paraguay signed an MOU (Memorandum of Understanding) on Biofuels. This set up cooperation in evaluation of different feedstocks' potential; technological development in the industry; analysis of infrastructure and transportation system needs; and investment in Paraguay. Paraguay is working with Embrapa (Brazilian agricultural research and extension agency) on this task (USDA, 2008).

However, it seems quite unlikely that a full embrace of Brazilian investors will ever take place. Based on a history of Brazilian domination, Paraguayans are understandably reluctant to base a promising new industry on strong levels of external dependency. Benitez's vision for the sector seems to reflect that of many in Paraguay, that the Brazilian road will not lead to employment or national value added effects. Hence he calls for "1000 micro distilleries producing 1 million litres of ethanol" through intense coordination of small farmers, the public sector, and private middlemen.

The possibility of capturing value added for local employment is a pressing concern and also reinforces the general notion and political fact that biofuels is seen in light of a more nationally-oriented development. The general desire for national development is further reinforced by the political fallout of the multinational-led growth in agribusiness in Paraguay, leading to large-scale consolidation of lands, and ubiquitous protests by those who want land reform, with soy being a particular target. We are not in a position to gauge the merits of such debates, though capitalization of agriculture and displacement have been universal developments. However, this brings us back to the problem of the lack of competitiveness of Paraguayan producers with Brazilian counterparts, meaning in the long-run a likely unsustainable situation of domestic protection and inability to realize exports. Indeed, during field research the smuggling in of cheaper ethanol from Argentina and Brazil was revealed as a source of concern for the government.

7.6 Sustainability certification and co-ops

One way around this situation suggested by an interviewee was to market Paraguayan production internationally as sustainable. The interviewee revealed that promising discussions were underway with the European Union for certifying Paraguayan production. This would give Paraguay the possibility to differentiate its product from Brazil. It seems plausible in the sense that the EU not only needs to import to meet its own biofuel targets, but in that Paraguay's production would not overwhelm its own producers as would Brazil's.[2]

However, there have been several major problems with this laudable idea. The first is that certification does not provide the resources to monitor and violations are rarely enforced. The second is that certification systems are not set up for local input (Friends of the Earth 2008). Brazil has already moved in this direction, thus could provide tech transfer, but this also means the potential dissipation of advantage based on differentiation.

Moreover, as noted above, Paraguay would clearly need institutional reform. It could set up local stakeholder councils that included growers, large consumers, public officials, workers,

---

[2] The US and EU markets remain protectionist limiting the potential international market for ethanol, though there are discussions about ways to open it up. For more information, see Hira forthcoming,

and local communities. These would begin to develop best practice standards in conjunction with attempting to follow certification standards as laid out in a number of European-based proposals. These are costly proposals and would require considerable up front investment. Paraguay would need assistance from the EU in financial and technical terms, as well as long-term contracts, to seriously consider such a move.

Another intriguing idea presented itself from field research. In Paraguay, there is a long history of very successful agribusinesses based on co-operatives. Some co-operatives have become dominant in areas such as dairy products, even achieving the ability to export. Co-ops by their nature are egalitarian with family farmers sharing capital goods such as tractors and helping each other through the frequent swings of agricultural markets and conditions. They would seem to be an ideal vehicle for avoiding the large concentrations of land and production in much of the modern sectors of Paraguayan agriculture, while still offering a vehicle for achieving Brazilian style economies of scale and possibly even co-op distilleries and transportation networks. Many Paraguayan co-ops have their own financial organizations. Some have even begun pilot projects to produce sugarcane.

However, there are severe obstacles to this solution as well. Most of the successful co-ops in Paraguay in terms of exports are religiously-affiliated communities of highly educated European and North American immigrants. The homogeneity of such communities does not lend itself to easy transfer to the landless and poorly educated groups of Paraguayans displaced from the growth of agribusiness. Interviewees familiar with co-ops in Paraguay suggested that it would be unlikely for existing co-ops to fully invest in the sector unless there was more stable and consistent support from the government to support the industry over the long-run. Still, it is interesting to note that the first major biofuels seminar organized by the Paraguayan government (Sintesis, 2007, 42) contains a conclusion with this very suggestion- that co-ops would be the way to preserve small farmers in the market.

8. Conclusion

While daunting, the problems for creating an ethanol sector in Paraguay are not insurmountable. Agricultural and industry requirements call for a true government commitment to the sector. Thus far, a full acceptance of ethanol as a target industry does not seem to be in the minds of Paraguay's leaders. Governance requirements are more mine-ridden, reflecting problems throughout Paraguay's economy. A program of training and reform in the state sector is direly needed. The possibilities for sustainable certification and some portion of co-op production are two very intriguing avenues for further exploration. However, the first and most important step remains the lack of a clear understanding for a long-term vision and plan for Paraguayan development and industrial policy. Most interviewees seemed sceptical about the prospects for an active industrial policy and wanted to see instead a "level playing field," with the state acting as a stabilizing force for the market. We are doubtful that such an approach would suffice to create a brand new industry, however we do understand the general reservations based on the historical weakness and current problems of the Paraguayan state. Our report shows, in sum, that a viable ethanol industry is feasible only if Paraguay is able to close the gap in terms of cost efficiencies with Brazil, and that it might do so in a way that is considerably more socially advantageous than Brazil, but it will require a strong and flexible state policy to achieve these goals. Our article also provides a precautionary note to other would be states about the very real challenges to developing a viable biofuels sector.

9. Acknowledgements

We gratefully acknowledge funding by the IDRC-AUCC (International Development Research Centre of the Government of Canada and the Association of Universities and Colleges of Canada), the Universidad Americana of Paraguay, and Simon Fraser University. We are highly appreciative of the cooperation given to us by the Government of Paraguay, the various private sector companies, and of the warm hospitality of innumerable individuals. The research design for this project was first presented at the University of Illinois Urbana Champaign biofuels studies workshops in Fall 2008, sponsored by the Sloan Foundation. I would like to thank Gale Summerfield and Jüergen Scheffran for their efforts in organizing those events. I would also like to thank colleagues who participated in the SFU biofuels roundtable in Fall 2009 for their comments on an early draft.

10. References

Interviews*

We include here the names of key interviewees during field research by Anil Hira and Plinio Torres. All interviews were conducted by Hira in July 2009 in Asunción, Paraguay. By prior agreement, we have anonymized any particular citations in this document to protect our interviewees. We gratefully acknowledge the willingness of participants to meet with us and their frankness, and particularly their generosity in time and in setting up site visits. We also conducted 3 site visits, to 2 alcohol distilleries and to Friesland cooperative.

-Cárdenas Nuñes, Justo Pastor, Diputado (Deputy), Congreso de Paraguay (Congress)

-Petropar, Office of the President

-Ministry of Industry and Commerce, Office of the Vice Minister, UTEPI (Unidad Técnica de Estudios para la Industria (macroeconomic and industrial policy analysis office)), REDIEX (public-private network for export promotion)

-Office of the Vice President

Asociacón Guyra Paraguay. (2009). Evaluación del Impacto de los Biocombustibles en la Matriz

Socio-ambiental del Paraguay y Medidas de Acción Futuras, Asociación Guyra Paraguay, Asunción

BNDES (National Development Bank and CGEE (Center for Strategic Studies and Management). (2008). Sugarcane-Based Bioethanol: Energy for Sustainable Development, BNDES, Rio de Janeiro

Benitez, H. B. (2007). Produccion de Etanol de Caña de Azucar en micro destilerias continuas. MA Thesis, Business Administration, Supervisor: Plinio Torres, Asuncion Paraguay, Universidad Americana

UN Economic Commission for Latin America and the Caribbean (CEPAL). (2008) Statistical Yearbook for Latin America and the Caribbean 2008. Santiago: CEPAL (2003). Ian Thomason, Ricardo Sanchez, and Alberto Bull. Estudio Preliminar del Transporte de los Productos de Comercio Esterior de los Países sin Litoral de Sudamérica, (March), CEPAL, Santiago

Coyle, W. (2008). The Future of Biofuels: A Global Perspective. Oregon Wheat, (Feb.), pp. 26-30

Delucchi, M.A. (2005). A Multi-Country Analysis of Lifecycle Emissions from Transportation Fuels and Motor Vehicles, Research report, Davis, CA, University of California

De Oliveira, M. E. D.: Vaughan, B.E. & Rykiel Jr., E.J. (2006). Bioscience, 55/7, (July), pp. 593-602.

Energy Information Administration (EIA). (2008). World Energy Outlook, US Dept of Energy, Washington, accessed Mar. 5, 09, Available from www.eia.doe.gov

Farrell, A. (2006). Ethanol can contribute to energy and environmental goals. Science, 17 311 (5760), pp. 506–508

FICCI (Agribusiness information centre). accessed Mar. 4, 09, Available from http://www.ficciagroindia.com/production-guidelines/field-crop/Sugarcane/climatic-req.htm

Friends of the Earth. (2008). Sustainability as a smokescreen: The inadequacy of certifying fuels and feeds, (April), FOE Europe

Goldemberg, J. (2007). Ethanol for a Sustainable Energy Future. Science, 315, pp. 808-81

Goldemberg, J.; Coelho, S.T.; Nastari, P.M. & Lucon, O. (2004). Ethanol Learning curve- the Brazilian experience. Biomass & Bioenergy, 26, pp. 301-4

Hanratty, D. M. & Meditz, S.W., (Eds.). (1988). Paraguay: a country study, Library of Congress, Washington

Hira, A. (forthcoming a). World Food Demand: Coping With Its Growth. In Advances in Food Science and Technology, Vol. 1. Nova Science Publishers

Hira, A. (forthcoming b). Sugar Rush: Prospects for a Global Biofuels Market. Energy Policy

Hira, A. & de Oliveira, G. (2009). No Substitute for Oil? How Brazil Developed Its Ethanol Industry. Energy Policy, Volume 37, Issue 6 (June), pp. 2450-2456

Inter-American Development Bank (IDB). (2004). Fernández Valdovinos, C.G. & Naranjo, A. M. Economic Growth in Paraguay, Economic and Social Study Series RE1-04-009, (May),

IDB, Washington (2008). Paraguay: the Bank's Country Strategy with Paraguay (2004-2008), IDB, Washington

International Energy Agency (IEA). (2008). World Energy Outlook 2008, IEA, Paris (2007). IEA Energy Technology Essentials, accessed 13 Aug. 2008, Available from www.iea.org

International Energy Agency/OECD. 2007a. World Energy Outlook. IEA/OECD Paris. (2007b). Key World Energy Statistics 2007. IEA/OECD, Paris.

IICA (Instituto Interamericano de Cooperación Para La Agricultura). (2007). El Estado del Arte de los Biocombustibles en el Paraguay, (May), IICA, Asunción

Larson, E.D. (2005).Liquid biofuel systems for the transport sector:a background paper. Proceedings of the Global Environmental Facility Scientific and Technical Advisory Panel Workshop on Liquid Biofuels, New Delhi, August 29– September 1

Lewis, P. H. (1991). Paraguay since 1930, In: The Cambridge History of Latin America: Volume VIII

Latin America since 1930, Spanish South America, Bethell, L. (Ed.), pp. 233-68, Cambridge U Press, NY

Lokey, E. (2007). How the next US president should slow global warming. Energy Policy. 35, pp.5399- 5402

Luo, L.; van der Voet, E. & Huppes, G. (2009). Life cycle assessment and life cycle costing of bioethanol from sugarcane in Brazil. Renewable and Sustainable Energy Reviews, 13, pp.1613-19

Mathews, J. (2006). A Biofuels Manifesto: Why biofuels industry creation should be 'Priority Number One' for the World Bank and for developing countries, Working paper, (Oct. 9), Macquarie University, Sydney, Australia

Molinas, J.; Pérez-Liñán, A. & Saiegh, S. with Montero, M. (2006). Political Institutions, Policymaking Processes and Policy Outcomes in Paraguay, 1954-2003, Research Network Working Paper #R-502, Inter-American Development Bank, Washington

Nickson, A. & Lambert, P. (2002). State Reform and the 'Privatized State' in Paraguay. Public Administration and Development, 22, pp.163-74

Paraguay Vende. (2004). Stevia from Paraguay, USAID, Washington

Penn World Tables (PWT). accessed Mar. 3, 2009. Available from: http://pwt.econ.upenn.edu

Prakash, A. (2007). Grains for Food and Fuel- At What Price? Proceedings of FAO meeting, Rome, April 11

Round Table on Biofuels (Mesa Rodonda de Biocombustibles). Reports no. 1, (2008); 2: (2008); 1, (2009). REDIEX, Asunción, Paraguay

S&T Consultants Inc. (2006). Issue Paper on Biofuels in Latin America and the Caribbean. Sept. 18, Prepared for Inter-American Development Bank

Sintesis del Primer Seminario de Biocombustible en Paraguay: Integración Productiva y Oportunidades de Negocios. (2007). Organized by Ing. Agr. Giovanna Facetti Doria. Asunción: Government of Paraguay, Brazilian Embassy, and Inter-American Development Bank

Souto, Guillermo. (2008) Paraguay explora su potencial en biocombustibles. Comunica. 4 (May-Aug): 44=52.

Ugarte Centurion, D. (1983). Evolución Histórica de la Economía Paraguaya, Editorial Graphis, Asunción

USAID. (2004). Diagnostico de los Principales Problemas Relacionados Con La Tierra Rural en Paraguay, USAID, Washington

USDA. (2008). Paraguay Bio-fuels Annual 2008. Foreign Agricultural Service GAIN Report, PA8005, Dec. 29, USDA, Washington

Van den Wall Bake, J.D.; Juninger, M.; Faaij, A.; Poot, T. & Walter, A. (2009). Explaining the experience curve: Cost reductions of Brazilian ethanol from sugarcane. Biomass and Bioenergy 33, pp.644-58

Vice President, Office of. (2009). Paraguay: Visión de Futuro. Presentation to Unión Industrial Paraguaya, May

Wikipedia. Entry on sugarcane, accessed Mar. 4, 2009, Available from: http://en.wikipedia.org/wiki/Sugarcane

World Bank. World Development Indicators (WDI) database, Accessed Mar. 3, 2009

(2006). ESMAP. Paraguay: Reforma del Sector Hidrocarburos, Doc. 319/06, March, World Bank, Washington

(2005). Republic of Paraguay: Institutional and Governance Review: Breaking with Tradition: Overcoming Institutional Impediments to Improve Public Sector Performance, Poverty Reduction and Economic Management Unit, Latin America and the Caribbean. Report No. 31763-PY, June, World Bank, Washington

Worldwatch Institute. 2007. Biofuels for transport: global potential and implications for sustainable energy and agriculture Sterling, VA: Earthscan.

7

Biofuels and World Agricultural Markets: Outlook for 2020 and 2050

Guyomard Hervé[1], Forslund Agneta[1,2] and Dronne Yves[1]
[1]*INRA,*
[2]*TAC, (previously INRA, France)*
France

1. Introduction

The possibility to produce biofuels from different agricultural feedstocks has raised huge interest during the last decade. This interest can be related to the parallel increase in fossil oil prices and the growing awareness about the need to reduce greenhouse gas (GHG) emissions worldwide. Biofuels are also seen by many governments as a means to contribute to the diversification of energy supply and sustain agricultural incomes by creating new outlets for several agricultural products, notably cereals, vegetable oils and sugar plants.

Ambitious public policies aiming at developing biofuel production and consumption in replacement of fossil fuels used in road transport have thus been set up all around the world. Policy instrumentation varies depending on the country. Altogether, policies aim at encouraging the supply of agricultural feedstocks used as raw materials for biofuels, the industrial production of biofuels and/or their domestic consumption by setting blending mandates and/or subsidizing biofuel use.

A period of keen interest was witnessed in the first years of the current decade which has led many countries such as the United States (US), the European Union (EU), Brazil and also several countries in Asia to set very ambitious policy targets for biofuels. But the boom in agricultural prices and the following food crisis in 2007-2008 have severely depreciated the public image of biofuels because of their potential negative impact on world food security in a context of land scarcity. Simultaneously, the issue of the impact of biofuel crops on GHG emissions due to induced land use changes has progressively emerged; it is today a matter of considerable controversy. In addition, concerns have risen about the relatively low energy yield of current biofuels and the budget cost of public policies aiming at encouraging their development. Initially, the debate about these interrelated issues has been confined to a narrow audience, mainly in the academic sphere. However, over the past three years, many stakeholders including environmental organizations, farmers' unions, the media, etc., have shown a considerable interest in the matter leading to a very lively debate worldwide, and more particularly in the EU.

First-generation (1G) biofuels produced from traditional food and feed crops are increasingly criticized for their adverse impacts on world food security and GHG emissions, essentially because they can divert land from food and feed, as well as land forest uses. As a result, hopes turn to a quick development of second-generation (2G) biofuels produced from various sources of biomass that do not directly compete with food and feed crops and, furthermore, are expected to be more efficient in transforming biomass into bioenergy. However, the fact that there is no direct competition does not mean the absence of indirect competition when land is required for growing biomass, even for 2G biofuels.

Our research objective is then to analyze to what extent the development of biofuels and, within this general framework more specifically the development of 2G biofuels in line with the first one, could affect world food security and the environment (GHG emissions and biodiversity protection). The chapter is structured as follows.

In Section 2, we present a general framework of GHG emissions and energy uses worldwide, 1G and 2G biofuels produced today and that could be produced on an industrial scale in the future. We summarize the potential benefits of 1G biofuels that are used for justifying public support, and we recall the main criticisms against them. We then analyze the theoretical arguments in favor of 2G biofuels, and again why there might be discrepancies between theory and reality.

Section 3 depicts the worldwide increase in 1G biofuel production and consumption over the last decade. The analysis distinguishes bioethanol obtained from cereals and sugar crops, and biodiesel obtained from vegetable oils. We also depict the current weight of agricultural products used for biofuel production as compared to total uses of agricultural products, and we analyze the current trend for these crops in terms of areas, yields and prices.

As far as prospects are concerned, two time horizons are considered, 2020 in Section 4 and 2050 in Section 5. For each horizon, we analyze the potential impact of several scenarios for biofuel development on world food security, focusing on cereals and oilseeds used for food, feed and energy. These scenarios differ in terms of assumptions concerning, firstly the total supply and demand in biofuels in 2020 and 2050, secondly the substitution rates of 1G biofuels by 2G ones, thirdly the yields of biomass used for biofuel. The co-products of the process of transforming biomass into biofuels are taken into account since 1G biofuels jointly produce large amounts of co-products that can be used for animal feed, which is not the case with 2G biofuels. Concerning the 2050 horizon, we also analyze the impact of biofuel development and the replacement of the first generation of biofuels by the second on world GHG emissions and biodiversity protection.

The 2020 analysis is based on original simulations performed using a world agricultural partial equilibrium model called OLEOSIM while the 2050 analysis is a review of literature. They both show that the development of 1G biofuels will have a negative impact on world cereal production used for food and feed. However this negative impact is partially alleviated by the production of co-products associated with the supply of 1G biofuels from cereals and oilseeds. The replacement of 1G biofuels by the second generation will alleviate this negative impact. However it will not suppress it, notably if large amounts of 2G biofuels have to be produced from dedicated energy plants that require land and thus, indirectly, compete with other land uses: food, feed, environment protection, urban and transport infrastructures, etc. In the same way, the partial replacement of the 1G of biofuels by the second will reduce the negative impacts of biofuels on GHG emissions and biodiversity linked to land use changes; but it will not eliminate them. More generally, the increase in agricultural production required for food, feed and fuel worldwide should associate an expansion in cultivated area and, more importantly, a significant improvement in yields, notably in world regions where they are very low and low today.

The challenge is then to develop agricultural practices, techniques and systems that make it possible to achieve high levels of land productivity and simultaneously protect the environment and preserve natural resources.

2. The general setting: why are first- and second-generation biofuels subject to criticisms?

2.1 Competition between food and non-food uses of agricultural products

Besides traditional uses, including food, feed, firewood and cooking, biomass can be used for energy production (bioenergy) and other industrial uses (bioproducts). Bioenergy comprises uses for transport, including biofuels, as well as for heat and electricity production. Bioproducts comprise biomaterials and biomolecules. Biomaterials include biodegradable polymers or biopolymers (bio plastics), fiber and composite materials (agrimaterials), paper and paperboard; biomolecules include surfactants, solvents, lubricants and cosmetics.

In 2005, total use of biomass represented about 13.4 billion metric tons (Bt) of dry matter produced from a total of 7.7 billion hectares (Bha). Out of these 7.7 Bha, about 5.2 (1.6 Bha of crops and 3.6 Bha of pastures) were used for food, directly or indirectly through the filter of animal feed; 2.6 Bha corresponding to forest areas, including 2.4 Bha of natural or semi-natural forests, were used for wood production and only a few tens of million hectares (Mha) were mobilized for the production of bioenergy and bioproducts (Wirsenius 2008). Beyond the uncertainties and inaccuracies in the figures due to missing or unreliable data, leading to difficulties in evaluating land surfaces dedicated to any particular use, orders of magnitude are robust: they clearly show the modest part of non-food uses of agricultural production compared to food uses, at least until 2005.

In 2007, global emissions of CO2 attributable to petroleum products and their use amounted to 10.9 giga metric tons (Gt). The transport sector with 6.6 Gt accounted for 60%of this total, most of which (4.8 Gt) for the road transport sector only (IEA 2009). In the short and medium terms, beyond energy savings and improved vehicle technologies, biofuels produced from biomass are seen as the major, if not the sole, alternative to the use of fossil oil in road transport.

More generally, the production of energy from renewable resources should grow sharply over the next decades in the context of both a rising energy demand and a gradual dwindling of non-renewable energy resources. Food demand will also increase due to population growth as our planet will host more than 9 billion people by 2050, nearly 2 billion more than today, economic growth and increased urbanization. These last two elements will result in a shift in food consumption at the expense of plant products - cereals, roots and tubers – and in favor of animal products that are less effective at converting solar energy into food calories. Hence, the question of the ability of our planet to simultaneously satisfy nutritional needs and non-food uses, mainly for energy production, is raised in a context where development must necessarily be sustainable from an economic, social and environmental point of view, at the very least much more sustainable than today. The challenge is huge but it is not insurmountable as long as all stakeholders join forces and act quickly (see, for example, Guyomard 2009; Guillou 2010).

In that general perspective, it is important to analyze the competition for land between food and non-food uses, particularly for energy. More specifically, we will assess to what extent the replacement of a part of 1G biofuels made from plant storage organs also used for human food and animal feed by 2G biofuels that use residues and waste, wood from forests or dedicated crops, could respond to two criticisms against 1G biofuels, namely a negative impact on food security and GHG emissions. But in a first step, we will recall why 1G biofuels and government policies aiming at encouraging their development are the subject of an increasing questioning and to what extent 2G biofuels could bring an answer to this questioning.

2.2 Promises of the second-generation of biofuels in response to criticisms of the first one

In 2008, global biofuel production amounted to 46 million tons of oil equivalent (Mtoe), slightly more than 2% of total fuel used in road transport, mainly in the form of bioethanol in the US and Brazil and biodiesel in the EU. This production used 320 million metric tons (Mt) of sugar crops (17% of world production), 100 Mt of cereals (5%) and 11 Mt of vegetable oils (9%). It mobilized 28 Mha, i.e., 3% of world surfaces in sugar plants, cereals and oilseeds (authors' estimations). These figures show that the use of agricultural commodities for biofuel production is still relatively modest today. However there has been a rapid development since the early 2000s, and acceleration in the more recent years. In 2008/09, the increase in the world cereal demand for bioethanol production was higher than the demand for food, respectively, 28 Mt and 13 Mt (authors' estimations from data of the United States Department of Agriculture).

2.2.1 From criticisms addressed to the first generation of biofuels…

First-generation biofuels are subject to criticism on the basis of two main arguments: firstly, they could be a threat for global food security; secondly, their energy, environmental and economic balance could be not as favorable as initially hoped.

First-generation biofuels are made from storage organs of terrestrial plant materials also used for food, directly or indirectly through animal feed: sugar crops (cane and beet), grains (mainly corn and wheat) and oilseeds (mainly soybean, rapeseed, sunflower and palm). The fear of excessive competition with food uses is therefore immediate: in the recent past, it reached a climax when farm prices soared at the end of 2007 and the beginning of 2008; it is still with us today in the previously recalled context of feeding 9 billion people by 2050 while respecting the environment and natural resources.

This Malthusian fear is doubled with the criticism that the environmental balance of 1G biofuels in terms of reducing GHG emissions can become negative once the changes in land use are recorded. The GHG balance of 1G biofuels is positive, with however large variations depending on the feedstock used as input, when the analysis is made for a given area. It is much less positive, and even can be negative, when the surfaces of sugar crops, grains and oilseeds involved in biofuel production are obtained from former grassland or by cutting down forest surfaces (Fargione *et al.* 2008; Searchinger *et al.* 2008).

The calculation for a given surface means in practice to record only the direct environmental benefit, i.e., the reduction of GHG emissions allowed by the use of one liter of biofuel instead of an equivalent volume of fossil fuel. It is of course necessary to complete this partial picture by taking into account the environmental cost. The latter is mainly related to land use changes, and more specifically to the loss of carbon storage in grassland and forests when the mobilization of one hectare of crops for energy requires, directly or indirectly, the "sacrifice" of one hectare of grassland or forest. This instantaneous effect is coupled with a dynamic loss of organic production from grassland and forests. Finally, the environmental dimension cannot be reduced to GHG emissions. The potentially negative effects of land conversion for energy objectives in terms of loss of biodiversity, use of fertilizers and pesticides beyond the absorptive capacity of agro-ecosystems, etc., should also be taken into account.

The energy efficiency of 1G biofuels per area unit used is also generally considered as modest, except for bioethanol produced from sugar crops, primarily sugar cane. Finally, production costs remain high today, especially when prices of raw plant materials used as resources are high too.

In this strongly questioning, if not critical, context, more and more voices are calling for a halt in the production of 1G biofuels, at least as long as we have not made sure that their development is not detrimental to food production, and would like 1G biofuel farming to be allowed only if it has been proved to have a positive effect on the environment, energy and economic balances. Beyond the scientific and technical progress that can be made on these three points, hopes for the longer term focus on later generation biofuels. In this chapter, we focus solely on 2G biofuels to the extent that those of the so-called third generation are still today at the research stage: they include, for example, hydrogen or oil production from macro- and micro-algae.

2.2.2 ...to the promises of the second generation

While 1G biofuels are produced from crop storage organs, 2G biofuels are produced from lignocellulose. The latter is the main component of plant cell secondary walls and is, therefore, the most abundant biomass constituent in terrestrial areas (Cormeau and Ghosse 2008).

Three main sources of lignocellulosic biomass can be mobilized for 2G biofuels, namely (i) waste and residues from agricultural, forestry, industrial, urban and/or household activities, (ii) forestry resources, i.e., wood from forests, and (iii) dedicated crops of annual plants (wheat, corn… used as a "whole plant") or perennial plants (forage plants like fescue, orchardgrass or ryegrass, herbaceous plants like miscanthus or switchgrass, and shrubs like short and very short rotation poplar, willow, eucalyptus or black locust plants harvested every three to ten years). Two ways of transforming lignocellulosic biomass can be used: a thermo-chemical process which consists in cracking molecules under the action of heat, and a biochemical pathway in which once the raw material has been disintegrated, the complex carbohydrates of lignocellulose are hydrolyzed into simple sugars which are then transformed into ethanol by fermentation. While the thermo-chemical process requires large facilities and significant investments in order to benefit from reduced costs, the biochemical pathway can use the facilities currently used for 1G biofuels. Second-generation biofuels are being experimented in research and demonstration platforms, with hopes for industrial applications and marketing in a decade's time.

Second-generation biofuels appear promising for at least three reasons: (i) the potential raw material is quite abundant, (ii) there is no direct competition with food crops when the resource is a waste, a residue, forest wood or a non-food dedicated crop, and (iii) their energy efficiency, and hence their economic efficiency, appears greater in terms of biomass yield per unit area used as well as in terms of conversion efficiency of this biomass into liquid energy. However the step from promises to reality should be cautiously taken.

When the raw material is a waste and/or a residue, the question of competition with food use does not arise. However it is important to note the potentially negative impact of removing agricultural and forestry residues on the microbiological and physical properties of soils. For example, Powlson *et al.* (2008) come to the conclusion that the energy savings associated with the burial of wheat straw in arable soils are greater than those generated by their removal to produce biofuels or electricity. In practice, the two interrelated questions raised by an energy-oriented use of residues and waste are the potential biomass availability and the cost of mobilizing this biomass (collection and storage costs).

But if the raw material is a dedicated culture (even if it cannot be used for food), the question of competition with food uses of land arises under the same theoretical terms as for

1G biofuels. Supporters of 2G biofuels derived from dedicated crops suggest growing them on "marginal" land, in some way unsuitable for food crops (including for economic reasons). But the potential for land to be mobilized in this way is uncertain because the need for a minimum profitability will likely require that, at least, some of these dedicated crops are located on sufficiently good land to obtain sufficient returns. In summary, the issue of allocating land to different uses (food and non-food, but also recreational, environmental, urban, etc.) also arises with 2G biofuels from dedicated crops.

3. World production of first-generation biofuels and impact on agricultural prices

3.1 World production of first-generation biofuels

3.1.1 The 2009 picture

In 2009, world production of 1G biofuels[1] is 51.8 Mtoe versus 45.9 Mtoe in 2008, that is an increase of 5.9 Mtoe or 13% in one year. Biofuel production is dominated by three countries: the US (22.0 Mtoe, 42% of world production), Brazil (13.9 Mtoe, 29%) and the EU (10.0 Mtoe, 18%). Other countries (5.9 Mtoe, 11%) are more recent players and some of them show very strong annual growth: China (1.3 Mtoe), Argentina (1.1 Mtoe), Canada (0.83 Mtoe), Thailand (0.69 Mtoe), Colombia (0.42 Mtoe) and India (0.35 Mtoe). The rest of the world corresponds to approximately 1.2 Mtoe (Figure 1).

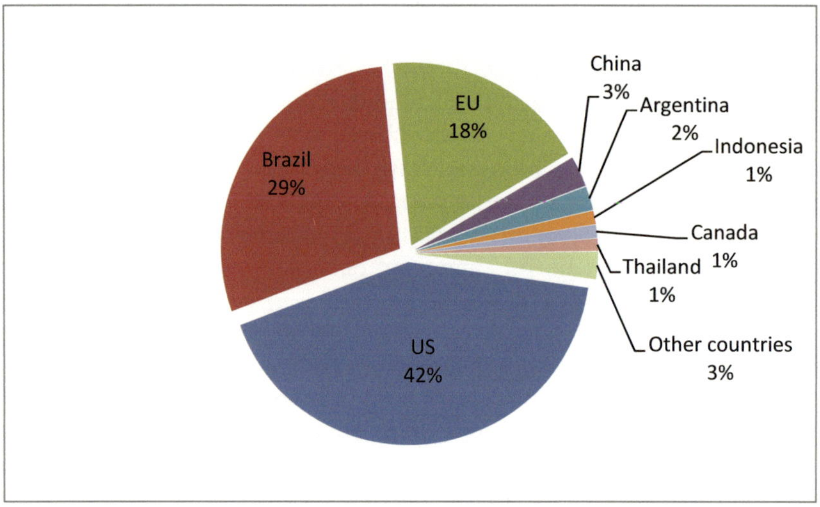

Fig. 1. Geographical distribution of world biofuel production in 2009, in Mtoe; Source: INRA from the Biofuels Platform

According to the Biofuels Platform, their production in 2009 is made of 74.0 billion liters (Bl) of bioethanol (approximately 37.7 Mtoe[2], 73%) and 17.9 Bl of biodiesel (14.1 Mtoe[3], 27%). In

[1] World production figures vary significantly according to the sources. In addition, comparison between sources is made difficult by the use of different measure units (liters, gallons, barrels, metric tons, tons of oil equivalent, etc.).

[2] The conversion factor is 0.51 toe for 1,000 liters of bioethanol.

[3] The conversion coefficient is 0.78 toe for 1,000 liters of biodiesel.

terms of production location, the structure of the world market for bioethanol is very different from that of biodiesel (Figure 2). The bioethanol market is overwhelmingly dominated by the US (40.1 Bl) whose production now largely exceeds that of Brazil (24.9 Bl). The third world actor, the EU, is a much more modest (2.9 Bl) and furthermore less dynamic producer than several other players. The rest of the world produces over 6 Bl; China, with 2.1 Bl, is an increasing player, followed by Canada, Thailand, Colombia and Australia. As far as biodiesel is concerned, world production is largely dominated by the EU (8.7 Bl out of a total 17.9 Bl, i.e., 49%). Several countries have also strongly developed their production in the more recent years, often for export: the US (2.1 Bl), Brazil (1.5 Bl) and Argentina (1.3 Bl). The rest of the world, mainly Thailand, China, Colombia and South Korea, produces 4.3 Bl.

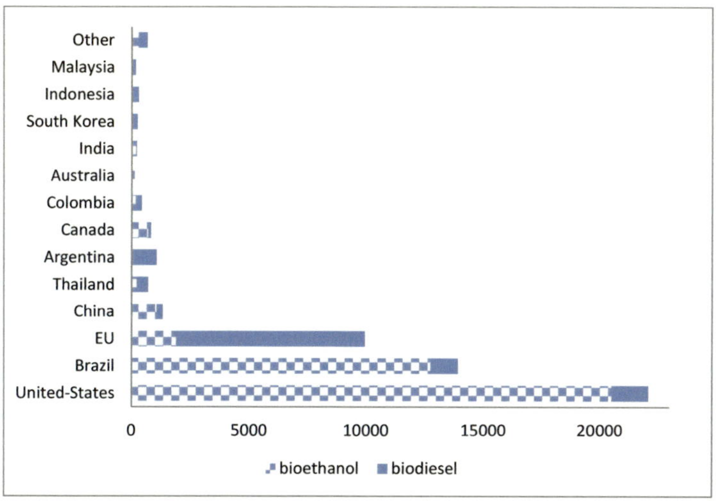

Fig. 2. Bioethanol and biodiesel world production in 2009, in toe; Source: INRA from the Biofuels Platform

In 2009, world biofuel production increased by 5.9 Mtoe (+13%), with 4.4 Mtoe for bioethanol (+13%) and 1.4 Mtoe for biodiesel (+12%). More than half of this overall growth came from the US (+2.7 Mtoe which corresponds to an increase of domestic production of +14%) and the EU (+1.6 Mtoe, +19%). Brazilian production increased more slowly, by 0.55 Mtoe (+4.1%). Production in the rest of the world increased sharply (+1.1 Mtoe, +22%) with very strong growth rates in Canada, China, Thailand, Colombia and South Korea. Among the main producing countries, only Indonesia experienced a decrease in production in 2009 with respect to 2008.

3.1.2 Evolution over the last twenty years

Biofuel production is influenced by both public policies aiming at encouraging their development and relative prices of fossil oil and agricultural products used for biofuel production. The importance of each factor varies according to countries, and, to some extent, the sub-periods considered over the last two decades.

World biofuel production started developing between 1975 and 1985, mainly in the form of bioethanol with Brazil as the main producer. It increased very slightly from 1985 to 2000 when it was equal to 9.4 Mtoe. Since then it has soared to 51.8 Mtoe in 2009. It is expected to reach 57.8 Mtoe in 2010. Biodiesel production was negligible until the beginning of this century (less than 7% of total biofuel production in oil equivalent in 2000). Since then its share has continuously increased to reach 26% in 2009. This share is expected to be 27% in 2010 (Figure 3).

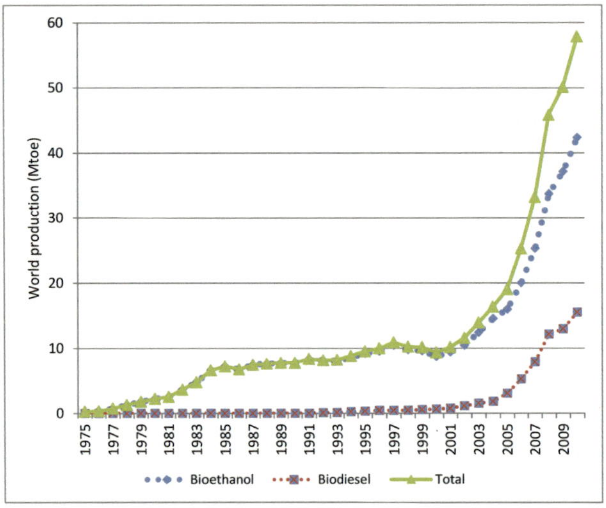

Fig. 3. World bioethanol and biodiesel production, 1975-2010, in Mtoe; Source: INRA estimations from various sources [4]

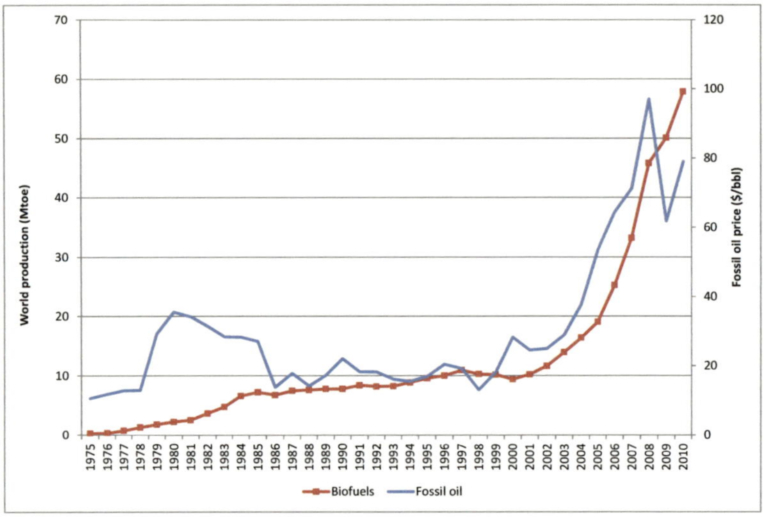

Fig. 4. Evolution of world biofuel production in Mtoe and fossil oil prices in $/bbl; Source: INRA estimations from various sources for biofuels (see footnote 4) and UNCTAD for fossil oil prices (mean Brent/Dubai/Texas)

[4] The main sources used for constructing Figures 3 to 12 are the Biofuels Platform, the European Biodiesel Board (EBB), the Earth Policy Institute, the FAO, the OECD, FO Licht, the International Energy Agency (IEA) and the Renewable Fuel Association (RFA).

As long as fossil oil prices remained close to US $20 per barrel (bbl), biofuel production stagnated around 10 Mtoe. When fossil oil prices started to soar as from 2000, biofuel production followed the same pattern. It is however noteworthy that the 2009 decline in fossil oil prices had no effect on the upward trend of biofuel production (Figure 4). This means that biofuel development worldwide cannot be explained by fossil oil prices only; other factors are playing, notably public policies.

Bioethanol

Brazil is the most ancient producer of bioethanol. In 1984, its production already reached 11.3 Bl. At that date, Brazil held 87% of the world bioethanol market, followed by far by the US but with 1.6 Bl only. In 2000, while Brazilian production had returned to 10.5 Bl after a peak at 15.4 Bl in 1997, the US production reached 6.2 Bl. Following the adoption of an ambitious mandate in 2004, the US production exceeded that of Brazil for the first time in 2005 (Figure 5).

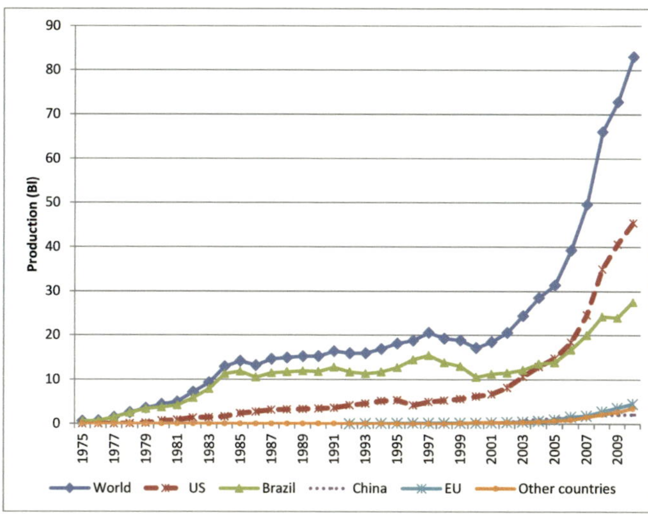

Fig. 5. Bioethanol production in Bl, 1975-2010, in the world and main producing countries; Source: INRA estimations from various sources

In the US, bioethanol production started in 1999 when fossil oil prices began to rise; since then, it has continued to do it regularly except in 2009 (Figure 6). As a result, it can be of some interest to consider in parallel three evolutions over the last decade, that of bioethanol production, that of fossil oil prices and that of corn prices (corn is the raw material used for bioethanol production in the US). More specifically, Figure 7 depicts the annual growth rate of bioethanol production and the evolution of the price ratio of petrol oil on corn. This ratio was low and stable over the 1990 decade. Since 2000, the two curves evolve in parallel with a time lag of two or three years which corresponds to the delay needed to build and start up new bioethanol production plants. The record progression of 10 Bl registered in 2008 despite a very high corn price (US $165/ton) can then be explained by the fact that two years before, the price of oil (US $404 for 1,000 l, i.e., 64$/bbl) was five times higher than the price of corn (US $79/ton). The corn price increase in 2008 and the resulting price ratio of 3.7 have been followed by lower increases in bioethanol tonnages in 2009 and 2010.

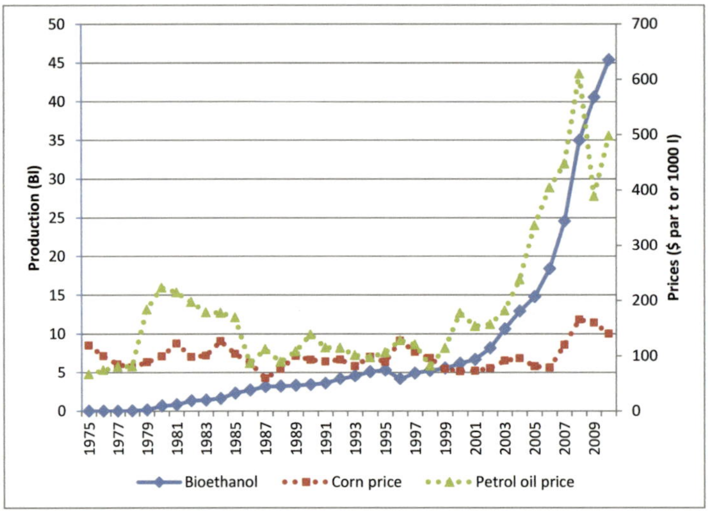

Fig. 6. Evolution of bioethanol production in the US in Bl, in parallel with fossil oil and corn prices; Source: INRA estimations from various sources for bioethanol, UNCTAD for fossil oil prices (in $ per 1,000 l) and USDA for corn prices paid to farmers (in $ per metric ton)

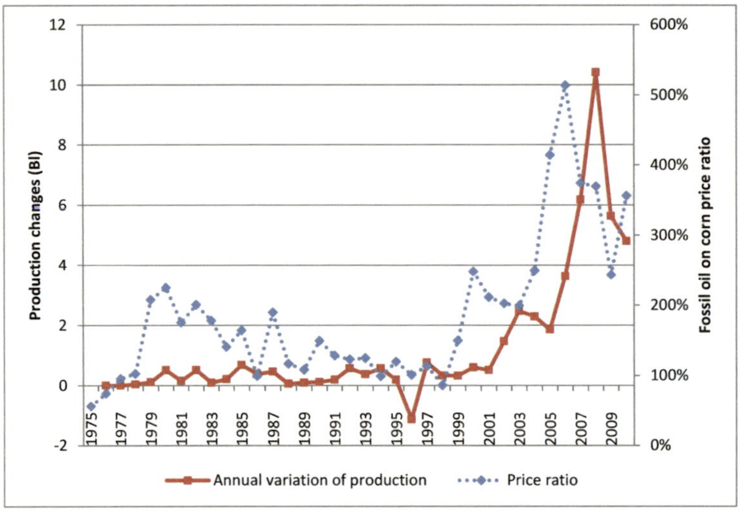

Fig. 7. Annual variation of bioethanol production in the US and parallel evolution of the petrol oil on corn price ratio; Source: INRA estimations from various sources for biofuel, UNCTAD for fossil oil prices (in $ per 1,000 l) and USDA for corn prices paid to farmers (in $ per metric ton)

In Brazil, bioethanol production increased sharply between 1975 and 1985, from 0.55 Bl to 11.8 Bl. This happened despite a very high price of sugar in 1980 and 1981 because of the simultaneous rise in fossil oil prices and also because of the biofuel development policy in place at that date (Pons 2007). Between 1985 and 2000, production increased more slowly because of a low fossil oil price and a relatively high sugar price for sugar. Ethanol production rose again in 2001 with both an increase in fossil oil price and a decrease in sugar price. It has continued to rise in the following years from that date. Production was multiplied by more than 2.5 between 2000 and 2010 (Figure 8).

As in the US where bioethanol production growth rates are influenced by the fossil oil on corn price ratio, Brazilian bioethanol production growth rates are related to the fossil oil on sugar price ratio (Figure 9). But contrary to what can be observed in the US, it appears that Brazilian bioethanol production annual changes precede those of the fossil oil on sugar price ratio by about one year. This can be explained by the dominant position of Brazil on the world sugar market: when an increased part of Brazilian sugar production is devoted to bioethanol production, sugar prices decrease the year after, and vice-versa.

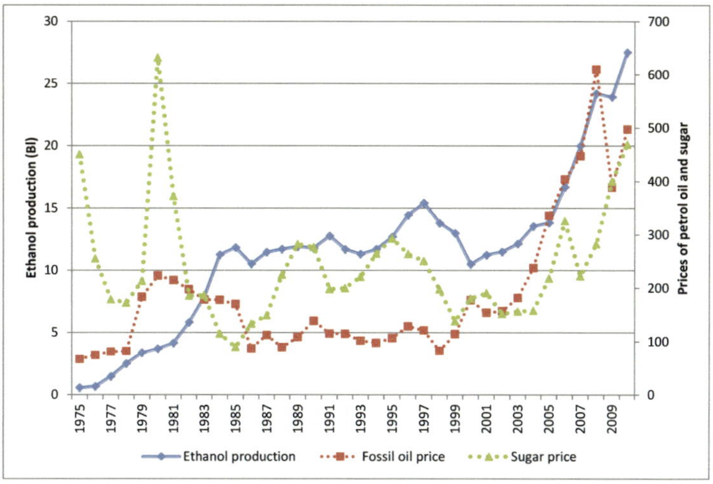

Fig. 8. Evolution of bioethanol production in Brazil in Bl, in parallel with fossil oil and sugar prices; Source: INRA estimations from various sources for biofuel, UNCTAD for fossil oil prices (in $ per 1,000 l) and sugar prices (mean of ATS Caribbean port prices in $ per 100 kg)

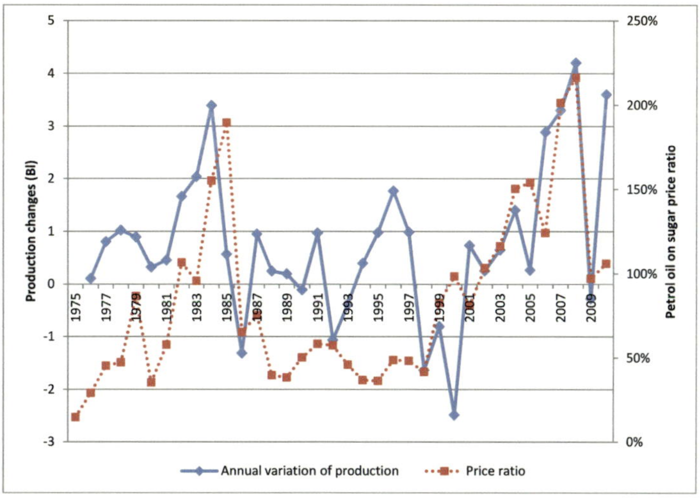

Fig. 9. Annual variations of bioethanol production in Brazil and evolution of the fossil oil on sugar price ratio; Source: INRA estimations from various sources for biofuel, UNCTAD for fossil oil prices (in $ per 1,000 l) and sugar prices (mean of ATS Caribbean port prices in $ per 100 kg)

Biodiesel

World production of biodiesel was equal to 0.8 Bl in 2000. It reached 4 Bl five years later and more than 16 Bl ten years later. Even if the EU is still the main producer with a market share around 55% today, several other countries did also record significant rises over the last five years: non-EU biodiesel production was equal to 0.16 Bl in 2004 and 7.7 Bl in 2009. Increases have been particularly marked in Argentina, Brazil and the US, a large part of production from these three countries being exported, notably towards the EU. Other bioethanol producers are Indonesia, Malaysia and Thailand. Because of the EU anti-dumping policy, US production and exports of biodiesel significantly decreased in 2009 and 2010 (Figure 10).

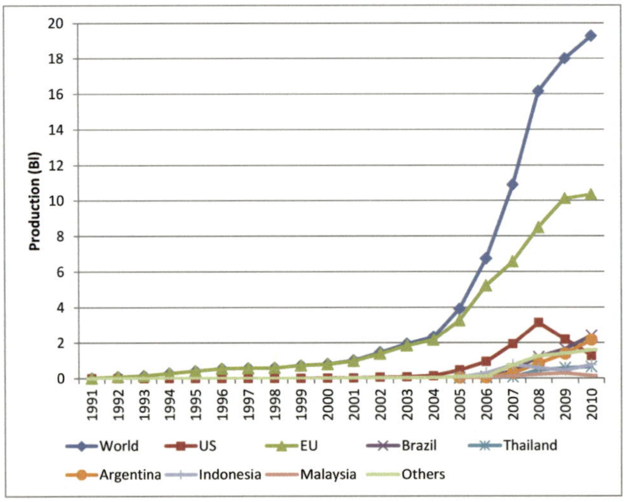

Fig. 10. Biodiesel production in Bl, 1991-2010, in the world and main producing countries; Source: INRA estimations from various sources

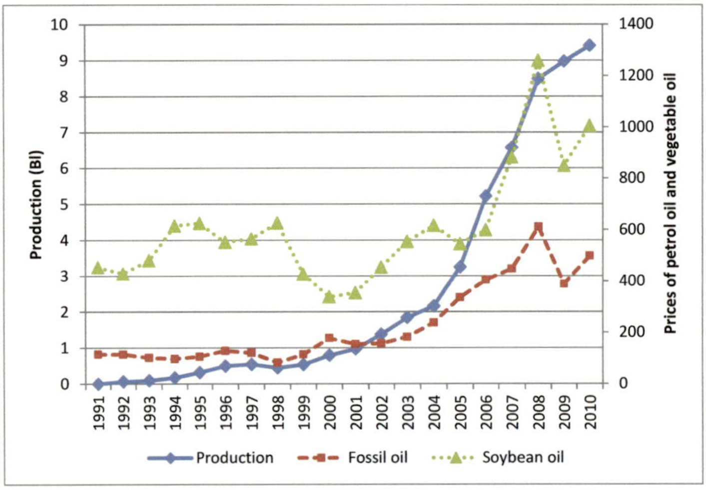

Fig. 11. Evolution of biodiesel production in the EU in Bl, in parallel with fossil oil and soybean oil prices; Sources: INRA estimations from various sources for biodiesel, UNCTAD for fossil oil prices (in $ per 1,000 l) and soybean oil prices (Dutch FOB ex Mill in $ per ton)

In the EU, biodiesel production started at the beginning of the 1990 decade (Figure 11). It increased at a very moderate rate during fifteen years to reach about 2 Bl in 2004. Annual growth rates were much higher during the 2005-2008 years. They were more modest in 2009 and 2010 (Figure 12). In 2010, EU biodiesel production was equal to 9.4 Bl.

EU biodiesel production is influenced by the domestic biofuel policy, the prices of fossil oil and vegetable oils and the fossil oil on vegetable oil price ratio[5]. From Figure 12, one sees that the high growth rates in EU bioethanol production observed from 2005 to 2008 corresponded to high fossil oil on vegetable oil price ratios. In 2009 and 2010, bioethanol production growth rates were lower although the price ratio remained at high levels: this can be explained by the fact that vegetable oil price levels were also high (Figure 11).

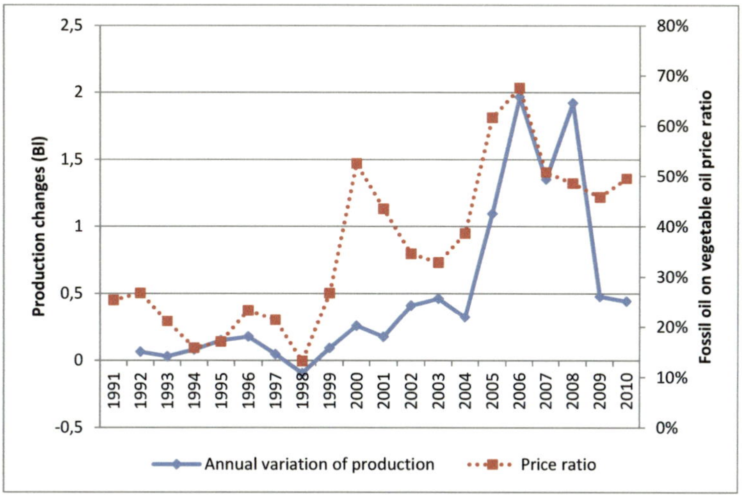

Fig. 12. Annual variations of biodiesel production in the EU and evolution of the fossil oil on soybean oil price ratio; Source: INRA estimations from various sources for biodiesel, UNCTAD for fossil oil prices (in $ per 1,000 l) and soybean oil prices (in $ per ton)

3.2 The role of first-generation biofuels in the 2007-08 agricultural price peak

When lower amounts of agricultural products are available under the effect of, for example, adverse weather conditions, this has a positive impact on prices. In the same way, any increase in demand has a positive effect on prices. As a result, we can state that the increase in 1G biofuel production is partly responsible for the soaring of cereal prices in the 2007-2008 period due to the expansion of bioethanol production in the US, mainly from corn, and for the soaring of vegetable oil prices due to the expansion of biodiesel production in the EU, mainly from rapeseed.

The story is far from straightforward however. It is essential to place the development of biofuels in perspective with all other supply and demand factors that influenced agricultural prices in 2007-2008. It is also important to follow the chronology of events and to differentiate between products by taking into account substitution and complementary effects between commodities, on both the supply and demand side.

---

[5] EU biodiesel production is essentially made from rapeseed oil. The price of this vegetable oil and the price of soybean oil are highly correlated. As a result, it is possible to use the soybean oil price as the "reference" price for all vegetable oils.

It would be dangerous to consider separately the corn-bioethanol situation in the US and the rapeseed-biodiesel situation in the EU for at least two reasons. Firstly, because the US have also developed a domestic biodiesel production and the EU a local production of bioethanol. Secondly, and more importantly, because the expansion of bioethanol-devoted corn crops in the US, particularly during the 2007/08 crop year, was carried out at the expense of domestic surfaces in soybeans. This in turn had an impact on oilseed prices and, because of substitution and complementary relationships between products, on the prices of all cereals and vegetable oils. Similarly, the development of EU biodiesel production did have an impact not only on rapeseed oil prices, but more generally on prices of all vegetable oils.
The global context of weak agricultural supply since the early 1990s and increased world food demand, notably in large emerging countries, led to decreases in cereal and oilseed stocks. The petrol oil price was rising since 2000 and the US dollar was depreciating with respect to a growing number of currencies. This was combined with increasing biofuel production, mainly in the US in the form of bioethanol (Figure 13). However, the use of stocks and the record harvest in 2004 contained cereal and oilseed price rises in 2003, 2004 and 2005; their international prices even slightly decreased in 2005.

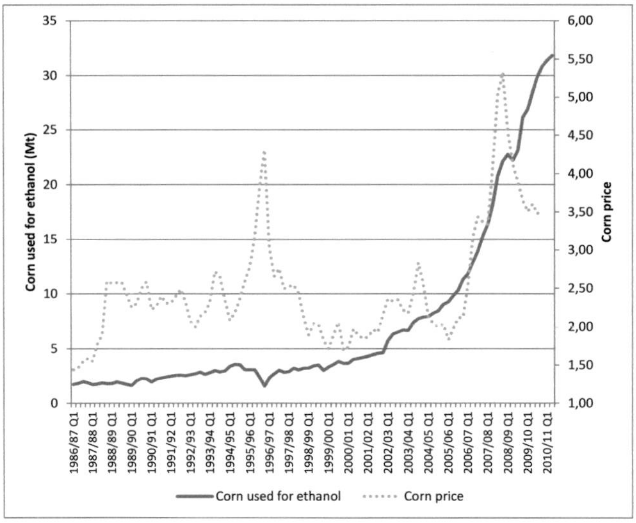

Fig. 13. Corn volumes used for bioethanol production in the US and corn prices paid to US farmers (quarterly data 1986/1987 to 2010/2011); Sources: INRA from USDA

In early 2004, the increase in fossil oil prices accelerated and the US dollar depreciated further. Cereal and oilseed stocks that had grown slightly in 2004 diminished again in 2005 and 2006, when, due to unfavorable weather conditions, world cereal production slightly decreased. From 2005, with low corn stocks and insufficiently dynamic domestic corn production, the US were unable to meet their bioethanol mandate from local feedstocks. They looked abroad, especially to Brazil. As a consequence, the international price of sugar rocketed, from 128.4 in June 2005[6] to 189.3 in December 2005 (+47% in six months), and to 254.3 in February 2006 (+34% in two months). In response to high sugar and ethanol prices, surfaces dedicated to sugar cane increased in Brazil. At the same time, US bioethanol

[6] According to the FAO monthly food price index (100 in 2002-2004).

producers gradually returned to the use of corn. As a result, the international price of sugar dropped from March 2006 and throughout 2007.

In a context of world stocks at their lowest level, the development of biofuels in the US and the EU played upward on the prices of cereals and vegetable oils as from the second semester of 2006.

In practice, the US sought to encourage bioethanol production from domestic plant resources, that is corn. Figure 14 depicts the growth of corn use for bioethanol production in the US over the period 2001/02 to 2010/11. In 2003/04, corn used for ethanol production was 30 Mt (11.1% of total US corn production). In 2006/07, the same use was 54 Mt (20.1%). In 2007/08, it reached 77.5 Mt (23.4%) and 8 Mha of corn of a total of 35 were used for the production of bioethanol in the US. These statistics suggest that the development of bioethanol production in the US is largely responsible for the upswing in the prices of corn in that country. As the US is, by far, the largest exporter of corn, the upward movement of US corn prices rapidly spread to world corn prices.

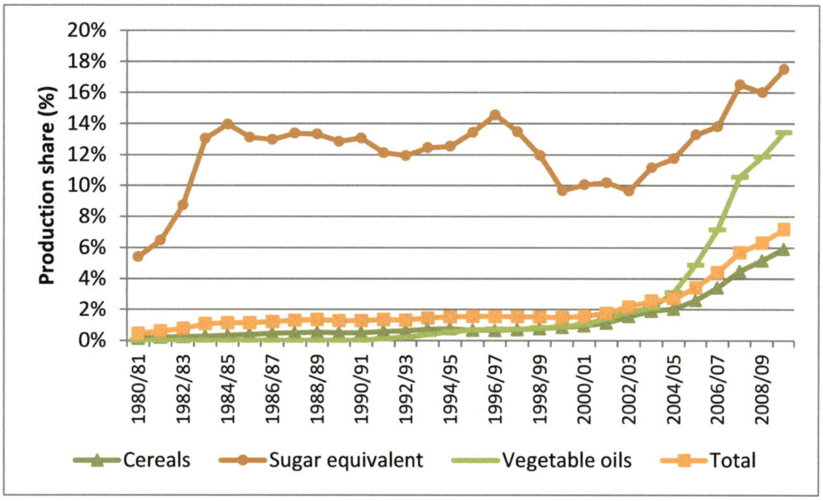

Fig. 14. Corn production in the US (Mt), corn used for bioethanol (Mt) and share of corn used for bioethanol (%); Source: INRA from USDA

The first impact of bioethanol production increase in the US was to raise the world price of sugar from June 2005 to February 2006. The second impact was to increase the international price of corn from spring 2006 and, incidentally, to cause the downward adjustment of the world price of sugar because, in a way, of an "excess supply" of sugar. The increase in the world price of corn peaked during the 2006/07 crop year. The high price of corn compared to other cereals and oilseeds prompted US and world producers to increase corn acreage, especially during the 2007/08 crop year. The world surface of corn, which decreased by 3% over the period 1996/97 to 2002/03, increased by 14% from 2002/03 to 2007/08; half of that increase took place during the 2007/08 campaign only, mainly at the expense of oilseeds (Abbott *et al.* 2008). Undoubtedly this decline in global oilseed surfaces in 2007/08, primarily in the US, contributed to exert additional upward pressure on prices of oilseeds and vegetable oils. Between June 2007 and June 2008, the FAO price index of vegetable oils rose by nearly 80%. By comparison, over the same period, world corn prices increased by slightly more than 20%, rice prices by nearly 50% and wheat prices by a little more than 90% (Figure 15).

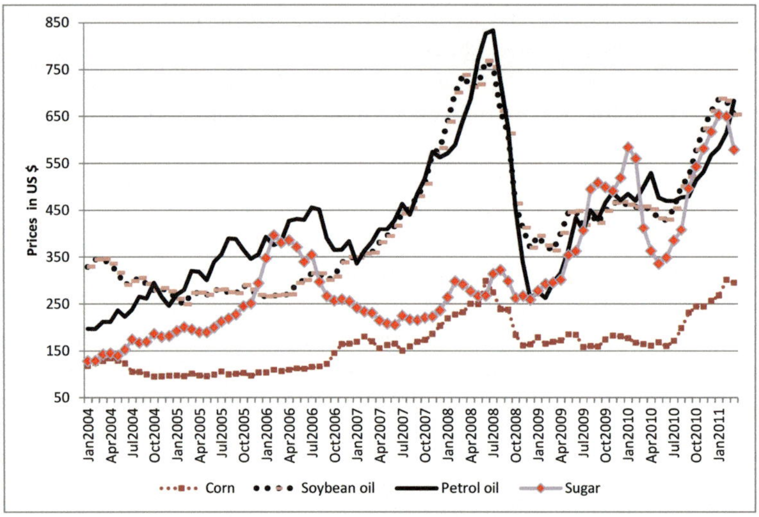

Fig. 15. Evolution of monthly prices for corn, soybean oil, sugar and fossil oil; Source: INRA from UNCTAD data; prices in US $/ton for corn, US $/500 kg for soybean oil, US $/10 kg for sugar and $/1,000 liters for fossil oil

The third impact of bioethanol development in the US was thus to increase the 2007/08 world corn area, including in the US, at the expense of the world sole in soybean in response to high corn prices and also in anticipation that demand for corn would remain firm in the years to come. This substitution effect on the supply side helped increasing the upward pressure on prices of soybeans and soybean oil, and by extension on prices of other vegetable oils that are highly substitutable on the demand side. This contagion effect on the prices of various cereals and vegetable oils was particularly strong because it took place in a context of medium- and short-term factors that played simultaneously in the direction of a general increase in agricultural prices (weak agricultural supply, disappointing harvests, dynamic food demand, low stocks, high fossil oil prices, weak US dollar, speculation on agricultural commodities, uncoordinated trade policies aiming at discouraging exports and encouraging imports of agricultural goods, etc.). In the specific case of vegetable oils, the concomitant development of the EU biodiesel production constituted an additional factor of upward pressure on prices, especially on the price of rapeseed oil.

Several studies have attempted to quantify the role played by biofuels in the 2007-08 increase in agricultural prices. Differences in terms of methodologies and models, data, product and country coverage, scenario definition or simulation horizons make it difficult to compare results. Nevertheless, according to these studies, it appears that the development of bioethanol production in the US can be held highly responsible for the rise in corn prices in the first place and then, through the play of substitutions in supply and demand, for the rise in the prices of the various cereals and vegetable oils (including, according to the IFPRI study based on the IMPACT model, the world price of rice even though rice is not used to produce bioethanol). Based on a survey of several studies and additional ad-hoc analyses, Collins (2008) concludes that 25 to 50% of the increase in the corn price in the US and worldwide between the 2006/07 and 2008/09 crop years can be attributed to the growth of the US bioethanol market.

4. Cereals, oilseeds and sugar: production and uses

4.1 The 2009/10 situation

In 2009/10, cereals, oilseeds and sugar crops were cultivated on 954 Mha worldwide (Figure 16). They produced about 2.9 Bt of primary products: 2.2 Bt of cereals (76%), 488 Mt of oilseeds (17%) and almost 200 Mt of sugar equivalent (7%). The bulk of oilseeds produced 229 Mt of oil cakes and 137 Mt of oil, palm oil included. The rest of oilseeds were used directly for food (60 Mt), feed (14 Mt) or seeds (11 Mt).

Food consumption is by far the largest use with more than 1.3 Bt of cereals, vegetable oils, oilseeds and sugar. These global amounts correspond to average consumptions per capita of 200 kilograms. Feed consumption arrives at the second rank with more than 1Bt (cereals, oil cakes, co-products of cereals and oilseeds, and others). Considering a production of 346 Mt of beef, pork and poultry meat and eggs, this feed use corresponds to an average consumption of 3.1 kilograms of annual crops per kilogram of animal products. The third

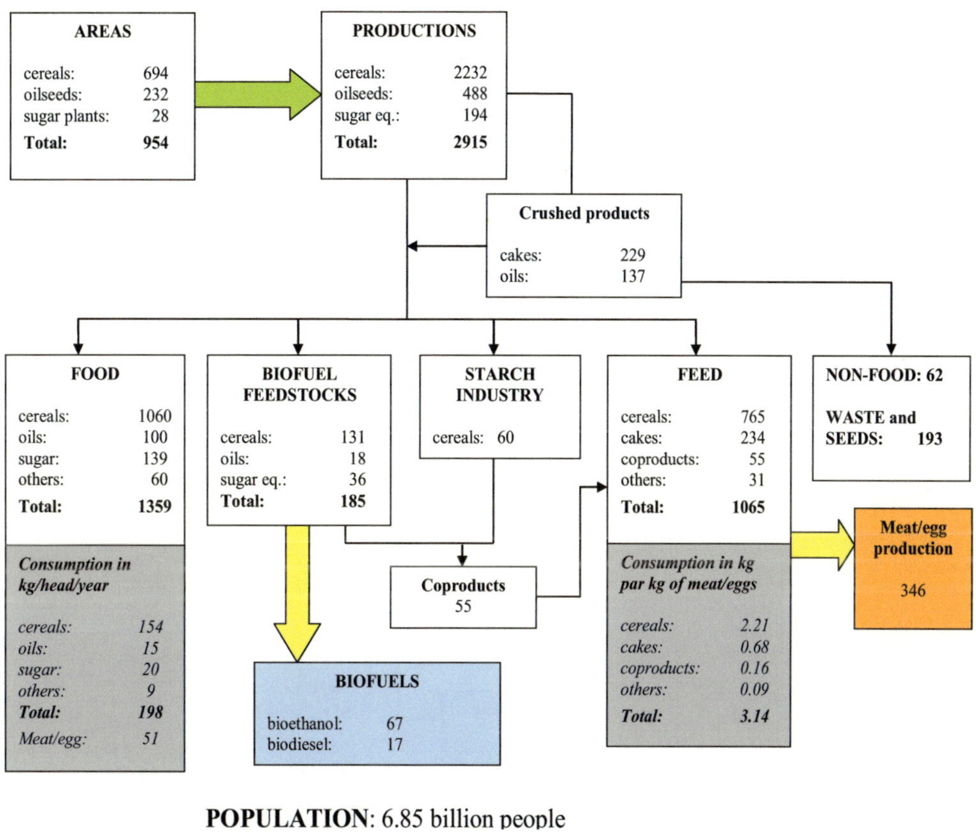

Fig. 16. Food, feed, fuel and other uses of cereals, oilseeds and sugar crops worldwide in 2009/10 (Mt). Source: INRA from various sources, notably the USDA/PSD database and FAO statistics[7]

[7] Concerning "crushed products", palm oil production is included. For the "starch industry", cereal use is estimated from data on corn-gluten-feed production (Oil World 2010) and bioethanol production.

use is biofuel with approximately 185 Mt of cereals, vegetable oils and sugar. Biofuels represent 6% of the world production of cereals, oilseeds and sugar, and amounts to 60 Mha of agricultural land. Other uses include starch production, oil for soap production, lubrication, paint and varnish, lipid chemistry, etc. Ethanol production from cereals and starch produce 55 Mt of co-products used for animal feed under the form of, for example, corn gluten feed (CGF) or dried distillers grains and solubles (DDGS). Co-products from biodiesel production amount to 10 Mt; they are included in the category 'total oil meal".

4.2 Changes between 1996/97 and 2009/10

When comparing the supply-demand balance for crops in 2009/10 with the same figure thirteen years earlier, in 1996/97, one notes first that the surfaces in cereals, oilseeds and sugar plants have increased by 5.3% (+48 Mha) while world population has increased by nearly 19% (+1 billion people). The area devoted to cereals has slightly decreased, the area of sugar plants is remained practically constant and the area in oilseeds has increased by more than 55 Mha. Despite area contraction, world cereal production has increased by 19% thanks to improvements in yields (+21%). World production of oilseeds has increased much more importantly (+75%) thanks to area expansion (32%) and improvements in yields (+32%). This increase in oilseed production is matched by a corresponding increase in vegetable oils, oil meals and oil cakes used for food, feed and fuel (Figure 17).

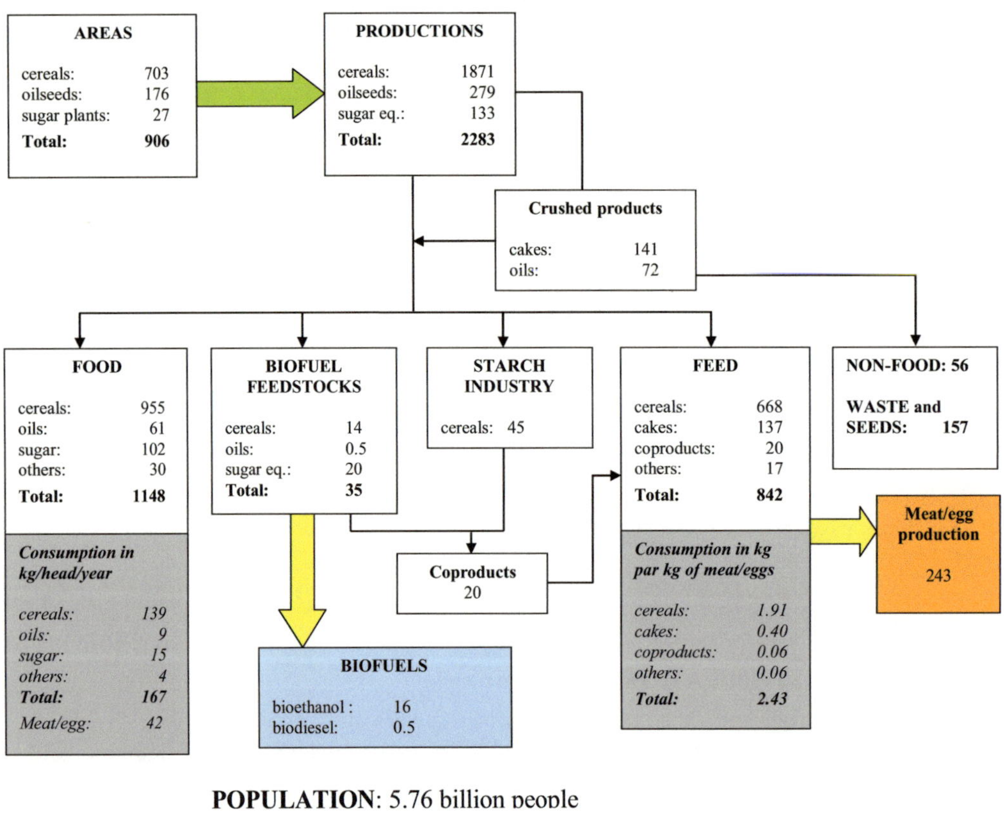

Fig. 17. Food, feed, fuel and other uses of cereals, oilseeds and sugar crops worldwide in 1996/97 (Mt). Source: INRA from various sources, notably the USDA/PSD database and FAO statistics

World production of cereals has increased by 361 Mt (+19%) between 1996/97 and 2009/10, out of which 115 Mt (32%) have been used for biofuels, 100 Mt (28%) for food, 100 Mt (28%) for feed and 46 Mt for other uses: it appears thus that the first outlet of additional cereals has been biofuel use. World production of vegetable oils has increased by 65 Mt (+90%), out of which 39 Mt (60%) have been used for food, 17 Mt (26%) for fuel and 9 Mt (14%) for other uses: by contrast with cereals, the first outlet of additional vegetable oils has been food uses. As a result, while cereal consumption per capita has increased by only 11%, from 139 to 154 kilograms, that of vegetable oils has increased by 66%, from 9 to 15 kilograms. World production of sugar has increased by 61 Mt (+46%), out of which 37 Mt for food and 16 Mt for fuel. Sugar consumption per capita has increased by 33%, from 15 to 20 kilograms. Coupled with significant increases in individual consumption of meat and eggs (from 42 to 51 kilograms/head/year), these figures show that average individual food consumptions have increased significantly over the thirteen-year period 1996/97 to 2009/10. They also show that increases have been heterogeneous between products, much more important for vegetable oils and sugar than for meat and eggs, and cereals. This means that biofuel production development over the period, from 17 Mt to 84 Mt, has had an impact of food security here defined in terms of cereals, oilseeds and sugar available for food consumption: this impact has been more pronounced for cereals than for oilseeds and sugar.

4.3 Areas, yields and production

Figures 18 and 19 display the evolution of surfaces devoted to the set of cereals, oilseeds and sugar plants from 1980/81 to 2010/11. One can distinguish two main sub-periods. From 1981 to 2003, upward movements phases have been followed by downward movements so that the area in 2003 is practically equal to that of 1981, about 880 Mha. One also notes that annual changes can be significant, +17 Mha in 1997 or -15 Mha in 2003. From 2004, the trend is clearly increasing: in eight years, the area of cereals, oilseeds and sugar plants has increased by nearly 80 Mha (+9%). The area increase was particularly important in 2004 (+27 Mha) partly in compensation for the decrease in 2003. The increase was also important in 2008 (+17 Ma) and 2009 (+14 Mha) in response to the 2007-08 agricultural price peak. The increase was negligible in 2009 in reaction to the 2008-2009 agricultural price decline and the 2008 financial crisis.

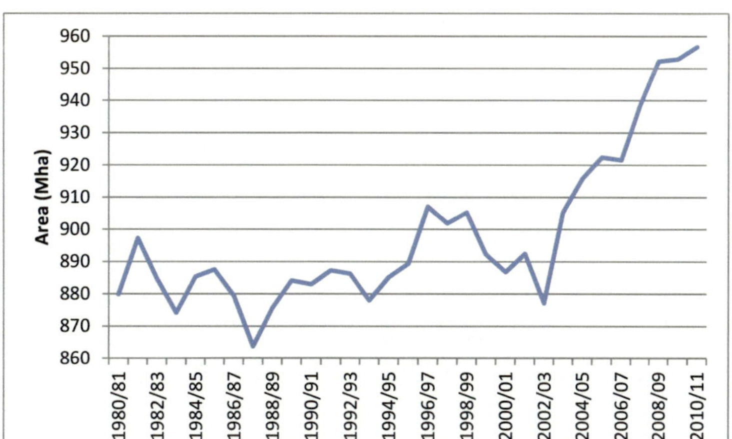

Fig. 18. World areas in cereals, oilseeds and sugar plants, 1980/81 to 2010/11, in Mha; Source: INRA from USDA/PSD and FAOSTAT data

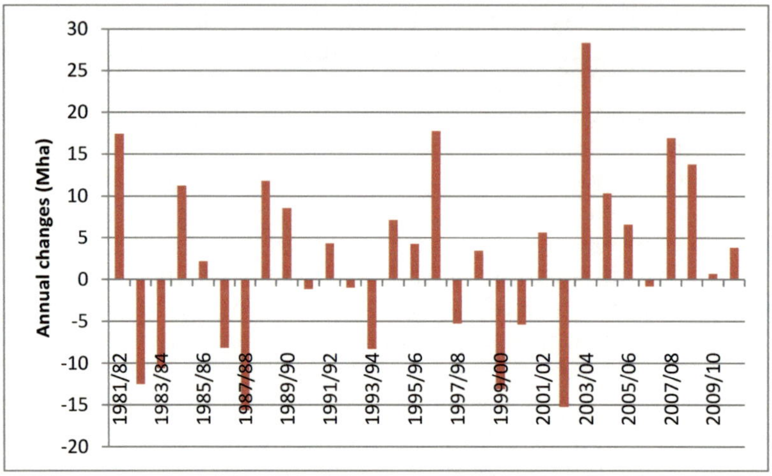

Fig. 19. Annual changes in the world area in cereals, oilseeds and sugar plants, 1981/82 to 2010/11, in Mha; Source: INRA from USDA/PSD and FAOSTAT data

While Figure 20 below depicts the evolution of areas, yields and production in cereals, oilseeds and sugar crops considered as one aggregate, the following figures display the same information for cereals (Figure 21), oilseeds (Figure 22) and sugar crops (Figure 23). From Figure 20, one notes that production and yields in cereals, oilseeds and sugar plants considered as one single aggregate have increased at the same rate than world population from 1981 to 2003. During this first sub-period, the contribution of area evolution to production growth was negligible; production increases were essentially the result of increases in yields. During the second sub-period, from 2004 to 2011, increases in yields and area expansion play together so that the production growth has outweighed that of world population. Finally, it is worthwhile noting the slowing down of growth in yields at the end of the period, from 2009.

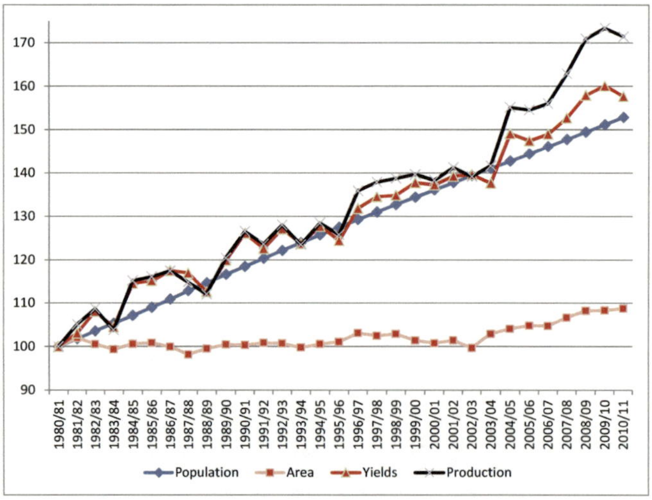

Fig. 20. World population and world areas, yields and production in cereals (including rice), oilseeds and sugar plants (in sugar equivalent), 1980/81 to 2010/11, base 100 in 1980/81; Source: INRA from USDA/PSD and FAOSTAT data

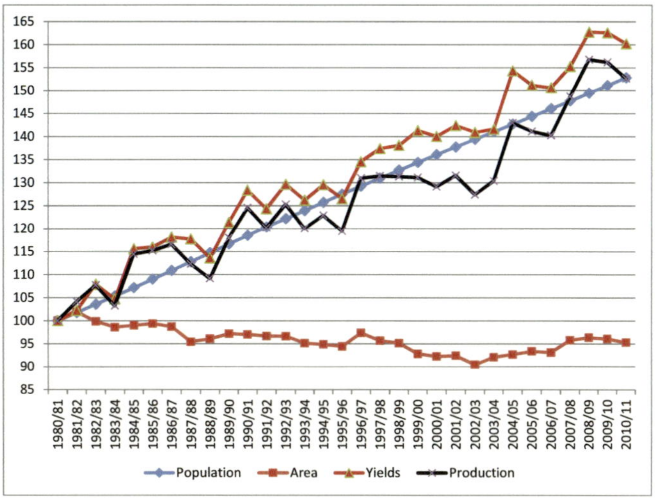

Fig. 21. World population and world areas, yields and production in cereals (including rice), 1980/81 to 2010/11, base 100 in 1980/81; Source: INRA from USDA/PSD and FAOSTAT data

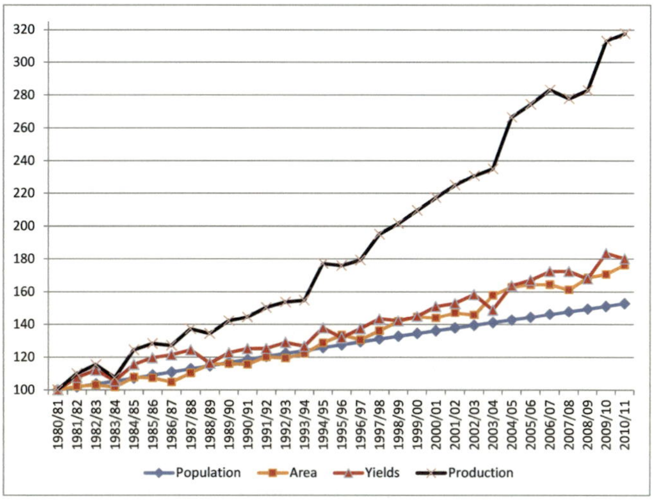

Fig. 22. World population and world areas, yields and production in oilseeds, 1980/81 to 2010/11, base 100 in 1980/81; Source: INRA from USDA/PSD and FAOSTAT data

This global picture masks huge differences between cereals on the one hand, oilseeds and sugar crops on the other hand. From Figure 21, it appears that cereal production growth was lower than that of world population from 1997 to 2008 because of the area contraction since yield growth rates were greater than those of population during this sub-period. The area devoted to cereals increased in 2008 and 2009 in response to the 2007-08 agricultural price peak. It decreased again in 2010 and 2011. One also notes the decrease in cereal yields in the very recent years. This contrasts with the case of oilseeds. As Figure 22 shows, for oilseeds, production growth exceeds that of world population from 1990 onward thanks to both improvements in yields and area expansion: for the two factors, growth rates exceeded that of population. In the same way, world sugar production is increasing at a much greater rate than world population again thanks to both improvements in yields and area expansion even if, by contrast with oilseeds, each determinant raises more slowly than population (except yields in the more recent years).

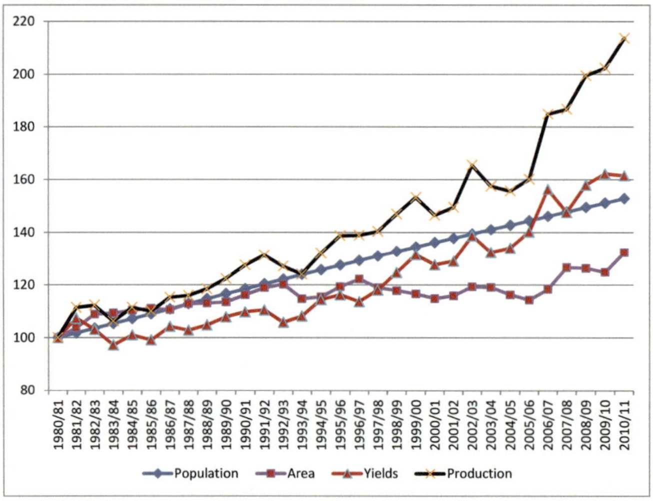

Fig. 23. World population and world areas, yields and production in sugar crops, 1980/81 to 2010/11, base 100 in 1980/81; Source: INRA from USDA/PSD and FAOSTAT data

4.4 Uses for biofuel production

The three main feedstocks used for 1G biofuels are cereals, oilseeds and sugar plants. Quantities used for biofuel were very low up to 2000/01. They have considerably increased over the last decade, from around 40 Mt in 2002/03 to more than 180 Mt in 2009/10 (Figure 24). By contrast with cereals and oilseeds, the share of world production of sugar used for biofuel was already significant in the 1990s, around 14%; after a decrease to 10% at the end of the nineties, it has increased over the 2000 decade to reach around 18% in 2009/10 (Figure 25). The share of cereals and vegetable oils used for biofuel were negligible until the years 2000 (around 1%). They have considerably increased from that date, first for vegetable oils (more than 13% in 2009/10), to a lesser extent for cereals (around 6% in 2009/10) [8].

5. Biofuel outlook by 2020 and 2050: to what extent second-generation biofuels could alleviate some of the negative consequences of first-generation biofuels?

The 2007-2008 peak of world agricultural prices was the result of a large combination of factors including 1G biofuels (see sub-section 3.2). In the structural context of strong food demand, weak agricultural supply and decreasing inventories over more than a decade,

[8] Data on cereals, oilseeds and sugar plants used for biofuel production are very partial. They are estimated using technical coefficients from data of biofuel production expressed in tons. They vary in function of feedstocks used. For bioethanol in Brazil, we have considered that 16 tons of sugar cane containing 10% of sugar are needed to produce 1 ton of ethanol; the technical coefficient is thus 1.6 to compare with a mean coefficient of 3.2 for bioethanol in other countries (Canada, China, EU, United States…). For biodiesel, we have estimated that 1.02 ton of vegetable oil is needed to produce 1 ton of biofuel in all countries.

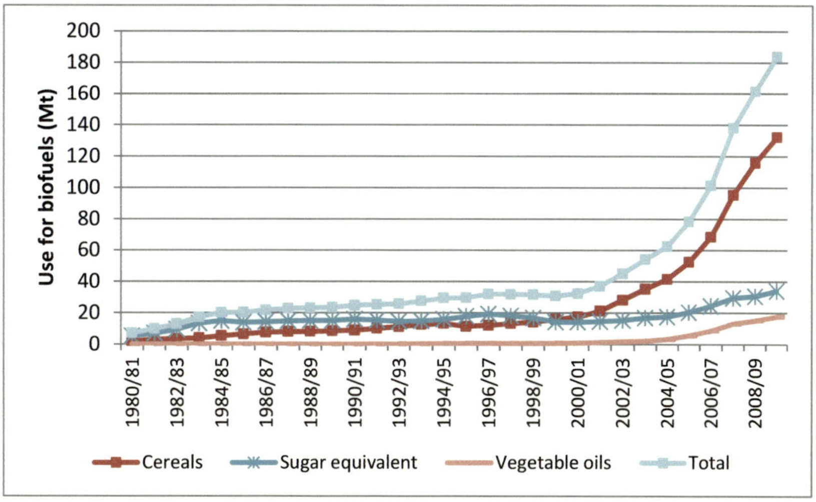

Fig. 24. Quantities of cereals, vegetable oils and sugar crops (in sugar equivalent) used for biofuel production, 1980/81 to 2009/10, in Mt; Source: INRA estimations from various sources

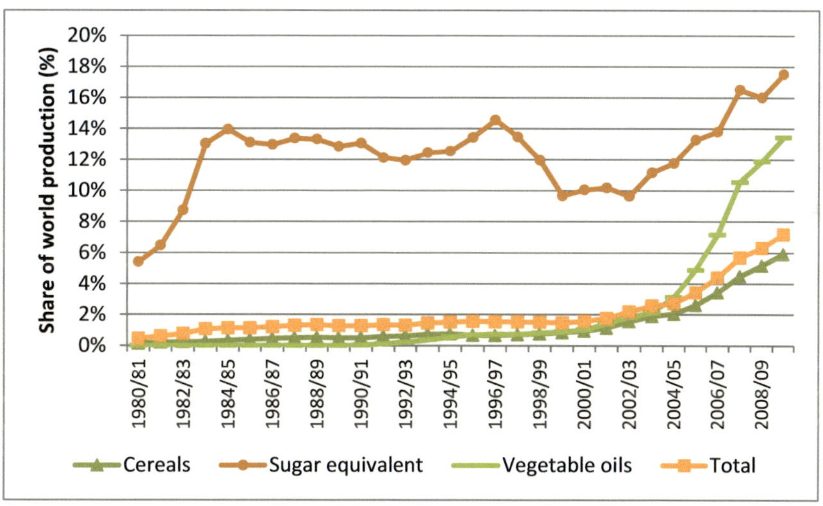

Fig. 25. Shares of world production of cereals, vegetable oils and sugar plants (in sugar equivalent) used for biofuel production, 1980/81 to 2009/10; Source: INRA estimations from various sources

some cyclical factors played negatively on quantities available for world markets (climatic accidents and restrictive export policies), others played positively on demand (biofuel production development, speculation, import encouragement policies and US dollar depreciation), and others played positively on agricultural production costs (fossil oil price increase). The very sharp correction that occurred from mid-2008 can then be explained by the fact that several factors that had pushed agricultural prices upwards in 2007-2008 returned: good weather conditions, US dollar appreciation, fossil oil price decrease, speculator exit from agricultural commodity markets because of the financial crisis. In

addition, the high agricultural prices of 2006 and 2007 encouraged producers to increase their acreage in field crops at the expense of fallow land or land previously devoted to forest or grassland, in Latin America, including Brazil, but also in Southeast Asia, sub-Saharan Africa, the United States and Europe. They also led farmers to seek the highest yields. In such a context of temporarily renewed dynamism in agricultural supply, the occurring downward adjustment of agricultural prices allowed stocks to (slightly) increase and countries to stop their exceptional import encouragement and export restriction policies. As the story is never ending, it is because several factors again return and exert a positive influence on prices that they are on an upward trend from the second part of 2010: bad weather conditions notably in Russia and Australia, US dollar depreciation, fossil oil price increase, unilateral restrictive export policies and speculation on agricultural commodities.
What was true yesterday will be true tomorrow. As a result, the issue of competition between food and non-food uses of agricultural products, and the induced questions of impacts of this competition on agricultural prices, food security or the environment, cannot be analyzed independently of the economic context. More specifically, the development of biofuels and uses of biomass for biofuel production cannot be analyzed solely in terms of growth perspectives and theoretical availability of land resources. The economic dimension must be taken into account insofar as non-food uses of biomass are ultimately determined by prices and incomes, which are directly influenced by policies, scientific progress and technical innovations, etc. The importance of the economic dimension can be illustrated by relating the increase in the world utilized agricultural area (UAA) in 2007 and 2008 in response to high agricultural prices (about +15 Mha each year) to the growth of the same area during the past thirty years (+2.5 Mha per year between 1976 and 2006).
First-generation biofuels are criticized on three main grounds: (i) their low energy efficiency in the current production conditions, (ii) the economic cost of policies put in place for promoting their development, and (iii) increasing doubts about their environmental impact, particularly in terms of GHG emissions when they involve land use changes (see sub-section 2.2.1). It is now suitable to examine the extent to which 2G biofuels can silence such criticisms, or at least dampen them. Let us first examine this question by 2020 on the basis of original simulations performed using the OLEOSIM model (Dronne *et al.* 2009). Before that, lest us remind the following point already outlined in sub-section 2.2.2. If the raw material used for 2G biofuel production is a residue and/or a waste, the question of competition with food uses does not arise. But when it is a forest resource and/or a dedicated energy crop, competition with food uses of land occurs under the same theoretical terms as for 1G biofuels: are then at stake the questions of land availability, including marginal land areas, productivity of forest and dedicated crops and efficiency of transforming this biomass into 2G biofuels.

5.1 Analysis in 2020

Dronne *et al.* (2009) analyze the impacts of biofuel development by the year 2020 for cereals, oilseeds and sugar crops. Three scenarios are considered. In the "baseline" scenario, world biofuel production is held constant at its 2006 reference level, that is 37.3 Mt (31.4 Mt of bioethanol and 5.9 Mt of biodiesel) or 25 Mtoe. In the second scenario called "1G", the political objectives of incorporation for the year 2020 are attained and satisfied by 1G biofuels only, at a level of 175 Mt (125.1 Mt of bioethanol and 49.8 Mt of biodiesel) or 143 Mtoe. In the third scenario called "X% 2G", the development of biofuels reaches the same level as in the "1G" scenario, but with varying shares of 2G biofuels replacing 1G biofuels: different blend rates and energy efficiency yields are considered defining seven "X% 2G" scenarios.

In the "baseline" scenario, the increase in cropland (+41 Mha, from 914 Mha in 2006 to 955 Mha in 2020) mainly favors cereal production despite a strong demand increase for oilseeds (oil and meals) for both food and feed. In the "1G" scenario, the arable land dedicated to the three considered crops is 960 Mha, that is +46 Mha relative to 2006 but only +5 Mha relative to the situation in the "baseline" scenario in 2020. On these 960 Mha, 68 Mha (7%) are used for biofuel production: 33 Mha of cereals (4% of world area in cereals), 27 Mha of oilseeds (13% of world area in oilseeds) and 8 Mha of sugar crops (28% of world area in sugar plants). In practice, the development of 1G biofuels leads to a sharp decline in cropland dedicated to other uses than biofuel production, including those dedicated to food: these are 893 Mha in 2020 which is 46 Mha less than in the "baseline" scenario without biofuel development. This means that the development of 1G biofuels by 2020 as simulated by Dronne *et al.* would have a doubly negative effect on world food security as it would result in a decrease in cropland surfaces dedicated to food (and feed) and an induced increase in crop prices, as compared to the situation in 2020 without biofuel development.

To what extent could the replacement of the first generation of biofuels by the second one change the picture briefly described above? In the seven "X% 2G" scenarios, 2G biofuels are gradually introduced from 2015 onward under different assumptions regarding blend rates in fossil fuel and energy efficiency yields of dedicated bioenergy crops. The first lesson is that the (partial) substitution of the first generation by the second should make it possible to limit the negative impacts of biofuel development on world food security. Crop areas devoted to food and feed are more important in the seven "X% 2G" scenarios when compared to the "1G" scenario in which only the 1G biofuels are available. The gain reaches 25 Mha in the most favorable "X% 2G" scenario in which 35% of 1G biofuels are replaced by 2G biofuels in 2020[9] and the yields of dedicated bioenergy crops are rather high (25 tons of dry matter per hectare). However the gain is limited to 10 Mha in a variant with a replacement rate of 1G biofuels by 2G biofuels of 20% at world level in 2020[10] and rather conservative assumptions regarding yields for 2G dedicated crops (12 tons of dry matter per hectare).

However, even in the most favorable scenario considered by Dronne *et al.*, the hectares devoted to cereals, oilseeds and sugar crops for food and feed uses in 2020 (910 Mha) are significantly lower than the corresponding surfaces at the same time horizon in the "baseline" scenario without biofuel development (939 Mha). This means that the development of 2G biofuels can only mitigate the negative impacts of biofuels on world food security, defined here in terms of crop surfaces for food and feed as well as world crop prices; it cannot eliminate them. Furthermore, the decline in the production of co-products associated with 1G biofuels penalizes the animal sector insofar as the increased availability in cereals and oilseeds induced by the replacement of 1G biofuels by 2G biofuels does not compensate for the more important decline in the availability of meals and co-products associated with the production of 1G biofuels.

---

[9] 35% is the 2020 world percentage for both bioethanol and biodiesel. In the EU, the replacement rate of 1G biofuels by 2G biofuels is much higher, that is 70% for both bioethanol and biodiesel. We acknowledge that the EU figure is unrealistic.

[10] The replacement rate is 25% in the EU.

5.2 Analysis in 2050

5.2.1 Estimation of the potential demand for biomass

The International Energy Agency (IEA) has attempted to estimate total primary energy demand by 2050 according to various effort levels aiming at reducing GHG emissions (IEA 2008). This demand ranges from 640 Exajoules (EJ), that is 153 Btoe, in the most constrained scenario ("Blue Map 2050") to 950 EJ, that is 227 Btoe, in the "business as usual" scenario ("Baseline 2050"). The "Blue Map 2050" scenario corresponds to a limitation of the atmospheric concentration in CO_2 at 450 parts per million (ppm) in 2050 which requires a halving of 2005 GHG emissions (and a 77% cut compared to the situation in 2050 in the "Baseline 2050" scenario). In this "Blue Map 2050" scenario, the total demand from energy-devoted biomass is 147 EJ; this corresponds to 23% of the total primary energy demand in 2050 and is three times the amount of biomass used for energy production in 2006. In the two other scenarios, the intermediate "Act Map 2050" scenario and the "Baseline 2050"scenario, the total demand from energy-devoted biomass also increases sharply compared to 2006, respectively +68 and +35 EJ (Figure 26).

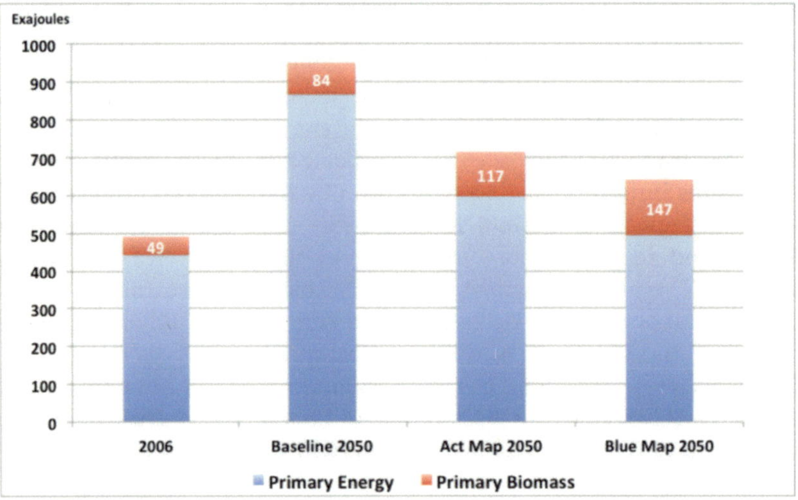

Fig. 26. Demand for primary energy and biomass used for energy production in 2050 in the three IEA scenarios, in EJ; Source: INRA estimations from IEA (2008)

In the "Blue Map 2050" scenario, 52% of reductions in GHG emissions in the transport sector are expected to be obtained through improved efficiency in the use of fuels, and 48% by replacing fossil fuels by alternatives including electricity, hydrogen and biofuels. Biofuels are expected to attain 30 EJ in 2050 (0.7 Btoe), that is 26% of the total fuel use at that date. To achieve this very ambitious result, the IEA assumes that 1G biofuels made from cereals and oilseeds will have disappeared by 2040-2045, with the exception of bioethanol produced from sugar cane for 3 EJ in 2050 (10% of biofuels at that date). The remaining, that is 90%, will be 2G biofuels made from ligno-cellulosic resources. In the intermediate "Act Map 2050" scenario, biofuels would reach 24 EJ (0.6 Btoe) in 2050. In the "Baseline 2050" scenario, the use of biofuels in 2050 would be significantly lower at 4.2 EJ (0.1 Btoe), but still substantially up compared to today.

To sum up, it appears that energy uses of biomass are expected to grow very significantly over the decades to come. The increase in energy uses of biomass will be highly correlated to the political mobilization aiming at reducing GHG emissions. After having assessed the demand for biomass by 2050, the next step consists in analyzing whether the supply of biomass will be sufficient to meet this demand.

5.2.2 Estimation of the potential supply of biomass

A literature review (Forslund *et al.* 2010) suggests that the potential supply of biomass for energy purposes is more than sufficient to meet the needs by 2050. According to some estimates, the potential supply would even be more than total energy demand in 2050. For example, Smeets *et al.* (2007) estimate that the potential supply of biomass for energy production will be more than 1500 EJ, out of which 1200 EJ from dedicated crops. Hoogwijk *et al.* (2003) quantify the same potential to slightly more than 1100 EJ, with again the largest share coming from dedicated crops. Two other studies, Fischer and Schrattenholzer (2001) and Berndes et al. (2003), are less optimistic with a potential of about 400 EJ, which is still quite a sizeable figure considering the current uses of biomass for energy purposes and total energy demand in 2050 in the different scenarios of the IEA.

The high variability in potential biomass supply estimates is mainly due to two key parameters, the land area that could be used for energy on the one hand, yields of energy crops on the other (in terms of both biomass yields per hectare and efficiency of transforming this biomass into energy). Land that could be used for energy depends on the total land area available and on land needs for other purposes than energy, first of all food and feed use, but also forestry, environmental, recreational, urban uses, etc.[11] Land needed for food depends on many parameters both on the demand side (demography, urbanization, incomes and prices…) and on the supply side (crop yields, importance of animal products relative to crops…). The equation has thus many unknowns, at least many uncertainty factors.

Just as there is variability about the total potential of biomass as an energy source according to studies, there is variability concerning the different types of biomass that could potentially be used: residues and waste, forest resources and dedicated crops. Nevertheless, the ranking of the three types of biomass is generally identical: first dedicated crops that represent up to 97% of the total potential for Hoogwijk et al. (2003) - 1098 EJ out of 1130 - and at least 44% for Schrattenholzer and Fischer (2001) - 200 EJ out of 450 -, second forest resources and residues and third residues of agricultural origin.

To sum up, despite the high variability of estimates about the biomass that could potentially be used for energy production, it appears that it would be sufficient to meet bioenergy demands in general, biofuel demands in particular, in 2050 and later on. Nevertheless, the

---

[11] According to FAO data, world emerged land is about 13.1 Bha. Excluding cropland (annual and perennial crops), forest areas, deserts and areas under human influence, that leaves a "surplus" of about 4.5 Bha (35%). Out of these 4.5 Bha, slightly more than 50% (2.4 Bha) are unsuitable for agricultural production (unproductive, marginal, steep and/or protected lands). That leaves 2.1 Bha of land potentially suitable for crops currently allocated to grass, shrubs and forest plantations outside forests. Out of these 2.1 Bha, researchers from the Land Use Change (LUC) program at the IIASA (International Institute for Applied Systems Analysis) consider that 60 to 70% of the biomass produced would be used for animal feed, which leaves a theoretical potential of 600 to 800 Mha for bioenergy production. This potential is only theoretical, first because it will be implemented only if it is economically profitable to do so, second because it will have to compete with increasing food uses in the years to come. For example, the FAO estimates that cropland should increase by about 120 Mha by 2050 in developing countries, essentially in Latin America and sub-Saharan Africa (FAO 2009).

potentially available biomass is estimated at a global scale without taking into account its geographical distribution and the necessary adaptation of local supply to regional demand at the scale of countries or groups of countries. The most widespread resources are not necessarily located in areas where consumption will be important. As a consequence, international trade will be indispensable. Moreover, it is also necessary to distinguish the potential supply of energy-devoted biomass from the prospects of biomass that should effectively be used for that usage.

These opportunities arise from supply-demand interactions on bioenergy markets. Supply and demand conditions are both under multiple influences (public policies, strategies of actors, scientific and technical progress…). Furthermore, the analysis cannot be restricted to bioenergy markets. It must be conducted taking into account, jointly and simultaneously, alternative uses of land and their corresponding applications (food and feed, environment, recreation, infrastructures, etc.). This is the purpose of the next sub-section that focuses specifically on assessing the impacts of bioenergy development on food security and the environment measured in terms of GHG emissions and biodiversity preservation.

5.2.3 Impacts of biofuel development on food security and the environment by 2050

As the main source of biomass for 2G biofuels is expected to be dedicated crops (see sub-section 5.2.2), the competition for land between food and non-food uses, already emphasized for 1G biofuels, is likely to remain a debated topic for 2050.

With this in mind, Fischer (2009) discusses the consequences of biofuel development in 2050 on world food security measured by food and feed uses, cereal prices and the number of people suffering from hunger. Fischer defines two main scenarios corresponding to two distinct blend rates of biofuels into fuels used for road transport, respectively 6% (225 Mtoe) and 11.3% (424 Mtoe). Both rates are implemented under three assumptions concerning the replacement of 1G biofuels by 2G biofuels at 26, 35 and 55%, respectively. Simulation results show that the development of biofuels would have a doubly negative impact on world food security, by lowering cereal quantities available for food and feed and by raising their international prices. As a result, the number of people suffering from hunger would increase in 2050 in all scenarios considered relative to the baseline scenario at the same date. Far from being negligible, these negative effects on world food security would increase in parallel with the total share of biofuels blended into fuels used for road transport and with the slowest rate of incorporation of 2G biofuels. In the "worst" scenario corresponding to a high incorporation rate of biofuels (11.3%) out of which only 26% are ligno-cellulosic biofuels, world cereal production would increase by 313 Mt and land dedicated to cereals by 48 Mha, relative to the baseline. But as the needs for energy cereals would be 446 Mt, the amount of cereals available for food and feed would shrink to 127 Mt, which, combined with rising prices (+27%), would have the ultimate effect of increasing the number of people suffering from hunger by more than 140 million in 2050, relative to the baseline. A faster development of ligno-cellulosic biofuels would reduce land needs for energy-devoted cereals, increase cereal quantities for food and feed and ultimately reduce the growing number of people suffering from hunger: in the same scenario of an 11.3% incorporation rate, but with 55% of ligno-cellulosic biofuels, the increasing number of people suffering from hunger would be "limited" to 70 millions in 2050, again relative to baseline.

For their part, Melillo *et al.* (2009) are particularly interested in environmental issues. More specifically, they compare the impacts of two scenarios for biofuel development aiming, at least in theory, at contributing to the same reduction in global GHG emissions. In 2050, all biofuels would be of the second generation and would all be produced from dedicated crops. In the first scenario called "deforestation", all surfaces can be mobilized for the production of biofuels as well as other uses, including agricultural production for food when this is economically beneficial. In the second scenario called "intensive", unmanaged land (e.g. tropical forests) can be mobilized only partially, that is by respecting the rates of land use changes observed in the past. In both scenarios, the economic development of 2G biofuels would be important and at an equivalent level (141 EJ in the "deforestation" scenario and 128 EJ in the "intensive" scenario, which in both cases is more than 10% of the total projected energy demand in 2050); the surfaces mobilized for that purpose would be significant and of similar magnitude (1.48 and 1.39 Bha, respectively).

In the "deforestation" scenario, cultivated agricultural land would increase by 1.73 Bha between 2000 (4.2 Bha) and 2050 (5.93 Bha) under the double pressure of increases in cropland for food and feed (from 1.61 to 2.0 Bha) and, more importantly, of increases in land areas devoted to dedicated energy crops (from 0 to 1.48 Bha) not compensated for by the very small decrease in grassland (from 2.58 to 2.45 Bha). The development of biofuels would have a negative impact on GHG emissions with a carbon debt[12] of 103 Pg C[13] over the 2000-2050 period mainly due to the deforestation of tropical forests in Latin America, sub-Saharan Africa and Southeast Asia. The impact on biodiversity would also be negative, particularly in Latin America (-520 Mha of natural and semi-natural forests and -60 Mha of other wooded land) and in sub-Saharan Africa (-310 Mha of natural and semi-natural forests and -120 Mha of other wooded land).

In the "intensive" scenario, cultivated agricultural land would be 4.98 Bha in 2050, that is 0.79 Bha more compared to the base year 2000 but 0.95 Bha less compared to the "deforestation" scenario in 2050: the sharp decline in grassland (from 2.58 to 1.79 Bha) is insufficient to offset the double increase in cropland for food (from 1.61 to 1.8 Bha) and in land devoted to dedicated energy crops (from 0 to 1.39 Bha). The reduction of biodiversity, measured in terms of reduction of natural areas now devoted to food crops, energy crops or pasture, is lower in the "intensive" scenario than in the "deforestation" scenario. However it is still significant with, for example, over 160 Mha losses of forest and wooded lands in Latin America and 270 Mha in sub-Saharan Africa. Furthermore, a more complete analysis would require taking into account, first the conversion of 800 Mha of grassland into food or energy crops, second the environmental consequences of the intensification of agricultural technologies and practices. The "deforestation" and "intensive" scenarios differ mainly in terms of carbon debt: from 2000 to 2050, the carbon debt would be more than three times lower in the "intensive" scenario (34 Pg C) than in the "deforestation" scenario (103 Pg C). This means that while one has to wait until the mid-century to observe the first net cuts in GHG emissions resulting from the substitution of fossil fuels by biofuels in the "deforestation" scenario, these become visible before 2035 in the "intensive" scenario.

---

[12] The decrease in terrestrial carbon stock, associated with biofuel development and induced land use changes, is commonly called the "carbon debt". During a lapse of time, and for unchanged land uses, the carbon debt decreases and can even be cancelled if GHG emissions linked to the production and the use of biofuels are lower than emissions from replaced fossil fuels.

[13] $1\ Pg = 10^{15}\ g = 1\ Gt = 10^{9}$ metric tons.

6. Conclusion

The objective of this chapter was to analyze the interactions between biofuel development worldwide and the agricultural markets for cereals, oilseeds and sugar plants in terms of quantities (supply and demand), prices and impacts on land use. More specifically, the aim was to determine the degree of competition between food and non-food uses, notably for biofuels, uses of agricultural land and products, and its consequences.

First-generation biofuels are criticized for their negative impact on world food security and their environmental performance, notably in terms of GHG emissions, that is often considered as negative when induced land use changes are taken into account. Furthermore, their technical efficiency is low and the cost of public policies aiming at encouraging their development is high relative to market and non-market profits. In this context, hopes are turning to second-generation biofuels produced from various wastes and residues, wood from forests or dedicated crops that do not compete directly with land uses for food and/or feed. Nevertheless, the absence of direct competition for land use does not mean that there is no indirect competition, as dedicated crops and forests, in the case of an expansion of forest areas to meet energy demands, require surfaces too.

Analyzing this indirect competition is all the more important as the demand for biomass for energy purposes should considerably increase over the coming decades and as the main source of biomass mobilized for that purpose should be dedicated crops. Our analysis shows that replacing first-generation biofuels by second-generation ones would only mitigate the adverse effects of the development of first-generation biofuels on world food security (analyzed in terms of agricultural products available for food and feed and in terms of agricultural prices) and the environment (notably in terms of GHG emissions and biodiversity preservation).

This mitigation will be stronger if the biomass used for second-generation biofuels is provided by dedicated crops grown on "marginal" land currently unoccupied by crops and forests. This point raises the related issues of quantifying these "marginal" land areas and of their sustainable exploiting. This is a vast domain with many parameters still unknown. Research and development should also focus on the transformation of ligno-cellulosic residues and waste from various sources in order to reduce collection and storage costs, and on the sustainable management of forests for enhancing their full potential in terms of uses of wood, surplus growth and residues for both energy and environmental features.

The food challenge (feeding 9 billion people by 2050) is associated with an environmental challenge (defining sustainable agricultural practices and systems) and an energy challenge related to the gradual depletion of fossil fuels. In connection with this third issue, it is more than likely that energy uses of terrestrial biomass will be significant in 2050. Can we quantify these energy uses? That is difficult, if not impossible, given the considerable uncertainties surrounding estimates of total energy consumption in 2050 (e.g. from simple to double, 550 to 1000 EJ per year, according to Clarke *et al.* 2007).

The food challenge requires actions on both the demand and supply side. As far as demand is concerned, developed countries (and rich households worldwide) need to reduce waste and losses at distribution and final consumption stages. They also need to change their consumption patterns to reduce overweight and obesity, and related diseases. On the supply side, in addition to reducing post-harvest losses, it will be necessary to increase crop yields, notably in regions where they are currently low, but in a sustainable way. Agricultural practices and systems used worldwide should radically change.

Quantifying the increase in agricultural production by 2050 is as difficult as quantifying the energy demand and its distribution among the different potential sources. In both cases, prices will ultimately determine supply and demand conditions. There is no doubt that these conditions are and will be influenced by policy measures. But in this area too, uncertainties are numerous. However there is a large consensus for recognizing that agricultural production will have to increase to satisfy food needs of an growing, increasingly urbanized and (on average) richer population. There is also a consensus that, as in the past, the increase in agricultural production volumes will be achieved mainly by increasing yields.

The required increase in yields across the world will necessarily be associated to sustainable farming practices and systems. This twofold aim calls for heavy investment in generic and systemic research, in farms' upstream infrastructure, notably in the objective of providing better access to machinery, water, fertilizers and crop treatment products for farmers in developing countries. It also calls for investment in downstream infrastructures, more specifically to storage and transportation in facilities to reduce post-harvest losses. It requires adopting a holistic approach based on an integrated management of agricultural ecosystems and the use of techniques aiming at conserving natural resources (Pretty *et al.* 2006). In this perspective, practices and systems will necessarily be diverse, adapted to local constraints and environmental resources, and capitalize on the knowledge and expertise of local actors supported by strong public policies.

Concerning biofuels, should we reject first- and second-generation biofuels because they could have a negative impact on world food security, because their environmental record in terms of GHG emissions and/or biodiversity preservation could be negative when induced land use changes are (too) important and/or because public promotion policies would be (too) costly with respect to market and non-market benefits? We do not think so, as the food and environmental challenge should not obscure the energy challenge related to the depletion of fossil resources. To meet the energy challenge, we must act on demand by promoting energy savings, and on supply by developing alternatives to fossil fuels as long as they are environmentally friendly and economically cost-effective, knowing that costs will vary depending on fossil oil prices. In other words, it is jointly and simultaneously necessary to examine our planet's capacity to meet the food, environment and energy challenges.

7. References

Abbott P., Hurt C., Tyner W.E., 2008 (updated 2009), What's Driving Food Prices? Farm Foundation Issue Report 75, July 2008 (updated March 2009).

Battjes J.J., 1994, Global Options for Biofuels from Plantations according to IMAGE Simulations. Interfaculty Vakgroep Energie in Milieukunde (IV), Rijksuniversiteit Groningen, Nijenborgh 4, 9747 AG Groningen, The Netherlands.

Berndes G., 2002, Bioenergy and Water: The Implications of Large-Scale Bioenergy Production for Water Use and Supply. *Global Environmental Change*, 12(4): 253-271.

Berndes G., Hoogwijk M., van den Broek R., 2003, The Contribution of Biomass in the Future Global Energy Supply: A Review of 17 Studies. *Biomass and Bioenergy*, 25: 1-28.

Biofuels Platform. http://www.biofuels-platform.ch/en/infos/eu-biodiesel.php.

Boyle S. (Greenpeace), 1993, Towards a Fossil Free Future - The Technical and Economic Feasibility of Global Phasing Out Fossil Fuel Use. Greenpeace International, London.

Bureau J.C., Guyomard H., Jacquet F., Treguer D., 2010, European Biofuel Policy: How Far Will Support Public Policy Go? Handbook of Bioenergy Economics and Policy, Khanna M., Scheffran J. and D. Zilberman editors, New York, Springer, 2010, p. 401-425.

Clarke L.E., Edmonds J.A., Jacoby H.D., Picher H.M., Reilly J.M., Richels R.G., 2007, Scenarios of Greenhouse Gas Emissions and Atmospheric Concentrations. Sun-Report 2.1A of Synthesis and Assessment Product 2.1 by the United States; Climate Change Science Program and The Subcommittee on Global Change Research, Department of Energy, Office of Biological and Environmental Research, Washington D.C.

Collins K., 2008, The Role of Biofuels and Other Factors in Increasing Farm and Food Prices: A Review of Recent Developments With A Focus on Feed Grain Markets and Market Prospects. Report Written as Supporting Material for a Review Conducted by Kraft Foods Global, Inc., the Current Situation in Farm and Food Markets, June 19, 2008.

Cormeau J., Ghosse G., 2008, Les biocarburants de deuxième génération : semer aujourd'hui les biocarburants de demain. In Demeter 2008, Economie et Stratégies Agricoles, pages 225-302.

Directive 2009/28/EC of the European Parliament and Council, April 23, 2009, On the Promotion of Energy from Renewable Sources.

Dronne Y., Forslund A., Gohin A., Levert F., 2009, Conséquences des politiques des biocarburants de l'UE et des grands producteurs à l'horizon 2020. Report for the French Ministry of Agriculture, INRA, UMR SMART, Rennes, France.

Dronne Y., Forslund A., Guyomard H., 2011, Les biocarburants de deuxième génération et la compétition pour l'usage des terres. *OCL*, 18(1):1-9.

European Biodiesel Board. http://www.ebb-eu.org/stats.php.

FAO, 2009, How to Feed the World in 2050? FAO, Issue Brief, Prepared for the High-Level Expert Forum, Rome, 12-15 October 2009;

http://fao.org/fileadmin/templates/wsfs/docs/expert_paper/How_to_Feed_the_World_in_2050.pdf.

Fargione J., Hill J., Tilman D., Polasky S., Hawthorne P., 2008, Land Clearing and Biofuel Carbon Debt. *Science*, 319: 1235-1238.

Fischer G., Schrattenholz L., 2001, Global Bioenergy Potential through 2050. *Biomass and Bioenergy*, 20: 151-159.

Fischer G., 2008, Implications for Land Use Change. Power Point Presentation for the FAO Expert Meeting on Global Perspectives on Fuel and Food Security, Rome, 18-20 February 2008.

Fischer G., 2009, World Food and Agriculture to 2030/50: How Do Climate Change and Bioenergy Alter the Long-Term Outlook for Food, Agriculture and Resource Availability? Paper prepared for the FAO Expert Meeting on How to Feed the World in 2050, Rome, 24-26 June 2009.

Forslund A., Gauvrit L., Guyomard H., Mora O., 2010, Les utilisations non alimentaires de la biomasse et la concurrence pour l'utilisation des terres : belles promesses ou vraies craintes ? INRA, Rapport RST de l'Académie des Sciences (en partenariat avec l'Académie d'Agriculture de France) sur Démographie, Climat et Alimentation Mondiale.

Global Renewable Fuel Alliance. http://www.globalrfa.org/pr_021111.php.

Guillou M., 2010, Peut-on nourrir la planète de façon durable ? Les défis de la recherche agronomique. *ParisTechReview*: 19 juillet 2010.

Guyomard H., 2008, Expliquer les évolutions des cours des matières premières agricoles : à l'impossible, nul n'est tenu ! *OCL*, 15(6): 364-377.

Guyomard H., 2009, Nourrir la planète de façon durable est possible, à condition que... *Politique Etrangère*, 2: 2-13.

Hoogwijk M., Faij A., van den Boek R., Berndes G., Gielen D., Turkenburg W., 2003, Exploration of the Ranges of the Global Potential of Biomass for Energy. *Biomass and Bioenergy*, 25: 119-133.

IEA (International Energy Agency), 2008, Energy Technology Perspectives 2008, Scenarios & Strategies to 2050.

IEA, 2008, World Energy Outlook 2008. IEA/OECD, Paris, 578 pages.

IEA, 2009, CO2 Emissions from Fuel Combustion: Highlights 2009.

INRA-CIRAD, 2009, Agrimonde, Agriculture and Food Worldwide in 2050: Scenarios and Challenges for Sustainable Development. Final Report of the INRA-CIRAD Foresight Study Agrimonde.

Lashof D.A. and D.A. Tirpak editors, 1990, Policy Options for Stabilizing Global Climate. Hemisphere Publishing Corporation: New York, Washington, Philadelphia, London.

Leemans R., van Amstel A., Battjes C., Kreileman E., Toet A., 1996, The Land Cover and Carbon Cycle Consequences of Large-Scale Utilization of Biomass as Energy Source. *Global Environmental Change*, 6(4): 335-358.

Melillo J.M., Gurgel A.C., Kicklighter D.W., Reilly J.M., Cronin T.W., Felz B.S., Paltsev S., Schlosser C.A., Sokolov A.P., Wang X., 2009, Environmental Unintended Consequences of a Global Biofuels Program. MIT Joint Program on the Science and Policy of Global Change, Report 168, January 2009.

Morton D.C., Defries R.S., Shimabukuro Y.E., Anderson L.O., Arai E., del Bon Espirito-Santo F., Freitas R., Morissette J., 2006, Cropland Expansion Changes Deforestation Dynamics in the Southern Brazilian Amazon. *PNAS*, 1003(39): 14637-14641.

Müller C., Bondeau A., Lotze-Campen H., Cramer W., Lucht W., 2006, Comparative Impact of Climatic Factors and Nonclimatic on Global Terrestrial Carbon and Water Cycles. *Global Biochemical Cycles*, 20.

Oil World Annual 2010. ISTA Mielke GmbH, Langenberg 25, 21077 Hamburg Germany.

Pons E., 2007, A Review of Policy Measures Supporting Production and Use of Bioenergy. OECD, TAD/CA/APM/WP(2007)24/FINAL OECD.

Pretty J.N., Noble A.D., Bossio D., Dixon J., Hine R.E., Penning de Vries F.W.T., Morison J.I.L., 2006, Resource-Conserving Agriculture Increases Yields in Developing Countries. *Environmental Science & Technology*, 40(4):1114-1119.

Powlson D.S., Riche A.B., Coleman K., Glendining M.J., Whitmore A.P., 2008, Carbon Sequestration in European Soils Through Straw Incorporation: Limitations and Alternatives. *Waste Management*, 28: 741-746.

Searchinger T., Heimlich R., Houghton R., Dong F., Elobeid A., Fabiosa J.F., Tokgoz S., Hayes D.J., Yu T., 2008, Use of United States Croplands for Biofuels Increases Greenhouse Gases Through Emissions for Land-Use Changes. *Science*, 319: 1238-1240.

Smeets E., Faij A., Lewandowski I., Turkenburg W., 2007, A Bottom-Up Assessment and Review of Global Bio-Energy Potentials to 2050. *Progress in Energy and Combustion Science*, 33: 56-106.

Sorensen B., 1999, Long-Term Scenarios for Global Energy Demand and Supply: Four Global Greenhouse Mitigation Scenarios. Roskilde University, Institute 2, Energy and Environment Group, Denmark.

Trostle R., 2008, Global Agricultural Supply and Demand: Factors Contributing To The Recent Increase in Food Commodity Prices. USDA, ERS, WRS-0801, May 2008 (revised July 2008).

Von Braun J., Torero M., 2008, Physical and Virtual Global Food Reserves to Protect the Poor and Prevent Market Failures. Policy Brief 4, IFPRI (International Food Policy Research Institute), Washington D.C.

Wirsenius S., 2008, Global Use of Agricultural Biomass for Food and Non-Food Purposes: Current Status and Future Outlook. Working Paper, Chalmers University of Technology, Department of Energy and Environment, Gothenburg, Sweden.

Worldwatch Institute, 2007, Biofuels for Transport. Global Potential and Implications for Sustainable Energy and Agriculture. Earthscan, United Kingdom.

8

The Potential for Production of Biomass for Biofuel by the Cultivation of Hybrid Poplar and Hybrid Aspen in the South of Sweden

Lars Christersson
Section of Short Rotation Forestry, VPE, Swedish University of Agricultural Sciences
Uppsala
Sweden

1. Introduction

More and more people are starting to realize that we must begin *now* to build a society that *does not* affect the environment (Lawson, 2008, Anonymous, 2010). What we are doing today is completely decisive for the development of the society of tomorrow. Two lines in the development of the society of the future are obvious. Either world leaders really start immediately to govern the development of the welfare states in an ecological and sustainable direction, so as to set an example and model for developing countries; or the politicians of the world do nothing and the society of the future will be created from the resultant chaos.

An example of what should be done now is what politicians did 40 years ago. As a direct consequence of the oil crisis in 1973, a new branch of the UN organization for food and agriculture (FAO) was set up in November 1974 on the initiative of Dr. Henry Kissinger, the former Secretary of State for Foreign Affairs in the USA (Anonymous, 2008). The name of this new organization was the International Energy Agency (IEA) and the three objectives were (1) to administrate the need for oil in the Western World and distribute the existing oil resources within this region, if the delivery of oil from the Middle East was interrupted. The new organization´s task was also (2) to execute a multilateral program for more effective use of energy and (3) to launch an advanced programme for research on new energy sources. The importance of IEA was strengthened by the next oil crisis in 1978.

Different countries became responsible for different parts of the organization. Sweden was chosen as Operating Agency for Energy from Biomass with Professors Gustaf Sirén and P-O Nilsson at the head, both at the Faculty of Forestry at the University of Agricultural Sciences (SLU) in Sweden (Christersson et al., 2008). Something similar *should* be done today, but compare this thinking with the results of the meeting in Copenhagen and Cancun, Mexico!

In Sweden at that time forestry was one of the most important industries with spruce (*Picea abies*) and pine (*Pinus sylvestris*) as totally dominant tree species. Professor Nilsson worked with the waste products from this forest industry, e.g. tops and branches from clear-cut areas, but also with sawdust from the sawmills, and black liqueur from the paper industry (Doherty et al., 2002).

But Professor Sirén started to think along new lines (Sirén, 1974, Sirén et al., 1983). He had previously battled against encroaching fast-growing seedlings of Salix (willow) and Populus (aspen), which suppressed and killed small, newly-planted seedlings of spruce and pine. Why do we not do the opposite and grow these fast-growing tree seedlings as short rotation coppice plantations, harvesting them early as biomass for energy and burn them for heat and electricity, Professor Sirén asked? He first concentrated his effort on willow plantations (Sirén et al., 1983), but later poplar also became involved when Professor Christersson in 1979 came on the scene (Christersson, 2010). In 1988 Dr. B. Ilstedt started breeding research with poplars (Ilstedt, 1996). So it was in Sweden at SLU that the concept "Energy Forestry" was created and developed (Dickmann and Kuzovkina, 2008).

Since the middle of the 1840s, European aspen (*Populus tremula*), the only Populus species growing in Sweden, had been used for matches and at the end of the century people concerned with this industry realized that there was not enough wood of this species for further development (Lundström, 1899). In the1890s, when the paper industry started to develop rapidly, the shortage of wood became apparent for this industry too. So scientists were asked to try to grow some more fast-growing species of Populus, originating from the USA and Canada, and its hybrids with the aim of producing more wood for matches and paper (Johnsson, 1953, Werner, 2010). In the match industries Sweden has been the leading country of the world since the beginning of the 1900s. However, when cigarette lighters came on the scene in the 1950s interest in using matches diminished.

Most of the imported poplar clones and hybrids to Europe in the old days originated from Oregon, Washington and Iowa, all states that are located at latitudes that correspond with France and Spain in Europe. Very few came from BC, Canada and Alaska, USA. This meant that most of the imported clones were damaged to some extent by the climate in Sweden (Christersson, 1996) so Swedish activities fluctuated over the years but finally decreased considerably.

However, when the oil crisis developed in the 1970s, interest in growing poplar and hybrid aspen for energy increased again. At first this was with the old poplar clones but later it was with new, imported poplar clones from BC and NWT in Canada and Alaska in the USA, i.e. from equivalent Swedish latitudes (Christersson, 1996, Karacic et al., 2003, Rytter and Stener, 2005).

The advantages of growing poplars instead of willows are that the poplar wood can be used for both energy and paper and even to some extent for construction, willows only for energy. Willows should be grown on very wet areas, poplars on wet areas. The cash flow is also better for those growing willow.

The oil crises hit Sweden particularly severely, because we have no oil, no coal and no gas. But we have district heating systems developed in every village, every town and city and many big combined heat and power plants (CHP) for production of heat and electricity. These mostly used oil as fuel at that time. That means that all the wood we can produce can be used directly to replace all the oil used in an already existing infrastructure. Thus 31. 5 % of all energy utilized in Sweden is currently from biomass (Andersson, 2010).

Biomass for energy has been used in Sweden for many years, but for many countries in the European Union this energy source is new. However, the European Union has now decided that the utilization of renewable energy sources should be 20 % by the year 2020 in all their member states. In Sweden the equivalent target is at least 50 % from renewable energy sources (Anonymous, 2009). This figure includes water power. The figure for Sweden for use of renewable energy sources is currently about 45 %.

These targets should be looked upon in the light of the situation that all developed regions of the world have so far been completely dependent on access to cheap sources of energy for their development. Until now, oil has served as such a source, but many people fear that coal will become the main energy source in the future, now that the price of oil is increasing (Azar, 2008).

In some countries people are using one hundred times more energy per capita than in others, and some countries, China for example, are building large rock storage chambers in their mountains, in order to stockpile currently cheap oil, bought from developing countries. However, in many countries, particularly the smaller ones there is increasing realisation that a more ecological approach towards energy policy is needed in order to be able to achieve sustainability.

Sweden has 300, 000 to 500, 000 ha of productive agriculture land available for growing biomass for biofuel (Lindroth and Åkerblom, 1984, Anonymous, 2006), and if the economics are good, many more hectares of fertile forest land can be used (Skogsstyrelsen, 2008). In many other countries the situation is the same. From the Ural mountains to the Atlantic there are 25 million hectares of set-aside, fallow or abandoned agriculture land not now used for food production (Andersson, 2010). One way of utilizing all this land area for producing biomass for energy is to cultivate fast-growing poplar, willow and hybrid aspen clones. These types of plantations have been shown to be one of the most efficient transformers of solar radiation energy to chemical energy in the form of wood (Christersson, 1987, 2006, 2010, Rytter et al., 2011).

In the literature (Stanton, 2009), growers of poplar are advised to plant them on mineral soils (Dickmann and Kuzovkina, 2008). But new results show that on organic soil with a pH above 6 poplars and hybrid aspen are doing very well (Christersson, 2010).

In the present investigation the results are given for cultivation of hybrid poplar and hybrid aspen plantations from four farms and from different soil types in the south of Sweden. It is a direct continuation of the investigation published by Christersson in 2010 and the results should also be valid for Denmark, Northern Germany, Poland and the Baltic States.

2. Materials and methods

On four farms in the south of Sweden and on different soil types, hybrid poplar and hybrid aspen were planted as bare-root seedlings in 1991 -1993 (Table 1). All the details are given in Christersson, 2010 and summarized in Table 1 - 3.

Location	Lat. °N	Long. °W	Altitude. m	Type of landscape	Annual precipitation mm	Year of plantation	Type of seedling
Kadesjö	55	13	30	level field	611	1991	bare-root
Näsbyholm	55	13	40	carex peat bog	623	1991	bare-root
Sturup	55	13	50	undulating field	650	1993	bare-root
Torup	55	13	60	slope	628	1993	bare-root

Table 1. Overview of the plantations

The four farms, Kadesjö, Näsbyholm, Sturup, and Torup, are located almost side by side in the southernmost part of Sweden, 10-20 km north of the Swedish south coast. The distance between the furthest apart is not more than 33 km. Further detailed information is given in Table 2 and 3. Some thinning was carried out at Kadesjö and Sturup.
All the poplar plantations are located on a bed rock of chalk. The soil is Baltic moraine clay of Scania. This is characterized by easily crushed rocks, almost without stones, often heavy clay (Lundegårdh et al., 1967). There were some differences between the soils of the four farms (Table 2).

Location	Bedrock	Soil type	pH (H_2O)
Kadesjö	**chalk**	**clay loam**	**6.4**
Näsbyholm	**chalk**	**clay, very rich in humus**	**6.1**
Sturup	**chalk**	**clay loam, with high humus content**	**5.9**
Torup	**chalk**	**heavy clay**	**6.3**

The soil analyses were performed by Dr Stig Ledin, SLU, Uppsala, Sweden)

Table 2. Bedrocks and soil types of the four farms

During the preceding 10 years the diameters of 43 -78 trees were measured along a line through the whole plantations; the height was measured of 10 randomized tree. The diameter measurements were accurate to one mm and the height to one metre, with the exception of the most recent year for Kadesjö and Näsbyholm, when the accuracy was one dm. For the height measurement it was difficult the most recent year to really see the top of the trees in the dense and very tall plantations. So in 2010 specialists were hired to make a more precise height measurement. The wood production was calculated by using the equation of Johnsson, 1953. The weight of one m^3 dry wood is 0.335 ton (Stener, 1998) and the weight of the top of the tree and the branches is estimated to 15 % of the stem weight (Karacic, 2005).
The Kadesjö plantation will be clear cut in 2011. In the Torup plantation some trees were felled and the annual height growths were measured for the last 11 years. The mean annual height growth was 1.6 m per year with a slight decrease in recent years.
The poplar plantations were fenced for the first 10 years at Kadesjö, Sturup and Torup but not at Näsbyholm. After that the fence was moved to other plantations in order to allow normal hunting. No damage by wild animals was observed.
It is planned to be harvest all the plantations after 20- 25 years when the price of the wood is favorable.

Location: South of Skåne, Sweden
General characterization: previously cultivated agricultural land
Latitude: 55° N
Longitude: 13° E
Altitude: 30-60 m asl
Mean annual temperature: 8.6° C
Mean temperature: May-August 14.2°C

Lowest temperature in winter: -26.3°C
Temperature sum of the year: 1547 daydegrees
Temperature sum of the growing season: 1132 daydegrees
Annual rainfall: 611 - 650 mm
Rainfall during the growing season: 426 mm.
Annual solar radiation: 1150 kWh/m^2
Solar radiation during growing season: 820 kWh/m^2
Start of the growing season (>5°C) day no 102 (12/4)
End of the growing season <5°C day no 319 (17/11)
Length of the vegetation period (>5°C-<5°C): 220 days
Size of the plantation: 3-10 ha
Soil type: see special table
Bed rock: limestone, chalk
pH: 5.9 – 6.4
Level of ground water: 0.2 - 1 m
Previous crops: wheat, oat, spring rape
Production: 5-7 ton grain/ ha, yr
Planting preparation: ploughing, harrowing
Species: h-poplar (OP 42), h-aspen (Ekebomix),
Planting method: by hand, spade
Time for planting: 1991-1993
Design: varied
Thinning: on Kadesjö and Sturup
Fencing: present for the first 10 years
Survival: > 90%
Weed management: none
Fertilisation: no
Irrigation: no
Rotation time: 20 years
Damage: wind throw
Utilisation of the wood: paper and energy

Table 3. Location, climatic and cultivation conditions for the four farms

3. Results

All the plantations are located very close together and at the same latitude; thus the climatic conditions are the same for all four. All are located on productive agriculture fields with a production capacity of 5 -7 ton grain of wheat per ha and year. The analysis shows that the pH of the soils varies between 5.9 and 6.4 (Table 2) with the highest value for Kadesjö and the lowest for Sturup. The highest clay content was in the field in Torup and the highest organic content was at Näsbyholm, which is a Carex peat bog. Näsbyholm also has the highest ground water level and part of it is flooded almost every winter. The field in Torup is on a slope with movable ground water. In Sturup the field is undulating with low areas with high organic content alternating with small sandy hills, which have a negative influence on the growth rates.

The highest wood biomass production was achieved at Näsbyholm for hybrid aspen (Table 5), probably because of the largest number of stems per hectare. However, the high growth rate is also dependent on very good water conditions on that field and on a nutrient- rich soil, caused among other things by the location of the plantation between intensively fertilized agriculture fields and a ditch. This plantation is something that we can call a 'vegetation filter' (Perttu and Kowalik, 1997, Elowson, 1999). There is a difference in growth rates between the hybrid poplar and the hybrid aspen clones but the difference is not significant. A future problem at Näsbyholm could be the high water table, which creates a loose soil with a risk of wind throw, particularly during winter when large areas are flooded, and because intense autumn storms are frequent in this part of Sweden. The production for both the poplar clone and the mixed hybrid aspen clones at this location reached 30 - 40 m^3/ha, yr, when there was canopy closure after some years. However, in some years with favorable climatic conditions the production reached over 40 m^3/ha,yr.

The poplar plantation at Kadesjö (Table 4) is interesting, because it is located on a very good and traditionally agricultural field. The conventional crops for this land would be wheat, sugar beet and rape with high to very high production rates, depending on the amount of the fertilization. The final woody biomass production of Kadesjö is slightly less than that of Näsbyholm but the differences are not significant.

The diameter growth at Näsbyholm (Table 5) of the hybrid poplar was always a little higher than for hybrid aspen, but the height of the trees was the same. The same trend for the diameter growth can be seen at Kadesjö (Table 4) initially, but it disappeared during later years. At Kadesjö the height of the hybrid poplar was always about 2 m more than that of the hybrid aspen clones. It is planned to harvest the Kadesjö plantation in 2011 so here it will be possible to compare the actual harvest with the calculated one and thus to evaluate the growth equation used.

At both Sturup (Table 6) and Torup (Table 7) only hybrid aspen clones were planted. The growth rate at Sturup was the same as for Näsbyholm and Kadesjö at the lower parts of the plantations, but much less on the hills. This is one of the reasons why the mean production rate is significantly lower here, even if we take into consideration that the plantation was two years younger. The other is that the thinning was applied very early so that canopy closure was delayed.

The lower results at Torup (Table 7) are caused by the very dense plantation.

All the four plantations were astonishingly free from damage. Only the wind throw on the loose soil at Näsbyholm was troublesome, and some of the same at Kadesjö. But it should be remembered that all the plantations were fenced, with the exception of Näsbyholm, for the first ten years and that the moose population in this part of Sweden is very low. But there was almost no damage caused by leaf rust or by insects (Steenackers, 1990.) Such damage can be devastating on many plantations in the rest of the world. At Torup the soil of the whole plantation was grubbed up by wild pigs, but whether this has a bad or good effect we do not know.

4. Discussion

Energy, water, and food are the most substantial and essential resources for human beings; all energy has its origin from the sun and food is a kind of energy. The energy radiates from the sun to the earth; it is absorbed by green plants, utilized by all living creatures and disappears out into the atmosphere as heat. Water, on the other hand, does not disappear

OP 42 (Populus maximowiczii x P.trichocarpa)										
Year	**2001**	**2002**	**2003**	**2004**	**2005**	**2006**	**2007**	**2008**	**2009**	**2010**
Age	**11**	**12**	**13**	**14**	**15**	**16**	**17**	**18**	**19**	**20**
Diameter, cm	**18.9**	**20.8**	**21.8**	**23.0**	**23.6**	**24.6**	**25.1**	**25.9**	**26.7**	**26.9± 4.5**
Height, m^3	**15**	**17**	**18**	**20**	**22**	**23**	**24**	**26**	**27**	**28.3±1.5**
Tree volume, m^3	**0.1890**	**0.2568**	**0.2974**	**0.3653**	**0.4207**	**0.4765**	**0.5164**	**0.5931**	**0.6531**	**0.6933**
No. trees/ha	**657**	**657**	**657**	**657**	**657**	**657**	**657**	**657**	**657**	**657**
Total prod./ha, m^3	**124**	**169**	**195**	**240**	**276**	**313**	**339**	**390**	**429**	**455**
Prod./year, m^3		**45**	**26**	**45**	**36**	**37**	**26**	**51**	**39**	**26**
Running 3 year period, m^3/yr				**39**	**36**	**39**	**33**	**38**	**39**	**39**

Calculated pulp production: MAI during 20 years= 23 m^3/ha,yr

Calculated energy wood production: 23 x 0.335 + 0.15 x 23 x 0.335= 9 ton TS/ha,yr

Hybrid aspen (P. tremula x P. tremuloides)										
Year	**2001**	**2002**	**2003**	**2004**	**2005**	**2006**	**2007**	**2008**	**2009**	**2010**
Age	**11**	**12**	**13**	**14**	**15**	**16**	**17**	**18**	**19**	**20**
Diameter, cm	**15.7**	**18.2**	**20.1**	**21.7**	**23.0**	**24.4**	**25.0**	**25.8**	**26.7**	**27.9± 3.0**
Height, m	**12**	**14**	**15**	**17**	**19**	**20**	**22**	**23**	**25**	**26.0±1.5**
Tree volume, m^3	**0.1065**	**0,1645**	**0.2134**	**0.2793**	**0.3480**	**0.4106**	**0.4716**	**0.5236**	**0.6068**	**0.6873**
No. trees/ha	**659**	**659**	**659**	**659**	**659**	**659**	**659**	**659**	**659**	**659**
Total prod./ha, m^3	**70**	**108**	**141**	**193**	**229**	**271**	**311**	**345**	**400**	**453**
Prod./year, m^3		**38**	**33**	**52**	**36**	**42**	**40**	**34**	**55**	**53**
Running 3 year period, m^3/yr				**41**	**40**	**44**	**39**	**39**	**43**	**47**

Calculated pulp production: MAI during 20 years= 23 m^3/ha,yr

Calculated energy wood production: 23 x 0.335 + 0.15 x 23 x 0.335= 9 ton TS/ha,yr

Table 4. Kadesjö, planted 1991

OP 42 (Populus maximowiczii x P. trichocarpa)

Year	2001	2002	2003	2004	2005	2006	2007	2008	2009	2010
Age	11	12	13	14	15	16	17	18	19	20
Diameter, cm	20.8	21.7	22.8	23.2	23.7	24.3	24.8	25.2	26.2	26.8±5.7
Height, m	14	15	17	18	20	21	22	24	25	26.2±2.8
Tree volume, m^3	0.2142	0.2484	0.3081	0.3364	0.3876	0.4266	0.4642	0.5204	0.5845	0.6393
No. trees/ha	718	718	718	718	718	718	718	718	718	718
Total prod./ha, m^3	154	178	221	242	278	306	333	374	420	459
Prod./year, m^3		24	43	21	36	28	27	41	46	39
Running 3 year period, m^3/yr				29	33	28	30	32	38	42

Calculated pulp production: MAI during 20 years 23 m^3/ha,yr

Calculated energy wood production: 23 x 0.335 + 0.15 x 23 x 0.335=9 ton TS/ha,yr

Hybrid aspen (*P.tremula x P. tremuloides*)

Year	2001	2002	2003	2004	2005	2006	2007	2008	2009	2010
Age	11	12	13	14	15	16	17	18	19	20
Diameter, cm	19.1	20.2	21.4	22.2	22.9	23.5	23.9	24.6	25.1	25.9±5.3
Height, m	14	15	16	18	20	21	23	24	25	26.1±2.1
Tree volume, m^3	0.1809	0.2156	0.2567	0.3083	0.3622	0.3992	0.4500	0.4962	0.5364	0.5953
No. trees/ha	868	868	868	868	868	868	868	868	868	868
Total prod./ha, m^3	157	187	223	270	314	347	391	431	466	517
Prod./year, m^3		30	36	47	44	33	44	40	35	51
Running 3 year period, m^3/yr				38	42	41	40	39	40	42

Calculated pulp production: MAI during 20 years 26 m^3/ha,yr

Calculated energy wood production: 26 x 0.335 + 0.15 x 26 x 0.335=10 ton TS/ha,yr

Table 5. Näsbyholm, planted 1991.

Hybrid aspen										
Year	2001	2002	2003	2004	2005	2006	2007	2008	2009	2010
Age	9 thin.	10	11	12	13	14	15	16	17	18
Diameter, cm	12.6	14.4	16.3	17.4	18.5	19.7	20.3	21.5	22.2	22.7± 4,8
Height, m	13	14	15	17	18	19	21	22	23	24
Tree volume, m^3	0.0749	0.1040	0.1413	0.1808	0.2152	0.2563	0.2990	0.3500	0.3890	0.4233
No. trees/ha	1100	729	729	729	729	729	729	729	729	729
Total prod./ha, m^3	82	76	103	132	157	187	218	255	284	309
Prod./year, m^3			27	29	25	30	31	37	29	25
Running 3 year period, m^3/yr					26	28	29	33	32	30
Calculated pulp production: MAI during 18 years =17 m^3/ha,yr										
Calculated energy wood production: 17x 0.335 + 0.15 x 17 x 0.335=7 ton TS/ha,yr										

Table 6. Sturup, planted 1993

Hybrid aspen (P. tremula x P. tremuloides)										
Year	2001	2002	2003	2004	2005	2006	2007	2008	2009	2010
Age	9	10	11	12	13	14	15	16	17	18
Diameter, cm	11.8	12.9	13.1	13.8	14.3	14.6	15.1	15.2	15.6	15.9±2.8
Height, m	9	10	12	14	16	18	19	21	22	23± 2
Tree volume, m^3	0.0466	0.0611	0.0748	0.0957	0.1165	0.1357	0.1525	0.1700	0.1870	0.2025
No. trees/ha	1792	1792	1792	1792	1792	1792	1792	1792	1792	1792
Total prod./ha, m^3	83	109	134	171	209	243	273	305	335	363
Prod./year, m^3		26	25	37	38	34	30	32	30	28
Running 3 year period, m^3/yr				29	33	36	34	32	31	30
Calculated pulp production: MAI during 18 years =20 m^3/ha,yr										
Calculated energy wood production: 20 x 0.335 + 0.15 x 20 x 0.335=8 ton TS/ha,yr										

Table 7. Torup, planted 1993

from the earth; it merely circulates from the earth (as water vapour) to the sky and back again (as rain). The problem is that human beings pollute it and destroy it for themselves and for all other living organisms and to clean all the waste water would cost a lot of energy Many scientists consider that fresh water is the Achilles heel of our society. Only 3 % of all water is fresh water. There is a lot of water on our planet, but it is salt water and energy is needed for desalination

Access to energy and pure water are the limiting factor for the development of a welfare state. Furthermore, most wars of to-day are actually caused by access to energy in the form of oil and many people fear that the next world war will be about access to fresh water. But even though the further development and well-being of the world are completely dependent on energy, there are few resources that exist in such large amounts as energy on our planet. Ten thousand times more energy radiates to the earth from the sun than is utilized (Cooper, 1975, Cannell, 1989). Expressed in another way, in half an hour the earth receives the same amount of energy as solar radiation than is used in a whole year. This fact is the greatest paradox of our time.

In this context let us try to analyze the result of the present investigation and concentrate on the energy effectiveness of a productive poplar plantation and compare the effectiveness of poplar with that of other tree species and food crops. The investigation has shown that the above-ground woody biomass production can reach a mean value of 10 ton dry matter per ha and year and that such results can be considerably higher in some very favorable years (Table 4-5). We estimate that the amount of leaves developed is 3 ton dry matter per ha and year, giving a total amount of produced biomass of say 20 ton as the absolute most. To be certain not to underestimate the amount of root biomass, we assume that the production of biomass in the soil is the same as above ground. So in the most favorable situation there may be a total biomass production of about 40 ton dry matter per ha and year.

The heat value of 40 ton dry biomass is 40 ton x 4.5 MWh = about 180 MWh per ha and year. The radiation from the sun during the growing season in the area is 820 kWh per m^2 (Table 3). One hectare is 10 000 m^2, giving a radiation per hectare during the growing season (May-August) of 8200 MWh. 180 MWh is little more than about 2 % of 8200 MWh The majority of the most productive agriculture crops are similar (Börjesson, 2007). All the remaining energy, about 98% of the incoming solar radiation, sooner or later radiates out into the atmosphere as heat. If the energy effectiveness of the plants can be increased by just a small amount, it will mean a lot for the energy and food supply of the world. This is the reason why ecophysiologic and genetic research is so important; particularly today when we know that the human population will increase by 50% by 2050, and when most people are aware that climate change cannot be prevented, only limited.

The discussion above is about the effectiveness of plants in absorbing radiation energy from the sun. The energy efficiency of a crop or a tree plantation is a completely different story. In this case we are calculating how much energy we must put in to produce so much biomass. If we do this calculation we find that spruce is the most efficient crop we can grow in Sweden (Figure 1) and that poplar and willows are second and third. Normal agriculture crops achieve only half these values.

However, spruce grows much more slowly than poplar and willow and has a rotation period of 50-60 years. The rotation period for willow is 2-4 years and for poplar 15-20 years. So if we have limited time and limited area to grow biomass for energy, poplar and willows are to be preferred (Christersson et al., 2008). Even the Cash flow is also more favourable for poplar and willow growers. However, new ideas and new plant materials are on their way with respect to plantation of different species of the genera Picea and Abies in the south of Sweden (Bo Karlsson, per. com). For conifers the problem is their influence of the fertility of the soil on which they grow and their effect on the biodiversity and appearance of the landscape. The advantage of poplar plantations over willow plantations is that poplar wood can be used both for energy and paper.

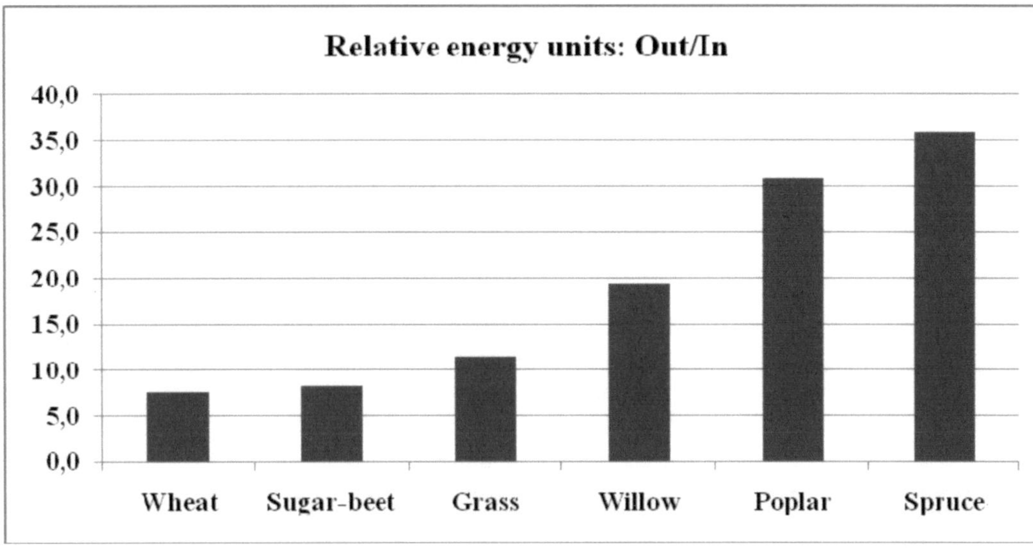

Fig. 1. Ratio between total energy output and input (energy balance quota) for some agricultural and forestry crops (data from Börjesson 2007, recalculated by K.Perttu).

The fertility of the soil of the four farms is similar. The soils are all clay but there are some differences in the amount of organic matter. The soil of Näsbyholm in particular is very rich in humus, which may explain the slightly higher production for hybrid aspen at that farm (Table 9). However wind throw is more frequent on this soil, because of its loser consistency. The small difference in the pH between the four farms is supposed to have only a slight effect. The lower woody biomass production of Sturup and Torup is partly because these two plantations are 2 years younger than the other two and partly because Sturup was thinned too early and the Torup plantation has very dense spacing.

	Poplar		H-asp	
	m^3/ha.yr.	ton/ha.yr.	m^3/ha.yr.	ton/ha.yr.
Kadesjö	23	9	23	9
Näsbyholm	23	9	26	10
Sturup			17	7
Torup			20	8

Table 9. Summary of the above- ground woody biomass production (MAI) of the four farms in ton/ha.yr. and in m^3/ha.yr.

The great disadvantage of growing poplar in the south of Sweden is the high risk of damage by browsing animals and by wind throws. Both moose and all types of deer will damage not only young plants but also thick stems by stripping the bark. Thus it is necessary to protect poplar plantations in some way. Currently the plantations are fenced, or every stem is netted, but these measures are very expensive. Development of some type of cheap protection is a prerequisite for introduction of poplar plantations on a large scale in Sweden.

In earlier investigations (Christersson, 2010) it has been found that damage from stem rot, caused by bacteria or viruses, seems to be mitigated by fertilization. Nor does this type of

injury occur on fertile soils. There are also large differences between clones in resistance to such diseases. So great care is needed in the choice of clones of poplar for plantations, particularly if the soil is poor.
Surprisingly little damage by fungi or insects has so far been found on poplar plantations in Sweden. It is only attack by leaf rusts, Melampsora and Massonina fungus that have had a negative effect on biomass production. However, such attacks have occurred only relatively late in the autumn and it is believed that the damage caused is only limited. Nevertheless, entomologists and mycologists warn of a forthcoming build up of large populations of both insects and fungi, if poplar plantations are introduced on a large scale (Steenackers, 1990).
The climate of Sweden is completely governed by the Gulf Stream and by winds from the south west. But sometime, even in summer, the wind direction changes and cold winds come to Sweden from the Arctic. This is a common phenomenon, particularly at the beginning of June. The temperature can then be as low as – 4 to - 6 degree Celsius, even in the most southern part of Sweden (Christersson,1987). At that time spring-planted seedlings of poplars are in full growth and large parts of the growing shoots are killed, sometimes the whole plant. But most often it is only in low-lying parts of the fields that such low temperatures occur, so such areas should be avoided for poplar. This problem can be solved by planting long cuttings (see below).

5. Visions

5.1 Future plantations

Concerning future energy supply and politics in Sweden, the Government has decided (Proposition 2008/09:162) that by 2020 the emission of greenhouse gases should have decreased by 40 % and that the amount of renewable energy used in Sweden in 2020 should reach at least 50%. That figure for 2008 was about 45 % (Andersson, 2010). The total energy use in Sweden of to-day is almost 400 TWh and a 5 % increase would be 20 TWh. With a production of 9 ton dry matter per hectare and year (Table 9), we need an area of 500 000 ha of poplar plantation to produce 20 TWh energy in the form of heat (9ton x 4,5MWh x 500 000 ha =about 20 TWh). In an inventory of available productive agriculture land for biomass production for energy, it was found that 3-500 000 ha is currently available and could be used for this type of cultivation (Anonymous, 2006). Furthermore, millions of hectares of agriculture land in Sweden have been planted with trees since 1950, particularly with spruce. It will soon be time to harvest these plantations; the next generation of forest on this land should be deciduous trees, because of problems with spruce root rot and with the fertility and pH of the soil.

5.2 Long cuttings

The trait of poplar-wood to form adventitious roots very readily is well-known and used in many nurseries and for establishing poplar plantations. Commonly, 10 - 30 cm or one-meter-long cuttings are used. In Sweden another method is now being developed: cuttings 3.5 m long are drilled one meter down into the soil in densely spaced plantations (2 x 2m) (Figure 2). The first harvest will take place after 15 years and the expected yield with newly bred plant materials will be 200 ton TS/ha (500 m^3). The second generation of such a plantation will consist of shoots from the stumps and from the roots of the first generation. This second generation will be harvested after another 15 years with the same or higher production. The plantation will be fenced for the first 10 years. The economy of such a plantation will not be much more than even for the first generation because of the cost of the fence. It is the second generation that will be profitable for the growers.

The second generation will not need to be fenced because the amount of new shoots from the stumps and from the roots is so enormous and because the shoots grow so fast that a normal population of moose and deer can be allowed to browse on the plantation without reducing wood production.

There are two advances in particular with this method of cultivation. The first is that it is possible to avoid a long establishment period of several years for a poplar plantation, always needed for half-meter long rooted plants or 30 cm cuttings. It is a question of achieving canopy closure as soon as possible. The second advantage is that even very fertile land areas subject to summer frosts can be used. These summer frosts are close to the soil surface and the growing points of the long cuttings, which are the most frost-sensitive, are located above the layer with temperatures below zero during a summer night frost.

Fig. 2. 3.5 m long cuttings, drilled one metre down into the soil. The plantation will be harvested for the first time after 15 years when the yield is expected to be 200 ton dry matter per hectare (500 m^3).

Other advantages are that no preparation of the soil is necessary and even in existing scrubby vegetation these long cuttings can be planted without clear-cutting the scrub. Even if the scrub is taller than the cuttings, only small light gaps need to be cleared. The growth rate of the poplar is so rapid that it soon suppresses the competing scrub.

There is one very important point to be remembered in producing the 3.5 m long cuttings. Normally two or three year old shoots are used. All the lateral branches should be removed and not more than three buds should be left at the apex. Thus most of the one year old top shoot should be cut off because at the beginning of the development of the cuttings it is important that the ratio of leaf surface to root is not too high. If it is, the cuttings will die from drought.

5.3 The pump

Most scientists agree that the increased carbon dioxide content of the air is caused by human activities and that this increase is the reason why the temperature of the atmosphere is rising (Johansson, 2009). This process is the so-called Anthropogenic Global Warming, AGW. But there are also different opinions (Karlsson, 2008) even if so far they are rather few and taciturn .

So far, the most important human activity in this respect is the utilization of oil, but it is expected that in the future it will be the utilization of coal and fossil gases that will be the major contributors to the increased carbon dioxide in the atmosphere (Azar, 2008). In some countries scientists are discussing and working to develop an economic method to store carbon dioxide securely deep in the underground (Johnsson, 2009). In a joint project, named BECCS (Bio-Energy with Carbon Capture and Storage), scientists in Norway and Sweden are collaborating to develop methods for a combination of biomass utilization and geological carbon dioxide storage (Karlsson et al., 2009). In one of the biggest enterprises in Sweden, Vattenfall , the possibility of trying to store carbon dioxide from burning coal in some deep clay layers is being investigated. Similar research is taking place in Austria and the USA.

If such underground storage can be expected to be able to hold the carbon dioxide permanently, an interesting situation will emerge. When we are growing biomass, as for example with fast-growing willows and poplars, the carbon dioxide of the air is bound in organic compounds in the leaves. In this way the carbon dioxide content of the atmosphere will decrease. The more fast-growing species there are, the more carbon dioxide is taken away from the atmosphere. The carbon dioxide is normally released again to the atmosphere, when we burn, eat or ferment the biomass or when it rots.

But if, in future, it becomes possible to pump the carbon dioxide, released from burning of biomass, to deep clay layers in the soil and retain the carbon dioxide there, we have created a pump of carbon dioxide from the air to the soil via the plants, and for the first time in history human beings will have developed a method to actively decrease the carbon dioxide in the air (Figure 3).

From the biomass energy we produce biofuels, electricity and /or heat and the wonderful situation arises that the more energy we use, the more biomass we must utilize, the more cars we drive, the more biomass must be grown, the more carbon dioxide disappears from the air and we can keep global warming, the AGW, to a minimum.

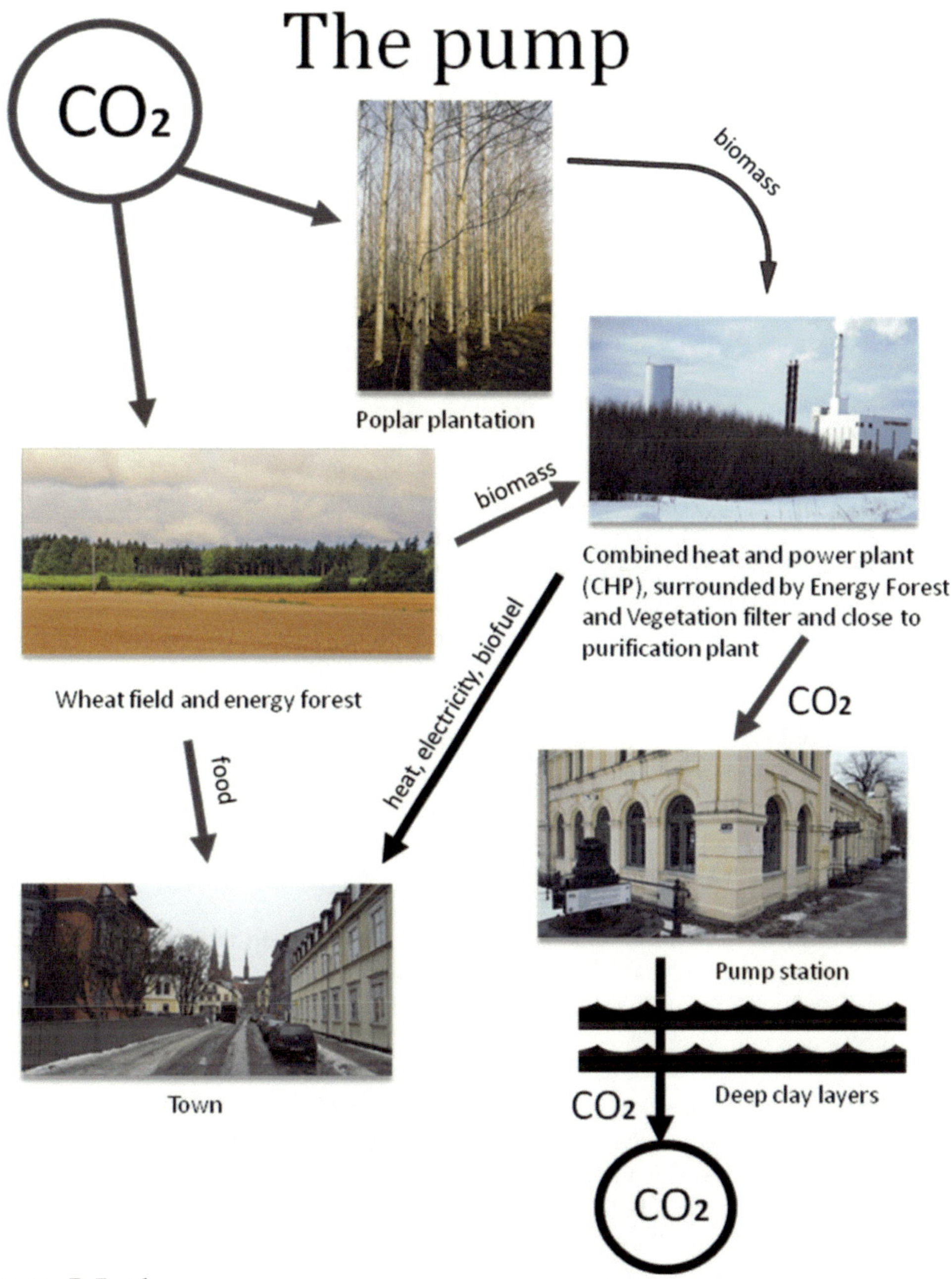

Illustrator: R. Reutler

Fig. 3. The pump. Poplar plantations, energy forest, vegetation filters and even agricultural crops such as wheat, take up CO_2 from the air and store the carbon molecules in the biomass. After harvest the biomass is transported to a combined heat and power plant (CHP) , the biomass is burned and the energy is transformed into biofuels, electricity and/or heat for the community. At the same time CO_2 is released and pumped deep down into soil layers that can retain the molecules permanently.

6. Acknowledgement

I am very grateful to technicians Jan Månsson, Göran Snygg and Torbjörn Andersson for their help with the measurement of the tree height at Kadesjö and Näsbyholm in 2010 and to Dr. Margaret Jarvis for correcting the language.

7. References

Andersson, K. 2010. Our view on ILUC and some comments on the Wibe report. Seminar at KSLA, Stockholm, Sweden, October the 27th, 2010.

Anonymous, 2006. På väg mot ett oljefritt Sverige. Kommissionen mot oljeberoende, Stockholm, 45 pp.

Anonymous, 2009. Energy in Sweden, 2009. Swedish Energy Agency, Eskilstuna, Sweden

Anonymous, 2008. Energy policies of IEA countries.---Review, Sweden, OECD/IEA

Anonymous, 2010. Climate challenge-the safety´s off.-Antologi, Formas Fokuserar.

Azar, C. 2008. Makten över klimatet.-A. Bonniers förlag, Stockholm, Sweden

Börjesson, P. 2007. Produktionsförutsättningar för biobränslen inom svenskt jordbruk.- Lunds Tekniska Högskola, Rapp. nr 61)

Cannell.M.G.R. 1989. Physiological basis of wood production: a review. Scand.J.For.Res. 4:459-490

Christersson, L. 1987. Damage to conifer seedlings by summer frost and winter drought. In Plant Cold Hardiness (ed. P. Li). 203-210. A.P.

Christersson,L. 1996. Future research concerning hybrid aspen, and hybrid poplar cultivation in Sweden. - Biom. & Bioenergi, 11, 2/3 109-113.

Christersson, L. 2006. Biomass production of intensively grown poplars in the Southernmost part of Sweden---observation of characters, traits and growth potential. Biom.Bioe. 30:497-508.

Christersson, L. 2008. Poplar plantations for paper and energy in the south of Sweden.--- Biom.Bioenr.32:997-1000.

Christersson,L 2010. Wood production potential in poplar plantations in Sweden. Biomass & Bioenergy, vol. 34, 9: 1289-1299

Christersson,L., Perttu,K., Ledin,S. and Elowson,S. 2008. Energiskog---odling av snabbväxande Salixarter.---In: SLU-tre decennier. Uppsala, Sweden

Cooper,J.P. 1975. Photosynthesis and productivity in different environments.(ed). Cambridge University. Press.Cambridge, London.

Dickmann,D. and Kuzovkina,L. 2008. Poplars and willows in the world.IPC Thematic Papers. FAO, Rome, Italy.

Doherty, S., Nilsson,P-O. and Odum,H. 2002. Energy evaluation of forest production and industries in Sweden.---Rep. no 1. Dept.Bioenerg. SLU, Sweden

Elowson,S. 1999. Willow as a vegetation filter for cleaning of polluted drainage water from agricultural land.---Biom. & Bioenerg. 16: 281-290.

Ilstedt, B.1996. Genetics and performance of Belgian poplar clones tested in Sweden.-For Gen 3:183-195.

Johansson, B. 2009. Climate challenge --- the safety´s off.---Formas Fokuserar 487 p.

Johnsson, H. 1953. Hybridaspens ungdomsutvecklingoch ett försök till framtidsprognos. Sv Skogsv för Tidskrift 51:73-96.

Johnsson, F. 2009. Carbon dioxide storage – an essential element in tackling global climate change. In " Climate challenge - the safety´s off". Formas Fokuserar, ed. Brigitta Johansson.

Karacic,A. 2005. Production and ecological aspects or short rotation poplars in Sweden Doctoral thesis no. 13, Faculty of National Resources of Agricultural Sciences, SLU, Uppsala, Sweden

Karacic,A., Verwijst,T. and Weih,M. 2003. Above-ground woody biomass production of short-rotation Populus plantations on agricultural land in Sweden. Scand.J.For.Res. 18427-437.

Karlsson,E. 2008. Jordens albedo och global medeltemperatur. Polarfront 133, 35:26-32

Karlsson,H., Byström,L. and Wiklund,J. 2009. BECCS (Bio Energy with Carbon Capture and Storage).--- Biorecro, Sweden.
www.biorecro.se

Lawson, Nigel. 2008. An Appeal to Reason.---SNS, Stockholm, Sweden

Lindroth,S. and Åkerblom,K. 1984. Areas for Energy forestry in Sweden. Dept. Forest Survey, SLU, Umeå, Sweden.

Lundegårdh,P., Lundquist,L. and Lindström,M. 1967. Berg och jord i Sverige. Almquist & Wiksell, Uppsala, Swedcen.

Lundström,C.F. 1899. Om poppelplanteringar till fyllande af den otillräckliga tillgången på asp.---P.A. Norstedt & Söner, Kungl. Boktryckeriet, Stockholm, Sweden

Perttu,K. and Kowalik,P. 1997. Salix vegetation filters for purification of waters and soils. Biom.&Bioenerg. 12(1):9-19.

Rytter,L.and L-G. Stener. 2005. Productivity and thinning effect in hybrid aspen stands in southern Sweden.--- Foret. 78:285-295

Rytter,L., Johansson,T., Karacic,A. & Weih, M. 2011. Investigation for a Swedish research program on the genus Populus. SkogForsk, Uppsala, Sweden

Skogsstyrelsen, 2008. Skogliga konsekvensanalyser 2008 - SKA-VB 08.Skogsstyrelsen, Rapport Nr 25, Jönköping, 146 s.

Sirén,G. 1974. Short-Rotation Forestry---a Chance to Bridge over the Imminent Pulp Wood Gap.---Sv.Skogsförb.Tidsk. 72:2, 315-325.

Sirén,G., Perttu,K., Eckersten,H., Linder,S., Christersson,L. and Sennerby-Forsse,L. Energy Forestry.---The National Swedish Board for Energy Sources Development (NE). 1983:11

Stanton,B. 2009. The domestication and conservation of Populus genetic resources.---Poplars and willows of the world. Chapter 4a. FAO Forestry Division Working Paper, IPC/9-4a. 92 pp.

Steenackers,V. 1990. Forty years of poplar research in Geraardsbergen.- Gaverstraat 4-9500. Belgium

Stener,L-G. 1998. Analys av fiberegenskaper för kloner av hybridasp.---In: SkogForsk, No. 387, Uppsala, Sweden

Werner,M. 2010. Föreningen skogsträdsförädling, från ursprunget 1936 och deförsta 70 åren--- Föreningen Skogsträdsförädling, Uppsa, Sweden.

9

Theoretical and Practical Evaluation of Jatropha as Energy Source Biofuel in Tanzania

Stelyus L. Mkoma and Faith P. Mabiki
Faculty of Science, Sokoine University of Agriculture
Tanzania

1. Introduction

Sustainable energy production and supply are strategic objectives for developed as well as developing countries. The energy sector plays a crucial role in attaining the United Nations Millennium Development Goals (Short, 2002), and the sustainability of modern economics is based in part on the capacity of countries to ensure their energy supplies. This is especially true for the transport sector, which consumes 30% of the world energy production, 99% of which is petrol-based (IEA, 2008). Global energy supply is currently mainly based on fossil fuels, which have many disadvantages. It is now widely agreed that more sustainable alternative energy sources will need to be developed. One potentially promising option consists of biofuels, since these are derived from biomass, have a closed carbon-cycle and do not contribute to the greenhouse effect. The biomass necessary for the production of biofuels can be derived from several sources; oil-producing crops are prominent among these. Due to relatively faster crop growth in the tropics, as well as the substantial land requirements for large-scale production, developing countries could potentially play a substantial role in the cultivation of oil-producing crops to yield major potential economic and environmental benefits for these countries. Notably, they could help to create additional income for the rural poor, and alleviate countries' balance of payments constraints by lessening oil import dependency or even by yielding export revenue. A gradual transition of the dominant energy regime in these countries from fossil fuels towards biofuels could thus have many advantages.

Current biofuels are actually based on traditional food crops such as maize, rapeseed or sunflower. A wide range of energy and global greenhouse gas budgets has been reported for them, although they are generally favourable compared with conventional fossil fuels like gasoline and diesel (Hill et al., 2006). However, these types of feedstock raise concerns because they are directly linked to food security issues? Also their cultivation is fuel-fertilizer and pesticide-intensive, with significant impacts on ecosystems. People living in the large part of the African, Asian and Latin America continents often lack access to energy sources in general. One approach to provide people with energy to increase living standards is to enable them to produce energy from local resources. The use of *Jatropha curcas* (Linnaeus), an inedible crop appears to be the promising alternative of local renewable energy source for people living in tropical and subtropical regions. When people use the term Jatropha, usually they refer to *Jatropha curcas* L. which is one of the 170 known species of this plant. Jatropha is a small wild plant belongs to the family Euphorbiaceae and is indigenous to Latin America and now found in all the tropical and subtropical zones (30°N; 35°S) of Africa and Asia. All parts of the plant, including seeds, leaves and bark, fresh or as a decoction, are toxic to humans and animals due to the presence of phorbol esters (Jongschaap et al., 2007).

Jatropha can be utilized for various purposes of which the application as fuel is probably the most interesting one from both an economical and ecological point of view. Nevertheless the other uses are worth mentioning as they provide insight in the total value chain of Jatropha products. Jatropha can be used for biogas production from its press cake formed during oil production. However, the press cake also has a wide variety of applications depending on local circumstances e.g. as a fertilizer (Kaushik et al., 2007). Jatropha can also be used to prevent and/or control soil erosion, to reclaim exhausted land (Benge, 2006), as a medicinal plant, be planted as a commercial crop, grown as a natural fence, especially to contain or exclude farm animals. In Tanzania the Swahili name for Jatropha is "Mbono Kaburi" (graveyard tree). It is known so because it was traditionally planted to mark a grave whenever someone dies. However, different domestic and traditional belief on the plant is excised in different parts of Africa and the world; these include guard against misfortunes reported in Brazil (Augustus et al., 2002).

Although Jatropha grows naturally in Africa, its cultivation on an economic scale is a recent venture for which little reliable scientific data exists either for environmental assessment or management. At present, the main agro-environmental impact studies in East African countries are largely qualitative, this include countries like Kenya (Kalua, 2008) and Tanzania (Eijck & Romijn, 2008). The Kenya biodiesel association created in 2008 to regulate and promote the production of Jatropha methyl ester in the country proposed to allow a 3% blending of biodiesel in conventional diesel fuel (Kalua, 2008). In Tanzania, the development of Jatropha biofuels is still in an early phase, and that its future is still unclear. Despite the favourable constellation of many contextual 'landscape' factors, there remain prominent barriers within Tanzania's existing energy regime (Eijck & Romijn, 2008).

This chapter reports on some important recent results from the Jatropha research in Tanzania. The chapter examines the *Jatropha curcas* L. as energy source biofuel in Tanzania. It also analyses the theory and practical evaluation of biofuel as renewable source of energy and Jatropha production, use and its application which have pronounced effects on the environment and economic aspects. Further, the chapter will concentrate on the production and energy potential of biogas from Jatropha press cake and provide an overview of the critical aspects, issues and best practice for sustainable exploitation of Jatropha in Tanzania.

2. Biofuel as renewable source of energy

An alternative to the oil-based energy production is provided by the biofuel, which can be produced from any biological material like plants, in contrast with fossil fuel like carbon, which is derived from long dead biological material. In developing countries, biofuel energy production could be a better alternative to solar energy system. Initial investment in respect to a solar installation and the cost of maintenance are limited. Moreover the solar-based energy production is not always suitable for rural regions and generally not easily reachable. Advantages of biofuels are mainly related to the fact that they are renewable sources of energy and can recover the use of diesel generators without further investments in new technologies. Also, biofuels are "carbon neutral" which means that the carbon released during the use of the fuel is reabsorbed and thus balanced by the carbon used by new plant growth if environmental conservation principles are observed.

Among different biofuels; the most important can be divided in ethanol based and oil based. The ethanol-based methods use sugar crops (sugar cane or sugar beet) and starch (corn or maize) to produce gas-fuel. Bioethanol fuel is mainly produced by the fermentation process, although it can also be manufactured by the chemical process of reacting ethylene with steam. There is also ongoing research and development into the use of municipal solid wastes to produce ethanol fuel. Globally renewable energy policies have promoted rapid growth in the biofuel economy. Currently, Brazil, Europe, and United States account for roughly 90% of total ethanol and biodiesel production (Coyle, 2007). Brazil is the first country where bio-ethanol is a real alternative to oil-based fuel (Brown, 2008). Since 1977 the Brazilian government made mandatory to blend 20% of bio-ethanol with gasoline, requiring just a minor adjustment on regular gasoline engines. Today the mandatory blend is allowed to vary nationwide between 20% to 25% bio-ethanol and it is used by all regular gasoline vehicles, plus three million cars running on 100% anhydrous bio-ethanol and five million of dual vehicles. The oil-based biofuels use oil plants are oil palm, soybean, algae or Jatropha. However, in United State of America the most common crop used for ethanol production is maize and for biodiesel feedstocks is soybean (as in Brazil) while in Europe rapeseed is the most common crop used for biodiesel all of which with large differences in energy efficiency between the production lines (Brown, 2008). In Brazil, soybean oil is a source that is already scaled up for biodiesel production. Nevertheless, other sources such as palm oil, algae, Jatropha, coconut, sunflower, peanut, cotton, and castor oil may be used in the near future once their cultivation achieve an economic up-scaling (Pinto et al., 2005). Oil plants-based biofuels can be a solution to the controversial question of "food vs fuel". In fact prices on a number of food types used for biofuel have doubled in the last couple of years even though biofuel is not the only cause. However, the impact of food price's increase is greater in poorer countries since demand for fuel in rich countries is now competing against demand for food in poor countries (Henning, 2003). For the mentioned reasons this chapter consider the suitability of the biofuel production using Jatropha plants.

3. Jatropha production, use and its application

Tanzania is globally placed to be the leader in biofuel production because of having ideal geographic and climatic conditions for growing a wide range of bio-fuel crops such as: sugar cane, palm oil, Jatropha, soy bean, and cotton. Of the 94 Mha total area of Tanzania 44.4 Mha is potential land available for agricultural investments. Jatropha is well promoted in Tanzania and investments have been reported to increase with strong political support. Currently, estimation for planted Jatropha in Tanzania is 17,000 ha which is 1.9% of the global cultivation and 14.4% of the total cultivation in Africa. Tanzania is considered very important for Jatropha cultivation sector with an estimate of up to 69,870 ha in 2010 to 620,110 ha in 2015 (GEXSi, 2008).

The plant *Jatropha curcas* (Linnaeus) belongs to the family Euphorbiaceae and is indigenous to Latin America and naturalized throughout tropical and subtropical parts of Asia and Africa. When people use the term Jatropha, usually they refer to this species which is one of the 170 known species of this wild plant (Figure 1) (Augustus et al., 2002; Akintayo, 2004; Jongschaap et al., 2007). Jatropha is a bush tree that is able to survive on marginal lands and can get up to 6 or 8 meters high. This perennial bush starts to grow fruits from the 2nd or

3rd year and can live up to 50 years. Under favourable conditions it can grow to a thick bushy fence of approximately one meter high in 6-9 months (Augustus et al., 2002). Jatropha grow under a wide range of rainfall between 200 and 1500 mm and specifically at 600 mm in moderate climates and 1200 mm in hot climates. In addition to water the plant needs nutrients which increase seed development and yield which varies significantly, with higher yields in fertile lands than marginal lands (Openshaw, 2000; Benge, 2006). Seeds from the Jatropha tree are known by many different names but "physic nut" is the most common and in Tanzania it is known as "Mbono". Studies have been conducted on the composition and properties of Jatropha seeds which also provide insight in the possibilities of using Jatropha oil for fuel purposes (Openshaw, 2000; Henning, 2003).

Fig. 1. Left: *The flowering Jatropha plant.* Right: *Representation of close-up of Jatropha fruits* (Photo taken at Sokoine University of Agriculture)

Jatropha being a woody plant can be used for a number of purposes as reported to in different parts of the world. In Tanzania the common uses of Jatropha include hedging, solid fuel, medicines, marking grave yards, supernatural beliefs, soap making, fertilizer and biogas production. These uses have been discussed in various sections in this chapter. Like in most rural areas in Africa, Jatropha plant in Tanzania is commonly used as a natural fence or live protection. The Swahili name for Jatropha in Tanzania is known so "Mbono Kaburi" (graveyard tree), because it was traditionally planted on graves. The reason behind this could be that Jatropha cuttings can be established in any season hence be used to mark a grave whenever someone dies. The Haya a tribe in Kagera region in Lake Victoria zone they call Jatropha "Mwitankoba" meaning a thunder killer tree. Traditionally every house was supposed to have at least one Jatropha tree near it to prevent the house from being destroyed during a thunder storm. In Tanga region (North eastern Tanzania) the Jatropha was used in supernaturally guided ordeals by the Sambaa community in Usambara mountains to determine the guilt or innocence of the accused. Accused persons had to consume the poison; the innocent vomited whereas the guilty died (Fleuret, 1980). Different traditional belief on the plant is excised in different parts of Africa and the world.

Jatropha plant can be used for medicinal purposes due to the medicinal properties of the bark and leaves which contain dye and latex. However, seeds, fresh or as a decoction are also used as traditional medicine and for veterinary purposes (Heller, 1996). Researches show that extracts from all parts of the Jatropha tree show different biological activities and pesticidal properties such as antibacterial, antifungal, insecticidal, fungicidal, molluscidal, blood coagulation and pain control (Heller, 1996; Jongschaap et al., 2007). The presence of the phorbol esters and a combination of lots of other compounds allow synergy of activity that have a broad and strong biological activity. The reported medicinal use of Jatropha in Tanzania include; the soap or oil used as skin antifungal applied external on infected part, the decoction of stem/root bark or leaves used to control hernia for boys, cough treatment and pain relief agents. The milky sap from Jatropha is used as a blood-clotting agent, in all these no scientific doses are seriously considered and it is prepared at locally through experience (Jongschaap et al., 2007).

Jatropha can be utilized for various purposes of which application as transport fuel is probably the most promising for the Jatropha seeds from both an economical and ecological point of view. Nevertheless the other uses as discussed earlier are worth mentioning as they provide insight in the total value chain of Jatropha products. The potential work that a fuel can do is determined by its energy content (Singh et al., 2008). The calorific values of solid and liquid Jatropha fuels as compared to those of conventional fuels are shown in Table 1. Most Jatropha related activities as well as this chapter are centred on liquid fuel (pure plant oil and biodiesel). Vegetable oil that has not been treated apart from filtering is often referred to as pure plant oil or straight vegetable oil. Jatropha oil is just one of a wide variety of pure plant oils. The main reason to use pure plant oil instead of converting it to biodiesel is the costs involved in the transesterification process. Compared to conventional diesel the use of pure plant oil in a diesel engine reduces the emission of sulphur oxides, carbon monoxides, poly-aromatic hydrocarbons, smoke, particle matter and noise. The main disadvantage of pure plant oil on the other hand is its high viscosity than normal diesel or biodiesel that leads to unsuitable pumping and fuel spray characteristics. The Jatropha oil is a potential material for fuel production with viscosity of 34-36 cST as compared to other plant oils such as soybean (31 cST), cottonseed (36 cST) and sunflower (43 cST) (Akintayo, 2004; Knothe & Steidley, 2005; Agarwal & Agarwal, 2007).

Parameter	Gross** calorific value (MJ/Kg)
[a]Jatropha seed	20.852 ± 0.08
[a]Jatropha press cake	18-25.1*
[a]Jatropha oil	37.832 ± 0.08
[a]Gasoline	47.127
[a]Crude oil	44.091
[b]Diesel fuel	46
[b]Kerosene	47
[b]Wood (15% water)	16
[b]Cooking coal (1-4% water)	35-37
[b]Coal (general purpose and 5-10% water)	32-42

Source: [a]Augustus et al., 2002 and [b]www.kayelaby.npl.co.uk
*depends on residue oil content in the press cake
**water formed and liberated during combustion is in the liquid phase

Table 1. Comparison of calorific value of Jatropha oil and seeds to conventional fuels

Biodiesel is a fuel obtained from mixtures, in different proportions, of fossil diesel and alkyl esters of vegetable oils or animal fats. Biodiesel is the fatty acids alkyl ester made by the transesterification of seed oils or fats, from plants or animals, with short chain aliphatic alcohols such as methanol and ethanol (Figure 2). Jatropha biodiesel have been pointed out to be among the twenty-six best suited fatty acid methyl ester mixtures (from a 75 species comparison) (Mohibbe et al., 2005). The important values for selection of fatty acid methyl esters for biodiesel are the cetane number (the ability of a fuel to ignite quickly after injection), iodine value (which indicates the degree of unsaturation) and other variables like linolenic acid value, boiling point and carbon chain length (Krisnangkura et al., 2006).

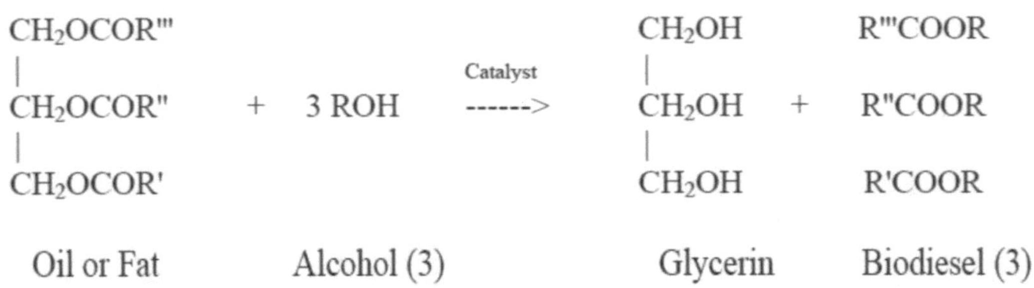

Fig. 2. The chemical reaction equation for catalyzed biodiesel production (Transesterification)

4. Jatropha biofuel by-products

The by-products of fresh Jatropha fruits include the shell, the hull, and the press cake. The fresh Jatropha fruit contains about 35–40% shell and 60–65% seed (by weight) of which 40–42% husk/hull and 58–60% kernels which consists of about 50% oil. The fruit shell is reported to contain about 34% cellulose, 10% hemi cellulose and 12% lignin and is good in minerals (Singh et al., 2008). In Tanzania is common to press the oil from peeled fruits and the shells are obtained after peeling the fruits leaving the seed with 100% hull to be pressed. Scientific investigation on the shell reveals its potentials for biological conversion and energy source as powdered briquettes and it high ash content (14.88%) a potential for fertilization of the soil. Using an up flow anaerobic digester filter, biological conversion of the shells after pre-treatment to remove fibres was possible producing biogas with 70% methane. Despite all that the shell briquettes still offers a more green energy opportunity for domestic and industrial fuel (Singh et al., 2008).

The Jatropha seed hull/husk is known to contain 3.97% ash, 71.04% volatile matter and 24.99% fixed carbon on dry weight basis, 10% moisture and the calorific value of 4044 $kcalKg^{-1}$. Laboratory gasification of the husks reached a maximum efficiency of 68.31% at a gas flow rate of 5.5 m^3h^{-1} and specific gasification rate of 270 $Kgh^{-1}m^{-2}$, at this point the calorific value of the gas is 1105 $kcalm^{-3}$. Biological conversion can be possible as well due to its high content of organic matter and this one contributes much to the Jatropha press cake organic matter because husks is left with the seed for pressing. Currently, the seed hull can not be counted separately as a source of energy since it part of the press cake.

The Jatropha press cake is obtained after separating the oil from the other seed contents chemically or mechanically. In Tanzania oil is obtained mechanically using a ram press or a screw press. Jatropha press cake in Tanzania is used as fertilizer, animal feed, source of biochemical and biogas production for energy source. Jatropha press cake can be used as organic manure, it nutrient content is richer than cow dung and neem cake and comparable to chicken manure. It contains macroelements nitrogen, phosphorus, potassium, sulphur and micronutrients Fe, Mn, Zn and Cu in a range from 800 to1000, 300 to 500, 30 to 50 and 18 to 25 $mgKg^{-1}$ of respectively. Jatropha press cake has also a nitrogen content of up to 6%, similar to that of castor beans and chicken manure. These increase the yield of Jatropha seeds up to 120% over the control without manure treatment. In Tanzania the Jatropha press cake is used as fertilizer without any scientific standards, research is needed in the area. The great hope behind many growers of Jatropha in Tanzania was to utilize the press cake for animal feed and selling to gain income. This has been a declined hope due to the fact that the Jatropha species grown in the country as of many parts of the world is toxic and yet the detoxification of cake is not of economically achievable and profitable at least at large scale. Therefore, the cake is only used as fertilizer and for energy source as briquettes or biogas production raw material (Janssen et al., 2005). Jatropha seed cake a source biochemicals and considered a best carbon sources among various carbohydrates, because it is pure, inexpensive and can be available in a mass supply. It has also been reported as good feed stock for production enzymes used by food industry (Muralidhara Rao et al., 2007).

5. Production and energy potential of biogas from jatropha press cake in Tanzania

Jatropha press cake is a good feedstock for biogas production because it is rich in organic matter; containing between 56%-64% crude protein (Benge, 2006). Biogas production from Jatropha press cake is about 60% higher as compared to the biogas generated from cow dung and had better calorific value as it had more methane (Lopez et al., 1997). More scientific investigations for anaerobic digestions of Jatropha press cake is required in order to draw a conclusion on its potentials to contribute as energy source.

5.1 Basics of the biogas process

Fermentation of organic matter by Microorganisms, allows the decomposition of organic matter aerobically or anaerobically. In aerobic decomposition the end products are carbon dioxide, heat and humus, during which most of the energy is lost as heat and production of new biomass. In anaerobic decomposition the main products are methane, carbon dioxide and peat or manure or sludge depending on the nature of process and the raw materials. When methane is produced under natural fossil conditions the gas produced is called natural gas. Biogas is normally produced through the process known as anaerobic digestion in swamps, marshes and intestinal tract of ruminant animals. Anaerobic digestion can be defined as the symbiotically stepwise process by which organic matter is digested to mainly methane (CH_4) and carbon dioxide (CO_2). It is a process due to concerted action of several metabolic groups of micro-organisms (Figure 3).

Biogas composed of a mixture of different gases including methane (50-75 Vol.%), carbon dioxide (25-50 Vol.%), hydrogen sulphide (50-6000 ppmv), Nitrogen (0-5 Vol.%), Ammonia (0-1 Vol.%), Oxygen (0-2 Vol.%), hydrogen peroxide (1-10 Vol. %) and water vapour (0-1 Vol.%) (Singh et al., 2008). The contents of the gases will depend on the type of the material composition of the organic matter used as feedstock and the process parameters and conditions during fermentation. The major difference between natural gas and biogas is the

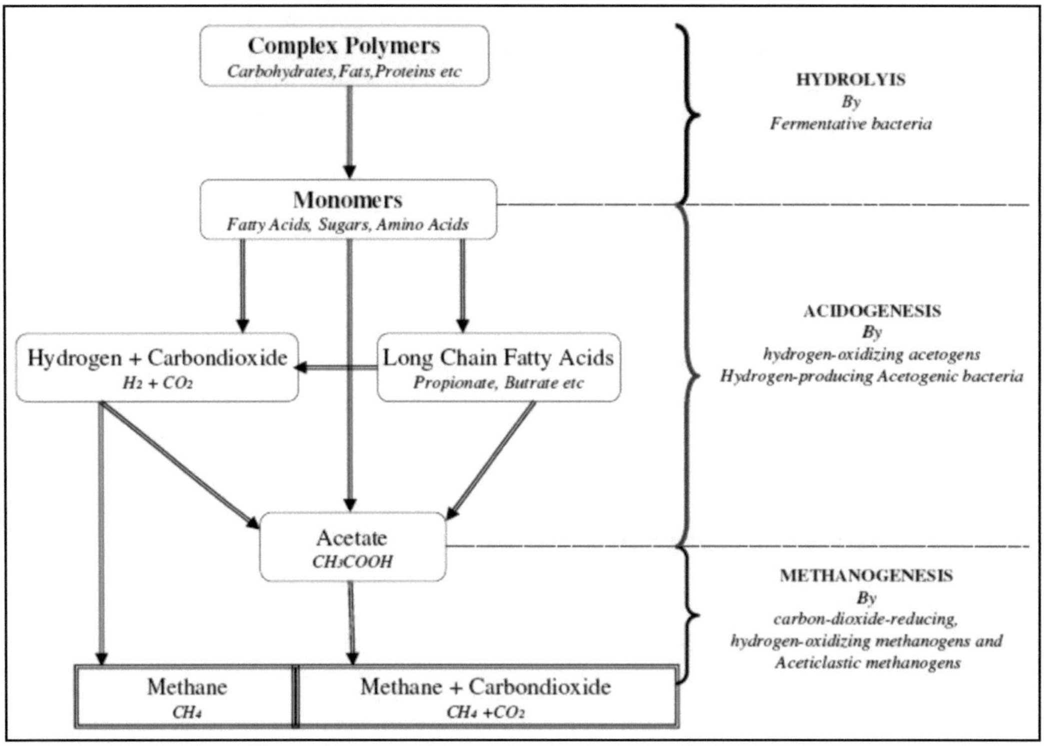

Fig. 3. Anaerobic digestion process

methane content and so it implies to their energy quality. The methane content is giving the energy quality of the gas and hence determines its application. The methane content of the biogas is usually measured at 50% to 80% by volume of the produced gas.

5.2 Process parameters and conditions

During the anaerobic digestion process parameters like pH, temperature and content of secondary gases in the digester must be controlled (Lopez et al., 1997; Wulf, 2005). Higher pH, value of the medium will allow gases such carbon dioxide and hydrogen sulphide to remain dissolved in the median and so methane yield is higher in the biogas produced. However, too high medium pH will again lead to accumulation of organic acids which can lower the pH of the medium. Because the methanogenic bacteria are pH sensitive, pH less than 6 and above 8 will limit the functioning and hence lower the production of methane most of the time, the buffer system is used to make the system stable throughout the process. Biogas producing can work at either psychrophilic (below 20 °C), or mesophilic (room temperatures/above 20 °C to 42 °C) or thermophilic conditions with temperatures between 42 °C -55 °C. However most of the bacteria work best at mesophilic conditions. At very low temperatures the microorganisms are very slow and the residence time can go up to 3 months. At thermophilic conditions the digestion is very fast (15-20 days) but the process becomes very temperature sensitive and promotes the volatile fatty acids formation. Sulphide content and ammonia gases content during digestion differs depending on substrate used, proteins high sources always can give higher sulphide and ammonia gases due to decomposition of protein, these can be controlled by pH stability and then the gas is purified or can be removed by additives, example sulphide can be removed by addition of iron salts in the medium.

The amount of biogas yield from anaerobic digester will depend on:

1. *The substrate composition*: Simplicity for degradation by type of microorganisms depends on structure components of the substrate ingredients, the values given in Table 3 are maximum values assuming that the substrate is fully converted usually 3% to 10% of the substrate is not available for biogas formation, these figures can realistically rise up to 30%-50% for proteins and fats respectively. Theoretically samples containing more fat/lipid contents will yield more gas and with high quality for different application (Table 2) relative to others under the conditions. These samples may include waste fat, grease and oils. The rate of digestion always depends on the structure complexity of the substrate and it should not be confused with the yield. Carbohydrates such as sugars are easily degradable than starch or protein so the protein will take more time to digest. On the other hand substrate like cellulose and hemicelluloses are even harder to degrade while the lignin is even not degradable.

Application	Energy type	Gas quality requirement
Heating/cooking	Heat	Low
Combustion in combined heat and power	Electric heat	
Compression and use in vehicles	Fuel	
Fuel cells	Electricity	
Compression and feeding in grid	Any	High

Source: Wulf, 2005.

Table 2. Showing biogas quality needed for different application

2. *Organic Total Solids (oTS)*: The amount of degradable material will determine the amount of biogas produced. In continuous process organic loading rate (KgoTS/m^{3}day) determines the yield. When the process can accommodate a high organic loading rate then this can imply higher yields. However on the other hand if the concentration of the organic total solids is too high, microoganisms can have a problem of adaptation and hence affects the yield and/or the retention time. Depending on the process and type of substrate, organic loading rate differs (Table 3).

Substrate	Theoretical biogas yield (N/kgoTS)	Theoretical biogas quality	
		Vol.% CH_4	Vol.% CO_2
Carbohydrates	746	50	50
Fats	1300	72	28
Proteins	800	60	40

Source: Wulf, 2005..

Table 3. Theoretical biogas yield and quality for different substrates

Biomass serves as the major source of energy in Tanzania providing about 90% of the total energy consumed, energy from oil and gas provides about 7% of the consumed energy (ProFOREST, 2008). Tanzania consumed as estimate of 25000 barrel of oil/day in the year 2005.

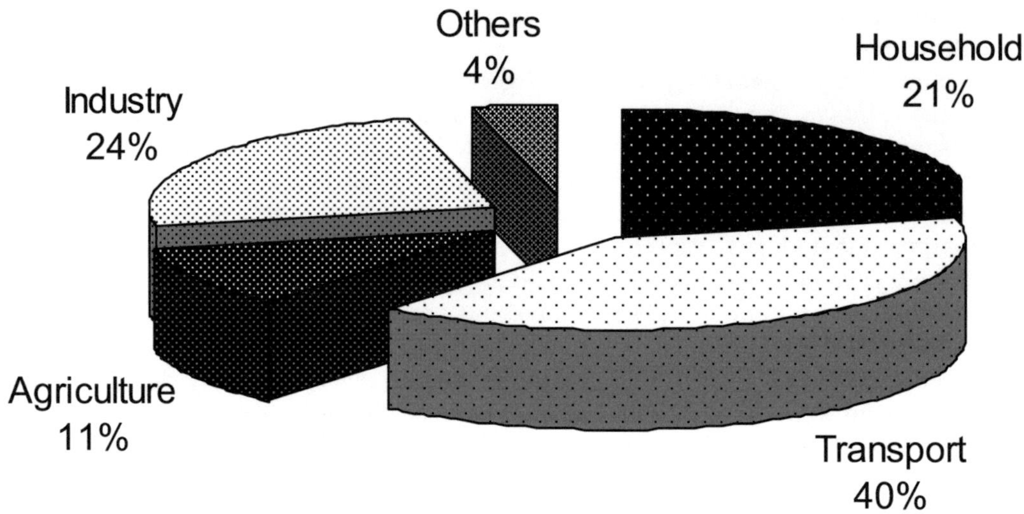

Fig. 4. Tanzania petroleum use pattern (Modified from Mgodo, 2008)

Two scenarios of Jatropha press cake availability are considered depending on the GXSI report on Jatropha Cultivation in Tanzania; experts' scenario as scenario 1 and GXSI scenario as scenario 2 (Table 4). In the scenario 1 the area estimated to be under cultivation is 11700, 34300 and 116000 for 2008, 2010 and 2015 respectively. For the scenario 2 the area identified under Jatropha cultivation is 17000, 69870, 620110 ha for 2008, 2010 and 2015 year respectively. Assuming that 6 T/ha constant yield of seeds per ha to 2015, 75% of the seed remain as press cake after pressing the oil, 4kg of seed gives 1L of oil and the 3Kg remains as press cake. The press cake is 100% available for biogas production, characterized by 92% oTS and 92% TS, biogas production is at 350 L/KgoTS with 65% methane. The proper technology is used resulting to efficient production and power losses during transmission are negligible.

Value added to 1L of oil pressed: One litre of oil is obtained by pressing 4 kg of Jatropha seed, leaving 3kg of press cake as a by-product. Biogas production from this cake will produce 578 L of biogas which is worthy 20.8 MJ of energy. Comparing to the energy value of 1 L of Jatropha oil which is 40.7 MJ, the value added is 51%. This will improve the Jatropha bioenergy system and contribute significantly in sustainability of Jatropha bioenergy system exploitation in Tanzania.

Contribution to National Energy Sector: The annual per capita electricity consumption in Tanzania was estimated to be about 80 kWh in 2005 while the annual energy consumption per capita was 29,300 MJ. Having these data as constant base values, with efficient exploitation of Jatropha press cake and use of the gas as energy source would contribute significantly to the energy sector in the country especially in rural areas with decentralization of the electricity distribution system. With scenario 1, there is a potential of 3.7×10^5 GJ existing currently which is mainly utilized as fertilizer and little biogas substrate in households, this potential is to be expected to increases up to 3.6×10^6 GJ in 2015. In scenario 2, the current potential is which 5.3×10^5 GJ which is 43% higher compared to the first Scenario. This as well is expected to increases up to 5.4×10^8 GJ in 2015 which is 500 times higher compared to the same in scenario 1. This energy amount can be used as source of heat for cooking at household level, or to generate electricity. The contribution of the energy in terms of Barrel of Oil Equivalent (BEO) in both scenarios ranges from 0.06 to 3.2 Million BEO per year which is equivalent to foreign exchange saving of 2-128 days due to importation of crude oil barrels. These energy potentials will therefore contribute to environmental conservation due deforestation, accelerate national development by increasing the number of Tanzanians with access to electricity; allowing small scale investments and business projects, these are directly linked to the millennium development goals and available national development strategies. However still this will need more research input in order to efficiently exploit the potentials.

Scenario	Year	Area under Jatropha cultivation (ha)	Amount of press cake (Tx10³)	Methane produced (m³)	Equivalent energy potentials		
					GJ	MWh	BEO
Scenario 1	2008	11,700	52	1.0×10^7	3.7×10^5	1.0×10^5	6.0×10^4
	2010	34,300	154	3.0×10^7	1.0×10^6	3.0×10^5	1.7×10^5
	2015	116,000	522	1.0×10^8	3.6×10^6	1.0×10^6	6.0×10^5
Scenario 2	2008	17,000	77	1.5×10^7	5.3×10^5	1.5×10^5	8.7×10^4
	2010	69,870	314	6.1×10^7	2.1×10^6	6.1×10^5	3.6×10^5
	2015	620,110	2790	5.4×10^8	1.9×10^7	5.4×10^6	3.2×10^6

Table 4. Energy potentials of the Jatropha press cake in two scenarios

For scenario 1, the potential for contribution to national annual electricity per capita is 134%, which is equivalent to 1.65% of the national total annual energy per capita in 2015. The corresponding values were 4% and 0.05% in 2008 and 15% and 0.19% in 2010. The energy can be exploited more efficiently at local levels to avoid power loss due to long distances and lack of capacity. In scenario 2, if the biomethane potentially produced is converted to electricity and fed in to the national electricity grid. It has a potential to contribute up to 25% in 2015 in the annual national electricity per capita, this contribute 0.31% of the total national annual energy per capita. The corresponding values were 3% and 0.003% in 2008 and 7% and 0.09% in 2010.

6. Critical aspects and issues on jatropha cultivation in Tanzania

The Government of Tanzania and international donors have identified biofuel as a priority growth sector and aimed at providing extensive support for investments in this sector. The hesitation is well revealed in the areas of land provision for biofuel crop cultivation though the country is politically stable and has favourable environment for business. On the other hand as the biofuel policy is not in place, government deferred new biofuel projects registration as it prepares a national policy on biofuels (Sawe, 2007). Though the biofuel issues are discussed very generally, it is worth noting that the pros and cons of the oil crops differ in many ways. The more focused biofuels crops in Tanzania include Jatropha, sugar cane, sisal and palm oil, this section focuses on the Jatropha. Billed as wonder crop, the establishment of Jatropha plantations on the ground in Tanzania has been far from successful, or, in some cases, ethical. Biofuel investment and production in Tanzania is a highly contentious issue. Biofuel investors have been doing business in Tanzania since 2000,

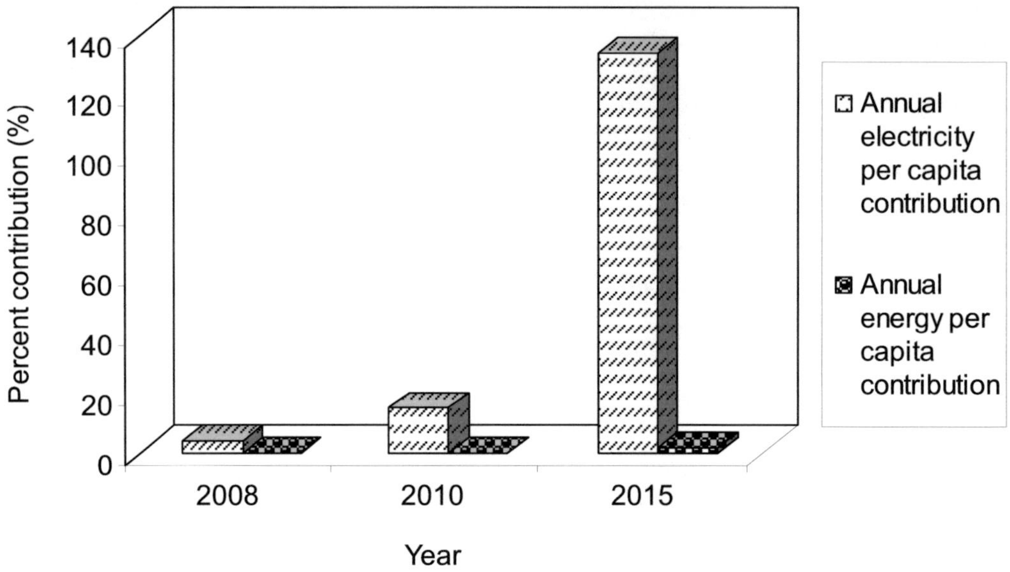

Fig. 5. Potential biogas contribution to Tanzania annual energy and electricity per capita in scenario 1

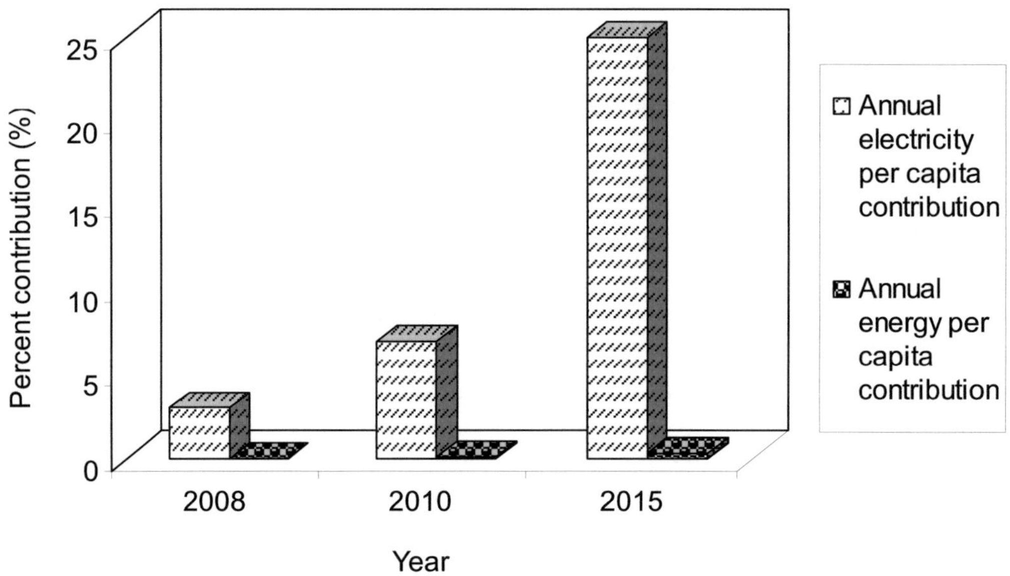

Fig. 6. Potential biogas contribution to Tanzania annual energy and electricity per capita in scenario 2

but business stepped up a gear after 2006. To date there are 17 investor companies, from UK, cane, sisal and palm oil, this section focuses on the Jatropha. Billed as wonder crop, the establishment of Jatropha plantations on the ground in Tanzania has been far from Germany, Sweden, the Nederland and America. This is a small number of investors compared to those in Brazil and Indonesia, but a number with clear motives. With over four million hectares requested by investors for biofuels only 650,000 hectares currently allocated, this is a sizeable potential earner for Tanzania (Sawe, 2007; Alweny, 2008).

Cultivation of Jatropha is going on and more investors and local farmers have shown interest of acquiring land for Jatropha cultivation. The Dutch based company Diligent Tanzania which has contracted about 5000 farmers for 10 years giving them support and advice on how to grow the Jatropha (Alweny, 2008). Another company the Sun Biofuels, a UK-based company, acquired 8,000 ha to grow Jatropha, part of a plan to expand capacity to 85,000 ha in the long-term. The company promises to pay workers $1095 per year for farming and harvesting and would devote an additional five percent of its budget towards "social infrastructure". PROKON Renewable Energy Ltd. Tanzania a German company in origin working in western Tanzania, with farmers cultivating more than 10,000ha of Jatropha supplied with Jatropha seed and extension services. Other Local institution and NGOs promoting Jatropha use in Tanzania include Tanzania Traditional Energy Development and Environmental Organization (TaTEDO), The Company For Technology Dissemination and Training (KAKUTE Limited-"Kampuni ya Kusambaza Teknolojia") and Tanzania Industrial Research and Development Organisation (TIRDO) (Sawe, 2007).

6.1 Water distribution and usage

Tanzania is surrounded by water bodies including the Indian ocean in the eastern part and great lakes, namely lake Victoria in the north, Nyasa in the south east and Tanganyika in the west of the country. There are rivers flowing throughout the year with many medium and small seasonal rivers. With this context Tanzania has good potential for irrigation of about 29.4 Mha, of which, currently only about 1% of the total potential have been developed. To date, Tanzania Jatropha cultivation is reported in areas with enough rainfall such as Arusha and Rukwa, and so no practical example from dry area such as those in central Tanzania. Large-scale systematic studies of the impact of Jatropha plantations on hydrological cycles is still lacking, however it is well documented that water use of Jatropha generally is low and that it is unlikely to have a negative effect on annual stream flow (ProFOREST, 2008; Jongschaap et al., 2007).

6.2 Food security

The decline of world food stocks particularly cereals and the rocketing of food prices in recent years have alarmed a threat to expansion of biofuels production (Janssen et al., 2005). Tanzania has currently not achieved food security, and therefore it is anticipated that converting main sites identified as suitable for growing food crops to produce biofuels crops is a threat to both accessibility and affordability of food. It has to be noted that food insecurity is not only caused by biofuel crops cultivation but poor farming practices and other factors which are not discussed in this chapter (FNT, 2008). To ensure food security the exclusive plantations cultivation should be avoided especially where it has to shift farmers and intercropping biofuel with food crops should be encouraged whenever possible. About 75% of Tanzanians are rural based and mostly engaged in crop production.

6.3 Environmental conservation and biodiversity issues

In order to suffice the external market during export production large scale biofuels production, monocultures cropping may be preferred over crop rotations or mixed farming. This can well be associated with deforestation and decrease in crop and farm biodiversity. Though the Biofuel policy is not in place the National Environmental Management Act is in place, under National Environmental Management Council (NEMC) ensures that all the investors in biofuels crop production carry out Environmental Impact Assessment (EIA) and adhere to recommended production practices. A number of research reports on the ability of Jatropha to conserve the environment in many ways including binding the soil, fertilizing, carbon conservation and air quality improvement However, these ability are subjected to more research to understand fully their effects if any (Soares Severion et al., 2007). The Massai plain near Arusha is very much endangered by erosion due to overgrazing. KAKUTE plants Jatropha trees to control erosion at Engaruka area in Arusha but failed because the field was near a water source for cattle and so they were destroyed. It is expected that other areas could also be affected hence research is needed.

6.4 Social impacts

Gender issues are imports to be looked on in Jatropha cultivation in Tanzania, who does the work, who gets the money, changes of the distribution of the workload, and changes of the social status. In Tanzania land is owned by the government which gives lease and permission to people owning it legally. Most of the Tanzanians especially those in rural areas do not have legal entitlement to their land, they own it by historical inheritance. The most people involved in production activities at lower levels are women, however their work is classifies as unpaid labour and so not counted in connection to the economic development of the nation. Women are the possible and cheap labour in many developing countries. It is reported that owners tend to prefer women workers, as they are able to pay them less than their male counterparts and find them a docile and dependent workforce, and are therefore more exploitable (Hurst et al., 2007). People without own farm land should have accessibility of seed, Jatropha in public forests for free collection. In Most part of Tanzania, there are wild and farm Jatropha trees. Jatropha trees planted as protection hedges, there is always an ownership of a family and only members of the family can allow collecting seeds. Other social issues include cultural and religious traditions or indigenous knowledge. In Tanzania, women are not allowed to own farm land and trees. But in both cases the responsible for the distribution of the land gave some plots to women groups to grow Jatropha there. Concerning indigenous knowledge the soap making with oil from different oil fruits is well known in Tanzania and has an old tradition.

6.5 Economic issues

In Tanzania the following economic evaluation of activities of the use of the Jatropha plant is based on experience of KAKUTE in its Jatropha project ARI-Monduli (Alternative Resources of Income for Monduli women). It is ascertained by KAKUTE that the collection of seeds and its sale gives the least added value as compared to oil extraction which also is not as good as soap making. This explains very clearly that the Massai women of Engaruka are not very much interested to sell seeds or oil, they want to gain the added value of the whole production chain and sell only soap.

7. Best practice for sustainable exploitation of Jatropha in Tanzania

Biorefineries are facilities that integrate biomass conversion processes that produce fuels, power, and chemicals from biomass. Biorefineries involves producing multiple products, taking advantage of the differences in biomass components and intermediates and maximize the value derived from the biomass feedstock. Using the concept of biorefinery therefore, it is possible to reduce amount of and increase safety of waste products, increase economic effectiveness of biomass source and valorisation of all plant material to useful and high value products (see Figure 7) (Zinoviev et al., 2007). There is clear indication that the Jatropha system in Tanzania can adopt the biorefinery concept as the industry is growing very fast with full support from investors, government and local people. The biorefinery concept could significantly reduce production costs of plant-based chemicals and facilitate their substitution into existing markets.

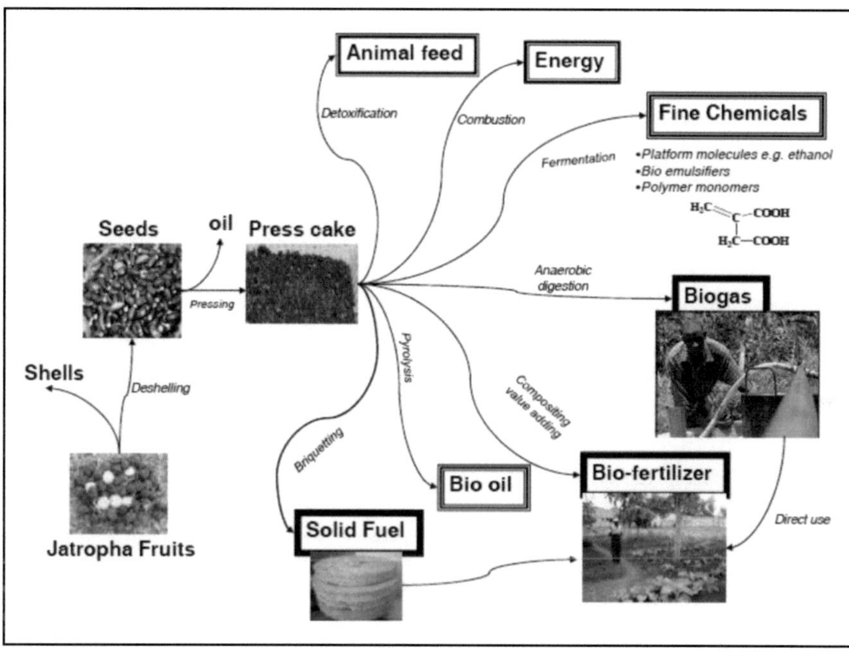

Fig. 7. The valorisation of Jatropha press cake for biorefinery potentials

Possible more use of the Jatropha by-products includes the utilization for biodiesel, production of chemicals, soap and lubricants and also utilization of Jatropha shells as an energy source, biogas production residues as fertilizers and biogas as fuel in engines, Jatropha press as substrate for enzyme production and platform chemicals (Figure 7). The biggest challenge in the exploiting the biorefinery concept are lack of experience and knowledge (best practices) on the Jatropha system i.e. still a young industry, lack of research to avail proper technologies and processes proper in the context of Tanzania, lack of capital, lack of biofuel policy and lack of local expertise.

The concept of people, planet and profit is very important to a country like Tanzania which still strives for the wellbeing of the people. The concept tries to guide the investor sand country governments to ensure the projects are sustainable for the middle and long term.

There should be no destruction of village or social structures, no infringement of common lands or traditional user rights and no displacement of people. There should be enhancement of local employment or income generation of local people, decent wages to be paid and preferably no dependency of a sole income source of people (risk avoidance). Planet is another criterion which takes care on what is real waste or idle land, minimal and no lasting environmental pollution in production by agro chemicals and fertilizers. There should be careful consideration of the sustainability of cropping. No selection of lands with high biodiversity importance and intercropping preferable, especially in the earlier years. Preparation of clear business plans, based on conservative/proven data, company profits preferably reinvested in the country, Jatropha should in first instance be used to supply internal markets. Local use is more energy efficient and there is always enough internal demand. Company profits sharing with farmers, and farmers decent payment and no excessive company profits.

8. Conclusion

The principle though originally focuses on risks for Jatropha large scale plantations but it fevers poverty reduction and the fulfilment of the millennium development goals (MDGs) in general (World Bank, 2006). Countries like Tanzania also aims to reduce poverty of her people but the environment the element of planet should be conserved and considered when biomass is utilized. On the other hand there is no business without profit then the investors have to be careful before and during investment. It is hereby highly recommended to take the precaution as the industry grows very fast. The ideas mentioned in the people, planet and profit concept are very important for sustainable exploitation of the Jatropha system products in Tanzania. In order to fully exploit the Jatropha products in Tanzania the following areas has to be seriously considered for advance:

- Speeding the formulation of biofuel policy by-products valorisation, oil production and blending. The biofuel policy will determine the investment environment and market for Jatropha but also pave a way for technology expansion and technology transfer.
- Research collaboration in the areas of technology and production with local institutions such Sokoine University of Agriculture (SUA) and other university and NGOs to fever the improvement of the local expertise and sharing of knowledge for the sustainability of the biofuel industry and specifically Jatropha system.
- Improvement of top down information and knowledge flow in order to facilitate the MDGs goals fulfilment and reduce conflict. Also improvement of technical know-how for mainstreaming and multiplication of technology

In Tanzania there are great opportunities for Jatropha biofuel industry and the biofuel industry in general. These include the land availability, supportive climatic conditions, water availability, political and peoples will and the need of the industry and its potentials towards attaining the MDGs. Already the short-term experience in Jatropha investment shows promising potentials in exploiting the products of Jatropha. Scenarios indicated positive contribution of the Jatropha bioenergy system to the energy sector in Tanzania. Jatropha press cake biogas alone has a potential to contribute 25% to 134% of the national annual electricity per capita by the year 2015, while adding more than 50% oil obtained after pressing of Jatropha seeds. Biorefinery of Jatropha products will also offer more energy contribution, wastes reduction and more value to Tanzania biofuel industry, particularly

Jatropha bioenergy system. Biofuel is an inter-sectoral industry; the success and sustainability of the industry in Tanzania will sorely depend on policy to be put in place.

9. Acknowledgement

The authors would like to acknowledge the International Centre for Science and High Technology of the United Nations Industrial Development Organization (ICS UNIDO) for sponsoring this work. The technical support form Faculty of Science, Sokoine University of Agriculture and the German Biomass Research Centre (DBFZ) is heartly appreciated.

10. References

Agarwal, D. & Agarwal, A.K. (2007). Performance and emissions characteristics of Jatropha oil (preheated and blends) in a direct injection compression ignition engine. *Applied Thermal Engineering*, Vol. 27, No. 13, (September 2007), pp. 2314-2323, ISSN: 1359-4311

Akintayo, E.T. (2004). Characteristics and composition of Parkia biglobbossa and *Jatropha curcas* oils and cakes. *Bioresource Technology*, Vol. 92, No. 3, (May 2004), pp. 307-310, ISSN: 09608524

Alweny, S. (2008). Potential boom for biofuels in Tanzania? Feature May 2008, Global facilitation unit for underutilized species, Available from http://www.underutilized-pecies.org/features/jatropha/salomes_jatropha_story.pdf

Augustus, G.D.P.S., Jayabalan, M. & Seiler, G.J. (2002). Evaluation and bioinduction of energy components of *Jatropha curcas*. *Biomass and Bioenergy*, Vol. 23, No. 3, (September 2002), pp. 161-164, ISSN: 0961-9534

Benge, M. (2006). Assessment of the potential of Jatropha curcas, (biodiesel tree) for energy production and other uses in developing countries, Available from http://www.ascension-publishing.com/BIZ/jatropha.pdf

Brown, L.R. (2008). Turning to renewable energy. *Chapter 12 in Plan B 3.0. Mobilizing to Save Civilization*, Elizabeth Doherty, (First Ed): Earth Policy Institute, W.W. Norton & Company, Inc., ISBN: 978-0-393-06589-3, New York, United States of America, Available from http://www.earth-policy.org/images/uploads/book_files/pb3book.pdf

Coyle, W. (2007). The future of biofuels: a global perspective. Amber Waves Nov 2007. USDA Economic Research Service Washington D.C., Available from http://www.ers.usda.gov/AmberWaves/November07/Features/Biofuels.htm

Eijck, J. & Romijn, H. (2008). Prospects for Jatropha biofuels in Tanzania: An analysis with Strategic Niche Management. *Energy Policy*, Vol. 36, No. 1, (January 2008), pp. 311-325, ISSN: 0301-4215

Fews Net Tanzania, FNT. (2008). Tanzania Food Security Update, Famine Early Warning Systems Network, Final report, January 2008, Available from http://www.fews.net/docs/Publications/Tanzania_2008_01_final.pdf

Fleuret, A. (1980). Non-food uses of plants in Usambara. *Economic Bot*any, Vol. 34, No. 4, (June 1980), pp. 320-333, ISSN: 0013-0001

Global exchange for Social investment (GEXSi) (2008). Global Market Study on Jatropha. Final Report 2008. Prepared for, the World Wide Fund for Nature (WWF) London/Berlin.

Heller, J. (1996). Physic nut *Jatropha curcas* L. Promoting the conservation and use of underutilized and neglected crops. Gatersleben: Institute of Plant Genetics and Crop Plant Research, and Rome: International Plant Genetic Resources Institute. Rome, Available from http://www.ipgri.cgiar.org/publications/pdf/161.pdf

Henning, R.K. (2003). *Jatropha curcas* L. in Africa. Baganì, Weissenberg, Germany.

Hill, J., Nelson, E., Tilman, D., Polasky, S. & Tiffany, D. (2006). "Environmental, economic and energetic costs and benefits of biodiesel and ethanol biofuels". PNAS Vol. 103 N° 30.

Hurst, P., Termine, P. & Karl, M. (2007). Agricultural workers and their contribution to sustainable agriculture and rural development. ISBN: 978-92-2-118709-7, Geneva, Switzerland, Available from http://www.ilo.org/wcmsp5/groups/public/---ed_dialogue/---actrav/documents/publication/wcms_113732.pdf

International Energy Agency, IEA. (2008). World Energy Outlook, executive summary. International Energy Agency/Organisation for Cooperation and Development, Paris, pp. 18.

Janssen, R., Woods, J., Sawe, E.N. & Pförtner, R. (2005). Liquid Biofuels for Transportation in Tanzania. Potential and Implications for Sustainable Agriculture and Energy in the 21st Century, Available from http://tatedo.org/cms/images/researchdocs/biofueltransport.pdf

Jongschaap, R.E.E., Corré, W.J., Bindraban, P.S. & Brandenburg, W.A. (2007). Claims and facts on Jatropha curcas L. Global Jatropha curcas evaluation, breeding and propagation programme. Plant Research International BV, Wageningen. Strichting Het Groene Woudt, Laren. Report 158, pp. 66.

Kalua, I. (2008). Stakes constraints of bio-fuels cultivation in East Africa. *Green Africa Foundation presentation, Vanatrop Workshop Non-food and energetic use of plants from Mediterranean and tropical areas, Agropolis International, Montpellier France*, October 22-24, 2008.

Kaushik, N., Kumar, K., Kumar, S., Kaushik, S. & Roy, S. (2007). Genetic variability and divergence studies in seed traits and oil contents of Jatropha (*Jatropha curcas* L.) accessions. *Biomass and Bioenergy*, Vol. 31, No. 7, (July 2007), pp. 497-502, ISSN: 0961-9534

Kearney, A.T. & van Eijck J. (2008). The Nairobichronicle "Biofuels growing in Tanzania, as Kenya lags behind" by AllAfrica.com, Available from http://nairobichronicle.wordpress.com/2008/08/09

Knothe, G. & Steidley, K.R. (2005). Kinematic viscosity of biodiesel fuel components and related compounds. Influence of compound structure and comparison to petrodiesel fuel components. *Fuel*, Vol. 84, No. 9, (June 2005), pp. 1059-1065, ISSN: 0016-2361

Krisnangkura, K., Yimsuwan, T. & Pairintra, R. (2006). An empirical approach in predicting biodiesel viscosity at various temperatures. *Fuel*, Vol. 85, No. 1, (January 2006), pp. 107-113, 0016-2361

Lopez, O., Foidl, G. & Foidl, N. (1997). Production of Biogas from J. curcas Fruit shells In: *Biofuels And Induslrial products From J. Curcas*, Gobitz, Mitielbach, Train (Ed): 118-122, Dvb-Verlag Für die Technische Universität Graz, ISBN: 3-7041-0242-3, Graz, Austria

Mngodo, J. (2008). Presentation at the Second FAO Technical Consultation on Bioenergy and Food Security (BEFS), Rome, 5-6 February, 2008. TACKLING THE NEXUS - Bioenergy and Food Security in Tanzania.

Mohibbe Azam, M., Waris, A. & Nahar, N.M. (2005). Prospects and potential of fatty acid methyl esters of some non-traditional seed oils for use as biodiesel in India. *Biomass and Bioenergy*, Vol. 29, No. 4, (October 2005), pp. 293-302, ISSN: 0961-9534

Muralidhara Rao, D., Jaheer Hussain, S.M.D., Pandu Rangadu, V., Subramanyam, K., Sivarama Krishna G. & Swamy, A.V.N. (2007). Fermentatative production of itaconic acid by Aspergillus terreus using Jatropha seed cake Short Communication. *African Journal of Biotechnology*, Vol. 6, No. 18, (September 2007), pp. 2140-2142, ISSN: 1684–531

Openshaw, K. (2000). A review of Jatropha curcas: an oil plant of unfulfilled promise. *Biomass and Bioenergy*, Vol. 19, No. 1, (July 2000), pp. 1-19, ISSN: 0961-9534

Pinto, A.C., Guarieiro, L.L.N., Rezende, M.J.C., Ribeiro, N.M., Torres, E.A., Lopes, W.A., Pereira, P.A.de P. & de Andrade, J.B. (2005). Biodiesel: An Overview. *Brazilian Journal of Chemical Society*, Vol. 16, No. 6B, (November/December 2005), pp. 1313-1330, ISSN: 0103-5053

ProForest (2008). Jatropha literature and perspectives review: Main potential social and environmental impacts arising from large scale plantings', ProForest Ltd., Oxford.

Sawe, E.N. (2007). Presenation by Director TaTEDO at the Eastern and Southern Africa Regional workshop on Biofuels, 28--29 June 29 June 2007, Nairobi, Kenya. Liquid Biofuels, Small Scale production and use, (issues, challenges and opportunities (UNEP).

Short, C. (2002). Energy for the Poor. Underpinning the Millennium Development Goals. Department for International Development (DFID), government of United Kingdom, London. Produced for DFID by Future Energy Solutions. DFID 1 Palace Street London SW1E 5HE. p. 32.

Singh, R.N., Vyas, D.K., Srivastava, N.S.L. & Narra, M. (2008). SPRERI experience on holistic approach to utilize all parts of Jatropha curcas fruit for energy. *Renewable Energy*, Vol. 33, No. 8, (August 2008), pp. 1868–1873, ISSN: 0960-1481

Soares Severion, L., De lourdes Sila de Lima, R., Bezerra Leoao, A. & Esberard de Macedo Betrao, N. (2007). Root system Charachterization of Jatropha curcus Plants propagated through five Methods. Expert Seminar on Jatropha curcus L. Agronomy and genetics. 26 -28 March 2007, Wageningen, Nertherlands.

The PROKON Renewable Energy Tanzania Ltd. Available from http://www.prokon-tanzania.com/tanzania.html

World Bank (2006). Agriculture and Achieving the Millennium Development Goals Washington, D.C.: Agriculture and Rural Development Department of the World Bank, Availbale from http://siteresources.worldbank.org/INTARD/Resources/Ag_MDGs_Complete.pdf

Wulf, S. (2005). Anaerobic digestion and biogas production, a Presentation at First Bonner Summer School on Sustainable Agriculture August 2005. p. 1-44.

Zinoviev, S., Sivasamy Arumugam, S. & Miertus, S. (2007). Background Paper on Biofuel Production Technologies Working document November 2007, Area of Chmeistry-UNIDO, Available from http://www.biofuels-dubrovnik.org/downloads

10

The Macro-Environment for Liquid Biofuels in the Brazilian Science, Mass Media and Public Policies

Edson Talamini[1] and Homero Dewes[2]
[1]*Federal University of Grande Dourados – UFGD*
[2]*Federal University of Rio Grande do Sul – UFRGS*
Brazil

1. Introduction

The economic interest for renewable fuels production and consumption has increased considerably in the last decade. In the liquid biofuels field, the main interest is being driven to biodiesel and ethanol production and consumption. Therefore liquid biofuels industry has become a new opportunity for investment allocation. Biodiesel and ethanol production has risen worldwide, mainly in United States, Brazil, Germany, France, Italy and Spain (International Energy Agency [IEA], 2006). In Brazil, the ethanol production was strongly supported by public policies from the middle 1970's to the early 1990's, when the market for ethanol was liberalized. On the other hand, biodiesel has been strongly supported by public policies since Lula's administration, remarkably by its inclusion in the National Plan of Agroenergy and National Program of Production and Use of Biodiesel - PNPB.
Regarding the government incentives for liquid biofuels production, the Brazilian ethanol production has increased significantly since the National Alcohol Program – PROÁLCOOL was launched to the present days. According to National Petroleum Agency – ANP (2001, 2010), the ethanol production increased from 10,7 billion liters in 2000 to more than 26,1 billion liters in 2009, an increasing rate superior to 140% in ten years. The biodiesel production, on the other side, went up and turned expressive in the last five years. In 2005 the Brazilian biodiesel industry comprised no more than eight biorefineries with a production capacity close to 85,3 million liters per year, but producing less than 0,75 million liters in that year. The scenario changed fast and five years later the biodiesel industry comprised 63 biorefineries online with a production capacity fifty times higher than in 2005 (National Petroleum Agency [ANP], 2006, 2010).
The previous results show that the Brazilian liquid biofuels industry is attracting even more investments along the production chain, from farming to processing and distributing stages. As stated by Stiglitz (2001), "the information affects the decision making in every context … and affects their [market participants] behavior". So, decision makers may be interested in scan the macro-environment for liquid biofuels properly, supporting their strategic planning and decisions on a structured scanning process which can return not just information, but organized, categorized, and assessed one. The macro-environmental scanning is a first and important stage in the strategic planning process through which the decision makers would

look out for the patterns and changes in the industry environment as a way to gather information which help them in the decision making process (Johnson et al., 2008; Grant, 2008; Wheelen & Hunger, 2008; Thompson et al., 2009; David, 2009).
The configuration of the macro-environment within it an industry or sector is embedded is a dynamic process once it may be the outcome of interactions between a wide range of stakeholders, mainly policy makers, scientists and journalists, along the public (society) and the industry actors. As a new field of interest and investments, the liquid biofuels sector demands for a set of particular public policies to regulate, to create incentives and/or to put some restrictions on it. In such case, the Brazilian government may play an important role in framing the liquid biofuels picture (Talamini & Dewes, 2009). On the other side, the scientific knowledge and scientists can bring in some influence on the macro-environment configuration interacting with policy-makers, journalists and society (Jasanoff, 1987; Sabatier, 1991; IPCC, 2004; Kalil, 2006; Nature, 2007). Journalists, on the other hand, may have a powerful ascendance on society and/or policy makers so that they can be able to lead a new frame for liquid biofuels (Gamson & Modigliani, 1989; Strömberg, 2001, 2004; Moirand, 2003; Kim, 2007).
Taking into account that, firstly, the Brazilian liquid biofuels industry has been attracting investments once the production and use of such kind of energy seems to be in expansion; and, secondly, scanning the macro-environment within the liquid biofuels industry is embedded could be relevant for decision makers as well as for scientists, journalists and policy makers, this chapter aims to answer the following questions: under which dimensions scientists, journalists and policy makers have configured the macro-environment for liquid biofuels in Brazil? Are the Brazilian public policies for liquid biofuels more science-based or mass media-based? How investments and liquid biofuels production in Brazil do react to the agenda of mass media, science and government? The main goal of this chapter is to identify the dimensions under which Brazilian scientists, journalists and policy makers have configured the macro-environment for liquid biofuels, correlating it with the investments (production) done in this industry.

2. Strategic planning and the macro-environmental scanning

In their widely-used book on strategic management, Johnson et al. (2008) identify the presence of three basic and required core elements for the strategic planning process: strategic position, strategic choice and strategy into action. The first core element concerns the evaluation of impacts of the external environment, internal resources and competences of firms, and expectation and stakes of interest groups on the strategy to be adopted. The analysis of the environment in which the organization is located is actually the first step of strategic planning for all reputed writers of strategy-handbooks, like Grant (2008), Wheelen & Hunger (2008), Thompson et al. (2009), David (2009), and Leidecker & Bruno (1984).
This prioritizing of the environment in strategic analysis goes back to Dill (1958). In his study, the author stated that the influences of the restrictions imposed by the environment were essential for the science of management, since the behavior of firms depend on autonomous environmental circumstances in which the firms are located, and how these are interpreted and turned into action by managers. Likewise, Terry (1977) states that the environment is the prime determinant of the form and behavior of an organization. Although criticized by adherents of the resource-based view in management (e.g. Rumelt, 1991, Barney & Hesterly, 2008) the environment is still the starting-point in almost all strategic planning.

But what is the environment? What are the variables that compose it? A widely accepted definition in the literature is that one proposed by Hall (2004), for whom the environment represents all those elements existing beyond the limits of the organization that may influence, directly or indirectly, the organization. In the management field, Thomas (1974) comments that the term environment should be understood as in the open systems approach. According to the writer, one should attach great importance to the idea that, since organizations exist in a dynamic environment, their resources are strongly affected by the forces of this environment.

As for the variables that make up a certain environment, it is necessary to identify which environment level is being analyzed. Thomas (1974), and in a similar manner Leidecker & Bruno (1984), proposes three different environment levels: general environment, operational environment and internal environment. In other words, the environment of a firm is made up of layers ranging from generic to specific. The general environment is made up of social, political, regulatory, economic and technological conditions existing in a national or global context.

With the purpose of utilizing macro-environment analysis for strategic planning, the variables found at the level of the general environment are usually grouped in factors or dimensions. When defining which dimensions or factors make up the general macro-environment, some writers use additional or supplementary dimensions to those originally proposed by Thomas (1974). A summary of the macro-environmental dimensions used by some writers is presented in Table 1.

As a conclusion about the dimensions that make up a given macro-environment, it could be said that a single or preferred set of dimensions does not exist. The variety and number of dimensions seem to depend on the line of business of a certain industry. However, a specific set of dimensions is found to be recurring among the consulted writers: the dimensions represented by the PESTEL acronym (see Table 1), as proposed by Walsh (2005) and Johnson et al. (2008), seem to cover most classifications of a macro-environment. We adopt this standard.

Following the determination of the concept and dimensions that make up the macro-environment of firms, it should next be understood how the macro-environment investigation process for the preparation of the strategic planning is carried out. Ginter & Duncan (1990) and Ginter et al. (1992) state that the macro-environmental analysis process is made up of four interrelated activities:

a. *Scanning* - scanning the macro-environment means to investigate the threat signals and possible opportunities that may affect the business;
b. *Monitoring* - the activity of monitoring the macro-environment is associated to the process of tracing the issues identified in the investigation process;
c. *Forecasting* - it is the process of estimating forecasts of directions, scope, speed and intensity of environmental changes on a plausible basis; and,
d. *Assessment* - the process of assessing the projected trends for the organization in terms of its relationship with the external environment.

These steps are also present in the strategic planning model proposed by Bates (1985)[1]. Based on these four steps in the macro-environmental analysis, it can be concluded that the scanning process is the first step and one of the main elements of strategic planning. This paper focuses on this scanning of the macro-environment. Therefore, it is important to

[1] The author defines the model as MAPing: *Monitor, Analyze* and *Predict*.

Writers	Macro-environment Dimensions
Thomas (1974)	Social, Political, Regulatory, Economic and Technological
Fahey & King (1977)	Economic, Political, Regulatory, Social, Cultural, Technological, Energy, Marketing/Industrial and Financial
Preble et al. (1985)	Legal, Economic, Political, Competitive, Technological and Cultural
Ginter et al. (1991)	Economic, Political, Social, Technological and Regional
Costa (1995)	Political, Economic, Social and Technological - PEST
Leonidou (1997)	Physical, Demographic, Sociocultural, Economic, Political/Legal and Technological
Fleisher & Bensoussan (2002)	Social, Technological, Economic, Ecological and Political or Legal - STEEP
Walsh (2005); Johnson et al. (2008)	Political, Economic, Sociocultural, Technological, Environmental and Legal - PESTEL

Table 1. The different macro-environment dimensions according to different writers

understand how firms carry out this scanning: What kind of information is usually assessed? What are the information sources? And, how is this information processed?

By studying the strategic planning processes, the first conclusion one reaches is that there are different levels for the deployment of scanning techniques by firms, and even the intensity of use by the same firm may vary over time (Fahey & King, 1977; Fahey et al., 1981; Stubbart, 1982). However, in an environment marked by speed and intensity of changes, macro-environmental scanning is highly recommended to monitoring the strategy of firms. Costa (1995) highlights some of the reasons why the deployment of a systematized analysis of the environment of the organization is relevant. In its essence, 'scanning' is a process based on the search and treatment of information on a certain macro-environment.

Ginter et al. (1992) illustrate clearly how the macro-environment scanning process captures and treats information. Figure 1 shows that, before scanning, the various pieces of information about the general macro-environment and the specific sectors of environment are dispersed. Although the information may be available, the identification of any pattern turns out to be a daunting task. After the scanning process, the result is a set of categorized, organized, accumulated and assessed information. The illustration demonstrates that scanning is a sorting filter for the pieces of information accessed by the firms or industry, so that, after its application, the macro-environmental patterns may be identified and assessed.

What are the different information's sources used for the purpose of the macro-environment scanning process? Figure 1 shows that information is the raw material which feeds the monitoring process. By analyzing different studies in this field, it was possible to detect the presence of two information sources: company internal and external, which in turn can be subdivided into: personal and impersonal sources. The studies of Keegan (1974), Ginter & Duncan (1990), Ngamkroeckjoti & Johri (2000) and Jogaratnam & Law (2006) show that the main information sources are still the people inside or outside the organizations.

However, evidently more recent studies in the scanning field have drawn the attention to the importance of the World Wide Web as an information source. The Internet has promoted a significant growth in the information volume available for decision-taking. Auster & Choo (1993, 1994), Choo (1994, 1999), Liu (1998), Liu et al. (2000) and Decker et al.

(2005) are just some of the studies conducted in recent years drawing attention on the importance of electronic information sources for business executives.

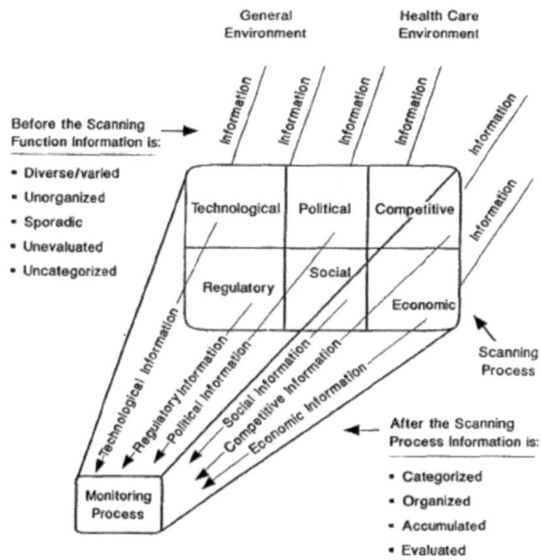

Fig. 1. Treatment of information through the 'scanning' process
Source: Ginter et al. (1992, p. 255)

If on the one hand access to information was made easier, we find that, on the other hand, there is the difficulty of analyzing such a large volume of data and information so as to extract the essential elements for planning the organizations' activities. The solution to this problem seems to be the combination of: (i) the theory and concept of scanning the macro-environment; and (ii) new information technologies (ITs) developed for electronic data treatment so as to extract a reduced and structured set of pieces of information. The studies of Liu (1998), Myers (1999), Wei & Lee (2004), Decker et al. (2005), Aasheim & Koheler (2006) are examples that the electronic 'scanning' of macro-environmental dimensions tends to be a powerful tool to better understand the current global environment, where information is in abundance, and in digital form.
Within the set of new techniques and technologies for macro-environmental scanning, the use of Text Mining is being discussed and presented as one of the intelligent techniques for the treatment of a large amount of information. In his book on this topic, Halliman (2001) discussed the use of Data Mining in depth and shows practical applications both to determine macro-environmental forces and to analyze scenarios. We will take up Data Mining, in line with Halliman (2001) and Lau et al. (2005) who used Text Mining as an analysis tool to identify macro-environmental dimensions associated to the business environment in the communication and hotel industries.

3. Method and procedures

The research method used in this research was based on the documental analysis of scientific articles published by Brazilian scientists, of mass media news and of official documents of Brazilian government. Karanikas & Theodoulidis (2002) e Hale (2005) assert that 80% of all available information occurs in a diverse range of written documents. To transform this information in knowledge this work used the concept of Knowledge Discovering in Texts - KDT, and Text Mining - TM techniques (Halliman, 2001). The procedures adopted for text mining were followed by a composition of phases, which came from the studies of Liddy (2000), Karanikas & Theodoulidis (2002), El Wakil (2002), Silva et al. (2004) e Hippner & Rentzmann (2006).

The selection of scientific papers, mass media news and governmental documents was made based on a key-words list, representative of the research object "liquid biofuels", both in English and Portuguese language, given the frequency in which these key-words have occurred in the literature on the issues related to bioenergy, bioeconomy, and biofuels. The key-words selected were: ETHANOL, ALCOHOL, BIOFUEL, BIOFUELS, BIO-FUEL, BIO-FUELS, BIODIESEL e BIO-DIESEL. The documents in which the term alcohol was related to an alcoholic beverage were discharged.

With all the key-words, searches were made in the following scientific papers websites: Scielo (www.scielo.br), *Portal de Periódicos CAPES* (www.periodicos.capes.gov.br) and Web of Science (http://portal.isiknowledge.com). Searching for mass media news two important newspapers with national circulation in Brazil were selected: *Folha de São Paulo* and *Valor Econômico*. Accessing the archives of such newspapers and by using the searching engine available in both newspapers, the news were retrieved and collected. The governmental documents were searched in the Brazilian government official web pages, from the Brazilian Government official website (www.brasil.gov.br). After that, all the web pages of the Ministries, Federal Offices and Autarchies were consulted using the browser on these pages and the documents were identified and downloaded. It is important to highlight that the governmental information which appeared in the "Press Room", available on most of the web pages accessed, were excluded as a way to distinguish official public policies and program from political discourse.

The search and collection of the textual documents from science, mass media, and government and the preliminary construction of the text-base had started on the first week of February 2007 and was concluded on the last week of following July. To analyze the performance of scientists, journalists and policy-makers along the time documents were collected within the timeframe 1997-2006. At the end of the browsing process, 219 scientific papers with at least one of authors from Brazilian scientific institutions, 4,121 mass media news, and 673 official documents of the Brazilian government were collected and archived into the text-base.

In the next phase, the electronic content of the documents were uploaded to a database built up with the help of QDA Miner® software, which prepares the documents for the text mining process. As the QDA Miner® software utilizes the *.RTF (Rich Text Format) extension to build its textual basis, 12 scientific articles and 49 governmental documents were lost because their .PDF files were protected or blocked. Then, the final set of the database were formed by 207 scientific papers, 4,121 mass media news, and 624 governmental documents.

To extract the knowledge from documents it was necessary to build an analytical structure able to extract the relevant information, because it was not found in the literature any methodology suitable to this purpose. Lists of appropriate key-words are frequently used, as can be seen in Vincent (2006), Crawley (2007) e Singh et al. (2007). In this direction, the first step to construct a specific structure to reach the proposed objectives was the definition of the macro-environmental dimensions to be used in this study. Accordingly to the literature about the macro-environment analysis, the most frequently used dimensions are those related to the "PESTEL" acronym, what stands for Politics, Economic, Socio-Cultural, Technological, Environmental and Legal (Walsh, 2005; Johnson et al., 2008). The number of dimensions and its label change from one study to another, depending upon the specific interest of the study, the environment and/or the activity researched, allowing some flexibility. For this study, nine dimensions were used, namely: Agronomical, Environmental, Cultural, Economical, Geopolitical, Legal, Political, Social, and Technological.

After the setting of the macro-environment dimensions, steps were carried out aiming at identifying the key-words which represent each dimension respectively, called here and onward "dimension-words" or shortly "d-words". The "d-words" are those relevant terms, which better discriminate a specific macro-environmental dimension. Therefore, nine different sets of "d-words" have been defined. The respective sets of "d-words" for each dimension were defined from the TF*IDF[2] index relevant to the words which occur in scientific texts published in journals specialized in the area of knowledge close related to the respective macro-environmental dimension. The number of "d-words" for each dimension was defined by using the percentile measure, selecting the "d-words" that quantitatively better discriminated each dimension. The average number of words selected was 14 "d-words" for each dimension. As low consistent, when single "d-words" were found in two or more dimensions, additional rules were added to the analytical structure for the knowledge extraction. The added rules took into account the simultaneous occurrence of defined terms in a same document. For the rules definition, the Jaccard's Coefficient was used (Chung & Lee, 2001).

The text mining was done using the textual basis in electronic format and the analytical structure for the knowledge extraction was made out of the macro-environmental dimensions and their respective "d-words". Using the WordStat module from the SIMStat® software it was possible to determine the frequency of each "d-word" in each set of documents and thereafter the frequency of use of the different macro-environmental frames by science, mass media, and government for liquid biofuels.

Absolute and relative frequencies of macro-environmental dimensions counted from documents by text mining were the main source of data. Data about ethanol and biodiesel production used in the analysis were mainly obtained from National Petroleum Agency – ANP. Cluster analysis using Jaccard's coefficient of agglomeration order was applied in analyzing the discourse in documents, paragraphs, and sentences. Granger causality tests were carried out to investigate how ethanol production reacts to scientific publication, mass media news, governmental documents, and the configuration of macro-environment as done by scientists, journalists, and policy-makers.

4. Results

This section is splitted up into three main sub-sections. The first one is dedicated to present and analyze the evolution of liquid biofuels industry along the last fifteen years, mainly how ethanol and biodiesel production increased (and also decreased) in response to internal and external phenomena. In the second sub-section, data on how Brazilian scientists, journalists and policy-makers have configured the macro-environment for liquid biofuels along the time are presented. Finally, in the third sub-section the results on how science', mass media'

---

[2]TF*IDF, Term Frequency (TF) multiplied by Inverse Document Frequency (IDF). For a more detailed review see Aizawa (2003) and Jing et al. (2002).

and government's agenda regarding liquid biofuels are correlated to the performance of liquid biofuels industry in Brazil.

4.1 Investments and liquid biofuels production

The ethanol is a relative well established market in Brazil since the National Program of Alcohol (*Proálcool*) was launched at the middle of 1970s. The *Proálcool* emerged as a series of governmental incentives for ethanol production and consumption. Such public policies were based on the perspective that ethanol could be an interesting and successful alternative for the raising in petroleum prices at international market. For almost two decades the ethanol production was supported by public policies which made compulsory the use o ethanol blended in gasoline, since there were no vehicles equipped with engine powered exclusively by ethanol. The public policies were also the main driver of changing in auto industry. The manufacture of vehicles with engines able to run exclusively on ethanol gave new impetus to *Proálcool*. Step by step the ethanol market has becoming a less government regulated market and more a price regulated market.

As such, the ethanol supply and demand were affected by a set of variables and balance changed frequently. At the end of 1990s the ethanol market was in decline, as proved by data shown in Table 2. But things were about to change the direction. The worldwide discussion about global warming has put new light on ethanol. The targets for reducing the Greenhouse Gases emissions as proposed in the Kyoto Protocol drove the governments to look for clean fuels alternatives. Then, the liquid biofuels appear as a potential solution for transportation fuel. Attentive to new demands, the automobile industry has invested in new technologies and launched hybrid cars. The production and sale of flex fuel vehicle (FFV) increases dramatically in Brazil since it was launched about 2004.

Year	Production (billion liters/year, anhydrous + hydrated ethanol)	Variation (%)
1995	12,746	-
1996	14,133	+10,88
1997	15,493	+9,62
1998	14,122	-8,85
1999	12,982	-8,07
2000	10,700	-17,58
2001	11,466	+7,16
2002	12,589	+9,79
2003	14,470	+14,94
2004	14,647	+1,22
2005	16,040	+9,51
2006	17,764	+10,73
2007	22,557	+26,98
2008	27,133	+20,28
2009	26,103	-3,79

Table 2. Ethanol production in Brazil
Source: ANP (2001, 2010)

The perspectives for ethanol market became better and the ethanol industry responded promptly. The sector reacted positively and new investments were done. New plants were installed and production capacity was expanded. During the 2000s years the production increased in almost all. From 2000 to 2008, the ethanol production increased from 10,7 billion liters to more than 27,1 billion liters, an increase rate higher than 150%. The decrease in ethanol production from 2008 to 2009 can be seen as an occasional phenomenon. For the future, the prospective scenarios are favorable to Brazilian ethanol market. Brazil has a cost-based competitive fuel produced mainly from sugarcane, a culture well adapted to soil and climatic conditions of major agricultural regions of the country and where there is still land to be used for agriculture purposes. The increasing demand for energy (renewable sources of energy also) around the globe may drive Brazil as an important global ethanol exporter. Taking into account the recent performance of Brazilian economy and the real increases of people income, the FFV's sales tend to grow even more demanding for more ethanol or blended gasoline/ethanol fuel.

If the Brazilian market for ethanol is already established to another liquid biofuel the things are still under development. The history's timeline of biodiesel is short in Brazil. The biodiesel production and consumption emerged from worldwide tendency of searching for renewable and clean sources of energy at the end of 1990's and beginning 2000's. Before 2000 the biodiesel industry was only a few experimental plants. As occurred with ethanol in years before, the biodiesel sector has changed its future from a set of public policies. Two major programs from Brazilian government were launched about 2005: the National Plan of Agroenergy and the National Program for Production and Use of Biodiesel. The National Plan of Agroenergy established objectives and goals for the agroenergy sector as a whole, while the National Program for Production and Use of Biodiesel defined the set of incentive policies to expand production and consumption of biodiesel. Following some lines of the past experience of *Proálcool*, the Brazilian government made mandatory the addition of biodiesel into the diesel fuel. At the beginning, was mandatory a blended fuel called B2 (2% of biodiesel added to diesel). Since 2010 the B5 is the mandatory blending. In the government of President Lula's view, the biodiesel could be also used as a "social fuel", promoting the social and economic inclusion of small farmers as raw material suppliers. The governmental support to biodiesel sector, via public policies, sounded well among investors, as shows Table 3.

Year	Biodiesel Refineries	Nominal Capacity (Million liters/year – B100)	Variation in Nominal Capacity (%)	Production (Million liters/year – B100)	Variation in Production (%)	Production/ Nominal Capacity (%)
2005	8	85,320	-	0,736	-	0,86
2006	19	638,620	+648,50	68,548	+9,213.59	10,73
2007	45	2,475.069	+287,56	402,176	+486,71	16,25
2008	62	3,315.339	+33,95	1,167.128	+190,20	35,20
2009	63	4,391.815	+32,47	1,608.053	+37,78	36,61
2010	68	6,198.268	+41,13	3,349.702	+108,30	54,04

Table 3. Biodiesel production in Brazil
Source: Primary data from ANP (2006, 2010)

From 2005 to 2010, the number of plants processing biodiesel (B100) grew from eight to sixty-eight, a growth rate of 750%. In the same period, the nominal capacity of producing B100 biodiesel grew more than 7,000%, jumping from just 85 million liters in 2005 to more than six billion liters in 2010. With the enlargement of biodiesel market by increasing mandatory blends (B2, B3, B4 and B5), the production also grew tremendously at the same time period. The future for biodiesel in Brazil is promising. Brazilian government has signalled it will raise levels of addition of biodiesel to diesel in the near future, generating a growth in demand for biodiesel. New biodiesel plants are in construction and the production can be doubled just with the nominal capacity currently installed. Research and technology development are being in course to improve the biodiesel industry competencies and productivity. New crops which can be used as raw material are studied and new production processes are continuously incorporated. The natural conditions presents in Brazil may lead the country to also be an exporter of biodiesel.

Summarizing, the market for two major liquid biofuels (ethanol and biodiesel) in Brazil have been affected by a set of phenomena. Economical and geopolitical events, like the price oil crisis and the power of the main petroleum producers; environmental issues, like global warming; social action, as the inclusion of small farmers; agronomic developments, like the adaptation of crops to the soil and climatic condition; cultural events, like decision-taking process of investors based on government signals and incentives; legal issues, like the official decrees defining the level of biofuels to be blended with fossil fuels; technological advances, like in the auto industry with the production of FFV. However, none of those dimensions was more important to the development of the market for liquid biofuels in Brazil than the political one. Policy-makers and the public policies proposed by them were decisive along the time and cannot be considered as incorrect to reserve to the Brazilian government the central role in such journey. At the same time, we cannot ignore the particular role of science and mass media in framing and configuring the macro-environment for liquid biofuels.

4.2 The macro-environment for liquid biofuels: science, mass media and government

In this section, the results on how the macro-environment for liquid biofuels has been configured by Brazilian scientists, journalists and policy-makers are presented and discussed. At first, we will present and discuss the results from Brazilian science. As shown in Table 4, Brazilian scientists have configured the macro-environment for liquid biofuels under three main dimensions: technological, environmental and agronomical. Economical and political dimensions have occurred with an intermediate frequency in Brazilian scientists agenda. Social, cultural and legal aspects were less frequently observed in the content of scientific publications.

Some changes in the main dimension under which the macro-environment for liquid biofuels have been configured can be seen. The environmental dimension, for instance, was the most frequently used by Brazilian scientists in 1998, 1999, 2001, 2003 e 2004, but it was overcome by economical dimension in 2002 and by agronomical in 2006. Such findings illustrate that the macro-environmental configuration is a dynamic process. In addition, the findings suggest that the interest of scientists can be affected by a dimension which may be more in evidence at a time. On the other hand, a trend analysis can also be carried out. In such case, it is possible to observe that among the most frequently macro-environmental dimensions used by scientists just technological dimension presents an increasing tendency along the time, especially in the last six years of the time series studied. Agronomical and environmental aspects related to liquid biofuels also presented a decreasing tendency regarding their presence in the scientists' agenda. On the other hand, geopolitical and political aspects have gained importance from 2001 and the trend analysis indicates an increasing importance of such matter in scientific field for the future.

Dimensions	1997	1998	1999	2000	2001	2002	2003	2004	2005	2006	Total
Agronomical	*103*	*190*	*323*	*508*	*183*	*254*	*465*	*534*	*3625*	*3826*	*10011*
	(43.6)	(11.1)	(18.6)	(27.0)	(24.0)	(6.6)	(21.6)	(20.3)	(21.8)	(23.4)	(20,9)
Cultural	*6*	*100*	*23*	*41*	*33*	*89*	*48*	*101*	*546*	*526*	*1513*
	(2.5)	(5.9)	(1.3)	(2.2)	(4.3)	(2.3)	(2.2)	(3.8)	(3.3)	(3.2)	(3,2)
Economical	*7*	*200*	*221*	*87*	*28*	*1247*	*239*	*183*	*1413*	*1984*	*5609*
	(3.0)	(11.7)	(12.7)	(4.6)	(3.7)	(32.5)	(11.1)	(7.0)	(8.5)	(12.1)	(11,7)
Environmental	*67*	*342*	*446*	*509*	*325*	*812*	*596*	*856*	*3627*	*2445*	*10025*
	(28.4)	(20.1)	(25.6)	(27.1)	(42.7)	(21.2)	(27.7)	(32.6)	(21.8)	(14.9)	(20,9)
Geopolitical	*16*	*248*	*225*	*90*	*55*	*199*	*203*	*169*	*1840*	*1945*	*4990*
	(6.8)	(14.5)	(12.9)	(4.8)	(7.2)	(5.2)	(9.5)	(6.4)	(11.1)	(11.9)	(10,4)
Legal	*5*	*66*	*58*	*31*	*14*	*104*	*34*	*43*	*622*	*653*	*1630*
	(2.1)	(3.9)	(3.3)	(1.6)	(1.8)	(2.7)	(1.6)	(1.6)	(3.7)	(4.0)	(3,4)
Political	*1*	*184*	*124*	*78*	*11*	*259*	*116*	*116*	*1288*	*1026*	*3203*
	(0.4)	(10.8)	(7.1)	(4.2)	(1.4)	(6.8)	(5.4)	(4.4)	(7.7)	(6.3)	(6,7)
Social	*0*	*69*	*24*	*3*	*6*	*36*	*29*	*36*	*205*	*187*	*595*
	(0.0)	(4.0)	(1.4)	(0.2)	(0.8)	(0.9)	(1.4)	(1.4)	(1.2)	(1.1)	(1,2)
Technological	*31*	*306*	*297*	*532*	*106*	*832*	*418*	*591*	*3463*	*3793*	*10369*
	(13.1)	(17.9)	(17.1)	(28.3)	(13.9)	(21.7)	(19.5)	(22.5)	(20.8)	(23.1)	(21,6)
Total	*236*	*1705*	*1741*	*1879*	*761*	*3832*	*2148*	*2629*	*16629*	*16385*	*47945*
	(100.0)	(100.0)	(100.0)	(100.0)	(100.0)	(100.0)	(100.0)	(100.0)	(100.0)	(100.0)	(100.0)

Note: Relative frequency in parenthesis

Table 4. Macro-environment configuration for liquid biofuels by Brazilian scientists

To better understand how Brazilian scientists have been using the set of dimensions in configuring the macro-environment for liquid biofuels an additional analysis of documents content was carried out. At this point, we were interested in identifying what are the main dimensions used in a conjoint way to compose the scientists' discourse. The mining process from scientific documents allow us group the macro-environmental dimension in clusters according its co-occurrence in three different levels of analysis: in a same document (a), in a same paragraph (b), and in a same sentence (c), as can be seen in Figure 2.

Taking the documents as units of analysis (Fig. 2, a) and clustering the macro-environmental dimensions in seven clusters, results show that the most frequently co-occurrence in Brazilian science discourse are shown by the topics related to agronomical and technological aspects of liquid biofuels production and consumption. Such aspects of liquid biofuels are followed by environmental matters. Departing from a more general (Fig. 2, "a", document) to a more specific level (Fig. 2, "b", paragraph and "c", sentence) of content analysis in scientific publications it is possible to identify what issues are highlighted by the scientists when reporting their studies and research on liquid biofuels. Thus, when two dimensions appear together in sentences of scientific documents, one can assume that the relationship between these two aspects of macro-environment are more revealing of the scientific approach than when such dimensions occur only at the document level. Although some minor changes can be perceived at a paragraph level, it is important to note what the configuration of a six-clusters analysis at sentences level revealed. In such case, it is possible to identify three main clusters: agronomical and environmental; political and legal; and, economical and technological. The findings reveal that Brazilian scientists are interested in agronomical, technological and environmental issues at a general approach, but to develop the knowledge on liquid biofuels they need to put particular emphasis on agronomical-environmental, political-legal, and economical-technological aspects.

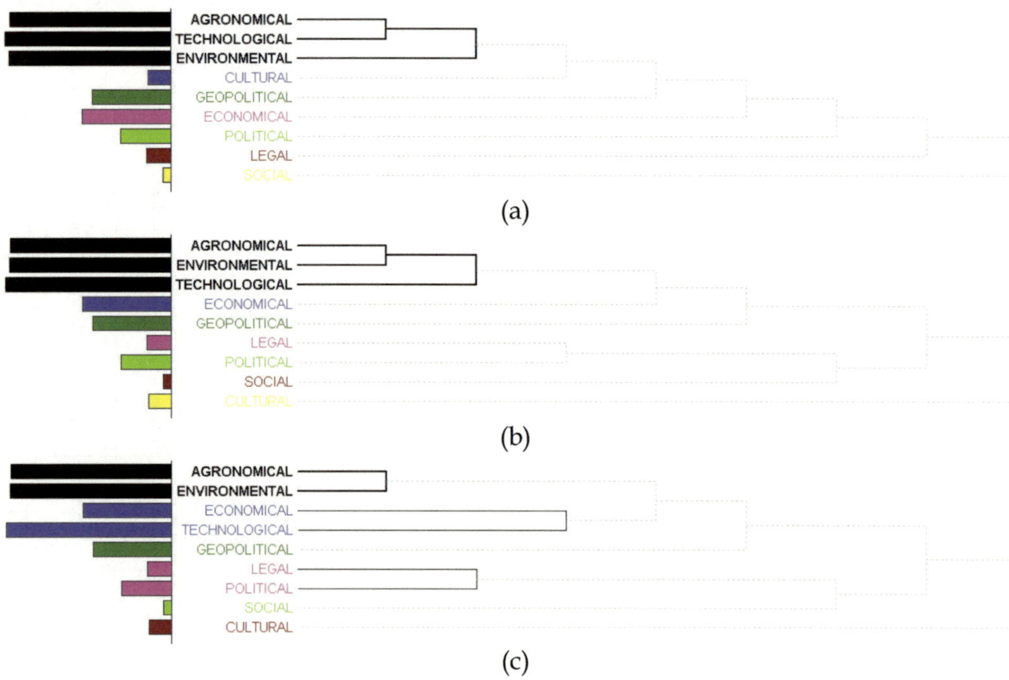

Note: "a", in the same document; "b", in the same paragraph; "c", in the same sentence

Fig. 2. Science document content analysis by dimensions agglomeration order using Jaccard Coefficient

The second group of stakeholders to be analyzed is composed by Brazilian journalists. As well established in the literature, the mass media plays an important role both in disseminating the knowledge produced by science, both to the lay public and to policy-makers, as in the discussion of relevant topics on public policies. As the liquid biofuels is a relative emerging sector which needs an intermediation between scientists, policy-makers and the lay public, and the mass media can properly performs this role, it is interesting to investigate how the macro-environment for liquid biofuels have been configured by Brazilian journalists along the time.

Compared to science, the macro-environmental configuration done by mass media is more stable. There is not much change between the most frequently used dimensions during the time. In general, Brazilian journalists have focused on four main dimensions when framing liquid biofuels: economical, technological, political and geopolitical (see Table 5). The agronomical and environmental dimensions have occurred in an intermediate frequency level, while cultural, social and legal can be considered as irrelevant for journalists regarding its low level of occurrence in the mass media news.

Despite the journalists present a more uniform framing of liquid biofuels over time, based on four main dimensions, trend analysis shows that those dimensions that were most frequently used have a trend of declining its relative importance in setting the macro-environment. In this direction, the geopolitical issues related to liquid biofuels have gained importance in the journalists' agenda, mainly from 2000 on. Other aspects of liquid biofuels production that journalists have driven an increasing attention at are those related to agronomical dimension, followed by technological matters. On the other side, economical, political and environmental issues are losing relative importance in setting the macro-environment made by journalists.

Dimensions	1997	1998	1999	2000	2001	2002	2003	2004	2005	2006	Total
Agronomical	*102*	*43*	*84*	*134*	*155*	*286*	*217*	*567*	*877*	*2133*	*4598*
	(3.8)	(2.6)	(3.3)	(4.8)	(8.8)	(8.3)	(6.6)	(8.5)	(9.4)	(11.3)	(8.7)
Cultural	*62*	*27*	*40*	*40*	*41*	*88*	*102*	*220*	*207*	*445*	*1272*
	(2.3)	(1.7)	(1.6)	(1.4)	(2.3)	(2.5)	(3.1)	(3.3)	(2.2)	(2.4)	(2.4)
Economical	*538*	*535*	*1094*	*1005*	*406*	*844*	*805*	*1573*	*1913*	*4119*	*12832*
	(20.1)	(32.7)	(42.6)	(35.8)	(23.0)	(24.5)	(24.6)	(23.6)	(20.5)	(21.8)	(24.2)
Environmental	*309*	*99*	*243*	*231*	*228*	*277*	*349*	*541*	*848*	*1273*	*4398*
	(11.5)	(6.1)	(9.5)	(8.2)	(12.9)	(8.0)	(10.7)	(8.1)	(9.1)	(6.7)	(8.3)
Geopolitical	*391*	*177*	*133*	*213*	*176*	*520*	*399*	*974*	*1388*	*3071*	*7442*
	(14.6)	(10.8)	(5.2)	(7.6)	(10.0)	(15.1)	(12.2)	(14.6)	(14.9)	(16.3)	(14.0)
Legal	*199*	*71*	*77*	*60*	*74*	*191*	*147*	*253*	*444*	*720*	*2236*
	(7.4)	(4.3)	(3.0)	(2.1)	(4.2)	(5.5)	(4.5)	(3.8)	(4.8)	(3.8)	(4.2)
Political	*498*	*386*	*396*	*607*	*270*	*556*	*512*	*876*	*1259*	*2487*	*7847*
	(18.6)	(23.6)	(15.4)	(21.6)	(15.3)	(16.1)	(15.6)	(13.1)	(13.5)	(13.2)	(14.8)
Social	*52*	*38*	*49*	*39*	*42*	*68*	*81*	*120*	*211*	*364*	*1064*
	(1.9)	(2.3)	(1.9)	(1.4)	(2.4)	(2.0)	(2.5)	(1.8)	(2.3)	(1.9)	(2.0)
Technological	*525*	*258*	*452*	*477*	*370*	*621*	*660*	*1544*	*2164*	*4261*	*11332*
	(19.6)	(15.8)	(17.6)	(17.0)	(21.0)	(18.0)	(20.2)	(23.2)	(23.2)	(22.6)	(21.4)
Total	*2676*	*1634*	*2568*	*2806*	*1762*	*3451*	*3272*	*6668*	*9311*	*18873*	*53021*
	(100.0)	(100.0)	(100.0)	(100.0)	(100.0)	(100.0)	(100.0)	(100.0)	(100.0)	(100.0)	(100.0)

Note: Relative frequency in parenthesis

Table 5. Macro-environment configuration for liquid biofuels by Brazilian journalists

Following the same analysis as done in scientific publications, also in mass media news further analysis is relevant to identify the discourse composition made by Brazilian journalists when publishing news on liquid biofuels. In Figure 3 are presented the results of clustering analysis taking into account the documents (a), paragraphs (b), and sentences (c) as units of analysis. The meaning and interpretation of these results have been discussed previously. As can be seen in Figure 3 ("a" and "b") there are no significant differences in the clusters when compared "documents" and "paragraphs". In both cases, the findings suggest that there are two main groups of mass media news. The most relevant is the group of mass media news that drives attention to a composition of political and geopolitical aspects of liquid biofuels sector. In the second group remains those news in which journalists have driven attention on economical and technological issues of liquid biofuels market.

As deeper the analysis goes, reaching the sentences as the unit of analysis (Fig. 3, "c"), more clear became the pattern in content of discourse done by journalists. The most relevant cluster of issues used in mass media is that composed by political, legal, and geopolitical aspects related to liquid biofuels. It is important to note that legal dimension occurs with a low relative frequency in the content of mass media news as a whole. However, such legal issues are relevant when composing the complete ideas proposed by journalists regarding liquid biofuels topic.

Last, but not least, we are interested in investigating how the macro-environment for liquid biofuels have been configured by Brazilian policy-makers along the time. Despite the central role of Brazilian Government in giving support to *Proalcool* first and recently to biodiesel production and consumption, the analysis of public policies on liquid biofuels is a fundamental subject. The general findings can be seen in the Table 6.

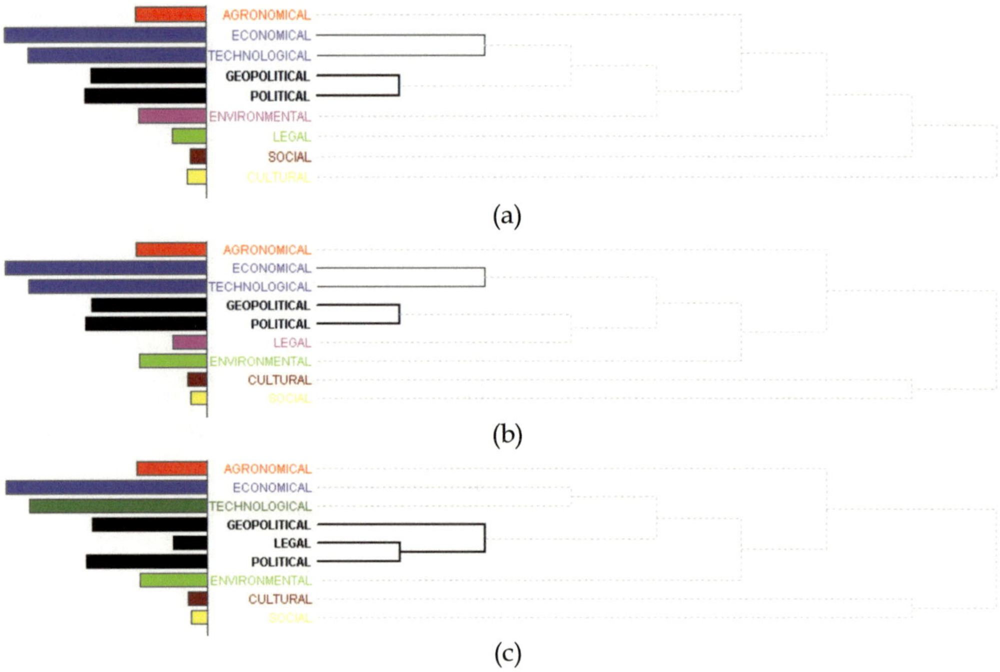

Note: "a", in the same document; "b", in the same paragraph; "c", in the same sentence

Fig. 3. Mass Media news content analysis by dimensions agglomeration order using Jaccard Coefficient

The findings obtained from text mining procedures on public policies documents have revealed that Brazilian policy-makers have focused mainly on technological aspects of liquid biofuels. The technological dimension have predominated absolutly along the ten years studied, not been surpassed by any other dimension in no time. The relative importance of technological dimension in Brazilian public policy may be related to Petrobrás, a public-private company which controls the liquid biofuels supply throughout the country. The second dimension most frequently observed in public policies is the geopolitical one, which also presents a trend of increasing its relative importance along the time. The economical, environmental, agronomical and political dimensions are in a set with intermediate relative frequency as observed on public policies documents. Legal, cultural and social dimension have presented the lowest frequency on public policies. Contrary to expectations, the social dimension did not appear as one of the main approaches of public policies, even in years after the inauguration of first President Lula's Government in 2002.

Looking for the future, the trend analysis reveals that liquid biofuels public policies in Brazil signal some change in the issues to be addressed in such policies. The most frequently used

Dimensions	1997	1998	1999	2000	2001	2002	2003	2004	2005	2006	Total
Agronomical	*16*	*94*	*1204*	*3284*	*372*	*1410*	*950*	*2407*	*2551*	*8960*	*21248*
	(2.1)	(5.2)	(8.2)	(8.6)	(8.8)	(7.7)	(11.7)	(7.4)	(12.0)	(14.1)	(10.5)
Cultural	*7*	*21*	*601*	*2023*	*104*	*992*	*219*	*1513*	*881*	*3166*	*9527*
	(0.9)	(1.2)	(4.1)	(5.3)	(2.5)	(5.4)	(2.7)	(2.7)	(4.1)	(5.0)	(4.7)
Economical	*112*	*185*	*2323*	*3850*	*501*	*2102*	*849*	*3844*	*2359*	*7486*	*23611*
	(15.0)	(10.3)	(15.8)	(10.1)	(11.8)	(11.4)	(10.4)	(10.4)	(11.1)	(11.8)	(11.6)
Environmental	*60*	*213*	*908*	*5146*	*538*	*3055*	*1545*	*2575*	*2454*	*6794*	*23288*
	(8.0)	(11.8)	(6.2)	(13.5)	(12.7)	(16.6)	(19.0)	(7.9)	(11.5)	(10.7)	(11.5)
Geopolitical	*93*	*148*	*1994*	*5793*	*524*	*2563*	*937*	*5083*	*2864*	*9708*	*29707*
	(12.4)	(8.2)	(13.6)	(15.2)	(12.4)	(13.9)	(11.5)	(11.5)	(13.5)	(15.3)	(14.6)
Legal	*110*	*150*	*1607*	*2426*	*253*	*1127*	*409*	*2747*	*2220*	*3269*	*14318*
	(14.7)	(8.3)	(10.9)	(6.4)	(6.0)	(6.1)	(5.0)	(5.0)	(10.4)	(5.2)	(7.0)
Political	*104*	*238*	*1740*	*3675*	*286*	*1681*	*816*	*3345*	*2216*	*6074*	*20175*
	(13.9)	(13.2)	(11.8)	(9.7)	(6.8)	(9.1)	(10.0)	(10.0)	(10.4)	(9.6)	(9.9)
Social	*19*	*69*	*658*	*1324*	*149*	*853*	*248*	*966*	*726*	*1616*	*6628*
	(2.5)	(3.8)	(4.5)	(3.5)	(3.5)	(4.6)	(3.0)	(3.0)	(3.4)	(2.5)	(3.3)
Technological	*228*	*681*	*3676*	*10508*	*1509*	*4628*	*2174*	*9981*	*4981*	*16322*	*54688*
	(30.4)	(37.9)	(25.0)	(27.6)	(35.6)	(25.1)	(26.7)	(26.7)	(23.4)	(25.7)	(26.9)
Total	*749*	*1799*	*14711*	*38029*	*4236*	*18411*	*8147*	*32461*	*21252*	*63395*	*203190*
	(100.0)	(100.0)	(100.0)	(100.0)	(100.0)	(100.0)	(100.0)	(100.0)	(100.0)	(100.0)	(100.0)

Note: Relative frequency in parenthesis

Table 6. Macro-environment configuration for liquid biofuels by Brazilian policy-makers

dimensions, like technological and economical, tend to lose their relative importance at the expense of inclusion of other aspects such as agronomical, geopolitical, and environmental. Such findings may indicate a set of public policies more aligned with the context of contemporary global energy sector. After being set the technological, economical, legal, and political standards, public policies tend to include other relevant aspects today, as: geopolitical issues, which involve the world's supply of fuel; and, environmental issues, in evidence due to global warming.

Following the same analysis pattern used in science and mass media, Figure 4 shows the results of a deeper inside document analysis. Taking into account the whole documents as unit of analysis (Fig. 4, "a") and a seven-cluster agglomeration, the findings are accurately in line with the configuration of macro-environment for liquid biofuels in Brazilian public policies. That is, the order of clustering of documents indicates that public policies have taken into account especially the technological, geopolitical, and economical issues. A second set of documents could be joined in a cluster of political-legal issues on liquid biofuels.

Using the paragraphs of public policies documents as unit of analysis (Fig. 4, "b"), the results reveal two main clusters of complete ideas used by Brazilian policy-makers. The cluster with higher Jaccard's Coefficient agglomeration order is that one composed by paragraphs which deal with political and social issues. Although the social dimension is one of the lowest frequency found in documents as a whole, the findings make sense regarding the recent public policies stated by Brazilian government on biodiesel. At last, the analysis using the sentences present in the public policies documents (Fig. 4, "c") reveals a main cluster composed by sentences which deals with technological-agronomical aspects, firstly, and adding economical ones, afterwards. It is important to note that the more close ideas expressed on sentences differ a bit from that general approach found in documents as whole, changing geopolitical issues by agronomical matters.

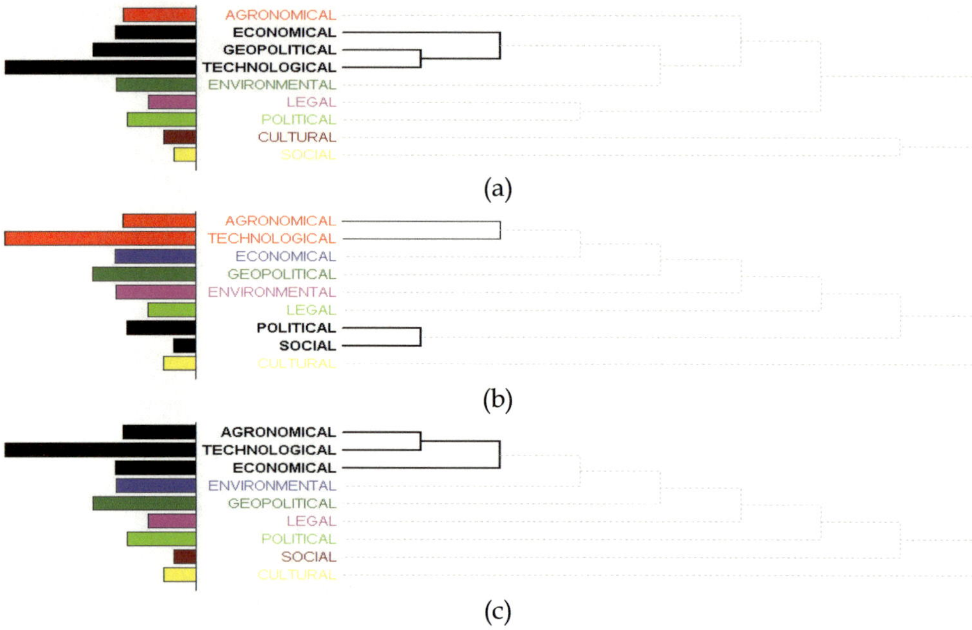

Note: "a", in the same document; "b", in the same paragraph; "c", in the same sentence

Fig. 4. Government documents content analysis by dimensions agglomeration order using Jaccard Coefficient

4.3 Correlating liquid biofuels production with macro-environment configuration

In this section we will search for some significant correlation between liquid biofuels production (ETP), number o scientific publication (SCP), mass media news (MMN), and governmental documents (GVD). As there is no sufficient data on biodiesel production to accomplish a correlation analysis, the Brazilian ethanol production was used as a proxy for investments on and liquid biofuels production in Brazil. The Pearson correlation values are shown in the Table 7.

	ETP	SCP	MMN	GVD
ETP	1.000			
SCP	0.761*	1.000		
MMN	0.765*	0.968**	1.000	
GVD	0.713*	0.918**	0.918**	1.000

Note: *correlation is significant at the 0.05 level
**correlation is significant at the 0.01 level

Table 7. Correlation matrix

As indicated by results the variables present a high level of correlation between each other. The higher correlation level was found between the number of scientific publication and mass media news. The number of governmental documents presents the same correlation value for both mass media news and scientific publications. Ethanol production presents higher correlation with mass media news, followed by scientific publications and governmental documents, although all values are significant at a same 0.05 level.

Taking into account that organizations react to macro-environmental configuration, more important than analyse the correlation between liquid biofuels and number of documents is to check for some correlation between production and macro-environmental configuration. In this direction in the Table 8 are presented the values for Pearson correlation between the amount of ethanol production and the total frequency that each macro-environmental dimension was counted in the scientific publications, mass media news, and governmental documents.

Macro-environmental dimensions	Ethanol Production - ETP
Agronomical by science	0.696*
Cultural by science	0.703*
Economical by science	0.592
Environmental by science	0.604
Geopolitical by science	0.718*
Legal by science	0.698*
Political by science	0.662*
Social by science	0.726*
Technological by science	0.679*
Agronomical by mass media	0.724*
Cultural by mass media	0.767**
Economical by mass media	0.688*
Environmental by mass media	0.777**
Geopolitical by mass media	0.771**
Legal by mass media	0.827**
Political by mass media	0.737*
Social by mass media	0.782**
Technological by mass media	0.761*
Agronomical by government	0.498
Cultural by government	0.282
Economical by government	0.367
Environmental by government	0.208
Geopolitical by government	0.325
Legal by government	0.293
Political by government	0.342
Social by government	0.146
Technological by government	0.290

Note: *correlation is significant at the 0.05 level
**correlation is significant at the 0.01 level

Table 8. Pearson Correlation – ethanol production and macro-environmental dimensions

In general, the higher Pearson correlation values were observed between ethanol production and macro-environmental dimensions used by journalists. Regarding this, the ethanol production is mainly correlated with the frequency that legal, social, environmental, geopolitical, and cultural dimensions were counted in documents during the text mining procedures. All of those correlations are significant at a 0.01 level. The correlations of ethanol production with other macro-environmental dimensions present in mass media were significant at the 0.05 level. The findings also indicate a medium correlation between ethanol production and the frequency that macro-environmental dimensions were counted in scientific publications. Just economical and environmental dimensions in science return a no significant correlation with ethanol production. All others correlations are significant at the 0.05 level. At last, there were no significant level of correlation between ethanol production and the frequency with which any macro-environmental dimension appeared in governmental documents. This is a somewhat surprising finding in spite of the historical importance that public policies on liquid biofuels have had over time. Summarizing, what we can learn from the correlation analysis is that the decision making process in the organizations is using more mass media as a source of information regarding its investments in ethanol production than governmental approach or scientific knowledge.

On the other hand, there may be a time lag between the macro-environmental setting made by scientists, journalists and politicians and the increasing in production of liquid biofuels. Searching for evidences in that direction pairwise Granger causality tests were accomplished with one and two years of time lag. Main findings can be seen in the Figure 5 ("a" and "b").

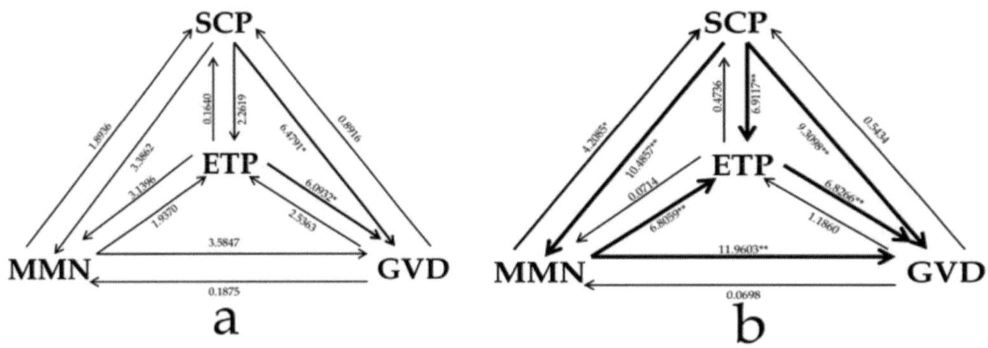

Notes: "a": two years time lag
"b": one year time lag
*significant at the 0.05 level
**significant at the 0.01 level

Fig. 5. Granger Causality Test (F-Statistic)

The Granger causality test is a statistical procedure to refuse or not the null hypothesis than a variable A does not cause a variable B. According to results presented in Figure 5 (a), we can accept the null hypothesis for almost all pairwise tests with a two years time lag. The exceptions are that: (i) the hypothesis that the number of science publications (SCP) does not cause the number of governmental documents (GVD); and, (ii) the hypothesis that the amount of ethanol production (ETP) does not cause the number of governmental documents (GVD), both cannot be refused. Also the hypothesis that the number of mass media news (MMN) precedes the ethanol production (ETP) and the number of governmental documents (GVD) cannot be refused.

On the other hand, the results change when just one year is used as time lag in Granger causality tests (Fig. 5, "b"). Many of the pairwise Granger causality test indicate that we cannot refuse the null hypothesis. Then, we cannot refuse the hypothesis, at a 0.01 level of significance, that the number of scientific publications precedes the number of mass media news, the governmental documents and also the ethanol production. Such findings suggest that science plays a leading role in prospecting the liquid biofuels trends, being followed by other stakeholders. The results also suggest that government reacts to ethanol production instead of the contrary, suggesting that the main role performed by government is regulating the market after its emergence and not to be central agent which stimulates the raising of liquid biofuels to a new industry. The Granger causality test revealed that government is preceded by all other stakeholders and does not precede any of them.

Additional analysis using Granger causality tests revealed that we cannot refuse the hypothesis that ethanol production is preceded mainly by macro-environmental dimensions used by journalists to frame liquid biofuels. Results indicate that ethanol production is preceded by agronomical, cultural, environmental, geopolitical, legal, political, social, and technological dimensions present in the mass media content. Ethanol production is also preceded by economical, environmental, and technological dimensions explored by scientists in producing knowledge on liquid biofuels. Once more, any macro-environmental dimension present in public policies precedes the ethanol production.

5. Conclusions

Three main questions were addressed in this paper regarding the macro-environmental configuration for liquid biofuels in the Brazilian science, mass media, and government and its implications for investments on and liquid biofuels production. From a content analysis accomplished by text mining procedures applied to scientific publication, mass media news, and governmental documents, some conclusions can be pointed out.

In general the conclusion is that Brazilian scientists, journalists, and policy makers have configured the macro-environment for liquid biofuels under different dimensions along the time. Scientists have emphasized more agronomical, environmental and technological aspects of liquid biofuels production and consumption. Journalists, as a characteristic of day-to-day agenda, have based their agenda mainly on economical and geopolitical issues. Technological standards and geopolitical issues are among the main topics present in the policy-makers' agenda. However, the relative importance of macro-environmental dimensions changes along the time. That means that investors in Brazilian liquid biofuels sector may find different macro-environments according to the source of information they use. So, a properly macro-environmental scanning practice should be encouraged before the accomplishment of strategic planning and decision-taking process.

As results suggested, the public policies for liquid biofuels in Brazil seem to be more a mass media-based process than a science-based one. Regarding the alignment between public policies and mass media news we could suggest that managers should use mass media ways of information (newspaper, magazines, broadcasts and so on) in scanning the macro-environment for liquid biofuels, once they will reproduce or influence the public policies in liquid biofuels field. On the other hand, despite the time lag between a scientific recommendation and its adoption as a public policy, we could suggest that managers should also look for information (scan the macro-environmental configuration) in science in advance, as a way to prospect possible public policies in the future.

The macro-environmental scanning in the liquid biofuels sector seems to be a useful tool in strategic planning process, agreeing with the results found by previous studies in the same direction. Regarding the public policy-making process in Brazil, the policy-makers and scientists seem not to work close to each other. It implies that in the liquid biofuels matter the Brazilian science seems to play a secondary role in public policy orientation. The results suggest that there is a distance between knowledge creation and its application by policy makers. Of course, there are many reasons for it. So, for scholars and for policy makers also, we can suggest a close cooperation and interaction in putting scientific knowledge creation in line with public interest, or conversely.

Taking into account the results obtained in this study, we can suggest some future studies in a sense to analyze this topic widely and deeper. For instance, to analyze the perception (positive, negative or neutral) of each stakeholder group on the liquid biofuels and correlate their perception with the investments and production of ethanol and/or biodiesel; to identify the gap between a scientific topic prescription and its real adoption by policy-makers including it into the public policies; and, future analysis should be done to understand how biodiesel production and consumption, specifically, react to macro-environment configuration.

6. Acknowledgments

The authors acknowledge financial support from FAPERGS, FINEP and the National Council for Scientific and Technological Development - CNPq for this project (Process: 400128/2009-8).

7. References

Aasheim, C. & Koehler, G. J. (2006). Scanning World Wide Web Documents with the Vector Space Model. *Decision Support Systems*, Vol. 42, pp. 690-699, ISSN 0167-9236.

Aizawa, A. (2003). An Information-Theoretic Perspective of TF-IDF Measures. *Information Processing & Management*, Vol. 39, pp. 45-65, ISSN 0306-4573.

ANP - National Petroleum Agency. (2001). Statistical Yearbook 2001, 28.10.2010, Available from:
http://www.anp.gov.br/?pg=15280.

ANP - National Petroleum Agency. (2006). Statistical Yearbook 2006, 28.10.2010, Available from:
http://www.anp.gov.br/?pg=8970.

ANP - National Petroleum Agency. (2010). Statistical Yearbook 2010, 28.10.2010, Available from:
http://www.anp.gov.br/?pg=31286.

Auster, E. & Choo, C. W. (1993). Environmental Scanning by CEOs in Two Canadian Industries. *Journal of the American Society for Information Science*, Vol. 44, No. 4, pp. 194-203, ISSN 1096-3480.

Auster, E. & Choo, C. W. (1994). How Senior Managers Acquire and Use Information in Environmental Scanning. *Information Processing & Management*, Vol. 30, No. 5, pp. 607-618, ISSN 0306-4573.

Barney, J. B. & Hesterly, W. S. (2008). *Strategic Management and Competitive Advantage. Concepts and Cases*. Pearson Prentice Hall, ISBN 978-0-132-33823-3, Harlow.

Bates, C. S. (1985). Mapping the Environment: an Operational Environmental Analysis Model. *Long Range Planning*, Vol. 18, No. 5, pp. 97-107, ISSN 0024-6301.

Choo, C. W. (1994). Perception and Use of Information Sources by Chief Executives in Environmental Scanning. *Library & Information Science Research*, Vol. 16, pp. 23-40, ISSN 0740-8188.

Choo, C. W. (1999). The Art of Scanning the Environment. *Bulletin of the American Society for Information Science*, pp. 21-24, ISSN 1550-8366.

Chung, Y. M. & Lee, J. Y. (2001). A Corpus-Based Approach to Comparative Evaluation of Statistical Term Association Measures. *Journal of the American Society for Information Science and Technology*, Vol. 52, No. 4, pp. 283-296, ISSN 1532-2882.

Costa, J. (1995). An Empirically-based Review of the Concept of Environmental Scanning. *International Journal of Contemporary Hospitality Management*, Vol. 7, No. 7, pp. 4-9, ISSN 0959-6119.

Crawley, C. E. (2007). Localized Debates of Agricultural Biotechnology in Community Newspapers: A Quantitative Content Analysis of Media Frames and Sources. *Science Communication*, Vol. 28, No. 3, pp. 314-346, ISSN 1075-5470.

David, F. R. (2009). *Strategic Management. Concepts and Cases*, Pearson Prentice Hall, ISBN 978-0-136-01570-3, Harlow.

Decker, R.; Wagner, R. & Scholz, S. W. (2005). An Internet-based Approach to Environmental Scanning in Marketing Planning. *Marketing Intelligence & Planning*, Vol. 23, No. 2, pp. 189-199, ISSN 0263-4503.

Dill, W. R. (1959). Environment as an Influence on Managerial Autonomy. *Administrative Science Quarterly*, Vol. 2, No. 4, pp. 409-443, ISSN 0001-8393.

El Wakil, M. M. (2002). Introducing Text Mining. *Information System Department*, Faculty of Computers and Information, Cairo University.

Fahey, L. & King, W. (1977). Environmental Scanning for Corporate Planning. *Business Horizons*, pp. 61-71, ISSN 0007-6813.

Fahey, L.; King, W. R. & Narayanan, V. K. (1981). Environmental Scanning and Forecasting in Strategic Planning - The State of the Art. *Long Range Planning*, Vol. 14, pp. 32-9, ISSN 0024-6301.

Fleisher, G. S. & Bensoussan, B. E. (2002). *Strategic and Competitive Analysis: Methods and Techniques for Analyzing Business Competition*. Prentice Hall, ISBN 978-0-130-888525-5, Upper Saddle River.

Gamson, W. A. & Modigliani, A. (1989). Mass media discourse and public opinion on nuclear power: a constructionist approach. *The American Journal of Sociology*, Vol. 95, No. 1, (July 1989), pp. 1-37, ISSN 0002-9602.

Ginter, P. M. & Duncan, W. J. (1990). Macroenvironmental Analysis for Strategic Management. *Long Range Planning*, Vol. 23, No. 6, pp. 91-100, ISSN 0024-6301.

Ginter, P. M.; Duncan, W. J. & Capper, S. A. (1991). Strategic Planning for Public Health Practice Using Macroenvironmental Analysis. *Public Health Reports*, Vol. 106, No. 2, pp. 134-141, ISSN 0033-3549.

Ginter, P. M.; Duncan, W. J. & Capper, S. A. (1992). Keeping Strategic Thinking in Strategic Planning: macro-environmental analysis in a State Department of Public Health. *Public Health*, Vol. 106, No. 4, pp. 253-269, ISSN 0033-3506.

Grant, R. M. (2008). *Contemporary Strategy Analysis*, Blackwell Publishing, ISBN 978-1-405-16309-5, Boston.

Hale, R. (2005). Text Mining: getting more value from literature resources. *Drug Discovery Today*, Vol. 10, No. 6, pp. 377-379, ISSN 1359-6446.

Hall, R. H. (2004). *Organizações: estruturas, processos e resultados*. 8 ed. Prentice Hall, ISBN 8587918761, São Paulo.

Halliman, C. (2001). *Business Intelligence Using Smart Techniques: environmental scanning using text mining*. Information Uncover, ISBN 978-0-967-49062-5, Houston.

Hippner, H. & Rentzmann, R. (2006). Text Mining. *Informatik Spektrum*, Vol. 29, No. 4, pp. 287-290, ISSN 1432-122X.

IEA - International Energy Agency. (2006). Statistics & Balances, 2006, 17.11.2009, Available from: http://www.iea.org/stats/index.asp.

IPCC - Intergovernmental Panel on Climate Change. (2004). 16 Years of Scientific Assessment in Support of the Climate Convention. WMO/UNEP, 13.04.2009, Available from: *www.ipcc.ch/pdf/10th-anniversary/anniversary-brochure.pdf.*

Jasanoff, S. S. (1987). Contested boundaries in policy-relevant science. *Social Studies of Science*, Vol. 17, No. 2, (May, 1987), pp. 195-230, ISSN 1460-3659.

Jing, L.; Huang, H. & Shi H. (2002). Improved Feature Selection Approach TF-IDF in Text Mining. Proceedings of the First International Conference on Machine Learning and Cybernetics, Beijing, China, November 4-5. pp. 944-946.

Jogaratnam, G. & Law, R. (2006). Environmental Scanning and Information Source Utilizations: exploring the behavior of Hong Kong hotel and tourism executives. *Journal of Hospitality & Tourism Research*, Vol. 30, No. 2, pp. 170-190, ISSN 1096-3480.

Johnson, G.; Scholes, K. & Whittington, R. (2008). *Exploring Corporate Strategy: Text and Cases*, Pearson, ISBN 978-0-273-71192-6, Harlow.

Kalil, T. (2006). Planning for US science policy in 2009. *Nature*, Vol. 443, (Oct. 19th, 2006), pp. 751-752, ISSN 0028-0836.

Karanikas, H. & Theodoulidis, B. (2002). Knowledge Discovery in Text and Text Mining Software, *Centre for Research in Information Management, Department of Computation*, UMIST, Manchester, UK, 26.04.2006. Available from: http://www.crim.co.umist.ac.uk.

Keegan, W. J. (1974). Multinational Scanning: a study of the information sources utilized by headquarters executives in multinational companies. *Administrative Science Quarterly*, Vol. 19, No. 3, pp. 411-421, ISSN 0001-8393.

Kim, H. (2007). PEP/SI: a new model for communicative effectiveness of science. *Science Communication*, Vol. 28, No. 3, (March 2007), pp. 287-313, ISSN 1075-5470.

Lau, K.; Lee, K. & Ho, Y. (2005). Text Mining for the Hotel Industry. *Cornell Hotel and Restaurant Administration Quarterly*, Vol. 46, No. 3, pp. 344-362, ISSN 0010-8804.

Leidecker, J. K. & Bruno, A. V. (1984). Identifying and Using Critical Success Factors. *Long Range Planning*, Vol. 17, No. 1, pp. 23-32, ISSN 0024-6301.

Leonidou, L. C. (1997). Finding the Right Information Mix for the Export Manager. *Long Range Planning*, Vol. 30, No. 4, pp. 572-584, ISSN 0024-6301.

Liddy, E. D. (2000). Text Mining. *Bulletin of the American Society for Information Science*, October/November, ISSN 0095-4403.

Liu, S. (1998). Business Environment Scanner for Senior Managers: towards active executive support with intelligent agents. *Expert Systems with Applications*, Vol. 15, pp. 111-121, ISSN 0957-4174.

Liu, S.; Turban, E. & Lee M. K. O. (2000). Software Agents for Environmental Scanning in Electronic Commerce. *Information Systems Frontiers*, Vol. 2, No. 1, pp. 85-98, ISSN 1387-3326.

Moirand, S. (2003). Communicative and cognitive dimensions of discourse on science in the French mass media. *Discourse Studies*, Vol. 5, No. 2, (May 2003), pp. 175-206, ISSN 1461-4456.

Myers, K. (1999). Technology for the Environmental Scanning Process. *Systemic Practice and Action Research*, Vol. 12, No. 4, pp. 409-424, ISSN 1573-9295.

Nature. (2007). A clear direction. Editorial, *Nature*, Vol. 447, (May 10th, 2007), p. 115, ISSN 0028-0836.

Ngamkroeckjoti, C. & Johri, L. M. (2000). Management of Environmental Scanning Processes in Large Companies in Thailand. *Business Process Management Journal*, Vol. 6, No. 4, pp. 331-341, ISSN 1463-7154.

Preble, J. F.; Rau, P. A. & Reichel, A. (1998). The Environment Scanning Practices of U.S. Multinationals in the Late 1980's. *Management International Review*, Vol. 28, No. 4, pp. 4-14, ISSN 0938-8249.

Rumelt, R. P. (1991). How much does industry matter? *Strategic Management Journal*, Vol. 12, pp. 167-85, ISSN 0143-2095.

Sabatier, P. A. (1991). Political-science and public-policy. *Political Science & Politics*, Vol. 24, No. 2, (June 1991), pp. 144-147, ISSN 1049-0965.

Silva, C.; Osório, F.; Vieira, R. & Quaresma, P. (2004). Mining Linguistically Interpreted Texts. In: *5th International Workshop on Linguistically Interpreted Corpora*, Geneva, 26.04.2006. Available from: www.coli.uni-saarland.de/conf/linc-04/silva.pdf.

Singh, N.; Hu, C. & Roehl, W. S. (2007). Text Mining a Decade of Progress in Hospitality Human Resource Management Research: identifying emerging thematic development. *International Journal of Hospitality Management*, Vol. 26, pp. 131-147, ISSN 0278-4319.

Stiglitz, J. E. (2001). Information and the change in the paradigm in economics. Nobel Prize Lecture, December 2001, 27.10.2010, Available from: http://nobelprize.org/nobel_prizes/economics/laureates/2001/stiglitz-lecture.pdf.

Strömberg, D. (2001). Mass media and public policy. *European Economic Review*, Vol. 45, No. 4-6, (May 2001), pp. 652-663, ISSN 0014-2921.

Strömberg, D. (2004). Mass media competition, political competition, and public policy. *Review of Economic Studies*, Vol. 71, No. 1, pp. 265-284, ISSN 0034-6527.

Stubbart, C. (1982). Are Environmental Scanning Units Effective? *Long Range Planning*, Vol. 13, No. 3, pp. 139-145, ISSN 0024-6301.

Talamini, E. & Dewes, H. (2009). Government and media in the configuration of the macroenvironment for liquid biofuels in Brazil. *Revista de Administração Pública*, Vol. 43, No. 2, (March-April, 2009), pp. 415-444, ISSN 0034-7612.

Terry, P. T. (1977). Mechanisms for Environmental Scanning. *Long Range Planning,* Vol. 10, June, pp. 2-9, ISSN 0024-6301.

Thomas, P. S. (1974). Environmental Analysis for Corporate Planning. *Business Horizons,* October, pp. 27-38, ISSN 0007-6813.

Thompson, A.; Strickland, A. J. & Gamble, J. E. (2009). *Crafting & Executing Strategy,* Mcgraw-Hill, ISBN 978-0-073-53042-0, Boston.

Vincent, R. C. (2006). A Comparative Study of Wsis News Coverage in North American and European Broadcast/Satellite, Newspaper and Wire Service Sources, 2001-2005. International Association for Mass Communication Research, July 23-29.

Walsh, P. R. (2005). Dealing with the Uncertainties of Environmental Chang by Adding Scenario Planning to the Strategy Reformulation Equation. *Management Decision,* Vol. 43, No. 1, pp. 113-122, ISSN 0025-1747.

Wei, C. & Lee, Y. (2004). Event Detection from Online News Documents for Supporting Environmental Scanning. *Decision Support Systems,* Vol. 36, pp. 385-401, ISSN 0167-9236.

Wheelen,T. L. & Hunger, J. D. (2008). *Strategic Management and Business Policy,* Pearson Prentice Hall, ISBN 978-0-132-38738-5, Harlow.

11

Sweet Sorghum as a Bioenergy Crop for the US Great Plains

Charles S. Wortmann and Teshome Regassa
Department of Agronomy and Horticulture, University of Nebraska-Lincoln, Lincoln, USA

1. Introduction

Sorghum (*Sorghum bicolor* L. Moench), including sweet sorghum, is widely adapted to diverse and often marginal crop production environments. Sweet sorghum stalks have high sugar content compared with other sorghum types and has potential for producing ethanol to be mixed with gasoline or for producing ethyl tert-butyl ether, an octane additive to gasoline. Sweet sorghum was introduced to the United States for syrup production in the 1850s (Winberry, 1980). Production peaked following sugar shortages during World War II at about 136 million L yr^{-1} of syrup in 1946 (Hunter & Anderson, 1997), but thereafter declined because of low sugar prices and inadequate production efficiency.

Sweet sorghum can be competitive with corn (*Zea mays* L.) and grain sorghum for ethanol yield when grain yield is less than 9 Mg ha^{-1}, and is comparatively efficient in nitrogen use (Smith & Buxton, 1993). Sweet sorghum can easily substitute for corn or grain sorghum in many cropping systems.

Currently, most ethanol produced in the U.S.A. is from the starch of corn grain with the support of federal subsidies. Energy gains with production of ethanol from grain are modest, typically ranging from 30 to 130% depending on N use efficiency, ethanol plant efficiency, and the efficient use of the distillers grain co-product. Sweet sorghum can be produced at less cost than corn, often with higher energy gains (Smith & Buxton, 1993). Rather than producing starch, sweet sorghum carbohydrates are stored in the stalk as sugar, with sugar concentrations of 8-20% (Rains et al., 1990). Conversion of sugar to ethanol requires less energy than starch as much energy is used to depolymerize the starch. Sweet sorghum has demonstrated potential to produce up to 6000 L ha^{-1} of ethanol in Iowa and Colorado U.S.A. (Smith & Buxton, 1993), equivalent to ethanol from approximately 20 Mg of corn grain. However, estimated ethanol yields were on average 33% more with grain of corn and grain sorghum compared with sugar of sweet sorghum for seven rainfed site-years in Nebraska U.S.A. (Wortmann et al, 2010). Seasonal availability, the need to transport and store much mass, and storability of sweet sorghum constrain sweet sorghum as a bio-energy crop.

In planning for bio-fuel production, long-term sustainability of cropping systems must be considered. Sustainability of a cropping system is very much dependent on production environment and resource availability. In one study comparing the sustainability of different bioenergy crops, sweet sorghum, along with oil palm (*Elaeis guineensis* L.) and sugarcane (*Saccharum* spp.) for biofuel, were found to be more sustainable in comparison to maize and wheat. This assessment considered efficiency in use of land, water, nitrogen and energy resources, and of pesticides, relative to net energy produced (Vries et al., 2010).

This chapter addresses sweet sorghum production for the U.S. Great Plains and other temperate production zones, harvest and processing issues, and energy and green house gas balances. An extensive literature review was conducted with most published papers reporting on research conducted in temperate zones.

2. Sweet sorghum production

2.1 Growth and sugar content

The agronomic principles and production practices for sweet sorghum and grain sorghum are similar (Hunter & Anderson, 1997). Reddy et al. (2005) reported much diversity among sweet sorghum genotypes with ranges in India of 13 to 24% for Brix (a measure of sugar and soluble starch in plant sap based on light refraction; a typical Brix measure for sweet sorghum sap is 85% sugar and 15% soluble starch), 7.2 to 15.5% for sucrose concentration in juice, 24 to 120 Mg ha^{-1} for fresh stalk yield, 36 to 140 t ha^{-1} fresh biomass yield, and 27 to 48 Mg ha^{-1} mill-ready stalk yield. Plant height can be as tall as 4.8 m (Freeman & Broadhead, 1973) and stalks can be more than 45 mm thick (Turhollow, 1994). Sweet sorghum has a range of maturity types, and is relatively well adapted compared with corn to water deficit stress, but yields are typically highest in deep, well-drained soils with good fertility. Sweet sorghum has the potential for producing a ratoon crop after harvest where the growing season is long enough.

Sweet sorghum growing degree days and thermal time are commonly calculated with a base temperature of 13° C (Barbanti et al., 2006; Ferraris & Charles-Edwards, 1986). Sugar yield is generally favored by early planting, but rapid emergence and vigorous seedling growth occur when soil temperature is above 18° C at planting (Lueschen et al., 1991). Sugar yield was increased with earlier planting and increased radiation during the reproductive stage (Ferraris & Charles-Edwards, 1986). Ricaud & Arenneaux (1990) reported mean stalk yields of 56 and 49 Mg ha^{-1} with 26 Apr and 25 May planting, respectively, in Louisiana U.S.A. Yield of stalk sugar in excess of 10 Mg ha^{-1} was observed for early sown crops and the sugar yield dropped to 3 Mg ha^{-1} for late-planted crops (Ferraris & Charles-Edwards, 1986). Juice yield was not affected by planting date in Mississippi U.S.A., but sugar yield was highest for early May planting (Broadhead, 1972) and similar for April and June planting (Broadhead, 1969). In another study conducted in the upper Midwest of the U.S.A., fermentable carbohydrate and ethanol yields were 13% more with earlier compared with later planting dates, and early planting of late-maturing sweet sorghum cultivars was recommended, despite a problem of lodging.

Sugar concentration of sweet sorghum increased as a function of the duration of growth, commonly peaking at the grain dough stage, and generally decreased with delayed planting irrespective of sampling stage (Ferraris, 1981; Geng et al., 1989). Planting of full season varieties commonly increases potential ethanol yield (Putnam et al., 1991; Zhao et al, 2009). The rate of sugar accumulation is nearly linear with growing time and with radiation intercepted. Early planting allows for a longer growing period and earlier canopy development for sunlight interception during the long days of June and July. The same studies found that production of highly-recoverable concentrated sugars was maximized with long season, tall- and thick-stalk sweet sorghum cultivars. Interception of radiation during the boot to early seed formation growth stage has been found to be very important to sweet sorghum sugar yield (Hipp et al., 1970).

2.2 Plant population and stand establishment

Uniform seedling emergence and vigorous stand establishment is important for sweet sorghum production but often challenging under unfavorable planting conditions. This is due to small seed size and often low germination rate and seedling vigor compared to grain sorghum. Once the crop has reached the fifth leaf stage, sweet sorghum growth is generally vigorous and competitive.

Several studies have addressed plant population and planting pattern. Across seven rainfed site-years in Nebraska, Wortmann et al. (2010) found similar harvestable stem number and sugar yield by sowing 7.5, 12.5, and 17.5 seed m^{-2} with 75-cm row spacing; increased harvestable tiller number compensated for the lower sowing rates. Sweet sorghum stalk yield was greater in Turkey with 15 plants m^{-2} compared with lower plant densities when planted with 65-cm row spacing (Turgut et al., 2005). In the northern Corn Belt of the U.S.A., sweet sorghum fermentable carbohydrate or ethanol yield was not affected by seeding rate (Lueschen et al., 1991). In a study of carbohydrate accumulation conducted in Australia, Ferraris & Charles-Edwards (1986) found lower early sugar concentration but slightly higher concentration with higher plant density. Broadhead and Freeman (1980) and Kuepper (1992), however, found reduced Brix, sucrose content, sugar yield, and juice content with increased plant density (Broadhead & Freeman, 1980; Kuepper, 1992).

Row spacing may be important. In Australia, Martin and Kelleher (1984) reported increased stalk and water soluble carbohydrate yield with 8 compared with 16 plants m^{-2} and when row spacing was reduced from 105 to 35 cm. They attributed the row spacing effect to greater photosynthetic productivity before anthesis and the production of taller, thicker stalks, the volume of which was closely related to post-anthesis carbohydrate accumulation. In Mississippi U.S.A., stalk yield and Brix were more by growing sweet sorghum in 52.5 cm row spacing compared with wider row spacing, but individual plant weight and juice content were more in wider rows and lodging was less (Broadhead & Freeman, 1980). In this study, however, stalk and sugar yield per hectare with 76-cm row spacing was similar compared with narrower row spacing and more than with 105-cm spacing.

2.3 Water use

Sweet sorghum has been observed to extract soil water to 270 cm depth in California U.S.A. (Geng et al., 1989). In this study, sweet sorghum had less yield loss compared to corn, sugarbeet (*Beta vulgaris* L.), and fodder beet (*Beta vulgaris* L.) under severe soil water deficit conditions. Water use efficiency of sweet sorghum was determined to be 310 compared to 370 kg water kg^{-1} dry matter for corn (Reddy et al., 2007). With adequate nutrient supply and irrigation, sweet sorghum hexose yield was 10.0 Mg ha^{-1} compared to 8.1 Mg ha^{-1} for corn (Geng et al., 1989), while under soil water deficit conditions sweet sorghum extracted more soil water and produced 29% more hexose compared with corn. In trials conducted between 40.8° and 42.0°N latitude in the U.S.A., total sugar and ethanol yield were similar, but total biomass yield was more with irrigated compared with rainfed production. Seasonal rainfall was not related to biomass or sugar yield in Nebraska U.S.A. where the cropping season rainfall ranged from 250 to 580 mm; median water productivity was 50 kg biomass and 8.1 kg of sugar per mm of seasonal rainfall (Wortmann et al., 2010); this did not account for stored soil water at one month before planting and available soil water remaining after harvest.

Most sweet sorghum research has been conducted under rainfed conditions. However, a study in Arizona U.S.A. on a sandy soil evaluated frequency of irrigation (Ottman & Miller,

2010). They found sweet sorghum to be responsive to irrigation under arid conditions but did not appear to be highly sensitive to frequency of irrigation. Water use was less and water use efficiency was greater when irrigating at 50 and 65% depletion of available soil water compared with irrigating at 35% depletion (Miller & Ottman, 2010).

2.4 Fertilizer use

Sweet sorghum response to applied nutrients varies with location. Dry plant yield in Louisiana U.S.A. was 40% more with 100 kg ha^{-1} N compared to no N applied; yield was not further increased with application of an additional 100 kg ha^{-1} N, but there was a 10% yield increase with addition of 90 kg ha^{-1} K (Ricaud & Arenneaux, 1990). They reported a 50% increase in total sugar yield by applying 100 kg ha^{-1} N, an additional 4% increase by increasing the N rate to 200 kg ha^{-1} N, and an additional 13% gain by adding 80 kg ha^{-1} K to the 100 kg ha^{-1} N. Nutrient uptake by sweet sorghum at the soft dough stage ranged from 109 to 214 with a median of 142 kg ha^{-1} for N, 11 to 31 with a median of 18 kg ha^{-1} for P, and 60 to 161 with a median of 113 kg ha^{-1} for K (Ricaud & Cochran, 1979). In a comparison with other potential bioenergy crops conducted in Kansas U.S.A., N and K removal in the above ground biomass was more with sweet sorghum compared with other crops, and P removal was less compared with maize and perennial grasses (Table 1) (Propheter & Staggenborg, 2010).

Sweet sorghum biomass yield in Turkey was increased by 16% and stalk diameter by 7% with application of 100 kg ha^{-1} N (Turgut et al., 2005). In California U.S.A., sweet sorghum used applied N much more efficiently than corn. Sweet sorghum required just 36% of the fertilizer N required by corn to maximize hexose yield, but produced 23% more hexose yield than corn (Geng et al., 1989). In Mississippi U.S.A., stalk yield was 24% more with the application of 45 kg ha^{-1} N compared to no N applied, but similar to the yield with application of 90 kg ha^{-1} N or with P application (Freeman & Broadhead, 1973). In other studies, fermentable sugar yield (Smith & Buxton, 1993), stalk dry matter yield at harvest (Barbanti et al., 2006), and fermentable carbohydrate and ethanol yield (Lueschen et al., 1991) were not affected by N application. In Texas U.S.A., total dissolved solids in juice decreased when a high N rate was applied (Wiendenfeld, 1984). Sweet sorghum did not respond to applied N when intercropped with alfalfa (Buxton et al., 1998). However, farmers producing sweet sorghum for syrup generally applied 34 to 56 kg ha^{-1} of fertilizer N (Kuepper, 1992). Some sweet sorghum cultivars have the capacity for associative N fixation with 0 to 18% of plant N determined, using the ^{15}N natural abundance technique, to be derived from the atmosphere (Yoneyama et al., 1998).

Sweet sorghum stalk dry matter and sugar yield were increased with application of 80 kg ha^{-1} N at only one of seven site-years in Nebraska U.S.A. while corn and grain sorghum grain yields were increased for all site-years with N application (Wortmann et al., 2010). Unpublished results from a related study in Nebraska U.S.A. found that total N uptake by sweet sorghum was similar to uptake by corn but the pattern of uptake differed. Nitrogen uptake by sweet sorghum was more gradual over a longer period of time than for corn and grain sorghum that had several weeks with a very high rate of uptake. Therefore, the high soil N supply needed by the grain crops compared with sweet sorghum at those critical growth stages likely accounted for the greater responsiveness of the grain crops to applied N. Sweet sorghum continued to take up N later into the season, allowing more time for soil organic N mineralization and for deeper root penetration and uptake of deep nitrate-N.

	Nutrient removal, kg ha^{-1}		
	N	P	K
Sweet sorghum	190	34	329
Maize	174	43	167
Forage sorghum†	152	34	292
Perennial grass†	43	48	52

†The forage sorghum values are means of three varieties including a photoperiod-sensitive sorghum. The perennial grass values are means of three species, including switchgrass (*Panicum virgatum* L.), big bluestem (*Andropogon gerardii* L.), and miscanthus (*Miscanthus giganteus*).

Table 1. Mean nutrient removal with the harvest of the above-ground biomass of four groups of bioenergy crops at two locations in Kansas U.S.A. (Propheter & Staggenborg, 2010).

Sweet sorghum biomass and juice yield increased with lime application when soil pH was low (Soileau & Bradford, 1985). Surface soil organic matter was 10 g kg^{-1} soil in this study and yields were depressed with N application in the absence of lime application.

3. Harvest, juice extraction, and transport

3.1 Sweet sorghum stalk harvest

Sweet sorghum stalk yield was not much affected by growth stage between flowering to physiological maturity but juice extraction efficiency decreased, and Brix and starch increased, with advancing maturity (Broadhead, 1974); sucrose yield was maximized during the dough stage. In other studies, syrup yield was maximized by harvesting during the late milk to hard dough growth stage (Broadhead, 1972; Tarpley et al., 1994). Stalk sugar concentration is often the lowest at boot stage and the highest at the soft dough stage (Lingle, 1987; Ricaud et al., 1979); the onset of sucrose accumulation was associated with the onset of the reproductive phase of growth and reduced acid invertase activity. Juice yield, per cent extracted, and purity were not affected by delaying stalk harvest until 3-4 weeks after physiological maturity, but Brix and sucrose were reduced by 6% and 4%, respectively, compared with harvest at or before physiological maturity (Broadhead, 1969). Sugar concentration of juice increased continuously until frost kill but thereafter declined (Nuese & Hunt, 1983). Sugar yield is dependent on length of growing season and the amount of radiation intercepted, with a linear increase in sugar yield at dough stage as photosynthetically active radiation increased from 20 to 80 MJ plant^{-1} due to earlier sowing and longer growing periods (Ferraris & Charles-Edwards, 1986). As long as the terminal meristem developed, the internodes increased in biomass and plant height increased, especially in late maturing cultivars (Coleman & Belcher, 1952). Sugar continued to accumulate in the fully-developed internodes well into seed development (Hunter & Anderson, 1997). In balancing potential ethanol yield with extending the harvest period, harvest of early maturing varieties may begin at about 20 days after anthesis (Zhao et al., 2009).

In traditional harvest for syrup production, sweet sorghum stalks were topped to remove the panicle and stripped of leaves before crushing for juice extraction because of effects on syrup taste (Winberry, 1980). Farmers staggered plantings over four weeks to prolong the harvest period (Broadhead, 1974). Juice was extracted with simple wooden or metal roller presses, a labor intensive procedure (Lamb, 1982). Juice extraction could be done without stripping stalks of leaves without syrup yield loss if: 1) the leaves were wilted before juice extraction; 2) juice was decanted after at least two hours of settling to remove sediment; and 3) alpha-amylase enzymes were used during preheating of the juice (Kuepper, 1992). Panicle and leaf removal is less important for ethanol production since taste is not an issue as it is for syrup produced for human consumption.

De-heading of sweet sorghum at anthesis resulted in more productive tillers and increases in main stalk diameter by 20%, juice yield by 30%, and sugar yields by 10% in India, but 5% less Brix, sucrose concentration, and juice purity (Rajendran et al., 2000). In another study, de-heading increased Brix and concentrations of sucrose and starch at the milk through physiological maturity growth stages while reducing plant lodging and increasing tillering, resulting in increased juice yield (Broadhead, 1973). Stalk water content was less with deheading but this did not reduce sugar yield (Broadhead, 1973; Hunter & Anderson, 1997).

Sweet sorghum produces much biomass and handling this biomass in the short harvest windows available in temperate zones poses a major challenge (Bennett & Anex, 2008). Modified forage harvesters that cut stalks into billets may be used for chopping and harvesting stalks before transporting to the juice expression site, but sugar loss before juice extraction is slower with intact compared to chopped stalks (Bennett & Anex, 2008).

In-field extraction of juice reduces the biomass to be transported, leaving the bagasse in the field for ground cover and nutrient cycling. A field harvester capable of expressing juice into large bladders for juice storage and fermentation has been proposed, but sugar extraction may be 30-40% less with current in-field extraction technology compared with larger stationary extraction equipment (Kundiyana et al., 2006).

Parameter†	Value
Sugar-to-ethanol yield	%
Stalk juice extraction	80
Brix to fermentable sugar	75
Fermentable sugar converted to alcohol	95
Grain and sugar conversion to ethanol	**L Mg^{-1}**
Maize or grain sorghum grain	423
Sweet sorghum sugar	665
Crop production and harvest, diesel-equivalent	**L ha^{-1}**
No-till production	4
Grain harvest, > 8 Mg ha^{-1} yield, L ha^{-1}	13
Sweet sorghum harvest and extraction, L M^{-1} of fresh stalks	0.3
Natural gas consumed to produce ethanol	**MJ L^{-1}**
Grain	5.44
Sugar	3.33

†Adapted from Wortmann et al., 2010. Other values used in calculations are reported in the BESS2008.3.1 User's Guide (www.bess.unl.edu; verified Mar. 24, 2011).

Table 2. Values used in calculations of ethanol yields and energy balance of maize, grain sorghum, and sweet sorghum with only the grain or sugar used for ethanol production in Nebraska U.S.A.

A self-propelled 4-row forage harvester adapted for sweet sorghum harvest was found to be economically competitive with other harvest alternatives; when the co-product value was included, the net farm-gate cost of fermentable carbohydrates ranged from \$7 to \$24 Mg^{-1} and less than the cost of fermentable carbohydrate of corn grain (Bennett & Anex, 2008). Mobile juice-extracting alternatives were not found to be economically competitive with stationary units, assuming reduced quality control and juice extraction efficiency. If the harvest area is near the juice extraction facility, the lower fermentable carbohydrate costs of sweet sorghum compared with corn grain were sufficient to offset increased costs of transporting the wet sweet sorghum biomass. However, processing costs were reduced by 50% with a processing plant of 379,000,000 L yr^{-1} compared to a small plant of 37,900,000 L yr^{-1} but requiring longer transport distances plus much storage capacity and much added cost (Bennett & Anex, 2009). Ensiled storage of wet sorghum stalks resulted in 20% loss of fermentable carbohydrates plus added costs, with the result that ethanol production from sweet sorghum was more costly than for maize grain. However, ethanol production from fresh sweet sorghum feedstock, even with the high transport costs, was more cost effective compared with grain of maize. In many studies, sugar yield is estimated based on Brix readings and expected efficiency of juice extraction. The relationship of Brix to sugar content and efficiency of juice extraction vary with the actual values dependent on numerous factors. It is important that the conversion factors be reported such as those reported in Table 2 in order that the results can be adjusted for the reader's conditions.

3.2 Stalk storage and juiced extraction

Delays in extracting juice with a stationary press following harvest of stalks often occur. Sugar loss from heaped intact stalks was just 3% in four days (Ricaud et al., 1979) and no significant sucrose inversion occurred during 24 hours after cutting. In another study, juice extraction and purity decreased by 3% and 5%, respectively, during 24 hours following intact stalk harvest, but sucrose and starch decreased by less than 1% during 48 hours after harvest (Broadhead, 1974). Temperature during storage appears to be important to losses with 20% of fermentable sugars lost in 3 days of storage at room temperature but no loss with refrigeration (Wu et al., 2010).

Chopped stalks can be stored as silage without appreciable sugar loss in fermentation if inhibited with an acrylic acid treatment (Hill et al., 1987), but a 20% loss in ensiled storage can occur without such treatment (Bennett & Anex, 2008).

There is evidence of an interaction of harvest growth stage and stalk storage time (Broadhead, 1974). Juice purity was not affected by harvest growth stage if the stalks were not stored, but juice purity was 70 and 73% less at 24 hours of storage for stalks harvested in the milk compared with the dough and physiological maturity stages, respectively.

Juice extraction efficiency can be improved by removing panicles and leaves (Lamb, 1982). Expression efficiency may be improved by repeated re-watering and re-expression and by maceration of the stalks before juice extraction by crushing, cutting, or shredding (Jankins, 1966). Stalk water content is important for juice extraction efficiency, with reduced efficiency when water content is less than 45% by weight. More sugar is extracted with repeated wetting and crushing following the initial expression. Juice extraction efficiency varies widely and it is important that the value used in estimating sugar yield from sweet sorghum be reported as in Table 2.

3.3 Sugar yield

In Louisiana U.S.A., stalk sugar concentration was 8.3 to 14.0% at flowering and 12.8 to 16.6% at soft dough, and total sugar yield was 4.3 to 8.5 Mg ha^{-1} at flowering and 6.6 to 11.7

Mg ha^{-1} at soft dough (Ricaud et al., 1979). Total sugar yield was 4.0 to 10.7 Mg ha^{-1} for several locations across the continental USA and up to 12 Mg ha^{-1} for Hawaii U.S.A. (Smith et al., 1987), equivalent to ethanol yields of 2129 to 5696 L ha^{-1} and comparable to ethanol yields with maize grain. Higher yields were reported for Florida U.S.A., ranging up to 17 Mg ha^{-1} (Vermerris et al., 2008). In the temperate U.S.A., sugar yields of sweet sorghum were as high as 6 Mg ha^{-1} with a sugar composition of 54% sucrose, 26% glucose, and 20% fructose (Smith & Buxton, 1993). Across seven site-years in Nebraska U.S.A. between 40.5 and 41.1° N latitude, sugar yield averaged 2.1 Mg ha^{-1} for a semi-arid site at 1300 m above sea level to 6.2 Mg ha^{-1} at lower altitude locations with a longer growing season.

Eventual commercialization of the conversion of cellulosic material to ethanol is likely to increase the value of sweet sorghum as a biofuel crop. In Kansas U.S.A., at 39.8° N latitude, calculated ethanol yields were 10,184, 6770, 7477, and 3073 L ha^{-1} for sweet sorghum, forage sorghum, maize, and perennial grass, respectively, when the total above-ground biomass was converted to ethanol (Propheter et al., 2010). Genetic improvement is also expected to result in increased productivity. Most research with sweet sorghum has been done with selected lines while significantly more yield potential was found in northern China with sweet sorghum hybrids (Zhao et al., 2009).

Sugar concentration along the length of sweet sorghum stalks is not uniform. The concentration of nonstructural carbohydrates in sweet sorghum was found to be 1.4 times higher in the upper and 2.7 times higher in the lower internodes compared with grain sorghum (Vietor & Miller, 1990). Sugar and sucrose concentration were found to be greater in the upper compared with the lower internodes at physiological maturity (Coleman, 1970). In another study, sugar concentration was highest at the seventh of 11 internodes (Krishnaveni et al., 1990). Stalk sugar concentration was usually higher at the middle stalk and least in the top 30-45 cm; the upper stem could be discarded in harvest without significant loss of sugar or juice yield (Janssen et al., 1930). Less concentration in older internodes may be due to less enzymatic activity compared with newer internodes, reducing their sink strength for sugar accumulation (Lingle, 1987). However, the mechanisms may be more complicated than enzymatic activities and sink strength and further investigation may be needed (Tarpley et al., 1994). Some cultivars partitioned a significant amount of carbohydrates to nodal tillers (Vietor & Miller, 1990).

3.4 Fermentation efficiency and ethanol yields

The theoretical yield of ethanol, which has a weight of 789 g L^{-1}, was determined to be 720, 646, 680, and 370 L Mg^{-1} for starch, glucose or fructose, sucrose, and maize grain, respectively (Smith & Buxton, 1993). They estimated that 5% of the sugar is used to produce microbial growth and non-ethanol products. Efficiency of maize grain conversion to ethanol has continued to improve. It was estimated at 417 L Mg^{-1} corn grain in 2005 (Wang et al., 2005). Dry-grind ethanol conversion of 423 L Mg^{-1} corn grain was common in the ethanol industry in 2009 (Table 2; Wortmann et al., 2010). Other energy yield estimates for comparison are 16 MJ kg^{-1} for biomass combustion, 18.5 MJ kg^{-1} for gasified wheat straw, ethanol yield of 0.36 L kg^{-1} of wheat, 18.3 MJ kg^{-1} for processing wheat to ethanol accounting for drying of the by-product, 7 MJ kg^{-1} of fresh bagasse of 50% dry weight, and 3.2 MJ kg^{-1} theoretical ethanol energy yield of fresh bagasse (Monti & Ventura, 2003), but these vary with conversion process and biomass composition (McAloon et al., 2000). Other conversion values are reported in Table 2.

Sweet sorghum juice can be converted to alcohol either by fermenting the juice or by fermenting the chopped stalks in a solid-state process (Rein, 1984). Alcohol conversion efficiency may be superior from chopped sweet sorghum than the corresponding juice. Fermentation efficiency can be improved by heating to 85° C and the addition of yeast at any temperature. Adding yeast reduced the temperature effect on fermentation efficiency and alcohol yield was maximized by heating juice to 60° C with addition of 0.25 g L^{-1} of yeast.

The U.S. Department of Energy estimated potential sweet sorghum ethanol yield to be 5590 L ha^{-1} (U.S. Department of Energy, 1979). Several sweet sorghum cultivars have the potential of producing greater than 25 Mg dm ha^{-1} $year^{-1}$ (Turhollow, 1994). In Iowa, ethanol yields of 11 sweet sorghum cultivars grown at six site-years ranged from 3850 to 4410 L ha^{-1} of ethanol production, assuming 95% extraction of sugars and 1.76 kg fermentable carbohydrate per liter of ethanol produced (Hunter, 1994). Other reported yields were 3050 to 4000 L ha^{-1} ethanol (Lueschen et al., 1991). Calculated ethanol yields were less in Nebraska U.S.A., averaging 1600 L ha^{-1} at a semi-arid location at 1300 m above sea level and ranging from 1800 to 4100 L ha^{-1} for locations with longer growing seasons and more precipitation (Wortmann et al., 2010).

3.5 Bi-product use

In addition to the ethanol produced by fermentation of sugar, other yield components of sweet sorghum may have biofuel or other value, including some grain yield and the bagasse remaining after juice extraction (Bennett & Anex, 2008). The grain and bagasse may be of value in animal feeding. The bagasse may be used in paper production, biofuel, or for soil application. Sweet sorghum hybrid varieties released in China have given biomass and grain yields of 25 and 5 Mg ha^{-1}, respectively, at a temperate latitude (Hong-Tu & Xiu-Ying, 1986).

3.6 Energy requirements and balances

Total energy yield, net energy yield, and the ratio of energy gained to energy input need to be considered in comparing biofuel sources and in comparing biofuel to fossil fuel (Table 2 and 3). The values used in these calculations vary and need to be reported in published works. The estimated crop production input of energy per liter of potential ethanol yield was 6.42, 5.25, 6.35, and 5.95 MJ L^{-1}, respectively, for maize, sweet sorghum, and sugar- and fodder beet grown in California U.S.A. (Reed et al., 1986). The respective theoretical ethanol yields are 4814, 5784, 7782, and 6886 L ha^{-1}. Estimated energy consumption for sweet sorghum compared to maize production in Nebraska U.S.A. was greater for fuel and transportation but less for N fertilizer and irrigation (Table 3; Wortmann et al., 2009). Energy required for converting the product to ethanol was not estimated.

The net energy gain was 17, 40 and 50% greater with sweet sorghum compared with fiber sorghum, wheat (*Triticum aestivum* L.) with no N, and wheat with N applied, respectively, assuming gasification of the crop residues (Monty & Venturi, 2003). The energy efficiency of ethanol production was estimated to be 90% compared with gasification.

The average energy output to input ratio was 2.83 for sweet sorghum across seven site-years in Nebraska U.S.A. compared to 2.13 and 2.21 for ethanol produced from grain of maize and grain sorghum, respectively (Table 3). Mean energy consumption for ethanol produced from sweet sorghum was approximately 3300 MJ ha^{-1} compared with 8900 and 5800 MJ ha^{-1} for maize and grain sorghum, respectively. These calculations were made using the BESS model

	Maize	Grain sorghum	Sweet sorghum
Grain or sugar yield, Mg ha^{-1}	7.94	6.24	2.85
N rate, kg ha^{-1}	107	50	0
Ethanol yield, L ha^{-1}	3361	2639	1892
Energy use rate, MJ L^{-1}	10.9	10.4	7.9
Energy yield, GJ ha^{-1} ††	78.3	60.9	39.9
Energy consumed, GJ ha^{-1}	36.6	27.5	14.5
Net energy yield, GJ ha^{-1} ††	41.6	33.4	25.3
Net energy ratio ††	2.13	2.21	2.70
Crop† energy use, MJ ha^{-1}	8932	5791	3294
Crop† CO_2 emission, kg Mg^{-1}	77.5	65.7	90.4
Crop† CH_4 emission, kg Mg^{-1}	0.080	0.070	0.073
Crop† N_2O emission, kg Mg^{-1}	0.38	0.26	0.98
Crop† CO_2e‡ emission, kg Mg^{-1}	192	144	385
Crop† CO_2e emission, g MJ^{-1}	21.5	16.2	27.4
Life cycle CO_2e emission, g MJ^{-1}	31.2	28.4	45.7
CO_2e reduction, %§ ††	66.1	69.1	48.8

† Values were calculated using the Biofuel Energy Systems Simulator (BESS; available at www.bess.unl.edu). Emission of N2O may be under-estimated for grain as the ethanol co-products were assumed to be fed to beef cattle, resulting in unnecessarily high protein rations with much excretion of urine-N that can be a significant source of N2O emission. This N2O emission was not considered in these calculations due to lack of good estimates.
‡ CO2e, total greenhouse gas emission expressed as CO2 equivalent.
§ This was calculated assuming 92 gCO2e emission MJ-1 for gasoline.
†† Grain crops included a standard energy and greenhouse gas co-product credit, while no co-product was included for sweet sorghum.

Table 3. Mean estimated yields, CO_2e emissions for grain and sugar produced, ethanol produced (g MJ^{-1}), and energy balances of maize, grain sorghum, and sweet sorghum determined over seven site-yr in Nebraska U.S.A. (adapted from Wortmann et al., 2010).

(Liska et al., 2009), and assumes processing in state-of-the-art ethanol plants and efficient use of the grain by-products in beef cattle feeding. In earlier work, the net energy ratio for sweet sorghum was estimated to exceed 2.0, with two units of energy recovered in the ethanol for each unit used for crop production and processing (Sheehan et al, 1978).

Mean net energy yield in the Nebraska U.S.A. study was 31 GJ ha^{-1} for sweet sorghum compared with 41 and 33 GJ ha^{-1} for maize and grain sorghum, respectively (Table 3; Wortmann et al., 2010). The mean reduction in greenhouse gas emission in replacing gasoline with ethanol produced from sweet sorghum as transportation fuel was 53%. The reduction may be greater because of uncertainty of the estimated N_2O emitted from decomposing bagasse, a major component of the greenhouse gas emission estimated on a carbon dioxide equivalent basis. In interpreting the results of comparing sweet sorghum with grain crops in Nebraska U.S.A., we must consider that grain crop production technology, including variety development, is much more advanced with the grain crops compared with sweet sorghum. Varietal differences indicated potential to increase productivity through genetic improvement. The potential of sweet sorghum hybrids compared with lines has been demonstrated (Zhao et al., 2009).

The cost ha^{-1} of sweet sorghum production was found to be greater than for maize in California U.S.A. because of high harvest costs (Geng et al., 1989), but hexose yield was greater with sweet sorghum and the costs of producing ethanol were very similar. In another study, the calculated cost of ethanol energy production was \$0.48, \$0.53, and \$0.58 L^{-1}, respectively, for maize, sugarcane, and sweet sorghum under best production potential scenarios in Florida U.S.A. (Rahmani & Hodges, 2006); processing and harvesting were major expenses for sweet sorghum. The cost of converting sugarcane juice to ethanol was estimated to be \$0.13 L^{-1} in Florida U.S.A. (Rahmani & Hodges, 2006); a similar cost may apply to converting sweet sorghum juice to ethanol.

4. Conclusion

There are several obstacles to the development of sweet sorghum as a competitive bio-energy crop for the U.S.A. Great Plains with greater challenges for the northern compared with the southern part of the region. The primary limitations of sweet sorghum for bioenergy in the U.S.A. Great Plains and other temperate climate zones include seasonality of harvest and large masses to be transported and stored. Fermentation of the expressed juice must be initiated quickly after harvest to avoid sugar loss. The loss of fermentable sugars from storing fresh juice at room temperature may be 20% after three days and up to 50% after one week, although losses were minimal with refrigerated storage (Wu et al., 2008; Wu et al., 2010). In temperate climates, the harvest window for sweet sorghum is limited by length of the growing season. Seed production is costly because of low seed yield and usually very tall plants. Few open-pollinated or hybrid cultivars are available for production, although there appears to be potential for significant increases in productivity in temperate zones with hybrid sweet sorghums (Zhao et al., 2009). Integration of distillation and distribution of sweet sorghum ethanol into existing grain-based ethanol processing systems would take advantage of existing infrastructure and reduce the challenges of transport and storage of sweet sorghum stalks or juice. Developing the means of stabilizing sweet sorghum juice to minimize sugar loss during storage would improve the feasibility of temperate zone sweet sorghum production. Profitable use of the bagasse such as with cellulosic ethanol production, without significant loss of nutrients for recycling in crop production, would add to the feasibility of sweet sorghum as a biofuel crop. Where bagasse is best returned to the land, combining small-scale processing technology, such as small-scale juice extraction, fermentation, and distillation linked with refinement at larger scale facilities, may reduce storage and transportation costs while enabling efficient nutrient recycling.

5. References

Barbanti, L., Grandi, S., Vecchi, A., & Venturi, G. (2006). Sweet and fiber sorghum (*Sorghum bicolor* (L) Moench), energy crops in the frame of environmental protection from excessive nitrogen loads. *Europ. J. Agron.*, Vol. 25, No. 1, (July 2006), pp. 30-39.

Bennett, A.S. & Anex, R.P. (2008). Farm-gate production cost of sweet sorghum as a bioehtanol feedstock. *Trans. of the ASABE*, Vol. 51, No. 2, pp. 603-613.

Bennett, A.S. & Anex, R.P. (2009). Production, transportation and milling costs of sweet sorghum as a feedstock for centralized bioethanol production in the upper Midwest. *Bioresource Technology*, Vol. 100, No. 4, (Feb. 2009), pp. 1595-1607.

Broadhead, D.M. (1969). Sugar production from sweet sorghum as affected by planting date, after-ripe harvesting, and storage. *Agron. J.*, Vol. 61, pp. 811-812.

Broadhead, D.M. (1972). Effect of planting date and maturity on juice quality of Rio sweet sorghum. *Agron. J.*, Vol. 64, pp. 389-390.

Broadhead, D.M. (1973). Effects of deheading on stalk yield and juice quality of Rio sweet sorghum. *Crop Sci.*, Vol. 13, pp. 395-397.

Broadhead, D.M. (1974). Effect of planting date and maturity on juice quality of Rio sweet sorghum. *Agron. J.*, Vol. 64, pp. 389-390.

Broadhead, D.M. & Freeman, K.C. (1980). Stalk and sugar yield of sweet sorghum as affected by spacing. *Agron. J.*, Vol. 72, No. 3, pp. 523-524, ISSN 0002-1962.

Buxton, D.R., Anderson, I.C., & Hallam, A. (1998). Intercropping sorghum into alfalfa and reed canarygrass to increase biomass yield. *J. Prod. Agric.*, Vol. 11, No. 4, pp. 481-486, ISSN 0890-8524.

Coleman, O.H. & Belcher, B.A. (1952). Some response of sorgo to short photoperiods and variations in temperature. *Agron. J.*, Vol. 44, pp. 35-39.

Coleman, O.H. (1970). Syrup and sugar from sweet sorghum. p. 416-440, In: *Sorghum production and utilization*, J.W. Wall & W.M. Ross (Eds.), pp. 416-440, AVI Publishers, Westport, CT U.S.A..

Ferraris, R. (1981). Early assessment of sweet sorghum as an agro-industrial crop. I. Varietal evaluation. *Aust. J. Exp. Agric. Anim. Husb.*, Vol. 21, No. 108, pp. 75-82.

Ferraris, R. & Charles-Edwards, D.A. (1986). A comparative analysis of the growth of sweet and forage sorghum crops. II. Accumulation of soluble carbohydrates and nitrogen. *Aust. J. Agric Res.*, Vol. 37, No.5, pp. 513-522.

Freeman, K.C., Broadhead, D.M. & Zummo, N. (1973). Culture of Sweet sorghum for syrup production. *USDA Agricultural Handbook No 441*, USDA.

Geng, S., Hills, F.J., Johanson, S.S., & Sah, R.N. (1989). Potential yields and on-farm ethanol production cost of corn, sweet sorghum, fodderbeet and sugarbeet. *J. Agron. Crop Sci.*, Vol. 162, No. 1, (Jan. 1989), pp. 21-29.

Hill, N.S., Posler, G.L., & Bolsen, K.K. (1987). Fermentation inhibition of forage and sweet sorghum silages treated with acrylic or maleic acid. *Agron. J.*, Vol. 79, No. 4 (Jul.-Aug. 1987), pp. 619-623, ISSN 0002-1962.

Hipp, B.W., Cowley, W.R., Gerard, C.J., & Smith, B.A. (1970). Influence of solar radiation and date of planting on yield of sweet sorghum. *Crop Sci.*, Vol. 10, No. 1, pp. 91-92 ISSN 0011-183X.

Hong-Tu, M. & Xiu-Ying, H. (1986). Breeding for sweet sorghum with high grain yield and sweet stalks. *Sorghum Newsl.*, Vol. 29, pp. 2-3.

Hunter, E.L. (1994). Development, sugar yield and ethanol potential of sweet sorghum. M.S. thesis. Iowa State Univ., Ames, U.S.A.

Hunter, E.L. & Anderson, I.C. (1997). Sweet sorghum. *Hort. Rev.*, Vol. 21, pp. 73-104.

Jankins, G. (1966). *Introduction to cane sugar technology*. Elsevier Science Ltd. ISBN-10 0444403191, Amsterdam, The Netherlands.

Janssen, G., McClelland, C.K., & Metzger, W.H. (1930). Sap extraction of Sorghum and the localization of juice and sugars in internodes of the plant. *Agron. J.*, Vol. 22, pp. 627-629.

Krishnaveni, S., Balasubramanian, T., & Sadasivam, S. (1990). Potentiality of sweet sorghum (Sorghum bicolor, Poaceae) for syrup preparation and alcohol production in India. *Econ. Bot.*, Vol. 44, No. 3, (July-Sept. 1990), pp. 355-359.

Kuepper, G. (1992). Sweet sorghum: production and processing. Kerr Center for Sustainable Agriculture, Poteau, OK U.S.A.

Kundiyana, D.D., Bellmer, D., Huhnke, R., & Wilkins, M. (2006). "Sorganol": Production of ethanol from sweet sorghum. ASABE Annual Meeting Presentation, Portland, OR, U.S.A.

Lamb, M. (1982). Mechanical expression of sweet sorghum for juice recovery. M.Sc. thesis, University of Nebraska-Lincoln, Lincoln NE U.S.A.

Lingle, S. (1987). Sucrose metabolism in the primary Culm of sweet sorghum during development. *Crop Sci.*, Vol. 27, No. 6, (Nov.-Dec. 1987), pp. 1214-1219, ISSN 0011-183X.

Liska, A.J., Yang, H.S., Bremer, V.R., Klopfenstein, T.J., Walters, D.T., Erickson, G.E., & Cassman, K.G. (2009). Improvements in life cycle energy efficiency and greenhouse gas emissions of corn-ethanol. *J. Ind. Ecol.*, Vol. 13, No. 1, (Feb. 2009), pp. 58–74.

Lueschen, W.E., Putnam, D.H., Kanne, B.K., & Hoversted, T.R. (1991). Agronomic practices for production of ethanol from sweet sorghum. *J. Prod. Agric.*, Vol. 4, No. 4, pp. 619-625, ISSN 0890 8524.

Martin, P.M. & Kelleher, F.M. (1984). Effects of row spacing and plant population on sweet sorghum yield. *Aust. J. Exp. Agric. Anim. Husb.*, Vol. 24, No. 126, pp. 386-390.

McAloon, A., F. Taylor, W. Yee, K. Ibsen, and R. Wooley. 2000. Producing ethanol from corn starch and lignocellulosic feedstocks. NREL/TP-580–28893. Natl. Renewable Energy Lab., Golden, CO.

Monti, A. & Ventura, G. (2003). Comparison of the energy performance of fiber sorghum, sweet sorghum and wheat monocultures in northern Italy. *Eur. J. Agron.*, Vol. 19, No. 1, (Feb. 2003), pp. 35-43.

Nuese, G.A. & Hunt, D.R. (1983). Field harvest of sweet sorghum juice. *Transaction of ASABE*, Vol. 26, No. 3, pp. 656-660.

Ottman, M.J & Miller, A.N. (2010). Irrigation frequency effects on growth and ethanol yield in sweet sorghum. *Agron. J.*, Vol. 102, No. 1, (Jan.-Feb. 2010), pp. 60-70.

Propheter, J.L. & Staggenborg, S. (2010). Performance of annual and perennial biofuel crops: nutrient removal during the first two years. *Agron. J.*, Vol. 102, No. 2, (Mar.-Apr. 2010), pp. 798-805.

Propheter, J.L., Wang, D., Wu, X., & Staggenborg, S.A. (2010). Performance of annual and perennial biofuel crops: yield during the first two years. *Agron. J.*, Vol. 102, No. 2, (Mar.-Apr. 2010), pp. 806-814, ISSN 0002-1962.

Putnam, D.H., W.E. Lueschen, B.K. Kanne, and T.R. Hoverstad. 1991. A comparison of sweet sorghum cultivars and maize for ethanol production. J. Prod. Agric. 4:377–381.

Rahmani, M. & Hodges, A. (2006). Potential feedstock sources for ethanol production in Florida. Retrieved Mar 29, 2011. Available from http://edis.ifas.ufl.edu/pdffiles/FE/FE65000.pdf

Rains, G.C., Cundiff, J.S., & Vaughan, D.H. (1990). Development of a whole-stalk sweet sorghum harvester. *Trans. ASAE*, Vol. 33, No. 1, pp. 56-62, ISSN 0001-2351.

Rajendran, C., Ramamoorthy, K., & Backiyarani, S. (2000). Effect of deheading on juice quality characteristics and sugar yield of sweet sorghum. *J. Agron. and Crop Sci.*, Vol. 185, No. 1, (July 2000), pp. 23-26.

Reddy, B.V.S., Ramesh, S.,Reddy, P.S., Ramaih, B., Salimath, P.M., & Kachapur, R. (2005). Sweet sorghum−A potential alternative raw material for bio-ethanol and bio-energy. *International Sorghum and Millets Newsl.*, Vol. 46, pp. 79-86.

Reddy, B.V.S., Kumar, A.A., & Ramesh, S. (2007). Sweet sorghum: a water saving bio-energy crop. International conference on linkages between energy and water management for agriculture in developing countries, IWMI, ICRISAT, Hyderabad, India.

Reed, W., Geng, S., & Hills, F.J. (1986). Energy input and output analysis of four field crops in California. *J. Agron. Crop Sci.*, Vol. 157, No. 2 (Aug. 1986), pp. 99-104.

Rein, B. (1984). Fermentation of sweet sorghum juice as affected by heating and yeast inoculums. Thesis. Lincoln, NE U.S.A.

Ricaud, R.B., Cochran, B., Arceneaux, A., & Newton, G. (1979). Sweet sorghum for sugar and biomass production in Louisiana. Manuscript report from the St. Gabriel Experiment Station.

Ricaud, R. & Arceneaux, A. (1990). Sweet sorghum research on biomass and sugar production in 1990. Manuscript report from the St. Gabriel Experiment Station.

Sheehan, G.J., Greenfield, P.F., & Nicklin, D.J. (1978). Energy economics, ethanol- a literature review. Alcohol Fuels Conf., Sydney, Australia.

Smith, G.A., Bagby, M.O., Lewellan, R.T., Doney, D.L., Moore, P.H., Hills, F.J., Campbell, L.G., Hogaboam, G.J., Coe, G.E., & Freeman, K. (1987). Evaluation of Sweet Sorghum for Fermentable Sugar Production Potential. *Crop Sci.*, Vol. 27, No. 4, (July-Aug. 1987), pp. 788-793.

Smith, G.A. & Buxton, D.R. (1993). Temperate zone sweet sorghum ethanol production potential. *Bioresource Tech.*, Vol. 43, No. 1, pp. 71-75.

Soileau, J.M. & Bradford, B.N. (1985). Biomass and sugar yield response of sweet sorghum to lime and fertilizer. *Agron. J.*, Vol. 77, No. 3, (May-June 1985), pp. 471-475, ISSN 0002-1962.

Tarpley, L., Lingle, S.E., Vietor, D.M., Andrews, D.L., & Miller, F.R. (1994). Enzymatic control of nonstructural carbohydrate concentrations in stems and panicles of sorghum. *Crop Sci.*, Vol. 34, No.2, pp. 446-452, ISSN 0011-183X.

Turgut, I., Bilgili, U., Duman, A., & Acikgoz, E. (2005). Production of sweet sorghum (*Sorghum bicolor* L. Moench) increases with increased plant densities and nitrogen fertilizer levels. *Acta Agriculturae Scandinavica*, Vol. 55, No. 3, (Sept. 2005), pp. 236-240.

Turhollow, A. (1994). The economics of energy crop production. *Biomass and Bioenergy*, Vol. 6, No. 3, pp. 229-241.

US Department of Energy. (1979). The report of the alcohol fuels policy review, Report No. DOE/PE-0012. Washington, DC U.S.A.

Vermerris, W., Rainbolt, C., Wright, D. & Newman, Y. (2008). Production of biofuel crops in Florida: Sweet sorghum. Retrieved mar 29, 2011. Available from http://edis.ifas.ufl.edu/ag298.

Vietor, D.M. & Miller, F.R. (1990). Assimilation, partitioning and nonstructural carbohydrates in sweet compared with grain sorghum. *Crop Sci.*, Vol. 30, No. 5, pp. 1109-1115 ISSN 0011-183X.

Vries, S.C. de, Giller, K.E., Ittersum, M.K. van, & Ven, G.W.J. van de. (2010). Resource use efficiency and environmental performance of nine major biofuel crops, processed by first-generation conversion techniques. *Biomass and Bioenergy*, Vol. 34, No. 5, (May 2010), pp. 588-601.

Wang, P., Singh, V., Xu, L., Johnston, D.B., Rausch, K.D. , and Tumbleson, M.E. 2005. Comparison of enzymatic (E-mill) and conventional dry-grind corn processes using a granular starch hydrolyzing enzyme. Cereal Chem. 82:734–738.

Wiendenfeld, R. (1984). Nutrient requirement and use efficiency by sweet sorghum *Energy Agr.*, Vol. 3, pp. 49-59.

Winberry, J. (1980). The sorghum syrup industry 1854-1975. In: *Agricultural History*, Vol. 54, No. 2, pp. 343-352.

Wortmann, C.S., Liska, A.J., Ferguson, R.B., Lyon, D.J., Klein, R.M. & Dweikat, I. (2010). Dryland performance of sweet sorghum and grain crops for biofuel. *Agron J.*, Vol. 102, No. 1, pp. 319-326.

Wu, X., Staggenborg, S., Propheter, J.L., Rooney, W.L., Yu, J., & Wang, D. (2008). Features and fermentation performance of sweet sorghum juice after harvest, Paper presented at the 2008 ASABE meeting, June 29- July 2, 2008, Providence, RI U.S.A.

Wu, X., Staggenborg, S., & Propheter, J.L, Rooney, W.L., Yu, J., & Wang, D. (2010). Features of sweet sorghum juice and their performance in ethanol fermentation. *Industrial Crops and Products*, Vol. 31, No. 1, (Jan. 2010), pp. 164-170.

Yoneyama, T., Terakado, J., & Masuda, T. (1998). Natural abundance of ^{15}N in sweet potato, pumpkin, sorghum, and castor bean: possible input of N2-derived nitrogen in sweet potato. *Biol. Fert. Soils*, Vol. 26, No. 2, pp. 152-154.

Zhao, Y.L., Dolat, A., Steinberger, Y., Wang, X,. Osman, A., & Xie, G.H. (2009). Biomass yield and changes in chemical composition of sweet sorghum cultivars grown for biofuel. *Field Crops Res.* Vol. 111, No. 1-2, (Mar. 2009), pp. 55-64.

12

Does Money Grow on Trees? People's Willingness to Pay for Cellulosic Wood Ethanol

Katherine Farrow, Mario Teisl, Caroline Noblet,
Shannon McCoy and Jonathan Rubin
University of Maine
USA

1. Introduction

According to the American Environment Research and Policy Center, America's dependence on fossil fuels, and the resulting global warming pollution, has been increasing both nationally and at the state level for decades (AERPC, 2009). The Intergovernmental Panel on Climate Change (IPCC, 2007) issued its Fourth Assessment Report in 2007, describing how atmospheric concentrations of carbon dioxide and other greenhouse gases have increased as a direct result of human activity for over one hundred years. Various implications of this increase in greenhouse gases include increases in average air and ocean temperatures, melting of snow and ice, and rising global average sea level. These environmental implications have important negative ecological and economic effects. Educational campaigns, policy initiatives and an increased public interest in alternative energies have led to the beginnings of a shift in this trend of increasing greenhouse gas emissions. Emissions declined in 17 states between 2004 and 2007 due to the use of cleaner and more efficient forms of energy (AERPC, 2009).

To continue this decrease in carbon emissions, it is in the interest of researchers and decision makers to expand the clean energy market, where doing so requires an understanding of the public's preferences and behavior regarding energy consumption. Attitudes are commonly linked to intentions and behavior, and as such, are believed to be an important component of the construction and implementation of various public policy initiatives (Krosnick, 1988; Ritchie & Spencer, 1994; Hini et al., 1995; Kaiser et al., 1999). Attitudes have been directly linked to behavioral change by Loudon and Della Bitta (1993), who state "behavioral change is a function of change in behavioral intentions…changes in behavioral intentions are related to change in attitude" (p.422), and by Bamberg (2003), who maintains that "degree of environmental concern has a direct strong impact on people's behavior" (p.4).

As a determinant of behavior, attitudes such as environmental concern are important to understand if we are to promote alternative energies like biofuels. Understanding whether or not environmental concern affects consumers' decisions to purchase biofuels will be of great use to policy makers and other groups interested in expanding the emerging biofuels market. On the other hand, consumer perceptions of biofuels are also likely to be important.

For example, Teisl et al. (2009) find that some people hold negative perceptions of ethanol (e.g., ethanol damages engines).

The goal of this study is to determine how much survey participants are willing to pay for cellulosic wood ethanol. Cellulosic ethanol made from wood has the potential to reduce carbon dioxide emissions and reduce oil imports while also reducing current conflicts between food and fuel production associated with some sources (corn, sugarcane) of ethanol (Solomon et al., 2007; Solomon & Johnson, 2009). However, an important consideration for market penetration by any product is the willingness of consumers to accept and use this product. Collantes (2010) reminds us that a consumer's value proposition (his/her perceived motivations for purchasing a certain product) for new technologies is best assessed in relation to existing mainstream technologies rather than in isolation. Therefore, the need to understand consumers *a priori* attitudes and beliefs regarding both gasoline and biofuels is of particular importance. Since the environmental impacts of biofuels differ across source material, we must become familiar with whether consumers are aware of, or sensitive to, these differences (Wegener & Kelly 2008). There may also exist subsets within the consumer base whose existing characteristics, attitudes or beliefs would incline them towards purchase of environmentally preferred products, including fuels, if appropriate messaging information could be presented.

2. Literature review

Though some willingness-to-pay (WTP) studies have been performed recently (Collantes, 2010; Jensen et al., 2010), literature regarding the acceptance of cellulosic ethanol is generally limited due to the pre-market nature of the product (Solomon & Johnson, 2009); insights into the factors that may impact consumer selection and acceptance of environmentally preferred products can be garnered from the 'green' behavior literature (see Clark et al., 2003; Carrus et al., 2008; Ek, 2005).

Attitudes have often been found to be a precursor to environmental behavior (e.g., Birgelen et al., 2009; Fraj and Martinez, 2007; Kaiser et al., 1999; Chan 2001); although often the effect is relatively weak (Fraj & Martinez, 2007). Attitude[1] towards a behavior is defined by Ajzen as "the degree to which performance of the behavior is positively or negatively valued" (Ajzen, 2006). Fraj and Martinez (2007) report that environmental psychologists have indicated two sets of environmental attitudes: one based on the actual eco-behavior under study, the other being a more general eco-attitude (i.e., an attitude toward the environment, not at a particular behavior).

Norms are shared beliefs about how people should act (Schwartz & Howard, 1982); social norms are generally defined as what the individual perceives as expectations on their behavior held by social groups important to the individual (e.g. peers, family or colleagues). These social expectations are assumed to be supported by real or perceived sanctions so that the individual has an incentive to adhere to the social norms (Ajzen, 1988). Personal norms are internal expectations held by the individual; e.g., a sense of obligation (Schwartz, 1977).[2] These norms have also been found to positively influence a person's eco-related behaviors

[1]In Ajzen's Theory of Planned Behavior model, attitudes are functions of beliefs.

[2]This is a common simple dichotomy; see Thøgersen (2006) for a rigorous taxonomy of norms.

(Ajzen et al., 2004; Ek & Söderholm, 2008; Hunecke et al., 2001; Thøgersen, 1999; Birgelen et al., 2009).

Perceived control reflects the degree to which a person views themselves as being able to perform a specific behavior. In other words, people exhibit a stronger willingness to change their environmental behavior when they feel they can individually have an effect on the quality of the environment (Roberts, 1996; Vermeir & Verbeke, 2006). Perceived behavioral control (Wall et al., 2008; Ajzen, 2002; Birgelen et al., 2009) is indicated to be a significant precursor to environmentally related behaviors, although sometimes there is no link found (Birgelen et al., 2009).

Willingness to pay (WTP) studies of biofuels to date have proffered mixed outlooks on the prospects of the market for biofuels in the U.S. Tatum (2010) found that the cross-price elasticity of gasoline and e85 derived from corn is near unity, indicating that the prices of corn ethanol and gasoline are closely linked, reducing its feasibility as a sustainable substitute. Evidence has also suggested that, given the high price, limited fueling availability, and reduced performance of cellulosic e85, owners of flex fuel vehicles have little incentive to discontinue using the current gasoline mix (Collantes, 2010). On a more optimistic note, Jensen et al. (2010) finds that average WTP estimates for cellulosic e85 range from 16.6 - 18.9 cents/gallon over e10 derived from corn, indicating an overall willingness to pay a premium for cellulosic ethanol.

3. Theory

Beginning with the economic assumption that demand is a function of price (P), other attributes of the product (A) and income (I) (Lancaster, 1971), we expand this theoretical model to include the decision maker's psychological characteristics (C). These characteristics include the standard theory of planned behavior variables: attitudes, beliefs, norms, and perceived behavioral control. The general theoretical model then becomes:

$$\text{Fuel choice} = f(\text{P, A, I, C}) \tag{1}$$

According to this model, choice (the decision to purchase fuel) is based on the attributes of the fuel (environmental, fuel security) and the individual's psychological characteristics - specifically their beliefs (perceptions) about the environment (e.g., the threat of global warming), fuel security (dependence on foreign sources of fuel), as well as attitudes towards new technologies and products.

Behavioral theories such as the cognitive hierarchy model and the theory of planned behavior suggest that attitudes are an important determinant of behavior. The cognitive hierarchy model, developed by Homer and Kahle (1988), asserts that people's behavior is the result of a cause-and-effect chain beginning with values, which shape beliefs, then attitudes and norms, and finally, determine the behavior itself. The theory of planned behavior has also serves as a common method for understanding the nature of the relationship between such beliefs and behavior. Daigle et al. (2002) studied environmental concern in the context of the theory of planned behavior, operating under the model that human behavior is guided by behavioral beliefs (attitudes towards a behavior), normative beliefs (beliefs about the normative expectations of others), and control beliefs (perceived ease or difficulty of performing the behavior).

4. Data and empirical model

4.1 Sampling and survey administration

During the summer of 2009 we administered a mail survey to a representative sample of 3,800 New England, USA residents[3] (500 residents per state, with an over sample of Maine residents - 800). The sample frame was purchased from *InfoUSA;* the *InfoUSA* database contains information about 210 million US residents.

The survey was administered with multiple mailings, including an introductory letter sent by post return-receipt requested to identify undeliverable addresses. In total 382 Maine residents and 958 New England (non-Maine) residents responded to the survey for a response rate of 52 and 38 percent, respectively yielding an overall response rate of 40 percent. The overall response rate is marginal, suggesting that individual survey results may not be a valid representation of the knowledge, practices and attitudes of the New England adult population. However, our purpose here is not to extrapolate our survey results to the aggregate population but to examine differences in attitudes, beliefs and fuel choice behaviors across different types of people.

4.2 Survey design

The survey instrument was informed by focus groups held in Maine and Massachusetts during the summer and fall of 2008 (Teisl et. al, 2009). The final survey instrument consists of six sections aimed at eliciting information regarding a consumer's environmental concern (in general, and regarding specific issues) including their experience with or knowledge of biofuels with a specific focus on cellulosic ethanol, consumer's driving habits, responses to environmental psychology constructs, a fuel choice experiment, current environmental behaviors and socio-economic characteristics.

The analysis of the fuel choice scenario is the basis of this chapter. Here, each respondent was asked to respond to one fuel choice scenario. In each scenario (Figure 1), respondents viewed information about three transportation fuels. One fuel represented their current fuel, one was a fuel that contained ethanol derived from wood and one fuel contained ethanol derived from corn. The fuels also differed in terms of price, environmental (greenhouse gases emissions) and level of fuel security (percent of fuel imported) attributes displayed. Respondents were told to assume that the products were exactly the same except for their prices and the information presented on the labels. Respondents were asked to assume they were purchasing one of these fuels in the near future (in a few months); this was to allow us to reasonably provide information about new fuels that were currently unavailable and allowed us to broaden the range of prices being used in the scenarios. Increasing the variation of the price variable in this way enabled us to better isolate its effect on the decision-making process. However, including prices significantly different than the actual market prices faced by the respondent may induce the respondent to reject the scenario altogether. Coupled with the fact that fuel prices vary significantly across the New England region, this problem is accounted for by the future frame of the question, making the choice seem reasonable.

The attributes displayed on the labels were chosen based on previous focus group research indicating that these attributes were the most important to consumers (Teisl et al. 2009). The

---

[3]The states included are Connecticut, Maine, Massachusetts, New Hampshire, New York, Rhode Island and Vermont.

actual values displayed on each label were generated from a normal distribution with predetermined means and standard deviations using Excel. Price per gallon for the consumer's usual fuel had a range of $1.50 to $4.50 with a mean of $2.50, while price for wood or corn based ethanol ranged from $1.30 to $4.65 also with a mean of approximately $2.50. The range of the greenhouse gases (GHG) levels presented to participants was based on carbon dioxide emission per gallon. Gasoline yields approximately 25 pounds of CO2 per gallon.[4] The range presented to participants was 15-25 pounds per gallon, with a mean of 20. Cellulosic ethanol yields 6.9 pounds per gallon to 12.5 pounds per gallon.[5] Respondents were given information on the percent reduction in emissions (where 7.5 is a 65% reduction) with a range of 40 to 80 percent and a mean reduction of 60 percent (approximately 8 pounds per gallon).[6] Information on corn-based ethanol's carbon dioxide emissions differ, however we based the scenario on an assumed 17 pounds per gallon (i.e. 20% reduction). Thus respondents were presented with a range of 5 to 60% reduction in GHG, with a mean of 23% reduction. Fuel import statistics for the scenario were based on Transportation Energy Data Book, Table 1.7 edition 27 (Davis et al., 2007). The price and attribute scores were then randomly assigned across respondents (i.e., each scenario is likely to have a unique price/attribute combination).

Although we provided price, environmental and fuel security attributes for each of the three fuels, we only provided information about ethanol and its source for two of the fuels. We did not provide this information for the respondent's current fuel because during the time of the survey administration, while all parts of the study area sold fuels containing ethanol, not all states where ethanol is sold required ethanol-containing fuels to be labeled on the fuel pump (Table 1). As a result, we included a question before the choice scenario explicitly asking the respondent if they currently use gasoline, gasoline mixed with 10 percent ethanol (e10), gasoline mixed with 85 percent ethanol (e85) or used diesel or other fuels. Surprisingly, 52 percent of the respondents claimed they were using gasoline without any ethanol even though e10 is the primary fuel sold in the region. Forty-five percent thought they were using e10 and the rest thought they were using e85, diesel or other fuels.[7]

That the sample of respondents was basically split in what fuel they thought they were currently using is problematic in that we need to define a status quo fuel to estimate a price premium for wood-based e10. To begin, we first used t-tests to examine whether the respondents who indicated they used only gasoline as their fuel (hereafter gas only respondents) were different than the e10 respondents.[8] In general we find that the two groups differ significantly in a number of areas. Gas only respondents are more concerned about global warming, hold more positive views of ethanol, and have more positive

[4]Based on 98.2 kg CO2e/mmBTU, U.S. Environmental Protection Agency, Regulatory Impact Analysis, Renewable Fuel Standard 2, p. 427.

[5]6.9 pounds per gallon is based on switchgrass at 27 kg CO2e/mmBTU, 12.5 pounds per gallon is based on the "advanced biofuel" requirements of 50% decrease in GHG emissions compared to 2005 baseline gasoline at 98 kg/mmBTU (U.S. Fed Register/vol. 75, no 56./Friday, March 26, 1020/Table V.C-4).

[6]Our framing numbers on the GHG intensity of fuels differs from current values used by the US EPA. These reflect a change in evolving science of lifecycle GHG accounting of the various fuels. The range given to survey participants is within the range of currently accepted values.

[7]Given only three percent citing the use of e85, diesel or other fuel, we dropped these respondents from further analysis.

[8]Contact second author for full results.

Assume that **in a few months** you went to your usual station to buy some fuel. In addition to the fuel you usually buy, you find two other types of fuel. The only difference between the fuels is what appears below. Note we have given you some information about your usual fuel.

YOUR USUAL FUEL	FUEL A	FUEL B
$1.52	$1.33	$1.44
	This product contains ethanol made from wood	This product contains ethanol made from corn
Each gallon of this product produces **22** pounds of green house gases	Each gallon of this product produces **4** pounds of green house gases	Each gallon of this product produces **20** pounds of green house gases.
63 percent of this fuel is imported	**50** percent of this fuel is imported	**57** percent of this fuel is imported

Which fuel would you purchase? (PLEASE CHECK **ONE** BOX)

- ❑ I WOULD CHOOSE MY USUAL FUEL
- ❑ I WOULD CHOOSE FUEL A
- ❑ I WOULD CHOOSE FUEL B

Fig. 1. Sample choice scenario; underlined numbers vary across respondents.

attitudes toward buying US-made products (Table 2). However, they are also less likely to try new products. In terms of their behaviors, gas only respondents drive less and use public transportation more often, but are less likely to belong to an environmental group or buy environmentally labeled products. *Importantly, they are twice as likely to choose the status quo fuel.* Given the above we determined that the model should allow for the status quo fuel to vary; in turn, the baseline fuel for gas only is modeled as not containing ethanol whereas the baseline fuel for e10 users is coded as containing corn-based ethanol. The choice to use corn-based e10 as the status quo fuel option is based upon our results that almost all respondents (97 percent) who have heard of ethanol as a fuel additive, cited corn is its primary source.

State	Labeling Required for Fuel Blends	Market Share 2009 of E10 (%)
Connecticut	Yes, at 1%	100
Maine	Yes, any blend	60
Massachusetts	Yes, at 1%	85
New Hampshire	No	85
New York	Yes, at 1%	90
Rhode Island	Yes, at 1%	95
Vermont	Yes, at 1.5%	65

Table 1. Northeast US E-10 Market Penetration and Labeling Requirements.
Sources: American Coalition for Ethanol, 2007 State by State Handbook; Fuel Testers

4.3 Estimated model

The goals of this study are to estimate respondents' willingness to pay a premium for wood-based ethanol and to determine the influence that different psychological factors may have on respondents' values for this fuel. Given the available data we operationalize the theoretical model (Equation 1) as:

$$C_{ik} = \alpha_1 PRICE_{ik} + \alpha_2 GHG_{ik} + \alpha_3(GHG_{ik} * GWIMP_i) + \alpha_4 IMPORT_{ik} + \alpha_5(IMPORT_{ik} * USBUY_i) + \alpha_6(IMPORT_{ik} * ECOBUY_i) + \alpha_7 WOOD\_GAS_{jk} + \alpha_8 CORN\_GAS_{jk} + \alpha_9 WOOD\_E10_{jk} + \alpha_{10} CORN\_E10_{jk} + \alpha_{11}(WOOD * TEKNO_i) + \alpha_{12}(CORN * TEKNO_i) + \alpha_{13}(WOOD * DAM_i) + \alpha_{14}(CORN * DAM_i) + \varepsilon \quad (2)$$

where C_{ik} is a dummy variable denoting individual i's choice of the kth fuel; 1 denotes the fuel was chosen, 0 otherwise. PRICE is the fuel's price as given in the scenario. GHG denotes the pounds of greenhouse gases produced per gallon of fuel and IMPORT denotes the percent of the fuel that is imported. WOOD_GAS and CORN_GAS are binary variables denoting whether the fuel contains wood- or corn-based ethanol, respectively, when the status quo fuel is gas only. WOOD_E10 and CORN_ E10 are binary variables denoting whether the fuel contains wood- or corn-based ethanol when the status quo fuel is corn-based e10. GWIMP, USBUY, ECOBUY, TEKNO and DAM are a set of psychological variables that were created through the use of factor analysis (explained below) which are meant to measure the individual's: concern about global warming; attitudes toward buying US-made and environmentally labeled products; aversion to trying new technologies or products; and perceptions of ethanol. These psychological variables are interacted with relevant product characteristics. ε is the error term. The model is estimated using conditional logit regression

4.3.1 Factor analysis

The survey contained a number of questions aimed at measuring individual's attitudes toward buying environmentally preferred products, US-made and new products, and their beliefs about ethanol. Two separate factor analyses were performed to confirm and develop these measures of attitudes and beliefs. All of the variables used to construct the factors (see Tables 3 and 4) used Likert rating scales (e.g. 1 = strongly disagree; 5 = strongly agree).

We use factor analysis on the above data to find a reduced set of factors that would help identify respondents by their psychological profile. Factor analysis is a data reduction technique used to investigate whether a group of variables have common underlying dimensions and can be considered to measure a common factor. Although the analysis can be used to summarize a larger number of variables into a smaller set of constructs; ultimately the analysis is not a hypothesis testing technique so it does not tell us what those constructs are (Hanley et al., 2005). In turn, the validity of naming the constructs is contingent upon researcher judgment and should be interpreted with some caution (Thompson & Daniel, 1996).

For the factor analysis we used principal components analysis followed by Varimax rotation. As is typical, factors with Eigen values less than one are dropped from further analysis as are variables with factor loadings of less than 0.6 as these are not considered statistically significant for interpretation purposes. To further verify the reliability of the factor analysis we compute Cronbach's alpha on the variables loading on each factor;

	'gasoline only'	'e10'
Psychological variables		
How concerned are you about the effect of global warming on the region	3.8	3.6
In general, I am hesitant to try new technologies	2.6	2.4
Reducing the amount I drive decreases global warming	3.8	3.6
It's worthwhile buying US-made products	4.3	4.2
Perceptions of ethanol		
Ethanol is cheaper than gas	2.8	2.5
Ethanol damages engines	2.8	3.2
Ethanol improves acceleration	2.7	2.5
Ethanol lowers fuel efficiency	2.9	3.2
Ethanol produces less pollution	3.5	3.2
Wood-based ethanol decreases dependence on foreign oil	4.2	4.0
Wood-based ethanol lowers the US trade deficit	3.8	3.6
Wood-based ethanol decreases global warming relative to gasoline	3.8	3.5
Wood-based ethanol decreases global warming relative to corn-based ethanol	3.5	3.3
Behaviors		
Miles of weekly driving	184	258
Percent belonging to an environmental group	21	27
Likelihood to buy eco-labeled products	2.9	3.1
Likelihood to use public transportation	1.7	1.5
Percent choosing the baseline fuel	20	10
Demographics		
Percent male	54	75
Age	55.5	55.5
Education (in years)	14.9	15.2
Average household income	80,700	91,000

Table 2. Sample of significant differences between the 'gasoline only' and 'e10' respondents.

aiming to have alphas greater than the minimum value of 0.70 suggested by Nunnally and Bernstein (1994).

The factor analysis on the first set of variables indicates that three factors (Table 3) explain respondent reactions toward buying environmental, new, and US-made products. Kaiser's overall measure of sampling adequacy is relatively high (0.85) indicating the factor model is appropriate; values greater than 0.80 are considered sufficiently high for analysis (SAS

	ECOBUY	TEKNO	USBUY
Buying greener products improves the environment	0.891	.	.
It's good to buy greener products	0.876	.	.
It's worthwhile to buy greener products	0.859	.	.
I improve the environment when I buy greener products	0.851	.	.
Reducing the amount I drive decreases global warming	0.736	.	.
I am hesitant to try new products	.	0.907	
In general, I am hesitant to try new technologies	.	0.886	
I am often skeptical about new products	.	0.860	
It's good to buy us-made products	.		0.867
It's worthwhile buying us-made products	.		0.876
Buying US-made fuel improves our economy	.		0.732

Note: Values less than 0.6 are not printed

Table 3. Factor analysis of individuals' attitudes toward buying environmental, new, and US-made products.

1994). We call Factor 1 ECOBUY because the variables loading highly on this factor reflect respondent's positive attitudes toward environmental purchasing. We call Factor 2 TEKNO as the variables loading highly on this factor mostly reflect people's hesitancy to try new products or technologies. We call Factor 3 USBUY because it reflects respondents' positive attitudes toward buying US-made products. Computation yields Cronbach's alphas of 0.91, 0.86 and 0.79, respectively; indicating our analyses have a relatively high degree of reliability.

Factor analysis on the second set of variables yields four factors (Table 4); here, Kaiser's overall measure of sampling adequacy is marginal (0.73). We call Factor 1 GWIMP as it relates to beliefs and concerns related to ethanol and global warming. The next three factors measure people's beliefs about the positive and negative aspects of ethanol or wood-based ethanol. Cronbach's alphas of 0.85, 0.52, 0.26 and 0.49, respectively; indicating only the first factor is relatively reliable. Given these result we drop Factors 2, 3 and 4 from further analysis.

Note that all factor loadings are positive (Tables 3 and 4) indicating that each of the factor scores are positively correlated to the variables originally used in their construction. In turn, although the factor scores are normalized to mean zero, the direction of each score is positively correlated to the direction of the original variables. Hence, higher (lower) factor scores indicate a higher (lower) level of importance for that factor.

We hypothesize that α_1, the respondents' reaction to price, will be negative. We anticipate that α_2 and α_4 will also be negative as greenhouse gases and fuel imports should be negative attributes. The parameters on the three greenhouse gas and fuel import interaction terms, α_3, α_5 and α_6 should also be negative as these parameters reflect the preferences of people who are more concerned about global warming, and have more positive attitudes toward buying US-made and environmentally preferred products. The signs of parameters on the

	GWIMP	Factor 2	Factor 3	Factor 4
Decreases global warming relative to gasoline	0.879	.	.	.
Decreases global warming relative to corn-based ethanol	0.868	.	.	.
How concerned are you about the effect of global warming on the region	0.828	.	.	.
Ethanol is cheaper than gas	.	0.742	.	.
Ethanol improves acceleration	.	0.715	.	.
Ethanol produces less pollution	.	0.650	.	.
Lowers the us trade deficit	.	.	0.877	.
Decreases dependence on foreign oil	.	.	0.819	.
Ethanol damages engines	.	.	.	0.813
Ethanol lowers fuel efficiency	.	.	.	0.802

Note: Values less than 0.6 are not printed

Table 4. Factor analysis of individuals' beliefs toward ethanol and wood-based ethanol.

four binary variables indicating the type of ethanol present in the fuel relative to the baseline fuel (α_7, α_8, α_9 and α_{10}) are indeterminate. The parameters on the four ethanol type interaction terms, α_{11}, α_{12}, α_{13} and α_{14} should all be negative as these parameters reflect the preferences of people who are hesitant to trying new products/technologies, or have more negative attitudes toward ethanol.

Note that the empirical model is consistent with our theoretical specification except we do not have good measures of norms and perceived control. In addition, we included several specifications of income in the model but income was never significant and so we decided to drop it from the final estimation.

4.3.2 Estimation of premiums

Estimates of the price premium garnered for wood-based ethanol are derived from the discrete choice model as follows and calculated as changes from the baseline:

$$\text{Premium} = (\mathbf{X}^*\boldsymbol{\alpha}^*)/\alpha_1 \qquad (3)$$

where $\boldsymbol{\alpha}^*$ and $\mathbf{X}^*$ denote the vector of parameter estimates and the vector of variables from (2) with the exception of the parameter estimate on price. Variables are coded such that product attributes reflect their average values. The psychological variables (GWIMP, USBUY, ECOBUY, TEKNO and DAM) are all coded to zero since factor scores representing the 'average' are already scaled to zero. The binary variables are coded to identify what ethanol is present in the chosen fuel relative to the baseline fuel. In turn, we will generate four premiums that reflect the combinations of e10 fuel (wood versus corn) and the type of respondent (those who think they currently buy e10 versus those who think they currently buy straight gasoline).

5. Results

As expected, the parameters on PRICE, GHG and IMPORT negatively impact purchase decisions. Also the parameter on the interaction term that measures how people's reactions to the greenhouse gas attribute changes with increased concern over global warming (GHG * GWIMP) is negative, indicating that global warming concerns increase the negative reaction to the greenhouse gas attribute. Surprisingly, the parameter on the interaction term that measures how people with more positive attitudes toward buying US-made products react to the fuel import attribute (IMPORT * USBUY) is insignificant; whereas the similar interaction that measures how people with more positive attitudes toward buying environmentally labeled products (IMPORT * ECOBUY) is negative.

The parameters on the four binary variables indicating how the 'average'[9] person reacts to wood- and corn-based ethanol, indicates that people who think they currently buy e10 are more likely to buy fuel containing either wood- or corn-based ethanol; whereas people who think they currently do not buy e10 are more likely to continue to buy gasoline without ethanol. Individuals who are less (more) likely to try new technologies/products are less (more) likely to buy wood-based e10, but have no special reaction for or against corn-based ethanol. Individuals who hold less (more) positive beliefs toward ethanol are less (more) likely to buy fuels containing ethanol.

	Parameter estimate	Standard error
Price per gallon (PRICE)	-7.942***	0.78
Pounds of greenhouse gases per gallon (GHG)	-0.041***	0.02
GHG * GWIMP	-0.037***	0.00
Percent of fuel that is imported (IMPORT)	-0.036***	0.01
IMPORT * USBUY	-0.001	0.00
IMPORT * ECOBUY	-0.019***	0.00
Wood-based e10: Base fuel is gasoline (WOOD_GAS)	-0.849***	0.21
Corn-based e10: Base fuel is gasoline (CORN_GAS)	-0.768***	0.23
Wood-based e10: Base fuel is e10 (WOOD_E10)	1.324***	0.43
Corn-based e10: Base fuel is e10 (CORN_E10)	0.758*	0.43
Wood-based e10 (WOOD) * Person avoids buying new technologies/products (TEKNO)	-0.269***	0.11
Corn-based e10 (CORN) * TEKNO	-0.175	0.12
WOOD * Person perceives ethanol damages engines (DAM)	-0.241**	0.10
CORN * DAM	-0.287***	0.11

Note: * denotes significant at the 10 percent level; ** denotes significant at the 5 percent level and *** denotes significant at the 1 percent level

Table 5. Discrete choice modeling results

[9]Here average means the person is average in their level of rejecting/accepting new technologies/products, and average in their positive/negative perceptions of ethanol.

The model parameters with appropriate variable coding provide estimates of the price premiums for the two types of e10 (wood-based versus corn-based) by the type of respondent (those who think they currently buy e10 versus those that think they buy gasoline without ethanol). Respondents who currently see themselves as buying e10 are willing to pay more for wood-based ethanol but not for corn-based ethanol (Table 6). The lack of a premium for corn-based ethanol is because these individuals are assumed to currently buy corn-based ethanol. As such, there are no greenhouse gas or fuel import improvements over the base fuel; however, there are greenhouse gas benefits for using wood-based ethanol (valued at about ½ cent for a 7 pound/gallon improvement in greenhouse gases). The other 7.5 cents appears to be driven by other unidentified benefits to using wood-based ethanol. Although not directly tested here, most respondents (70 percent) rated 'increases local employment' as an important benefit of producing and using ethanol made from trees.

Respondents who currently perceive themselves as using gasoline without any ethanol are also willing to pay more for wood-based ethanol (albeit less than the respondents above) but not for corn-based ethanol. Although these respondents respond positively to the larger greenhouse gas and fuel import reductions available with either ethanol relative to gasoline, they generally reject ethanol. We are not clear why this group rejects ethanol since the estimates here control for their perceptions of ethanol and their stronger aversion of trying new technologies/products.

	Wood-based e10	Corn-based e10
Estimate for respondents thinking they buy e10	$0.08	$0.00
Estimate adjusted for non-response	$0.03	$0.00
Estimate for respondents thinking they do not buy e10	$0.03	$0.00
Estimate adjusted for non-response	$0.01	$0.00
Share-weighted average premiums		
Based on unadjusted estimates	$0.06	$0.00
Based on estimates adjusted for non-response	$0.02	$0.00

Table 6. Per gallon premiums for two types of e10, by type of respondent and by share-weighted average

The above estimates reflect the responses of those interested enough in the survey topic to return a completed survey, and; thus reflects an upper bound estimate for each type of respondent. Only 40 percent of the sample actually returned the survey; if we were to assume non-respondents would not be willing to pay any price premium for ethanol, then a more representative willingness to pay estimate for wood ethanol would be $0.03/gallon premium for ethanol buyers and $0.01/gallon premium for non-ethanol buyers.

To develop a population estimate we weight the two wood-based premiums by the percent of respondents who fall into the two respondent types (i.e., 53 percent are currently gas buyers; 47 percent are currently e10 buyers). This weighted average premium for wood-based ethanol would range from $0.02-0.06 per gallon, with no premium for corn-based ethanol.

6. Discussion

We find a small but significant premium for wood-based ethanol in the New England market. However, these premiums would only exist in the market if the different fuels were labeled and consumers were educated about these differences (e.g. through marketing). One key result is that about half of the respondents in our sample do not think they currently buy e10 fuel even though e10's market penetration in the region was around 94-97 percent (the lower percent removes New Hampshire from the calculation as it did not require e10 labeling at the time of the survey).

7. Acknowledgements

Funding provided in part by the Northeast SunGrant Initiative, United States Department of Transportation, and by the Maine Sustainability Solutions Initiative, National Science Foundation Grant EPS-0904155 to Maine EPSCoR at the University of Maine. Maine Agriculture and Forest Experiment Station No. 3148

8. References

AERPC (American Environment Research and Policy Center). (2009). Too much pollution: State and national trends in global warming emissions from 1990 to 2007. http://www.environmentamerica.org/home/reports/report-archives/global-warming-solutions/global-warming-solutions/too-much-pollution-state-and-national-trends-in-global-warming-emissions-from-1990-to-2007

Ajzen, I. (1988) Attitudes, Personality, and Behavior. Milton-Keynes, England: Open University Press.

Ajzen, I. (2002) Perceived behavioral control, self-efficacy, locus of control, and the theory of planned behavior. *Journal of Applied Social Psychology*. 32(4), 665-683.

Ajzen, I. (2006) *TPB Diagram*. Retrieved from http://people.umass.edu/aizen/tpb.diag.html. Accessed March 17th, 2011.

Ajzen, I., Brown, T.C., & Carvajal, F. (2004) Explaining the discrepancy between actions and intentions: the case of hypothetical bias in contingent valuation. *Personality and Social Psychology Bulletin*. 30(9), 1108-1121.

Ajzen, I. & Fishbein, M. (1977) Attitude-behavior relations – theoretical analysis and review of empirical research. *Psychological Bulletin*. 84(5), 888 – 918.

Bamberg, S. (2003) How does environmental concern influence specific environmentally related behaviors? A new answer to an old question. *Journal of Environmental Psychology*, 23, 21-32.

Birgelen, M., Semeijn, J., & Keicher, M. (2009) Packaging and Proenvironmental Consumption Behavior Investigating Purchase and Disposal Decisions for Beverages. *Environment and Behavior*. 41(1), 125-146.

Carrus, G., Passafaro, P., & Bonnes, M. (2008) Emotions, Habits and Rational Choices in Ecological Behaviours: The Case of Recycling and Use of Public Transportation. *Journal of Environmental Psychology*. 28(1), 51-62.

Chan, R.Y.K. (2001) Determinants of Chinese Consumers' Green Purchase Behavior. *Psychology and Marketing*. 18(4), 389-413.

Clark, C.F., Kotchen, M.J., & Moore, M.R. (2003). Internal and external influences on pro-environmental behavior: Participation in a green electricity program *Journal of Environmental Psychology* 23, 237–246

Collantes, G. (2010) Do green tech policies need to pass the consumer test? The case of ethanol fuel. *Energy Economics*. 32(6), 1235-1244.

Daigle, J.J., Hrubes, D., & Ajzen, I. (2002) A comparative study of beliefs, attitudes, and values among hunters, wildlife viewers, and other outdoor recreationists. *Human Dimensions and Wildlife*, 7, 1 – 19.

Davis, S.C., Diegel, S.W. & Boundy, R.G. (2007) Transportation Energy Data Book: Edition 27, ORNL-6981 (Oak Ridge, TN, 2008), Table 1.7, web site http://cta.ornl.gov/data/index.shtml.

Ek, K. (2005) Public and Private Attitudes Towards "Green" Electricity: The Case of Swedish Wind Power. *Energy Policy*. 33(13), 1677-1689.

Ek, K. & Söderholm, P. (2008) Norms and economic motivation in the Swedish electricity market. *Ecological Economics*. 68(1-2), 169-182.

Fraj, E. & Martinez, E. (2007) Ecological Consumer Behavior: An Empirical Analysis. *International Journal of Consumer Studies*. 31(1), 26-33.

Hanley, A.J.G., Meigs, J.B., Williams, K., Haffner, S.M., & D'Agostino, R.B. (2005) Re: "(Mis)use of factor analysis in the study of insulin resistance syndrome." *American Journal of Epidemiology*. 161(12), 1182-1184.

Hini, D. , Gendall, P., & Kearns, Z. (1995). The link between environmental attitudes and behaviour. *Marketing Bulletin*. 6, 22-31.

Homer, P.M. & Kahle, L.R. (1988) A structural equation test of the value-attitude-behavior hierarchy. *Journal of Personality and Social Psychology*, 54, 638 – 646.

Hunecke, M., Blobaum, A., Matthies, E., & Hoger, R. (2001) Responsibility and environment - Ecological norm orientation and external factors in the domain of travel mode choice behavior. *Environment and Behavior*. 33(6), 830-852.

IPCC (Intergovernmental Panel on Climate Change). (2007) Summary for Policymakers. In: *Climate Change 2007: The Physical Science Basis. Contribution of Working Group I to the Fourth Assessment Report of the Intergovernmental Panel on Climate Change* [Solomon, S., D. Qin, M. Manning, Z. Chen, M. Marquis, K.B. Averyt, M.Tignor and H.L. Miller (eds.)]. Cambridge University Press, Cambridge, United Kingdom and New York, NY, USA.

Jensen, K.L., Clark, C.D., English, B.C., Menard, R.J. Skahan, D.K. and Marra, A.C. (2010) Willingness to pay for E85 from corn, switchgrass, and wood residues. *Energy Economics*. 32(6), 1253-1262.

Kaiser F.G., Wofling, S. & Fuhrer, U. (1999) Environmental attitude and ecological behavior. *Journal of Environmental Psychology*. 19(1), 1-19.

Krosnick, J.A. (1988) The role of attitude importance in social evaluation - a study of policy preferences, presidential candidate evaluations, and voting behavior. *Journal of Personality and Social Psychology*. 55(2), 196-210.

Lancaster, K. (1971) Consumer demand: A new approach. Colombia University Press: New York.

Loudon, D.L. & Della Bitta, A.J. (1993) *Consumer Behaviour: Concepts and Applications* McGraw Hill: Auckland.

Nunnally, J. C. & Bernstein, I. H. (1994) *Psychometric Theory (3rd Ed.)*. New York: McGraw-Hill.

Ritchie, J. & Spencer, E. (1994) *Qualitative data analysis for applied policy research*. In, Bryman, A. and Burgess, R.G. (eds.) Analyzing Qualitative Data. London: Routledge.

Roberts, J.A. (1996) "Green consumers in the 1990s: profile and implications for advertising", *Journal of Business Research*. 36(3), 217-231.

SAS Institute Inc. (1994) "Measuring the Internal Consistency of a Test, Using PROC CORR to Compute Cronbach's Coefficient Alpha." *SAS Communications*. 20:4, TT2–TT5.

Schwartz, S.H. (1977) "Normative Influences on Altruism." In Advances in Experimental Social Psychology, Volume 10. Berkowitz, Leonard (Ed). New York: Academic Press.

Schwartz, S.H., & Howard, J.A. (1982), "Helping And Cooperation: A Self-Based Motivational Model" in Cooperation and Helping Behaviour: Theories and Research. Derlege, V.J., and Gizelack, J.(Eds). NewYork: Academic Press.

Solomon, B.D., Barnes, J.R., & Halvorsen, K.E. (2007) Grain and Cellulosic Ethanol: History, Economics, and Energy Policy. *Biomass & Bioenergy*. 31(6), 416-425.

Solomon, B.D. & Johnson, N.H. (2009) Valuing Climate Protection Through Consumers Willingness to Pay for Biomass Ethanol. *Ecological Economics*. 68, 2137-2144.

Tatum, S.W., Skinner, S.J., & Jackson, J.D. (2010) On the economic sustainability of ethanol e85. *Energy Economics*. 32(6), 1263-1267.

Teisl, M.F. Noblet, C., Knox, A. and Rubin, J. (2009) *Consumer Biofuel Knowledge and Preferences: Results of Focus Groups*. School of Economics Staff Paper - SOE – 580, University of Maine, Orono ME, USA.

Thøgersen, J. (2006) Norms for Environmentally Responsible Behavior: An Extended Taxonomy. *Journal of Environmental Psychology*. 26(4), 247-261.

Thøgersen, J. (1999) Spillover processes in the development of a sustainable consumption pattern. *Journal of Economic Psychology*. 20(1), 53-81.

Thompson, B. & Daniel, L.G. (1996) Factor analytic evidence for the construct validity of scores: A historical overview and some guidelines. *Educational and Psychological Measurement*. 56(2), 197-208.

Vermeir I, Verbeke W. (2006) Sustainable food consumption: exploring the consumer "attitude–behavioral intention" gap. *Journal of Agricultural & Environmental Ethics*. 19, 169 – 94.

Wall, R., Devine-Wright, P. and Mill, G.A.. (2008) Interactions between perceived behavioral control and personal-normative motives - Qualitative and quantitative evidence from a study of commuting-mode choice. *Journal of Mixed Methods Research.* 2(1), 63-86.

Wegener, D.T. & Kelly, J.R. (2008) Social Psychological Dimensions of Bioenergy Development and Public Acceptance. *Bioenergy Research.* 1(2), 107-117.

Part 3

Biofuel Technical Analysis

13

Biodiesel from Non Edible Oil Seeds: a Renewable Source of Bioenergy

Mushtaq Ahmad, Mir Ajab Khan,
Muhammad Zafar and Shazia Sultana
*Biofuel Lab., Department of Plant Sciences, Quaid-i-Azam University Islamabad
Pakistan*

1. Introduction

1.1 Background justification for using non edible oil seeds as source of bioenergy

Environmental pollution and diminishing supply of fossil fuels are the key factors leading to search for the alternative sources of energy. Today, 86% of the world energy consumption and almost 100% of the energy needed in the transportation sector is met by fossil fuels (Dorian et al., 2006). Since the world's accessible oil reservoirs are gradually depleting, it is important to develop suitable long-term strategies based on utilization of renewable fuel that would gradually substitute the declining fossil fuel production. In addition, the production and consumption of fossil fuels have caused the environmental damage by increasing the CO2 concentration in the atmosphere (Westermann et al., 2007).

Currently the most often-used type of biodiesel fuel is vegetable oil fatty acid methyl esters produced by transesterification of high quality vegetable oil by methanol. Biodiesel derived from vegetable oil and animal fats is being used in USA and Europe to reduce air pollution and dependency on fossil fuel. In USA and Europe, their surplus edible oils like soybean oil, sunflower oil and rapeseed oil are being used as feed stock for the production of biodiesel (Ramadhs et al., 2004 ; Sarin and Sharma, 2007).

Since more than 95% of the biodiesel is synthesized from edible oil, there are many claims that a lot of problems may arise. By converting edible oils into biodiesel, food resources are actually being converted into automotive fuels. It is believed that large-scale production of biodiesel from edible oils may bring global imbalance to the food supply and demand market. Recently, environmentalists have started to debate on the negative impact of biodiesel production from edible oil (Butler, 2006). They claimed that the expansion of oil crop plantations for biodiesel production on a large scale may increase deforestation in countries like Malaysia, Indonesia and Brazil. Furthermore, the line between food and fuel economies is blurred as both of the fields are competing for the same oil resources. In other words, biodiesel is competing limited land availability with food industry for plantation of oil crops. Arable land that would otherwise have been used to grow food would instead be used to grow fuel (Anonymous, 2004). In fact, this trend is already being observed in certain part of this world. There has been significant expansion in the plantation of oil crops for biodiesel in the past few years in order to fulfill the continuous increasing demand of biodiesel. Fig. 1 shows the trend in global vegetable oil ending stocks due to the production of biodiesel in the years 1991–2005 (Anonymous, 2006). Although there is continuous increase in the production of vegetable oil; however, the ending stocks of vegetable oils are continuously decreasing due to increasing production of biodiesel. Eventually, with the implementation of biodiesel as a substitute fuel for petroleum-derived diesel oil, this may lead to the depletion of edible-oil supply worldwide.

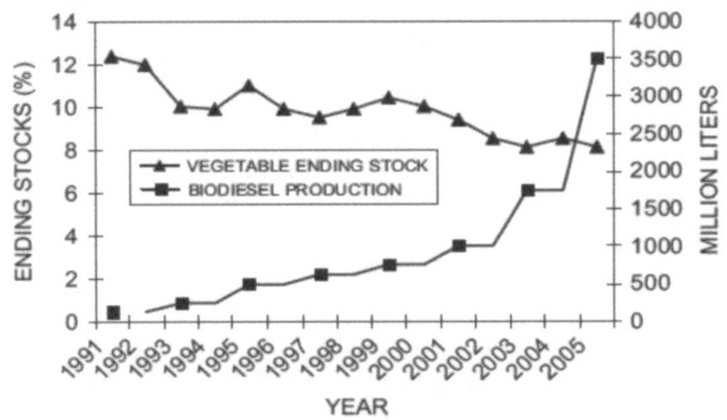

Fig. 1. Global vegetable oil ending stock and biodiesel production

1.2 Non edible oil seed crops for biodiesel

In order to overcome this devastating phenomenon, suggestions and research have been made to produce biodiesel by using alternative or greener oil resources like non-edible oils. The non-edible vegetable oils such as *Madhuca indica, Jatropha curcas* and *Pongamia pinnata* are found to be suitable for biodiesel production under the experimental conditions (Meher et al, 2006 ; Sent et al., 2003). Meher et al., (2006) found that the yield of methyl ester from karanja oil under the optimal condition is 97–98%. Oil content in the Castor bean, Hemp and Pongame seed is around 50, 35 and 30-40 % respectively. Neem seed contains 30% oil content. Biodiesel is the pure, or 100%, biodiesel fuel. It is referred to as B100 or ''neat'' fuel. A biodiesel blend is pure biodiesel blended with petrodiesel. Biodiesel blends are referred as Bxx. The xx indicates the amount of biodiesel blend (i.e., a B80 blend means 80% biodiesel and 20%petrodiesel).

1.3 Objectives

Extensive work has been done on the transesterification of non edible oils; however, no significant work has been done on the optimization, oil characterization and fuel analysis of most of the non edible oil seeds. An optimization study on biodiesel production from castor bean, hemp, neem and pongame was done in detail with one-step alkali transesterification process along with the fuel property analysis of these oils and their blends.

2. Experimental work

The seeds of castor bean, hemp, neem and pongame were used as raw material for biodiesel production. Seeds of these plants were expelled by using electric oil expeller (KEK P0015-10127), Germany. Methanol 99.9% purity, sodium hydroxide (NaOH) and anhydrous sodium sulphate (Na_2SO_4) were of analytical grade obtained from Merck (Germany).

2.1 Transesterification

The formation of fatty acid methyl esters (FAME) through transesterification of seed oils requires raw oil, 15% of methanol & 5% of sodium hydroxide on mass basis. However, transesterification is an equilibrium reaction in which excess alcohol is required to drive the reaction very close to completion. The vegetable oil was chemically reacted with an alcohol in the presence of a catalyst to produce FAMEs. Glycerol was separated as a by-product of transesterification reaction (Rao et al., 2008). The transesterification process was carried out using two litres round bottom flask equipped with reflux condenser, magnetic stirrer, thermometer and sampling outlet. One liter crude oil firstly filtered and heated up to 120°C to remove the moisture. The transesterification reaction performed at 6:1 molar ratio of methanol/oil, by using 0.34%, 0.67% and 1.35% (w/w) NaOH as catalyst. The temperature and the reaction time was maintained at 60°C and stirred for 2 hr with stirring velocity of 600 rpm. The resultant mixture was cooled to room temperature for the separation of two phases (Plate 1-4). The upper phase contained biodiesel and lower phase contained glycerin (by-product). Crude biodiesel contains the excess methanol, the remaining catalyst together with the soap formed during the reaction and some entrained methyl esters and partial glycerides.

2.2 Purification

After separation of the two layers, the upper layer of biodiesel was purified by distilling the residual methanol at 60°C. The remaining catalyst was removed by successive rinsing with distilled water by adding 1-2 drops of acetic acid to neutralize the catalyst. The residual can be eliminated by treatment with anhydrous sodium sulphate (Na_2SO_4) followed by filtration. Transparent blackish liquid was obtained as the final product.

2.3 Acid value

The acid value of the reaction mixture in the first stage was determined by the acid base titration technique (ASTM, 2003).

2.4 GC-MS method

The fatty acid methyl esters (FAMEs) contents were determined by gas chromatography, model GC–6890N coupled with mass spectrometer, model MS–5973 MSD (mass selective detector). Separation was performed on a capillary column DB-5MS (30 m ×0.32 mm, 0.25μm of film thickness). The carrier gas was helium with flow rate of 1.5 mL/min. The column temperature was programmed from 120-300 °C at the rate of 10 °C/min. A sample volume of 0.1μL HOB in chloroform was injected using a split mode, with the split ratio of 1:10. The mass spectrometer was set to scan in the range of m/z 50-550 with electron impact (EI) mode of ionization.

2.5 FT–IR method

Biodiesel samples were characterized by FT–IR, using a Bio-Rad Excalibur Model FTS3000MX in the range 4000 - 400 cm^{-1}. The resolution was $1cm^{-1}$ and 15 scans.

2.6 NMR method

NMR analyses were performed at 7.05 T using Avan CE 300 MHz spectrometer equipped with 5mm BBO probes. Deuterated chloroform and tetramethylsilane were used as solvent and internal standard respectively. ^1H (300 MHz) spectra were recorded with pulse duration of 30°, a recycle delay of 1.0 s and 8 scans. The ^{13}C (75 MHz) spectra were recorded with a pulse duration of 30°, a recycle delay of 1.89 s and 160 scans.

2.7 Fuel properties determination

Various fuel properties of COB, HOB, NOB, POB and their blends B100, B50. B20 and B10 were determined and compared with ASTM.

3. Results and discussion

3.1 Determination of Free Fatty Acid (FFA) content

The FFA has significant effect on the transesterification of glycerides with alcohol using catalyst (Goodrum, 2002). The high FFA content (>1%w/w) will cause soap formation and the separation of products will be exceedingly difficult, and as a result, low yield of biodiesel product would be obtained. It is important to first determine the FFA content of oil. The free fatty acid (FFAs) number of hemp crude oil was 1.76%, neem 2.5 % and pongame 2%. According to Anggraini *et al.*, (1999) if the free fatty acid content is more than 3% then the conversion efficiency decreases gradually.

3.2 Biodiesel yield

To achieve optimum yield of biodiesel from non edible oil seeds, alkali based transesterification was carried out. Alkali-catalyzed transeterification is much faster than acid-catalyzed and is used in commercial production of biodiesel. Even at ambient temperature, the alkali-catalyzed reaction proceeds rapidly usually reaching 95% conversion in 1-2 h (Rachmaniah et al., 2006). As a catalyst in the process of alkaline methanolysis, mostly sodium hydroxide or potassium hydroxide have been used, in concentration from 0.4 to 2% w/w of oil (Meher et al., 2006). Yield of methyl esters of HOB, NOB, POB and COB were investigated by changing catalyst concentrations while molar ratio (6:1) methanol / oil and temperature (60°C) was kept constant.The highest conversion rate was obtained with the catalyst concentration of 0.7 g, under these conditions the biodiesel yield was 74.36%, 70%, 77.16% and 70% for HOB, NOB, POB and COB respectively. When the catalyst concentration was doubled to 1.6 g, it was observed that the ester formation decreased with the increase in sodium hydroxide concentration and soap formation was increased. This is because the higher amount of catalyst may cause soap formation (Attanatho et al., 2004). Soap formation reduces catalyst efficiency, causes an increase in viscosity, leads to gel formation and makes the separation of glycerol difficult (Guo and Leung, 2003). The Conversion efficiency decreased when the catalyst concentration was reduced to half (0.4 g), as low concentration of catalyst suppress biodiesel.

3.3 Fuel properties

The experimental investigation was carried out for different fuel properties and the performance was evaluated according to ASTM D-445, D-1298, D-93, D-1298, D-1500, D-4294 and compared with diesel in Table 2.

3.3.1 Viscosity

High viscosity is the major problem preventing the use of vegetable oils and animal fats directly in diesel engines as it affects the flow of fuel and spray characteristics (Hossain and

Sr.No.	Plants	Concentration of catalyst	Amount of catalyst NaOH (g)	Biodiesel (%)	Glycerin (%)	Soap (%)
1-	HOB	Normal	0.7	74.36	15.62	0
		Double	1.4	60.62	44.53	0.4
		Half	0.35	14.42	6.44	0
2-	NOB	Normal	0.7	70	8.10	0.0
		Double	1.4	68	16.67	2.40
		Half	0.35	40	16.56	0.40
3-	POB	Normal	0.7	77.16	12.73	0.0
		Double	1.4	43.77	38.21	1.91
		Half	0.35	62.78	24.20	0.0
4-	COB	Normal	0.7	70	25.77	0.0
		Double	1.4	60	13.80	19.65
		Half	0.35	0.0	0.0	0.0

Table 1. Biodiesel yields of HOB, NOB, POB and COB by using NaOH as catalyst

Davies, 2010). High Speed Diesel (HSD) has viscosity of 1.3-4.1 @40 C where as the viscosities of COB and POB are 5.67 and 5.5 respectively which is slightly higher than the viscosity of HSD. While the viscosities of HOB and NOB are 3.83 and 4.81 which were closer to HSD. It shows that viscosities of these non edible oil seeds biodiesel were comparable to HSD.

3.3.2 Density

The density of COB, HOB, POB and NOB at 15 °C was found to be 0.88 Kg/L, 0.81 Kg/L and 0.87 Kg/L which are closer to the density of diesel (0.83) 0.86 kg/lit can be used as an alternative fuel.

Properties	NOB	COB	POB	HOB	HSD
Kinematic viscosity @ 40°C cSt	4.81	5.67	5.5	3.83	1.3-4.1
Density @15°C Kg/L	0.87	0.88	0.86	0.81	0.83
Flash Point (°C)	124	96	90	120	60-80
Sulphur % **wt**	Nil	0.0008	0.008	Nil	0.05
Color comparison	2.5	2.0	2	2.5	2.0
Pour point (°C)	9	-9	3	6	-35 to 15
Cloud point (°C)	12	-6	6	-25	-15 to 6

Table 2. Comparative analysis for fuel properties of NOB,COB,POB and HOB with HSD

3.3.3 Flash point

Flash point is the temperature that indicates the overall flammability hazards in the presence of air; higher flash points make for safe handling and storage of biodiesel (Hossain and Davis 2010). The Flash Points of biodiesel of NOB, COB and POB were 124 °C, 120 °C, 96 °C and 90°C which are higher than that of HSD (60-80 °C) (Fig. 2). For non-edible based seeds oils flash point are higher than fossil diesel (Anonymous, 2007; Pramanik, 2003; Ziejewski et al, 1992).

3.3.4 Sulphur contents

The most valuable result is the reduction and absence of percentage of total sulphur contents in NOB, COB, HOB and POB that will result in reduction of Sox in exhaust gases which is one of the reason of acid rain. Sulfur content of petrodiesel is 20–50 times higher than biodiesels (Shay, 1993).

3.3.5 Cloud point and pour point

Cloud point is the temperature at which a cloud of wax crystals first appear in the oil when it is cooled. The pour point is the lowest temperature at which the oil sample can still be moved. These properties are related to the use of biodiesel in colder region (Arjun et al., 2008). The cloud points of NOB, COB, POB and HOB were 12, -6, 6, -25 °C and the pour point are 9,-9,3 and 6 °C respectively. Pour point and cloud point of all oils were almost within the specified range. Lee *et al* (1995) argued that the cloud points were affected by the presence of monoglycerides while the pour points were not affected.

3.4 Biodiesel blends

Blending oils with diesel fuel was found to be a method to reduce chocking and extend engine life. Zhang and Gerpen (2006) investigated the use of blends of methyl esters of soybean oil and diesel in a turbo-charged, four cylinder, direct injection diesel engine modified with bowl in piston and medium swirl type. They found that the blends gave a shorter ignition delay and similar combustion characteristics as diesel (Orchidea et al., 2007). Table 3 presents data pertinent to the mixture of petroleum diesel and biodiesel in the following fashion:

Petroleum diesel (90%)- biodiesel (10%) : B10
Petroleum diesel (80%)- biodiesel (20%) : B20
Petroleum diesel (50%)- biodiesel (50%) : B50

The fuel properties of biodiesel blends NOB, COB, POB and HOB were compared with HSD. It was found that viscosity was slightly higher as the proportion of biodiesel in the mixtures increased. However, this event does not affect the atomization characteristics. Viscosity of B20 is very close to the viscosity of diesel. So that the biodiesel of B10, and B20 blends can be used without any heating arrangement or engine modification.

The density of different blends of methyl esters were increased with increase in blend percentage. The blends of B10, and B20 of NOB and HOB were closer to the density of diesel. The flash points of different blends of methyl esters are increased with increase in methyl ester percentage. It is also observed that the flash points of pure biodiesel were in comparison with HSD. Thus, it can be used as a fuel without any fire accidents.

3.5 Biodiesel analysis

The chemical composition and characterization of COB, HOB, NOB & POB based on GC-MS, NMR & FTIR analysis were shown in Fig 3-6.

Fuel Properties	PLANTS	B 100 %	B 50 %	B 20 %	B 10%	HSD
Kinematic viscosity @ 40°C cSt	NOB	4.81	3.80	3.30	3.10	1.3-4.1
	COB	5.67	4.43	4.48	3.52	1.3-4.1
	POB	5.532	6.23	4.1849	3.7959	1.3-4.1
	HOB	3.83	6.331	5.113	4.223	1.3-4.1
Density @15°C Kg/L	NOB	0.8785	0.8578	0.8476	0.8450	0.8295
	COB	0.8873	0.88722	0.88720	0.88111	0.8343
	POB	4.086	0.92	0.5072	0.1639	0.8295
	HOB	0.8195	0.8145	0.8224	0.8116	0.8343
Flash Point (°C)	NOB	124	72	64	65	60-80
	COB	96	86	83	77	60-80
	POB	90	83	72	70	60-80
	HOB	120	108	91	75	60-80

Table 3. Fuel properties of Biodiesel Blends and HSD

3.5.1 Gas Chromatography and Mass Spectrometry (GC-MS)

The composition of FA were analyzed by gas chromatography after converted into their corresponding methyl esters (FAME) (Knothe, 2001). The use of mass spectrometer would eliminate any ambiguities about the nature of eluting materials since mass spectra unique to individuals compounds would be obtained. Fatty acids (FA) components of biodiesel produced from COB, NOB, POB and HOB oil obtained by GC is presented in Table 4.

FA	Carbon number	COB	NOB	POB	HOB
Myristic Acid	C14:0	0.7	-	-	-
Palmitic Acid	C16:0	0.9	14.9	10.6	15
Stearic Acid	C18:0	2.8	14.4	6.8	65
Oleic acid	C18:1	90.2	61.9	49.4	15
Linoleic Acid	C18:2	4.4	7.5	19.0	5
Linolenic Acid	C18:3	0.2	-	-	-
Arachidic Acid	C20:0	-	1.3	4.1	-
Eicosenic acid	C20:1	-	-	2.4	-
Docosanoic acid	C22:0	-	-	5.3	-
Tetracosanoic acid	C24:0	-	-	2.4	-

Table 4. Fatty acid content of the methyl esters from non edible oil seeds

Oleic acid was the most common FA found in Castor oil, Pongame oil and Neem oil. COB was found to have 90.2 % oleic acid (18:1), while NOB have 61.9 % oleic acid (18:1). Myristic acid was present only in COB, Eicosenic acid, Docosanoic acid and Tetracosanoic acid were absent in all oils except for pongame oil. HOB was found to contain 65% stearic acid (18:0), 15% oleic acid (18:1), 5% linoleic acid (18:2),15% palmitic acid (16:0), Stearic and Palmitic acids are the major saturated FA found in HOB. It contains approximately 20% unsaturated fatty acids. COB was found to contain 90.2 % oleic acid (18:1),4.4% linoleic acid (18:2), 0.9%

palmitic acid (16:0) and 2.8% stearic acid (18:0). Oleic acid is the major unsaturated fatty acid found in COB. It contains approximately 94% unsaturated fatty acids. POB was found to contain 49.4% oleic acid (18:1), 19.0% linoleic acid (18:2), 10.6% palmitic acid (16:0) and 6.8% stearic acid (18:0). It contains approximately 70% unsaturated fatty acids. NOB was found contain 61.9 % oleic acid (18:1), 7.5% linoleic acid (18:2), 14.9% palmitic acid (16:0) and 14.4 % stearic acid (18:0). Oleic acid acid is the major Unsaturated fatty acids found in NOB. It contains approximately 69% unsaturated fatty acids.

The amount and type of fatty acids in the biodiesel determines the viscosity, one of the most important characteristics of biodiesel. Due to the presence of high amount of long chain fatty acids, POB may have a slightly higher viscosity compared to HOB, COB and NOB. The FAMEs of these species also meet the specification of 90/95% boiling point limit of 360°C specified in ASTM-D6751 and in other biodiesel standards. Generally, the FAMEs, which are mainly comprised of carbon chain lengths from 16 to 18, have boiling points in the range of 330–357°C; thus the specification value of 360°C is easily achieved. Besides, the concentration of linolenic acid and acid containing four double bonds in FAMEs should not exceed the limit of 12% and 1%, respectively (Pasto et al., 1992). FAMEs of all species are within the specified limit, so they are suitable for the production of biodiesel.

3.5.2 FT-IR analysis for fatty acid methyl esters

The FT-IR spectra in the mid-infrared region have been used to identify functional groups and the bands corresponding to various stretching and bending vibrations in the samples of oil and biodiesel. The position of carbonyl group in FT-IR is sensitive to substituent effects and to the structure of the molecule (Safar et al., 1994). The methoxy ester carbonyl group in HOB, NOB, COB and POB was appeared at 1743 cm^{-1} , 1741 cm^{-1}, 1742cm^{-1} and 1743 cm^{-1} respectively. The band appeared at 3465 cm^{-1} showed the overtone of ester functional group (Gelbard et al., 1995). The C-O stretching vibration in HOB and COB showed two asymmetric coupled vibrations at 1119 cm^{-1} and 1171 cm^{-1} due to C-C(=O)-O and 1017 cm^{-1} due to O-C-C (Table 5).

Sample type	Catalyst	C=O (cm^{-1}) Ester Carbonyl	C-O(cm^{-1})
NOB	NaOH	1741.1	1014.2
COB	NaOH	1742	1017
POB	NaOH	1740	1014.4
HOB	NaOH	1743	1017

Table 5. Chemical Composition of non edible seed biodiesel based on FT-IR analysis

3.5.3 ^{1}H NMR

Biodiesel of HOB, COB, POB and NOB were characterized by ^{1}HNMR spectroscopy as summarized in Table 6. Gelbard *et al.*, (1995) reported that the spectroscopic determination of yield of transesterification reaction utilizing HNMR depicting its progressing spectrum (Gerlbard et al., 1995). The characteristic peak of methoxy protons was observed as a singlet at 3.669 ppm, 3.64 ppm, 3.5 ppm and 3.67 ppm for HOB, POB, COB and NOB and a triplet of α–CH_2 protons at 2.31 ppm, 2.8 ppm, 2.24 ppm and 2.28 ppm for HOB, POB, COB and NOB.These two peaks are the distinct peaks for the confirmation of methyl esters present in biodiesel.

Sample type	CH_3-Methoxy proton (ppm)	α CH_2 Proton (ppm)	CH=CH Unsaturation (ppm)
NOB	3.64	2.28	5.32
COB	3.5	2.247	5.247-5.487
POB	3.674	2.816	1.263-1.310
HOB	3.65	2.31	5.31-5.39

Table 6. Chemical Composition of non edible oil seed biodiesel based on ^{1}HNMR analysis

3.5.4 ^{13}C NMR

The spectrum of ^{13}C NMR of the HOB, POB, COB and NOB were shown in (Table 7) which shows the characteristic peaks of ester carbonyl (–COO–) and C–O at 174.2 and 51.4 ppm for HOB, 173.28 ppm and 51.45 for POB, 174.19 and 51.32 for COB and 174.23 and 51.37 for NOB respectively.

Sample type	-COO- (ppm)	C-O (ppm)
NOB	174.23	51.37
COB	174.19	51.32
POB	173.28	51.45
HOB	174.21	51.35

Table 7. Chemical Composition of Biodiesel in ^{13}CNMR

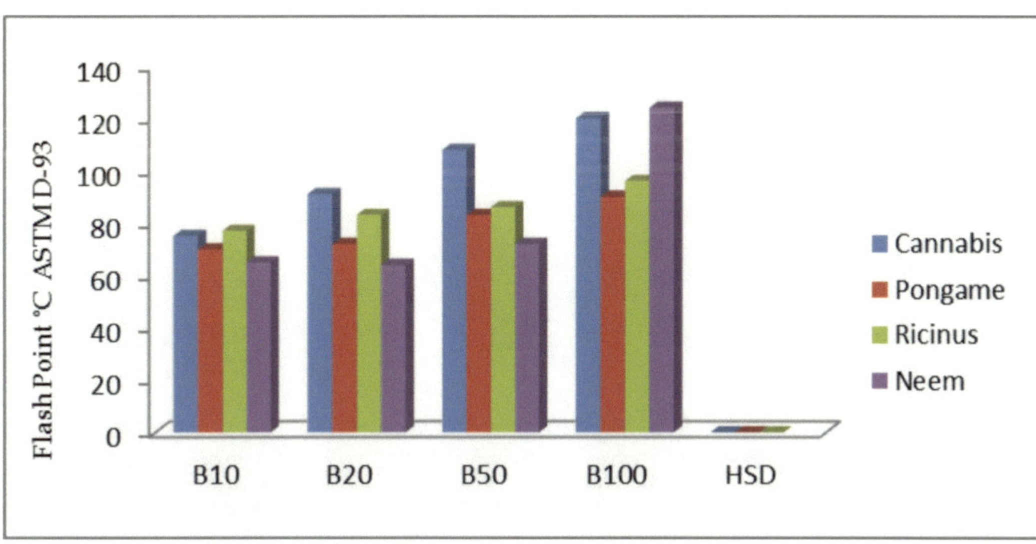

Fig. 2. Flash Point of Biodiesel blends

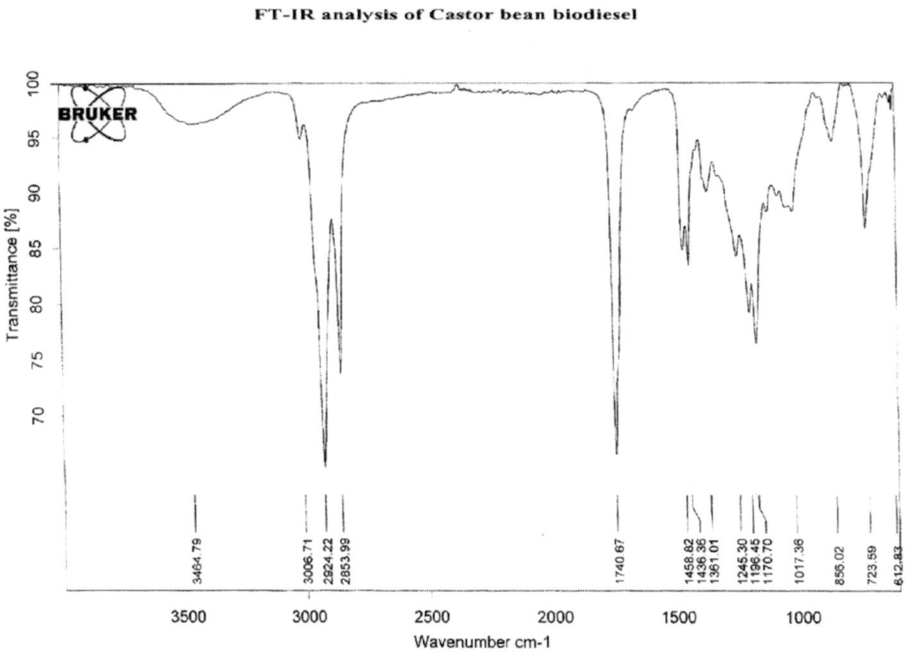
FT-IR analysis of Castor bean biodiesel
BRUKER
Transmittance [%]
100
95
90
85
80
75
70
3464.79
3006.71
2924.22
2853.99
1740.67
1458.82
1436.36
1361.01
1245.30
1196.45
1170.70
1017.36
856.02
723.59
612.83
3500
3000
2500
2000
1500
1000
Wavenumber cm-1

(a)

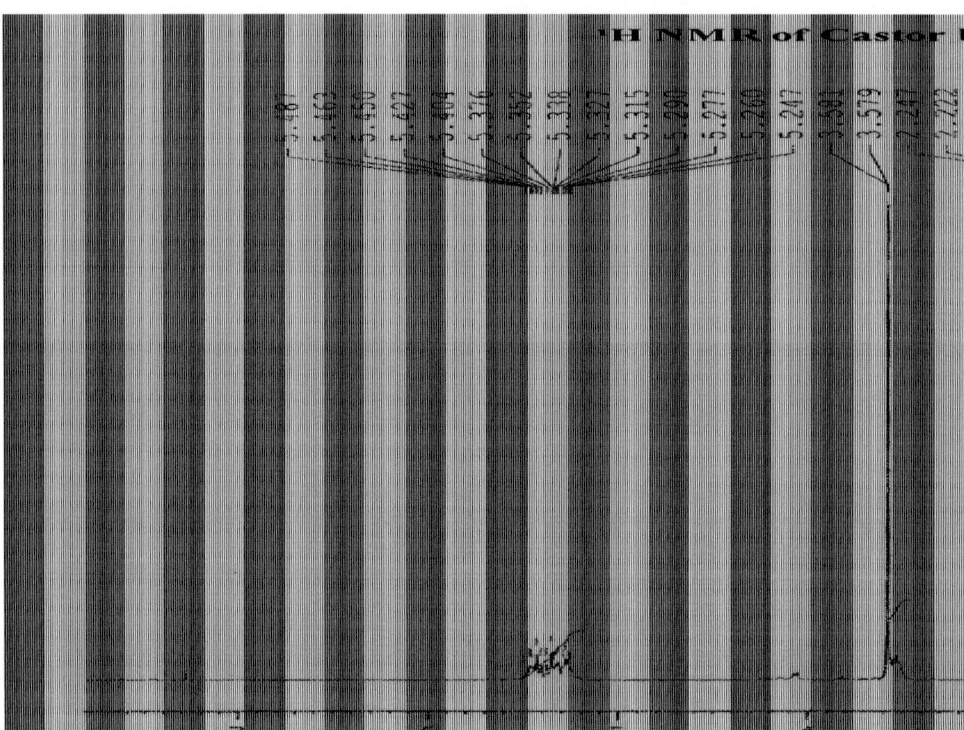
5.487
5.463
5.450
5.427
5.404
5.376
5.352
5.338
5.327
5.315
5.290
5.277
5.260
5.247
3.581
3.579
2.247
2.222
7
6
5
4

(b)

GC-MS analysis of Castor bean FAMEs

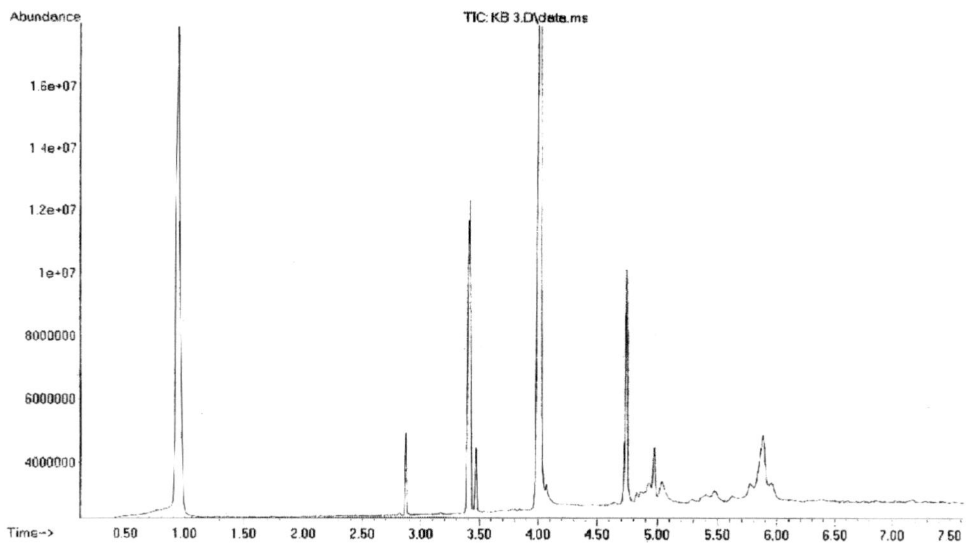

(c)

^{13}C NMR of Castor bean biodiesel

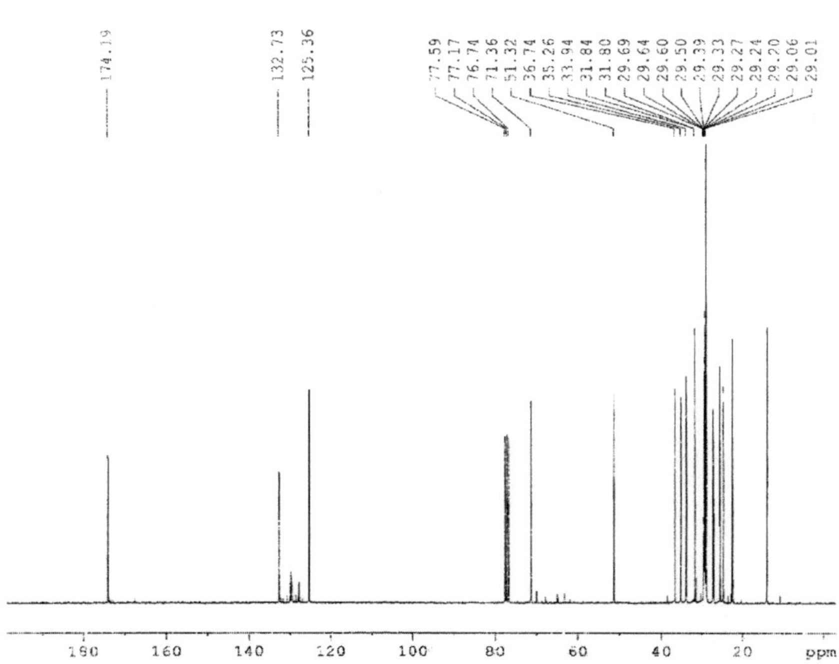

(d)

Fig. 5. COB chemical profile: A- FT-IR spectrum, B- 1H NMR spectrum, C- Gas chromatogram & D- 13C NMR spectrum

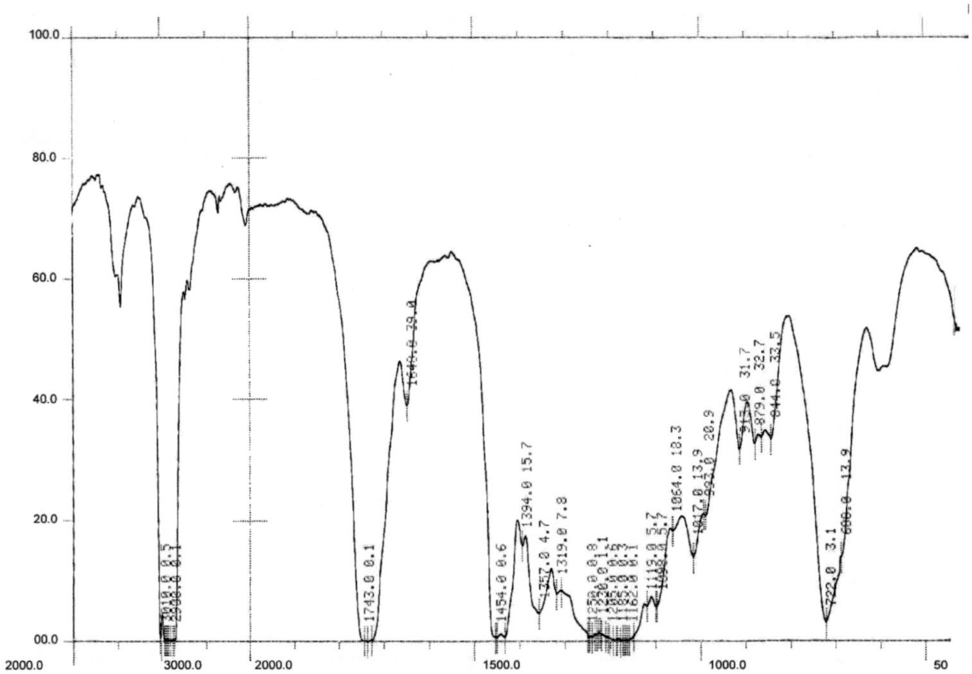
100.0
80.0
60.0
40.0
20.0
00.0
2000.0
3000.0
2000.0
1500.0
1000.0
50

(a)

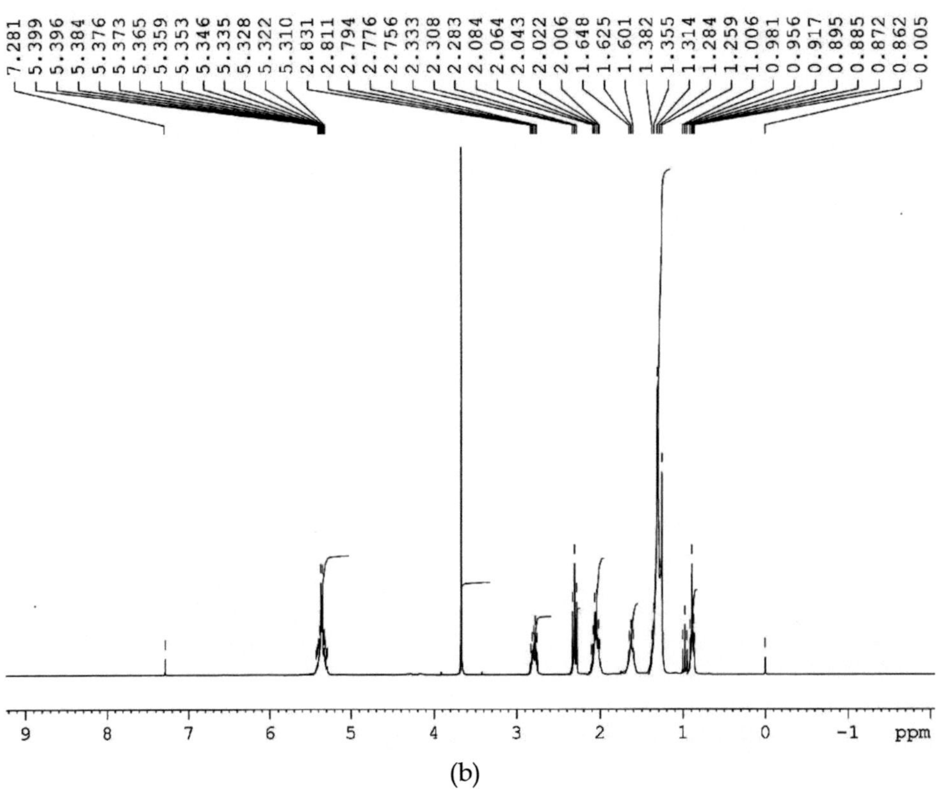
7.281
5.399
5.396
5.384
5.376
5.373
5.365
5.359
5.353
5.346
5.335
5.328
5.322
5.310
2.831
2.811
2.794
2.776
2.756
2.333
2.308
2.283
2.084
2.064
2.043
2.022
2.006
1.648
1.625
1.601
1.382
1.355
1.314
1.284
1.259
1.006
0.981
0.956
0.917
0.895
0.885
0.872
0.862
0.005
9
8
7
6
5
4
3
2
1
0
-1
ppm

(b)

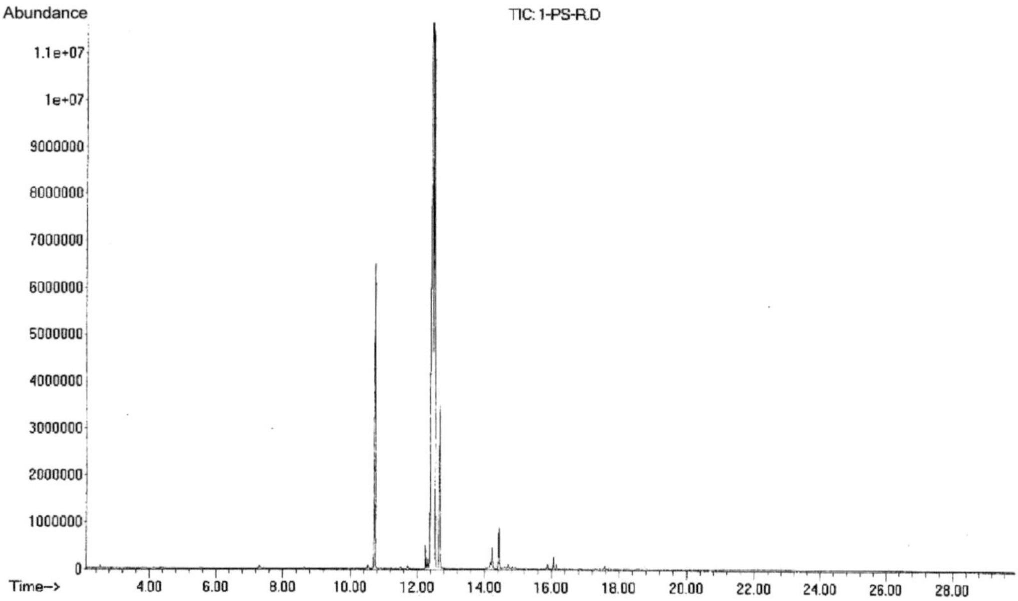

(c)

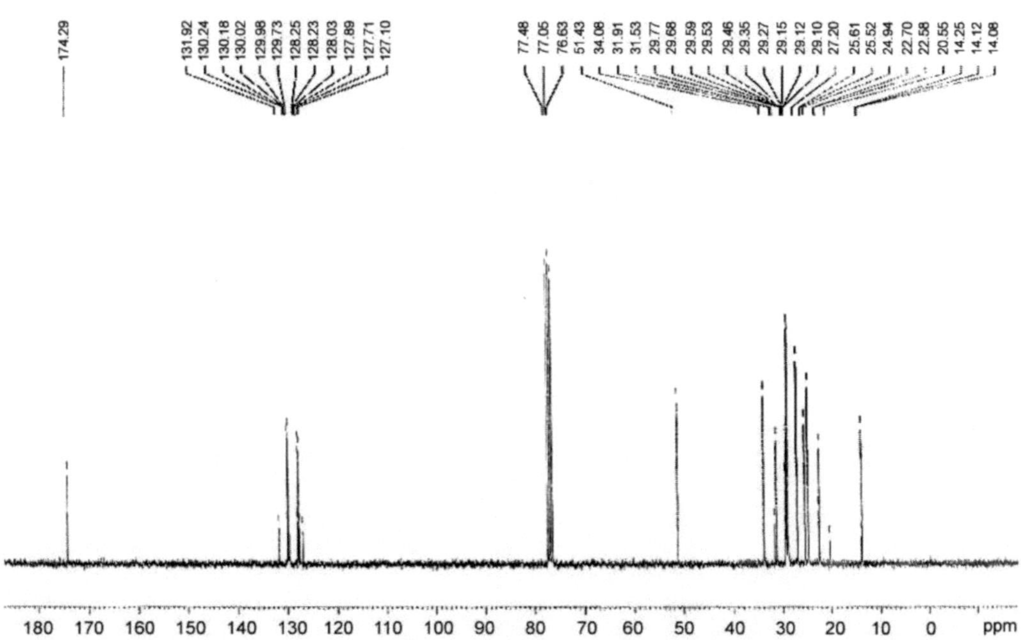

(d)

Fig. 6. HOB chemical profile: A- FT-IR spectrum, B- 1H NMR spectrum, C- Gas chromatogram & D- 13C NMR spectrum

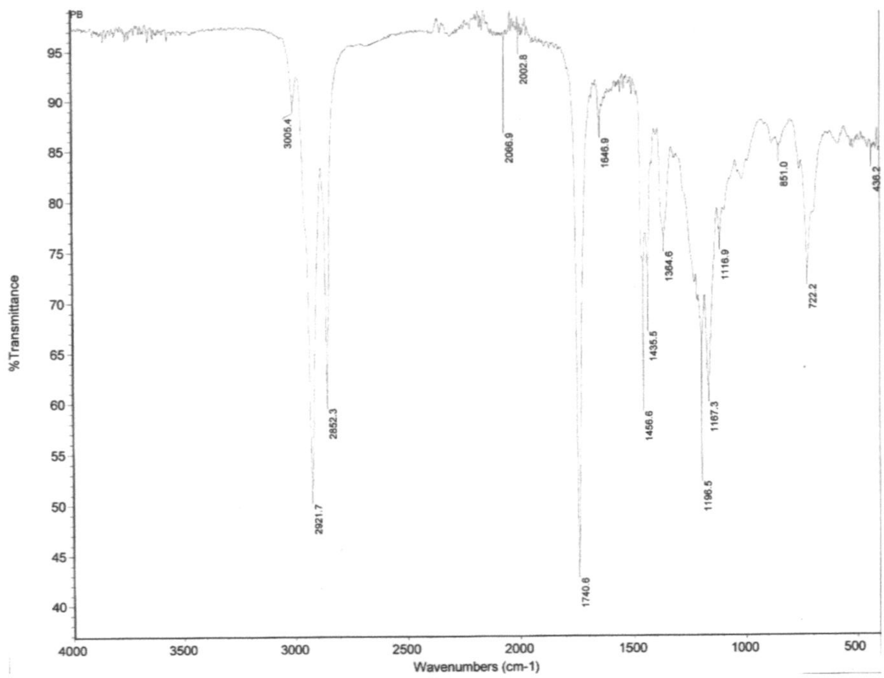
%Transmittance
Wavenumbers (cm-1)
95
90
85
80
75
70
65
60
55
50
45
40
4000
3500
3000
2500
2000
1500
1000
500
3005.4
2921.7
2852.3
2066.9
2002.8
1740.8
1646.9
1456.6
1435.5
1364.6
1196.5
1167.3
1116.9
851.0
722.2
436.2

(a)

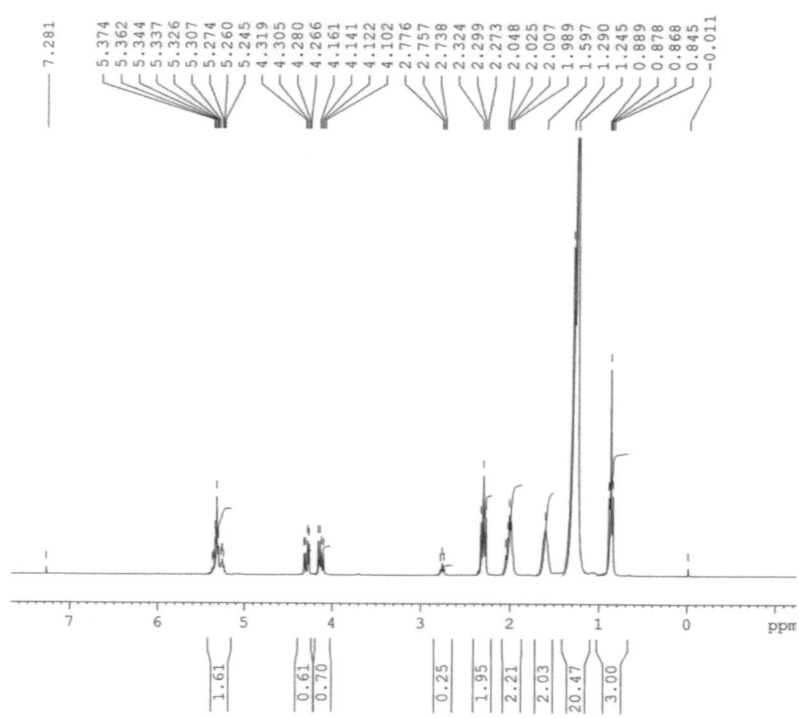
7.281
5.374
5.362
5.344
5.337
5.326
5.307
5.274
5.260
5.245
4.319
4.305
4.280
4.266
4.161
4.141
4.122
4.102
2.776
2.757
2.738
2.324
2.299
2.273
2.048
2.025
2.007
1.989
1.597
1.290
1.245
0.889
0.878
0.868
0.845
-0.011
7
6
5
4
3
2
1
0
ppm
1.61
0.61
0.70
0.25
1.95
2.21
2.03
20.47
3.00

(b)

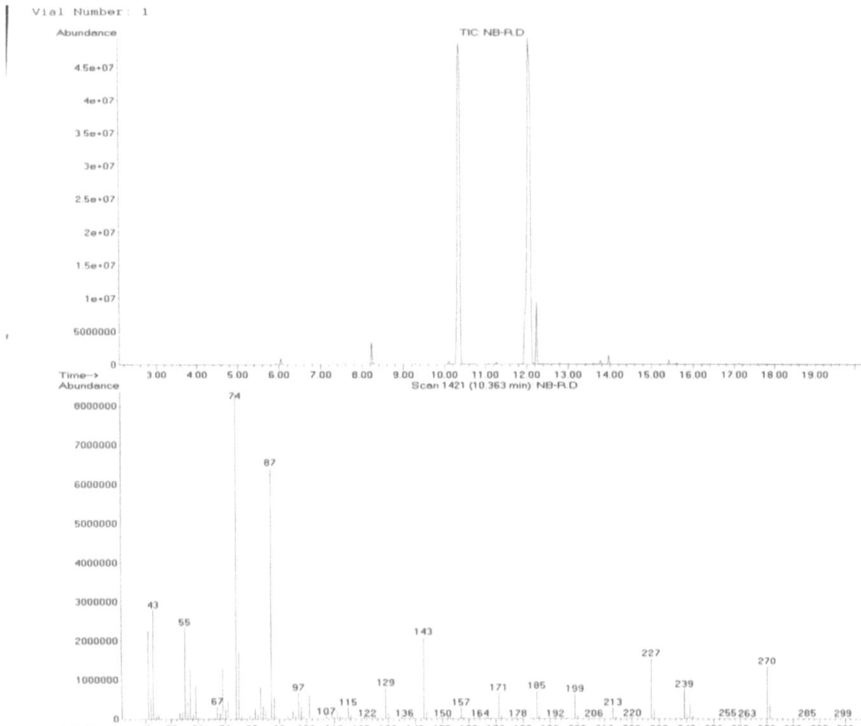

(c)

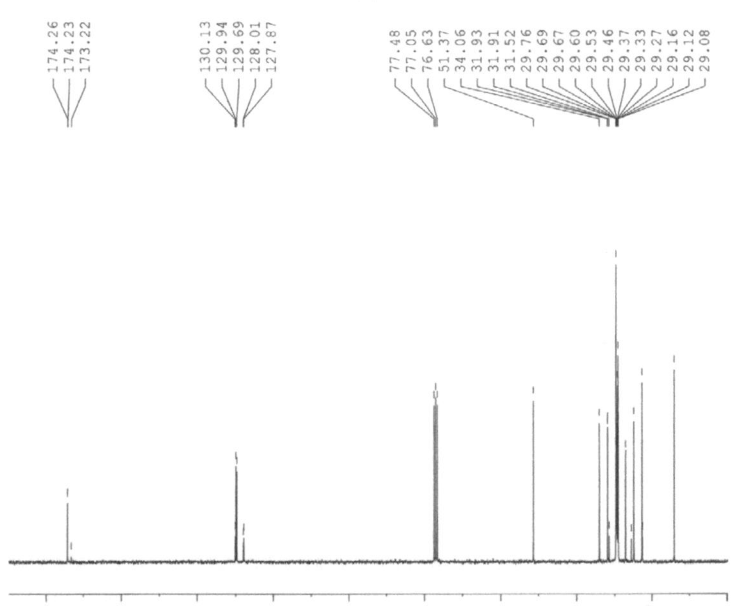

(d)

Fig. 7. NOB chemical profile: A- FT-IR spectrum, B- 1H NMR spectrum, C- Gas chromatogram & D- 13C NMR spectrum

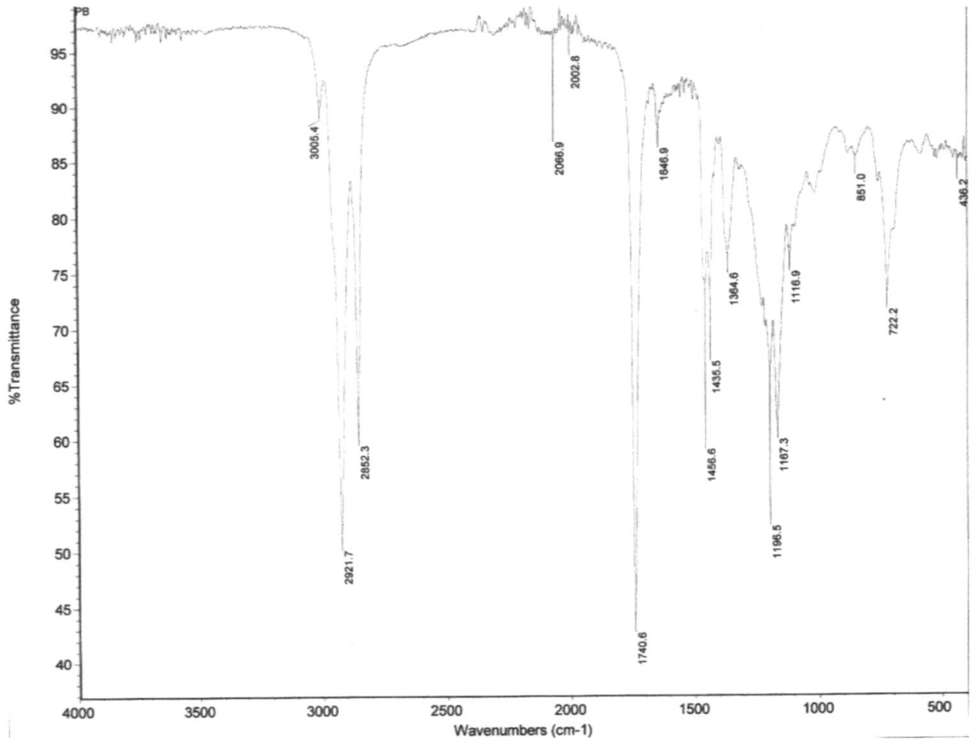
%Transmittance
Wavenumbers (cm-1)
3005.4
2921.7
2852.3
2066.9
2002.8
1740.6
1646.9
1456.6
1435.5
1364.6
1196.5
1167.3
1116.9
851.0
722.2
436.2

(a)

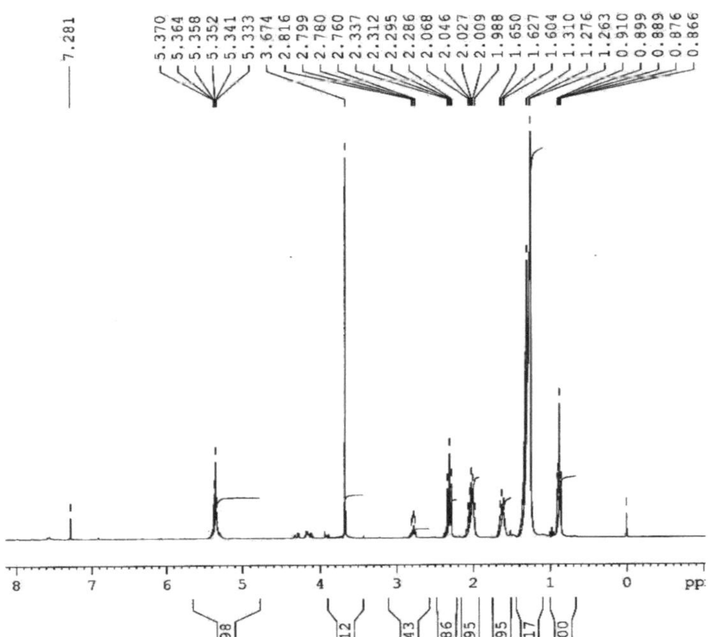
7.281
5.370
5.364
5.358
5.352
5.341
5.333
3.674
2.816
2.799
2.780
2.760
2.337
2.312
2.295
2.286
2.068
2.046
2.027
2.009
1.988
1.650
1.627
1.604
1.310
1.276
1.263
0.910
0.899
0.889
0.876
0.866
1.98
2.12
0.43
1.86
2.95
1.95
19.17
3.00

(b)

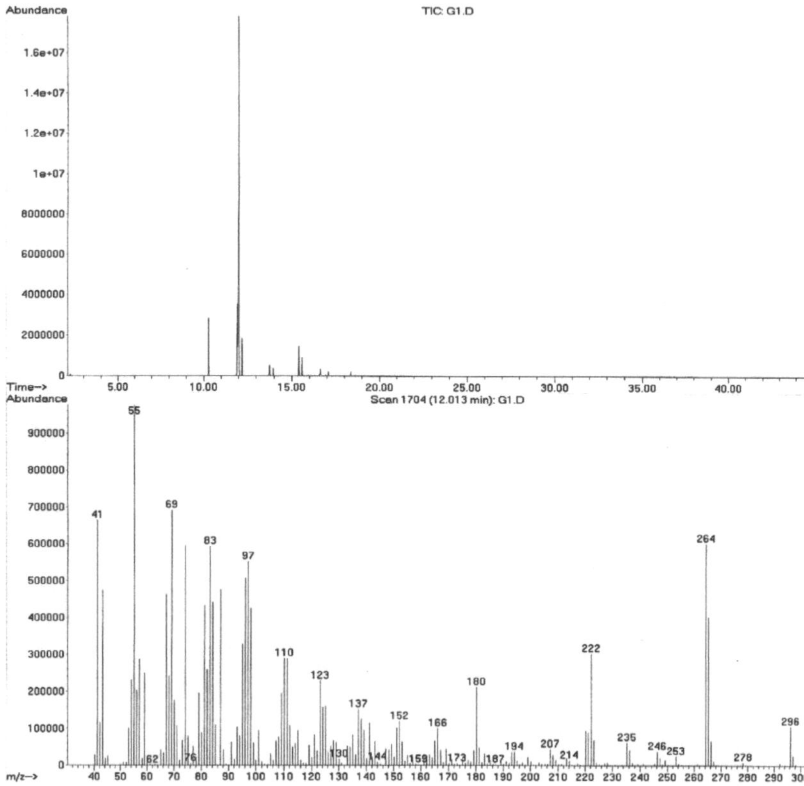

(c)

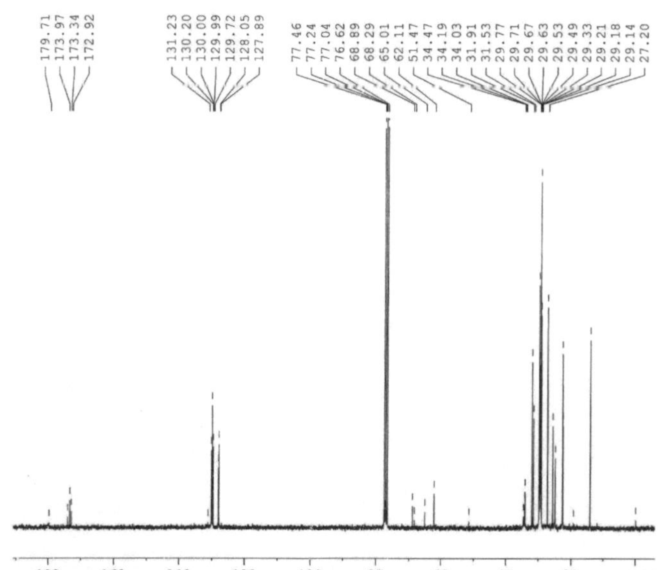

(d)

Fig. 8. POB chemical profile: A- FT-IR spectrum, B- 1H NMR spectrum, C- Gas chromatogram & D- 13C NMR spectrum

Plate 1. Extraction of oil by electric oil expeller

Plate 2. Filtration of crude oil

Plate 3. Mixing of sodium methoxide catalyst

Plate 4. Separation of glycerin as by product

4. Conclusion

In this study an optimized protocol for biodiesel production from non edible seeds of castor bean (*Ricinus communis* L.), hemp (*Cannabis sativa* L.), neem (*Azadirachta indica* A. Juss.) and pongame (*Pongamia pinnata* (L.) Pierre.) converted into fatty acid methyl esters (FAME) through base catalyzed transesterifiction using an optimum ratio of 1:6 (Oil : Methanol) at 60 ^{o}C. The fuel properties of biodiesel blends i.e. B100, B50, B20, B10 were compared with ASTM standards. Biodiesel from these sources was analyzed for qualitative and quantitative characterization by using H^1NMR, C^3NMR, GC-MS and FT-IR techniques. Based on qualitative and quantitative analysis of Biodiesel and their byproducts, it is concluded that the bioenergy from these species can be feasible, cost effective, environment friendly, if mass plantation of such resources may initiated in suitable places at global perspective.

5. List of abbreviations

ASTM	=	American Society for Testing and Materials
B100	=	100 % biodiesel
B20	=	20 % biodiesel + 80 % high speed diesel
B10	=	10 % biodiesel + 90 % high speed diesel
B5	=	5 % biodiesel + 95 % high speed diesel
B50	=	50 % biodiesel + 50 % high speed diesel
C^{13}NMR	=	Carbon - Nuclear Magnetic Resonance
COB	=	Castor Bean Oil Biodiesel
FA	=	Fatty Acid
FAME	=	Fatty Acid Methyl Esters
FFA	=	Free Fatty Acid
FT-IR	=	Fourier Transfer – Infra Red
GC-MS	=	Gas Chromatography – Mass Spectrometer
H^1NMR	=	Hydrogen – Nuclear Magnetic Resonance
HOB	=	Hemp Oil Biodiesel
HSD	=	High Speed Diesel
MSD	=	Mass Selective Detector
NOB	=	Neem Oil Biodiesel
POB	=	Pongame Oil Biodiesel

6. Acknowledgment

The financial support for biodiesel project of Higher Education of Pakistan is highly appreciated. The project confined to the identification, production and characterization of biodiesel from plants based resources.

7. References

Dorian, J. P.; Franssen, H.T.; & Simbeck, D.R. (2006). Global challenges in energy. *Energy Policy*. 34: 1984– 1991.

Westermann, P.; Jorgensen, B; Lange, L.; Ahring, B.K.; & Christensen, C. H. 2007. Maximizing renewable hydrogen production from biomass in a bio/catalytic refinery. *Int. J. Hydrogen Energy* 32: 4135- 4141.

Anonymous (2005). Carbon dioxide level highest in 650, 000 years. <http://news.mongabay.com/2005/1124-climate.html> (October 2007).

Anonymous, 1998. Austrian Biofuels Institute. 1998. Review on Commercial production of Biodiesel World- wide. *Report for the International Energy Agency*. Vienna, Austria.

Ramadhs, A. S.; Jayaraj, S.; & Muraleedharan, C. (2004). Use of Vegetable Oils as I.C engine Fuels : A Review. *Renewable Energy*. Vol. 29: 727-742.

Sarin, R & Sharma, M. 2007. Jatropha Palm biodiesel blends: An optimum mix for Asia. *Fuel*. Vol. 86: 1365-1371.

Butler, R. A. (2006). Why is oil palm replacing tropical rainforests? Why are biofuels fueling deforestation? April 25, 2006. See also: /http://news.mongabay.com/2006/0425-oil_palm.htmlS.

Anonymous (2004). Feeding cars, not people. 23 November 2004. See also: http:// www.monbiot.com/archives/2004/11/23/feeding-cars-not-people/S

Anonymous (2006). Bio-diesel demand disrupts vegetable oil balance. Special report, December 12, 2006. Chicago, Illinois: Fox Investments. See also: http://www.nsfutures.com/pdfs/beanoil_20061212.pdfs.

Pramanik, K. (2003). Properties and use of *Jatropha curcas* oil and diesel fuel blends in compression ignition engine. *Renew Energy*. 28: 239–48.

Meher, L.; Kulkarni, M.; Dalai, A.; Naik, S. (2006). Transesterification of karanja (Pongamia pinnata) oil by solid catalysts. *Eur J of Lipid Sci Technol*.108: 389–397.

Senthil, K. M.; Ramesh, A. & Nagalingam, B. (2003). An experimental comparison of methods to use methanol and jatropha oil in a compression ignition engine. *Biomass Bioenergy*. 25: 309–318.

Meher, L.; Dharmagadda, V.; Naik, S. (2006). Optimization of alkali-catalyzed transesterification of pongamia pinnata oil for production of Biodiesel. *Bioresour Technol*. 97: 1392–1397.

Rao, T. V.; Rao , G. P. & Reddy, K. H. C. (2008). Experimental Investigation of Pongamia, Jatropha and Neem Methyl Esters as Biodiesel on C.I. Engine. *Jordan Journal of Mechanical and Industrial Engineering. All rights reserved* 118 - *Volume 2, Number 2 (ISSN 1995-6665)*

ASTM, American Standards for Testing of Materials., (2003). D 189-01, D 240-02, D 4052-96, D 445-03, D 482- 74, D 5555-95, D 6751-02, D 93-02a, D 95-990, D 97- 02.

Goodrum, J.W. (2002). Volatility and boiling points of biodiesel from vegetable oils and tallow. *Biomass Bioenergy*. 22: 205–211

Anggraini, A. A. and V. G. Wiederwertung. (1999). Speiseolen/fetten im Energetisch Technischen Bereichein Verfahren und Kassel: Witzenhausen, Germany, p 193.

Rachmaniah, O.; Ju, Y. H; Vali, S. V; Tjondronegoron J. & Musfi, A. H. A. (2006). Study on Acid- Catalyzed Transesterification of Crude Rice Bran Oil for Biodiesel Production.Department of Chemical Engineering, Institut Teknologi Sepuluh Nopember, Kampus ITS Sukolilo.

Meher, L. C.; D. V. Sagar & S. N. Naik. (2006). Technical aspects of biodiesel production by transesterification—a review. *Renewable and Sustainable Energy Reviews*.10: 248–268.

Attanatho, L.; Magmee, S. & Jenvanipanjakil, P. (2004). The Joint International Conference on 'Sustainable Energy and Environment', 1-3 December 2004, Huan Hin, Thailand.

Guo, Y. & Leung, Y. C. (2003). Analysis on the biodiesel production using grease trap oils from restaurants., Macro Review. Japan. *Jpn. Macro Eng. Soc* 16: 421-426.

Hosasain, A. K. & P. A. Davies. (2010). Plant oils as fuels for compression ignition engines: A technical review and life-cycle analysis. *Renewable Energy* 35: 1–13.

Hossain, A. K. & P. A. Davies. (2010). Plant oils as fuels for compression ignition engines: A technical review and life-cycle analysis. *Renewable Energy.* 35: 1–13

Anonymous, (2007). Renewable - global status report, http://www.ren21.net [accessed July 2008].

Pramanik, K. (2003). Properties and use of Jatropha curcas oil and diesel fuel blends in compression ignition engine. *Renewable Energy* 28: 239–48.

Ziejewski, M. & Goettler, H. J. (1992). Design modifications for durability improvements of diesel engines operating on plant oil fuels. SAE 921630. USA: SAE; 1992

Shay, E.G. (1993). Diesel fuel from vegetable oils: status and opportunities. *Biomass and Bioenergy.* 4: 227–242.

Arjun, B., I. Chhetri, K. C. Watts and M. R. Islam. (2008). Waste Cooking Oil as an Alternate Feedstock for Biodiesel Production. *Energie.* 1: 3-18; DOI: 10.3390/en1010003

Lee, I.; Johnson, L.A. and Hammond, E.G. (1995). Use of branched-chain esters to reduce the crystallization temperature of biodiesel. *JAOCS.72*: 1155-1160.

Zhang, Yu, Jon, H. & Van Gerpen. (2006). Combustion Analysis of Esters of Soybean Oil in a Diesel Engine. SAE paper 960765.

Orchidea R.; Yi-Hsu, J;, Shaik R. V.; Ismojowati T.; & Musfil A. S. (2007). A Study on Acid-Catalyzed Transesterification of Crude Rice Bran Oil for Biodiesel Production Department of Chemical Engineering, Institut Teknologi Sepuluh Nopember, Kampus ITS Sukolilo

Knothe, G. (2001). Analytical methods used in the production and fuel quality assessment of biodiesel. *Trans ASAE.* 44(2): 193–200.

Biodiesel standard, ASTMPS121, USA (1999).

Pasto, D.; C. Johnson & Miller, M. (1992). Experiments and techniques in organic chemistry, first ed., Prentice Hall, New Jersey.

Safar, M., D. Bertrand, P., Robert, M.F. Devaux and C. Genut. 1994. Characterization of edible oils, butters and margarines by Fouier Transform Infrared Spectroscopy with attenuated total reflectance. *J. Am oil Chem. Soc.* 71: 371-377.

Gelbard, G.; Bres, O.; Vargas, R. M.; Vielfaure, F. & Schuchardt, U. (1995). ^{1}H Nuclear magnetic resonance determination of the yield of the transestterification of rapeseed oil with methanol, *J. Am. Oil Chem. Soc.* 72: 1239-1241.

14

The Development of the Multi - Fuel Burner

Anna Maiorova, Aleksandr Sviridenkov, Valentin Tretyakov,
Aleksandr Vasil'ev and Victor Yagodkin
Central Institute of Aviation Motors named after P.I. Baranov
Russia

1. Introduction

For modern propulsion engineering the trend to use of a wide spectrum of liquid fuels – both oil, and alternative is characteristic. The physical properties of various liquids corresponding to the Russian and international standards, are resulted in table 1. As we see from table. 1, a range of change of fuel properties, especially viscosity, is wide enough. One of the most pressing problems at present is creation of combustion chambers for engines and gas-turbine plants which can operate on fuel with as low as the increased viscosity at preservation of low level of toxic species emissions.

Liquid	Density, kg/m^3	Kinematic viscosity$\cdot 10^6$, m^2/s	Surface tension coefficient $\cdot 10^3$, N/m
Distilled water	998.2	1.003	72.75
Ethanol	788	1.550	22.3
Kerosene TS1	≥780	≥ 1.3	24.3
Rapeseed oil	916	88.62	33.2
Summer diesel	≤ 860	3.0-6.0	28.9
Winter diesel	≤ 840	1.8-5.0	27.8
FAME (biodiesel)	877-879	8.0	31.4

Table 1. Physical properties of liquids.

Maintenance of the majority of requirements shown to the combustion chamber directly depends on the chosen scheme of spraying system. Fuel ignition at an engine or gas-turbine plant start-up, stability and efficiency of combustion, levels of toxic species emissions are connected with fuel atomization and its mixing with air in atomization system. Several injectors of various types and a whole number of air swirlers various on a design can form modern atomization system. The review of various types of sprayer units and the analysis of the conclusions made in works (Lefebvre, 1985;Vasil'ev, 2007) shows, that the most perspective direction of researches is the development of the device with the pneumatic scheme of an atomization. The main lack of such scheme of an atomization is insufficient droplet's fineness on wake-up modes for assured lighting in combustor. To reach comprehensible droplet's fineness on low modes it is possible to use pressure-swirl injectors

with comprehensible maximum injection pressure (nearby 2 MPa). However at such injection pressure it is impossible to provide all range of operation modes with injectors of this scheme.
The most promising direction of researches is the development of the dual-orifice (on fuel) burner of the combined centrifugal-airblast scheme. The aims of this work are scheme selection, designing, test and research as injectors, and the burner as a whole for low-emission combustors on fuels as usual, as increased viscosity (kerosene, ethanol, diesel, biodiesel)

2. The selection of the sprayer unit scheme

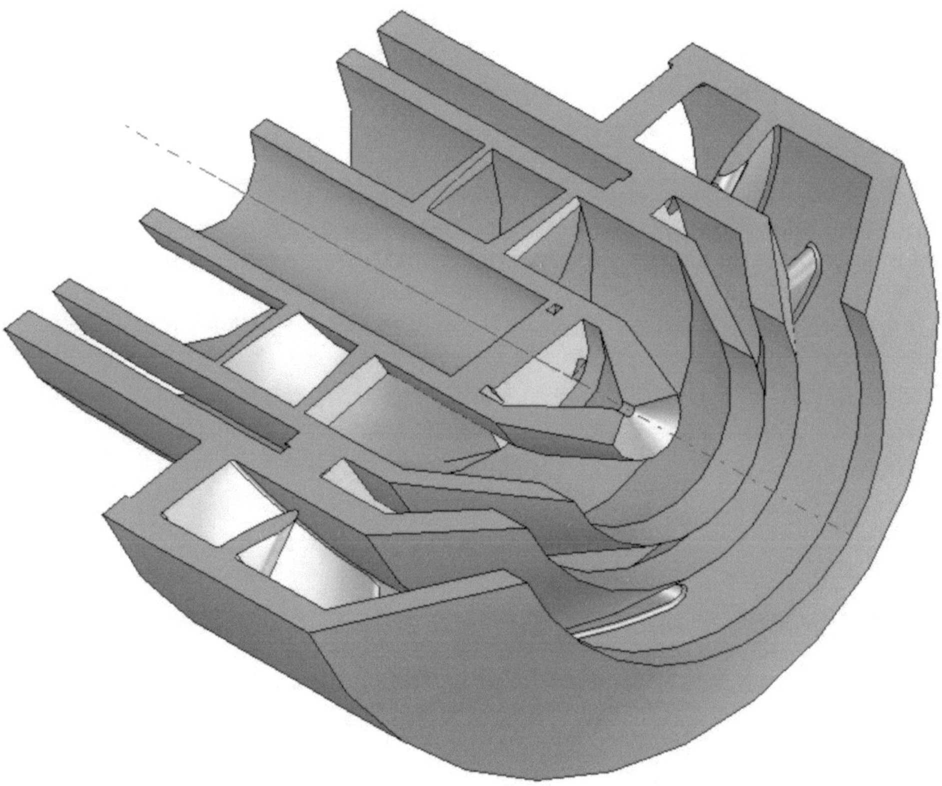

Fig. 1. A scheme of the sprayer unit

The shape of the designed sprayer unit is resulted on fig. 1. Nozzles of injectors place concentrically. The low-rate pilot channel (pressure swirl nozzle) is installed on a burner axis. Confidently to ignite the chamber, it is necessary to provide hit of a quantity of fuel droplets into plug discharge zone. Hence the atomizer should have a large fuel spray angle $2\theta_R$ (80-100°). The main channel - airblast nozzle, is located between two air swirlers for the best crushing of a liquid film and fuel spray stabilization.
Sprayer unit basic elements are shown on fig. 2. The sprayer unit consists of a casing 1, outer air swirler 2 and injector shaft 3. The casing forms an outer air nozzle 4 and the cowling with basic spraying edge 5. Blades of an outer air swirler 2 are disposed on a shaft of a fuel atomizer 3 which in turn contains the channel 6 of main fuel injection with

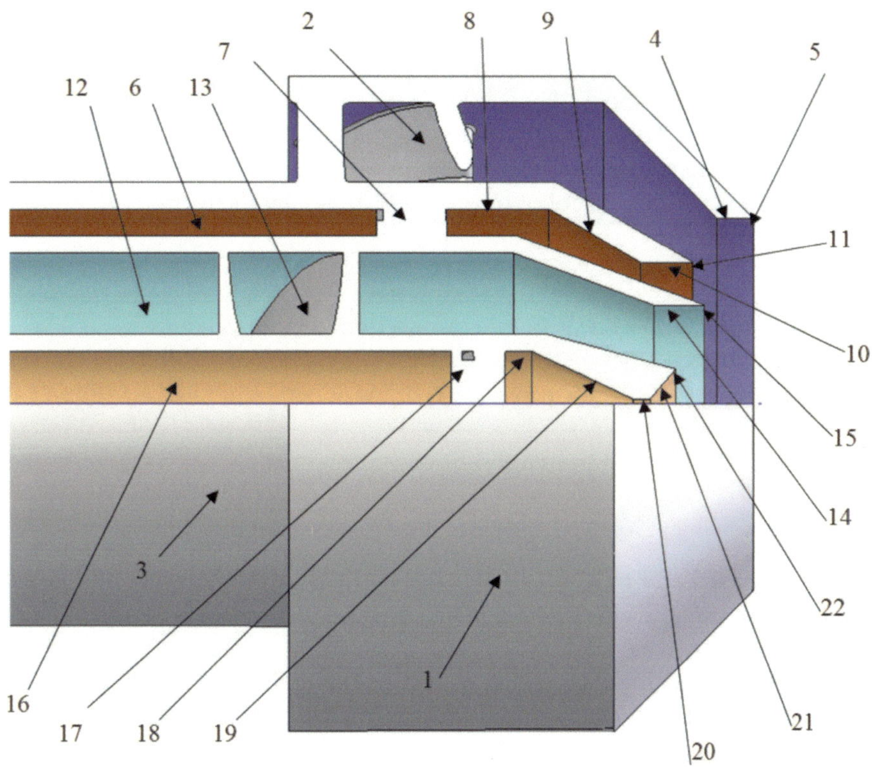

Fig. 2. Basic elements of sprayer unit

disposed in it consistently: fuel auger 7, swirl chamber 8, conic section 9, fuel nozzle 10 of main channel and cowling with spraying edge 11. Concentrically in the channel 6 of main fuel injection there is the channel 12 for air input into the central air passage with the central air swirler 13, the air nozzle 14 and an edge 15 disposed in it. On an injector axis installation of the pilot fuel channel 16 with auger 17, the swirl chamber 18, conical section 19, the nozzle 20, an expansion face 21 and an the cowling with edge 22 disposed in it is possible.

The sprayer unit works by a following principle. The pilot injector works on all power conditions, including provides start. It is offered to execute chamber start at fuel injection pressure of an order 0.5 MPA. On an intermediate mode injection pressure on the pilot channel can even be a little reduced for submission greater fuel shares through the main channel for the purpose of maintenance of the best uniformity of concentration in fuel-air spray. On a wake-up mode fuel moves only in the pilot channel 16 (working as pressure swirl atomizer). There the fuel passing through auger 17 is swirled , merging on length of the swirl chamber of 18 from discrete sprays in a fuel flow. In a conic section 19 fuel flow increases a tangential velocity and reaching a maximum at passage of the nozzle 20 reveals in a conic film. The face 21 helps to increase spray angle on a wake-up mode. . Breaking from edge 22 fuel film disintegrates on drops under the influence of internal (hydrodynamic) forces or to be atomized by external streams of air, depending on the relation of fuel and air momentums. On higher modes the fuel moving in the main channel 6 (operating as airblast atomizer) is swirled in the fuel auger 7, breaks from a spraying edge 11, and the fuel film to be atomized between two swirl streams of air formed by swirlers 2 and 13.

3. The design of the burner

At designing dual-orifice injector device it is necessary to know flow hydraulic characteristics on each of channels. Hydraulic characteristics of an injector are, first of all, the flow rate characteristic: $G_f = G_f(\Delta P_f)$, where G_f - mass flow rate of a liquid, ΔP_f - injection pressure, and the factors connected with it.
Definition of a discharge coefficient of injector C_d is given by the formula

$$C_d = G_f / (f_c \sqrt{2\rho_f \Delta P_f}), \tag{1}$$

where f_c - the nozzle area on a shear (in narrow section), ρ_f - liquid density
For the pressure swirl channel also it is necessary to calculate assumed drop sizes and a fuel-air spray angle, especially for a wake-up mode where spraying air doesn't yet exert essential influence on a spray.
The root angle θ_R is determined from a condition:

$$tg(\theta_R) = u_\varphi / u_x, \tag{2}$$

where u_x and u_φ - axial and tangential velocities of a liquid in the centre of a liquid film on an exit from the nozzle. The effective angle θe corresponds to an actual corner of projection of drops or a fluid spray cone angle.
After the spray angle and the film width w_e estimation Sauter Mean Diameter of droplets SMD was determined in calculations by Lefebvre formula (Lefebvre A.H., 1989)

$$SMD = 4.52\left(\frac{\sigma_f \mu_f^2}{\rho_f \Delta P_f^2}\right)^{0.25} (w_e \cos\frac{\alpha_e}{2})^{0.25} + 0.39\left(\frac{\sigma_f \rho_f}{\rho_A \Delta P_f}\right)^{0.25} (w_e \cos\frac{\alpha_e}{2})^{0.75} \tag{3}$$

Here μ_f, σ_f - - liquid dynamic viscosity and surface tension coefficient, ρ_A - air density.
For selection of fuel rates relation on engine operation modes and the main geometrical characteristics of fuel channels designing hydraulic calculations of pilot and main channels (tab. 2, 3) are carried out. The program *fnozzle*, based on a technique (Dityakin at al., 1977) with some refinements of authors was used. In designing calculation the ideal diesel was considered as fuel (ρ_f = 845 kg/m^3, ν_f = 4.17 mm^2/s, σ_f = 28.1 mN/m at T = 293 K). The injector device was designed on a landing place of 48 mm. On the basis of hydraulic designs the main geometrical characteristics of fuel channels have been chosen. For example, fuel nozzle diameter of a airblast injector has made 22 mm.

№	ΔP_f, MPa	G_f, kg/s	$2\theta_R$ °	SMD, mkm
1	0.0131	0.0008	76.2	177.8
2	0.1068	0.0025	86.0	72.0
3	0.1578	0.0030	87.6	61.2
4	0.3054	0.0041	89.6	47.1
5	0.4713	0.0050	90.9	39.7

Table 2. Hydraulic design of pilot fuel channel

From table 2 it is visible, that on a prospective wake-up mode (injection pressure an order 0.5 MPa) calculated drop size reaches 40 microns and a root fuel-air spray angle of an order 90° (a line 5). On a 100 % mode (a line 4) injection pressure is nearby 0,3 MPa, drop size makes 47 microns, and a root angle - 90°. The reached values should be enough for assured firing of the combustion chamber, taking into account that pressure and temperature in the combustor were considered as the normal.

№	ΔP_f, МПа	G_f кг/с	$2\theta_R$, °	SMD, мкм
1	0.0116	0.00250	135.9	300.0
2	0.1039	0.01350	144.9	103.3
3	0.1488	0.01700	146.7	85.0
4	0.3531	0.02500	148.6	58.6

Table 3. Hydraulic design of main fuel channel

Calculation of the main channel is resulted, basically, for the purpose of obtaining of the flow rate characteristic on fuel. As this channel works by a principle of a pneumatic atomization the real sizes of drops will essentially less. Fuel-air spray angle practically will depend completely on a direction of motion of air streams.

Analyzing the calculations carried out , it is possible to choose the fuel rate relation on injector channels for the main operating modes of the combustor (wake-up, underload, mode 100 %). Let's consider, that the maximum mass flow rate on one injector makes 29.1 g/s of diesel fuel. On a wake-up mode the pilot channel with the injection pressure corresponding to a line 5 in table 2 works only. It provides a fuel rate through the injector 5 g/s. On an underload mode, fuel is supplied in the main and pilot channels with identical pressure difference of an order 0,15 MPa (a line 3 in tables 2 and 3). Passing through the pilot channel 3 g/s, through the main - 17 g/s, we will receive the fuel mass flow rate through the sprayer unit 20g/s. On a 100 % mode fuel injection pressure in channels increases to 0.3 MPa (a line 4 in tables 2 and 3). Passing through the pilot channel – 4.1 г/с, through the main - 25 г/с, we will receive the fuel mass flow rate through the device accordingly 29.1 г/с. Thus, the fuel supply in both channels can be carried out by one pump with use of one simple valve. The size of diesel fuel droplets thus, taking into account gas recompression in the combustor, should not exceed 25-40 microns. Such values should provide high combustion efficiency of the diesel fuel moving through pilot and main channels of the burner , on all operational modes of low-emission combustor.

To obtain characteristics of airflows, reverse zone size and to select swirl scheme to the beginning of detail design 3D calculations of device, established in circular pipe, have been conducted by a technique (Patankar, 1980). Air pressure upon an exit from calculation area makes 0,1 МПа. Calculations of the gas flow are based on numerical integration of the full system of stationary Reynolds equations written in Euler variables. To find the coefficients of turbulent diffusion, use is made of the Boussinesq hypothesis on the linear dependence of the components of the tensor of turbulent stresses on the components of the tensor of deformation rates of average motion and two equations of transfer of turbulence characteristics. Details of the calculation technique one can find in the research (Maiorova at all, 2010).

Researches were conducted on 2 versions of devices: with airflows swirling in opposite directions (variant 1) and in one direction (variant 2). Calculated air flow rate G_A through the sprayer unit and swirlers are resulted in table 4.

Swirlers	G_A , g/s	
	swirling in opposite directions (variant 1)	swirling in one direction (variant 2)
Total	16.3	15.7
Outer	12.0	11.6
Central	4.3	4.1

Table 4. Air mass flow rate - preliminary design

The difference in flow rates values it is possible to explain by the absence of developed reverse zone in variant with opposite swirling and, as consequence, smaller outlet back pressure. Calculations show that, at opposite directions swirling, it is possible to receive higher intensity of turbulence and accordingly the best spray fineness. However thus there is no stable zone of reverse flow. The underpressure area is formed of the device exit behind the central body because of more axial velocity in comparison with variant 2. It will negatively affect stable combustion limits in combustor. More uniform pressure field in a cross-section direction on distance of 20 mm from a nozzle edge, is received in calculation 2.. It should positively affect boundary lines of ignition and lean blowout.
Thus it is possible to conclude, that the scheme with swirling in one direction is more preferable to continue researches and detail burner design.
The calculations carried out revealed the necessity of further air mass flow rate increase. It's necessary for the provision of reliable start and high combustion efficiency.
A number of calculations has been conducted to investigate the interaction of streams from the central and circumferential air swirlers at various swirl parameters. It has been received that the best performance is provided at use of two swirlers with an identical blade angles - on 45° to a device axis In the designed burner the following percentage of air mass flow rates through swirlers has been received: 33 % - in the central, 67 % - in outer (table 5).

Swirlers	G_A, g/s
Total	37.3
Outer	24.9
Central	12.4

Table 5. Air mass flow rate – final design

At designing of low-pressure injectors a collapse of a fuel bubble is essential danger, especially on low modes. For prevention of this phenomenon, already at a design stage, it is necessary to carry out the calculation of the shape of a fuel film. Fundamental theory of a calculation method was stated in (Chuec S. G., 1993). In research (Vasil'ev at al, 2010) this mathematical model has been applied to calculate the film form generated downstream of dual-orifice pressure-swirl atomizer. It has been shown, that the calculated film shape is in satisfactory agreement with the shape obtained in the experiment. With the specified geometrical parameters of an atomizer, these shapes are determined by the flow rate of liquid through the atomizer.
The simplified system of mass and momentum conservation equations in the coordinate system connected with a film surface taking into account gravity forces was solved by numerical method. Initial data about a film thickness, spray angle, longitudinal and

tangential liquid velocities were set from hydraulic design of an injector. The calculations carried out have shown that on operational modes there is a confident deployment of a fuel bubble. It will allow to provide hit of enough of fuel droplets into plug discharge zone and, as consequence, sufficient area for assured firing of the combustion chamber.

4. Results of the combined burner analysis

4.1 Flow rate characteristics of the burner

Testing of the burner manufactured begins with measurement of flow rate characteristics of injectors in tests without participation of air and their comparison with calculated ones (fig. 3.a). Measurements of the liquid mass flow rate were conducted by firm KROHNE flowmeter (a measurement error <1 %). The air mass flow rate was measured by PROMASS flowmeter. Fluid injection pressure was transduced by AZD pressure sensors. The comparison of experimental and calculated characteristics by air is presented on fig. 3.b. It is possible to consider the concurrence received as comprehensible (taking into account possible errors of manufacturing).

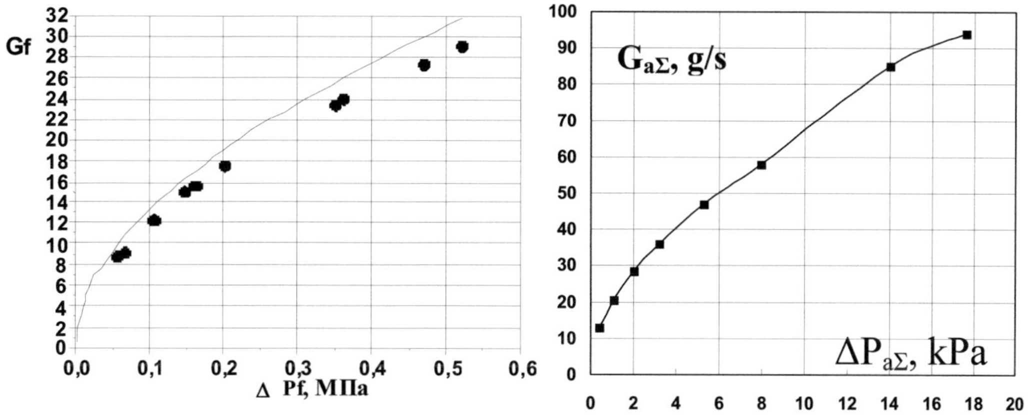

Fig. 3. Mass flow rate characteristics of an injector on kerosene (the outer channel) and air (total); lines - calculation, points - experiment

4.2 The investigation of fuel films without supply of airflows.

As is well known, the form of liquid exposed to an atomization, in an appreciable measure influences the quality of the aerosol received, basically on such parameters as fuel droplets distribution in cross section and a spray angle. In this connection, the complex of cold tests has been continued by investigation of the form of fuel films without supply of airflows. In experiences it was spent laser visualization of a stream. The flow of kerosene film at the outlet from the atomizer nozzle was recorded using a Canon XL-H1 three-matrix color video camera. Photos of the expiration of a fuel film at various mass flow rates and measured spray angles for injectors investigated are resulted on fig. 4 and 5. Operational modes for a pressure-swirl atomizer begin with flow rates corresponding to a photo 4b and above. In this range of flow rates it was possible to reach good stability of a fuel film angle. The photo on fig. 4a visually shows high uniformity of a fuel sheet even on lower modes, usually hard-hitting. For a pressure-swirl atomizer a target range of spray angles - 90-95° and high uniformity of injection are reached without of supply of an airflow.

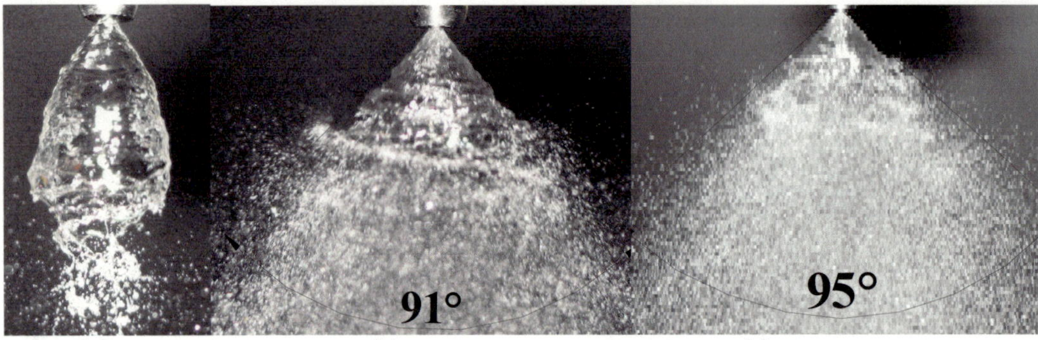

(a) (b) (c)

Fig. 4. Photos of the expiration of a kerosene film at various mass flow rates throw the pressure swirl nozzle; a) G_{f1} = 1.77 g/s, ΔP_{f1} = 60 кPa; b) G_{f1} = 2.7 g/s, ΔP_{f1} = 150 кPa; c) G_{f1} =3 .5 g/s, ΔP_{f1} = 286 кPa.

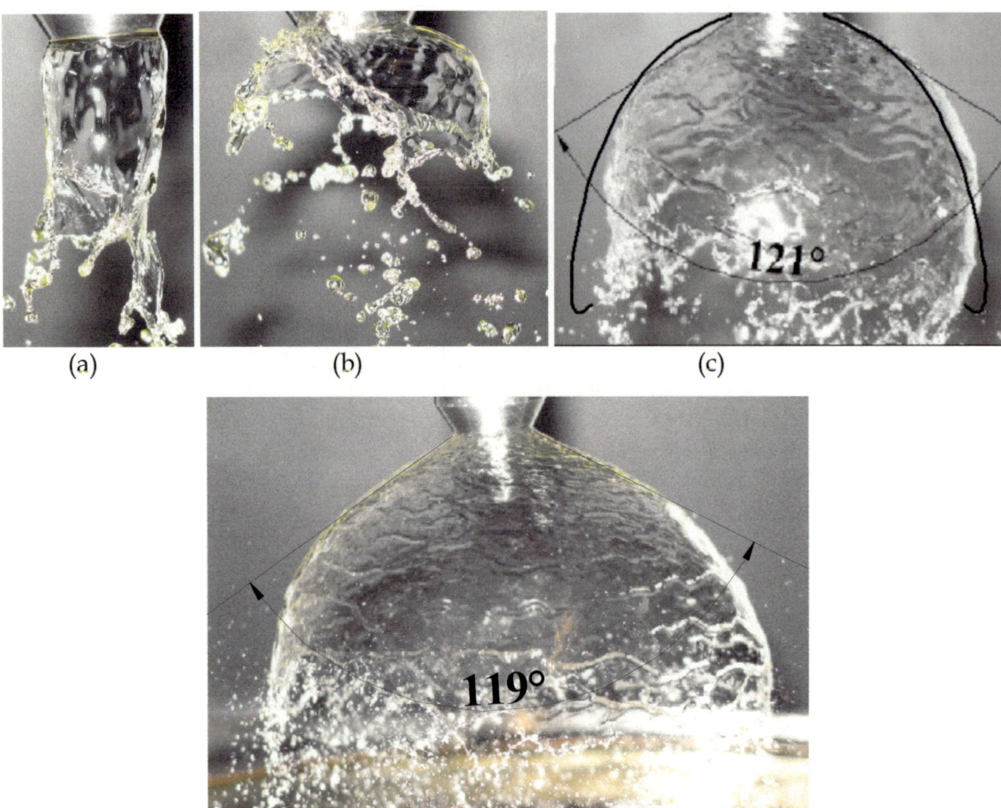

(a) (b) (c)

(d)

Fig. 5. Photos of the expiration of a kerosene film at various mass flow rates throw the airblast nozzle; a) G_{f1} = 9.2 г/c, ΔP_{f1} = 66 кPa; b) G_{f2} = 12.3 г/c, ΔP_{f1} = 105 кPa; c) G_{f1} =1 7.5 г/c, ΔP_{f1} = 202 кPa; line - calculation; d) G_{f2} =24,0 g/s, ΔP_{f2}=362 kPa

Photos of the expiration of a fuel film at various flow rates throw the airblast atomizer are presented on fig. 5. The comparison on Fig. 5c shows that the computational technique describes well the experimental data on the configuration of the fuel. Fig. 5c corresponds to underload mode, 5d - mode 100 %, thus spray angle - an order 120°. Substantial growth of fuel film diameter till the moment of its contact to swirled airflows provides reduction of its thickness in a zone of pneumatic spraying. As consequence, the fineness of atomization improves essentially.

Thus for the fuel channel of an airblast injector the spray angle without airflow submission, and the small thickness of a fuel film are received stable on modes. This allows to improve considerably the fineness of atomization even on low engine power settings.

4.3 The comprehensive investigations of the burner with air submission

Comprehensive test of the burner in open space with air submission were carried out. Measurements were conducted on a bench of laser diagnostics. The important parameters characterizing quality of the device performance - value and intensity of paraxial reverse zone were optimized. As a result of tests geometrical parameters of sprayer unit were updated. Axial rules of injectors and swirlers, and also blades angles of swirlers varied. As a result of optimization following blade angles have been chosen: 60° for the central swirler , 45 ° for the circumferential one. The difference of optimum angles from received in predesign, is connected possibly with distinction of calculation and experimental areas.

When carrying out aerodynamic 3-D calculations of optimized flame sprayer the total air flow through the burner (37.3 g/s), with a following percentage ratio of mass flow rates is received: 33 % - in the central swirler (12.4 g/s), 67 % - in outer one (24.9 g/s).

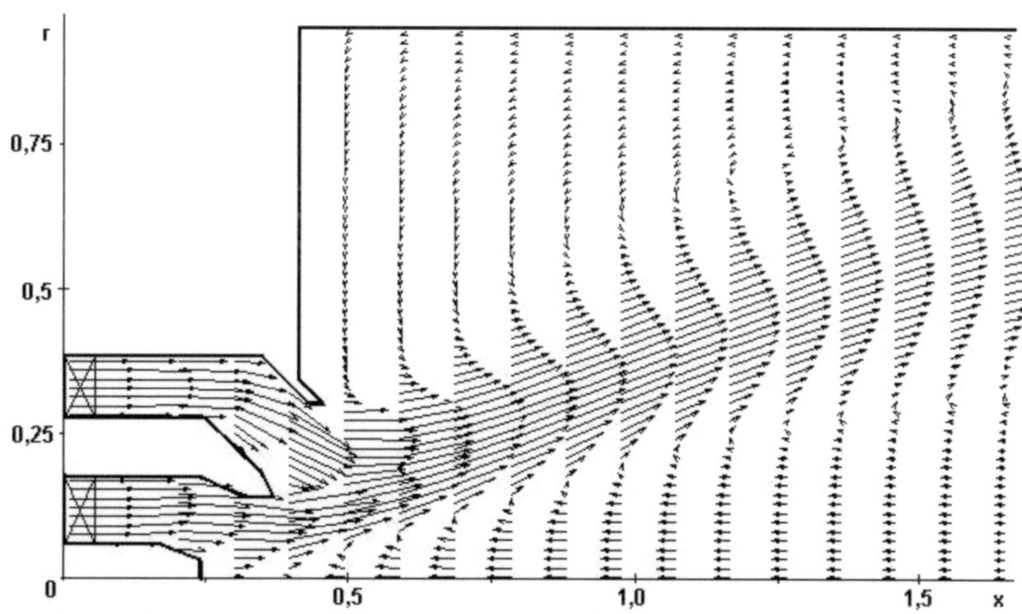

Fig. 6. Calculated vector velocity field behind the burner

Calculated flow pattern in the tube after the burner is given in fig. 6. As one can see from this fig, near the burner axis the advanced zone of reverse flow is formed that should promote the stability of combustion process.
The researches carried out have allowed to conduct comparison on radial distribution of axial and tangential velocities at distance 30 mm behind the burner. Results of imposing of experimental velocity profiles on calculated curves are shown on fig. 7. From the graphs presented on fig. 7, it is possible to draw a conclusion on satisfactory concurrence of calculation and experiment. Let's notice, that the calculated velocity profile is received for air, and experimental for a drop-forming phase formed by the pressure-swirl atomizer. In this connection the experimental velocity curve is a little bit wider then calculated.

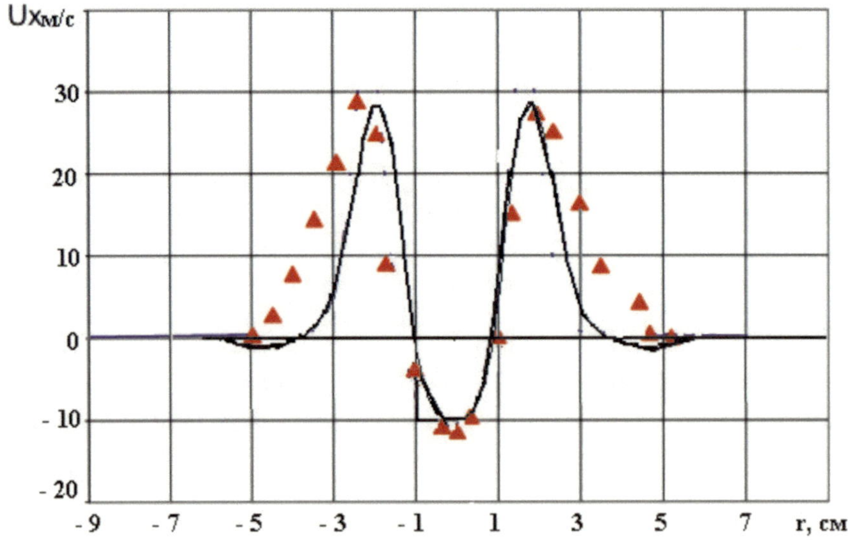

(a)

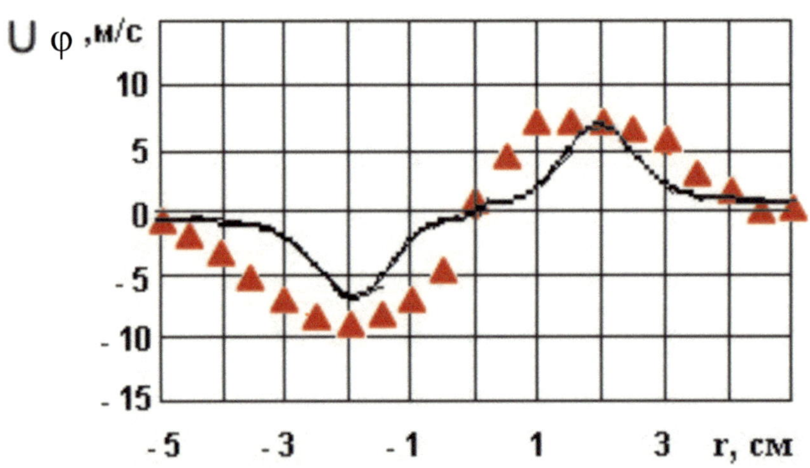

(b)

Fig. 7. Distribution of axial (a) and tangential (b) velocities on diameter of a spray; lines – calculation, points – experiment (PDPA measurements)

4.4 Comparative researches of burner performance on different hydrocarbon fuels

Let's estimate now possibility of using of the burner for working on various hydrocarbon fuels - oil and alternative. Calculated performance of a pilot injector at the wake-up mode are given in table 6. The performance of a main injector at the mode 100 % are presented in table 7.

fuel	C_D	ΔP_f, MPa	$2\theta_R$, °	SMD, mkm
Ethanol	0.140	0.893	102.0	19.5
Kerosene TS1	0.146	0.836	100.6	21.7
Ideal diesel	0.186	0.471	90.9	39.7
FAME (biodiesel)	0.195	0.415	86.3	52.8

Table 6. Calculated performance of a pilot injector at the fuel rate 5 g/s

fuel	C_D	ΔP_f, MPa	$2\theta_R$, °	SMD, mkm
Ethanol	0.003	0.335	151.3	41.4
Kerosene TS1	0.003	0.336	152.0	39.8
Ideal diesel	0.003	0.353	148.6	58.6
FAME (biodiesel)	0.003	0.356	144.3	81.4

Table 7. Calculated performance of a main injector at the fuel rate 25 g/s without air supply

As one can see from table 7, flow rate characteristics of the big main injector practically do not depend on fuel viscosity. Discharge coefficients are identical for all fuels and the difference of injection pressure is determined only by a difference of density. Values of spray root angles are very close for all fuels, and real sizes of droplets will be mainly air streams dependent. The pilot injector performance (table 6) to a greater extend depend on a fuel kind. The deviation of discharge coefficients from one for a diesel is within 24 %, the dispersion of spray angles - within 12 %. As a whole, however, the injector provides comprehensible performance on regimes on wake-up and underload modes for all kind of fuel observed.

Let's consider also results of calculation of the fuel film shape for various fuels. The film thickness on an exit from the main injector is essentially more than on an exit from the pilot injector. The fuel kind can exert the greatest influence on the deployment of a fuel bubble in the case of the main injector.

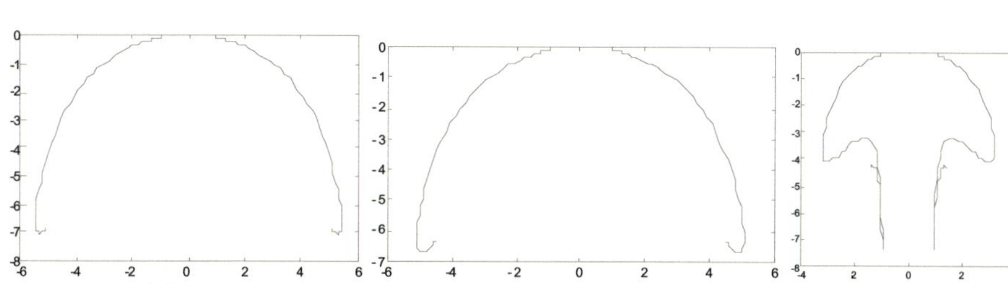

(a) (b) (c)

Fig. 8. Calculated shape of fuel film without supply of air; G_f=17.5 g/s; a - ethanol, b - kerosene, c - biodiesel

Calculated shapes of fuel film downstream of main atomizer without supply of air and with supply of air are given in fig. 8 and 9 respectively. Fuel mass flow rate corresponds to underload mode. The nozzle radius is assumed as characteristic dimension.

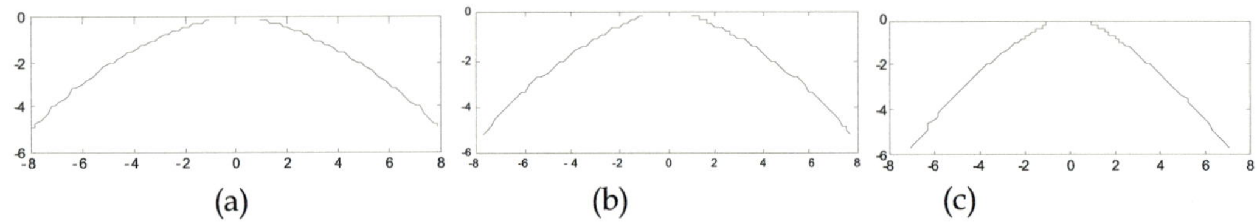

Fig. 9. Calculated shape of fuel film with supply of air; G_f=17.5 g/s; a - ethanol, b - kerosene, c - biodiesel

As one can see from fig. 8, in the absence of air supply for the regime considered the shape of fuel spray depend on fuel kind. In case of most viscous of them - a biodiesel - the fuel spray does not deploy. However, during air injection the shape of fuel film is determined, basically, by the airflow. Apparently from fig. 9, confident disclosing of a fuel spray for fuels as with normal, and the raised viscosity is observed in this case.
In fig. 10 - 12 the results of comparative test of the burner at normal conditions on different hydrocarbon fuels (PDPA measurements) are presented during air injection for wake-up mode (centrifugal nozzle works only) and underload mode (both nozzles work). At the underload mode measurements for mix of diesel fuel with rapeseed oil in ratio 50% - 50% were carried out too. Physical properties of this mix are: ρ_f = 867 kg/m^3 , ν_f = 12 mm^2/s. Apparently from Fig. 10-11, the difference of the droplets sizes or diesel fuel and kerosene is appreciable weakly. Diesel-oil droplets are large-scale near the axis. However their sizes are in the range of target values. Values of SMD average along the cross section are 40-60 mkm. Values of volumetric concentration are neighbour for all fuels. The flow structure as it is visible from fig. 12, is self-similar. This result shows, that at flow rates ratio used the atomization is determined mainly by an airflow.

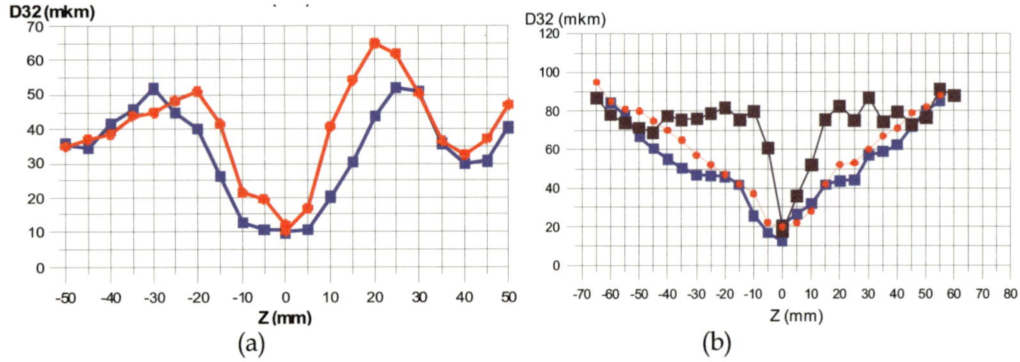

Fig. 10. Distribution of the droplets sizes on diameter of a spray; a) wake-up mode, ΔP_a=3 kPa, G_f=5 g/s; b) underload mode, ΔP_a=3 kPa, G_f=20 g/s; -•diesel; -■ kerosene; —♦ diesel - oil mix

The photo of fuel-air spray for underload mode is given in fig. 13. The spray angle for diesel and kerosene practically does not depend on a fuel kind that proves to be true also concentration structures (fig. 11), and makes an order 90°. It corresponds to target value of this parameter. Apparently from Fig. 13b the spray angle makes more than 80° even for such viscous fuel, because the spraying is determined by air streams.

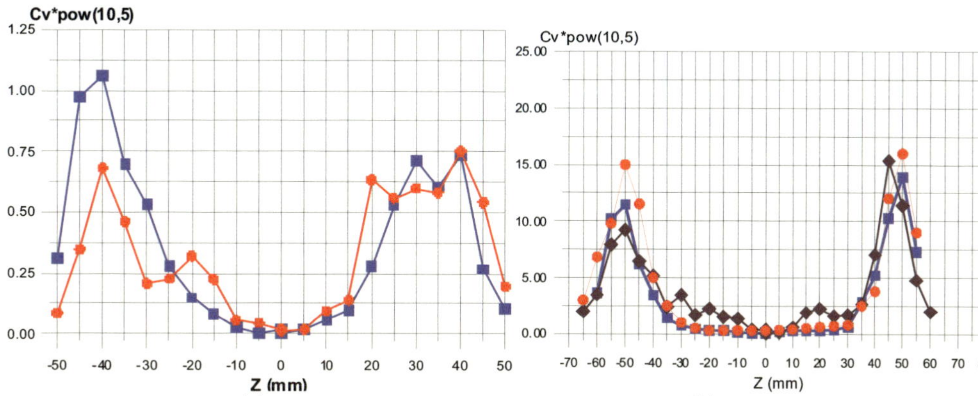

(a) (b)

Fig. 11. Distribution of the volumetric concentration of a liquid fuel on diameter of a spray; a) wake-up mode, ΔP_a=3 kPa, G_f=5 g/s; b) underload mode, ΔP_a=3 kPa, G_f=20 g/s; –•diesel; –▪kerosene; —♦ diesel -oil mix

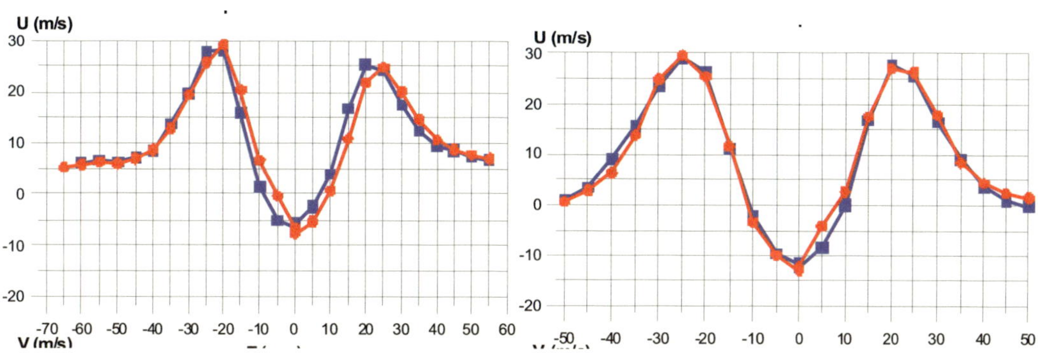

(a) (b)

Fig. 12. Distribution of the axial velocity on diameter of a spray; a) wake-up mode, ΔP_a=3 kPa, G_f=5 g/s; b) underload mode, ΔP_a=3 kPa, G_f=20 g/s; –•diesel; –▪ - kerosene

(a) (b)

Fig. 13. The photo of fuel-air spray; ΔP_a=3 kPa, G_f=20 g/s; a - diesel, kerosene; b - diesel -oil mix

Results of the present section show that the designed dual-orifice atomizer can be used for different fuels, both for oil, and for alternative. In addition injection valve modernization can be necessary only.

4.5 Tests of a burner with the low-emission combustion chamber compartment

For fire tests of a burner with the combustion chamber compartment the flame tube with permeability 4281 mm^2 was used. The kerosene TS1 was used as fuel. Boundary lines of the flame blowout (fig. 14) were determined only on one pilot channel. It is possible to assume, that connection of the second channel of a burner will allow to expand a zone of a stable running of the chamber even more. The received blowout boundary line shows, that the chamber steadily works in a range coefficient of air excess α_C from 1 to 6.5 and air volume flow rate Q_C up to 0.45 м^3/s at underpressure in chamber P_C = 0.08 MPa. This operation mode corresponds to altitude of an order of 2 km.

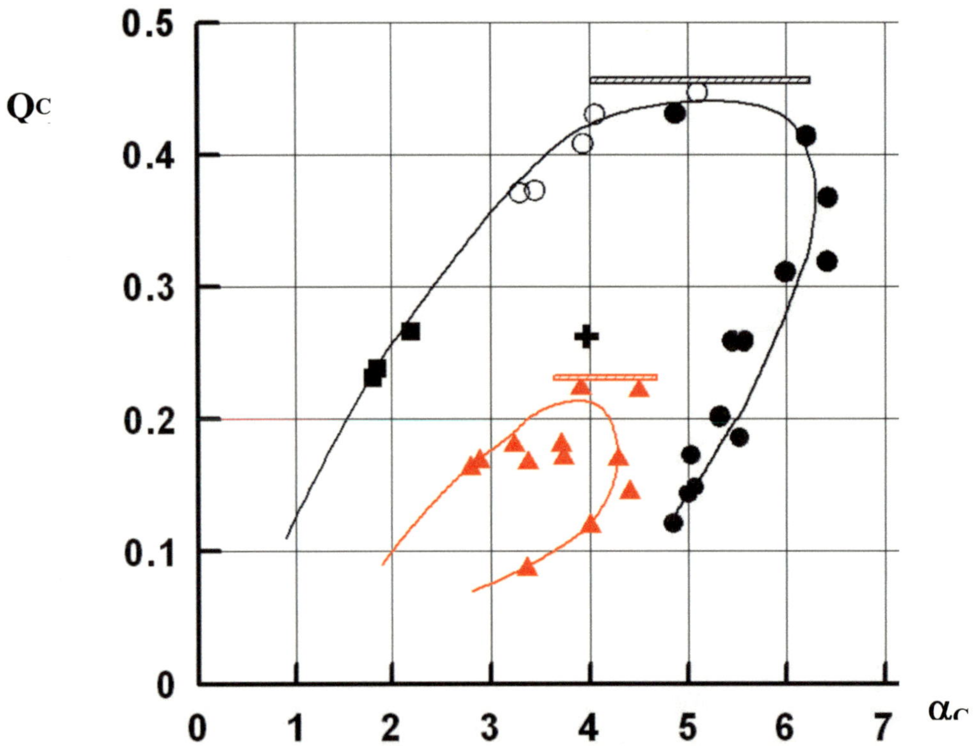

Fig. 14. Boundary lines of ignition and blowout in the combustion chamber compartment, the plug 2 J, H ≈ 2 km, T_C^* = 280 K, • – lean blowout, O - rich blowout, ▲-wake-up, ■ - there is no rich blowout, ✚ - the point of temperature field taking-out

The area boundary reaches satisfactory values on α_C, and comprehensible values on Q_C. The ignition domain boundary is sufficient on the square for assured firing of the combustion chamber. The given result allows to assert, that blowout characteristics in earth conditions will appear at least not worse received. Flame photos at various α_C are shown in fig. 15.

Also the temperature fields behind an exit from transition liner in a pipe of diameter 110 mm have been taken out under various α_C . The temperature field received has a symmetric appearance and small non-uniformity on value of temperature - the minimum value differs from maximum on 70 K. The temperature distribution on one radius is resulted in fig. 16.

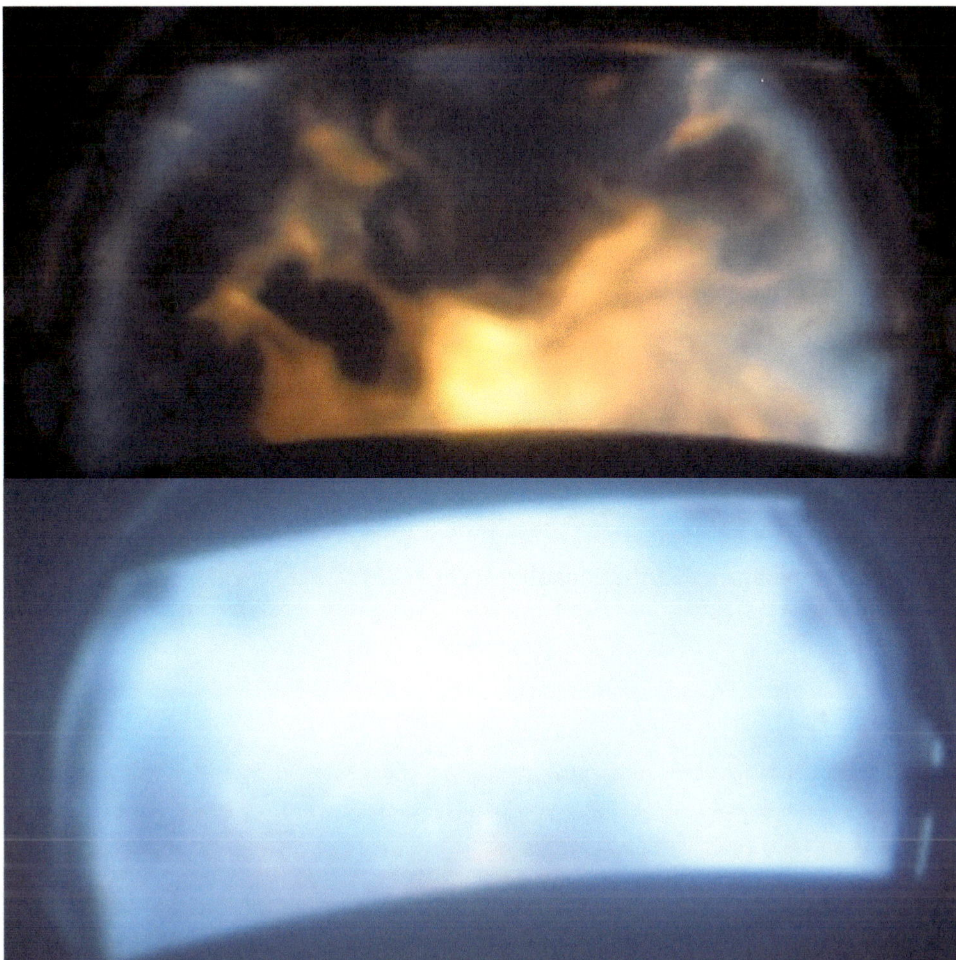

Fig. 15. Flame photos at various α_C.

Integration of this curve allows to receive mass average value of temperature Tav = 575K.
The dependence of combustion efficiency and average temperature behind the transition liner on air excess coefficient is presented in fig. 17. The combustion efficiency was calculated on value of temperature according to the work (Kulagin, 2003).
On the basis of the spent experiments it is possible to assert, that the burner developed has shown comprehensible characteristics. In particular wide side-altars of the stable combustion, assured firing of the combustion chamber, uniform enough field of gas temperature on an exit and satisfactory combustion efficiency, taking into account that tests occurred on a regime close to earth wake-up mode.

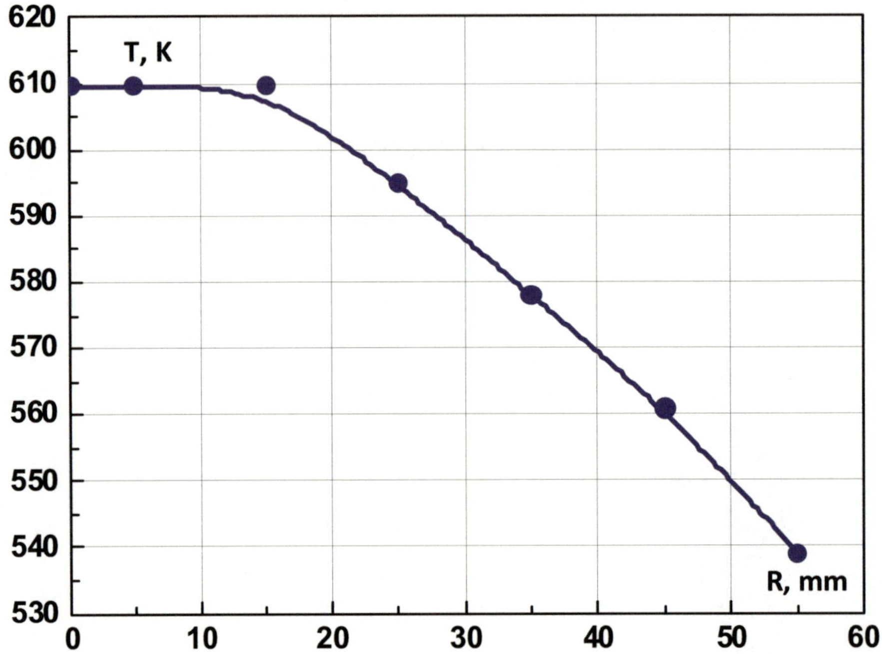

Fig. 16. Radial temperature distribution; $\alpha_C = 4$; $Q_C = 0.28$.

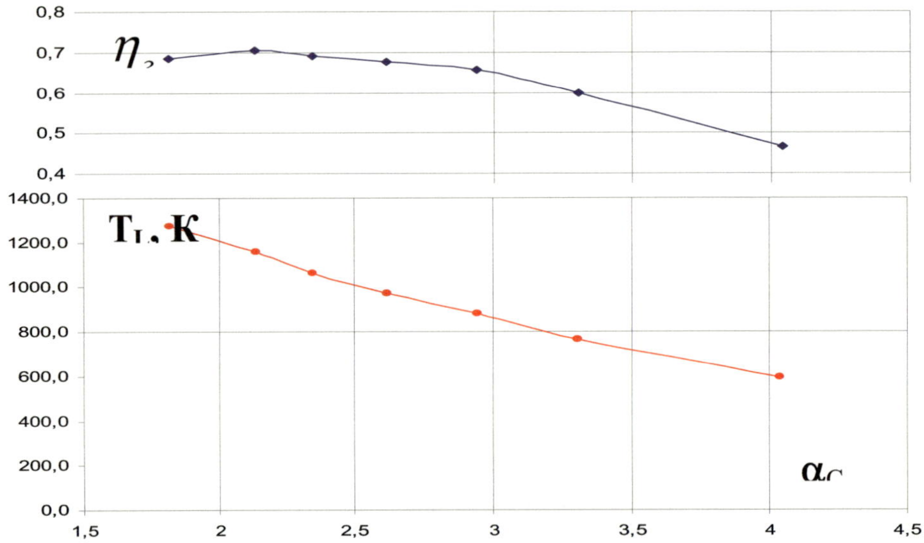

Fig. 17. The dependence of combustion efficiency and average temperature behind the transition liner on air excess.

5. Summary

The designing, manufacturing and test of individual injectors and the burner as a whole for low-emission combustion chambers of gas-turbine engine or gas-turbine plant is executed. The present work represents a complex of target researches on design-experiment basing of shape of the sprayer unit for low-emission combustors on fuels as usual, as of the increased viscosity (kerosene, ethanol, diesel, biodiesel).

The dual-orifice (on fuel) burner of the combined centrifugal-airblast scheme is proposed. Nozzles of atomizers place concentrically. The low-rate pilot channel (pressure swirl nozzle) is installed on a burner axis. The main channel – airblast nozzle is located between two air swirlers.

Hydraulic design of pilot channel and numerical 3D modeling of air channels of the burner are carried out. Geometrical parameters of the burner and blades angles of swirlers were chosen. These parameters were optimized during comprehensive test of the burner in open space with air submission.

On the basis of calculation researches two heads of injectors are designed and made: centrifugal and airblast for the combined burner.

The investigation of fuel films without supply of airflows is carried out. For a pressure-swirl atomizer a target range of spray angles - 90-95° and high uniformity of injection are reached. For the fuel channel of an airblast injector the spray angle without airflow submission, and the small thickness of a fuel film are received stable on modes. This allows to improve considerably the fineness of atomization even on low engine power settings.

The comparative researches of burner performance on different hydrocarbon fuels are carried out. It's shown that at fuel-air flow rates ratio used the atomization is determined mainly by an airflow. Schemes of devices worked out provide the adjacency of aerosol characteristics for combustibles investigated. Values of Sauter Mean Diameter average 40-60 mkm. The spray angle when both injectors working with air supply makes an order 90°.

Results of the research show that the designed dual-orifice atomizer can be used for different fuels, both for oil, and for alternative.

Fire tests of a burner with the low- emission combustion chamber are conducted. It is possible to assert, that the burner developed has shown comprehensible characteristics. In particular wide side-altars of the stable combustion, assured firing of the combustion chamber, uniform enough field of gas temperature on an exit and satisfactory combustion efficiency.

Results of present work are protected by the Patent of the Russian Federation (Vasil'ev at al. 2009).

6. Notation

C_d - discharge coefficient of injector

C_v - volumetric concentration of a liquid fuel, kg/m^3

D_{32} - droplet mean Zauter diameter, average along the circumference, m

G – mass flow rate, kg/s

P – pressure, Pa

Q - air volume flow rate, m^3/s

ΔP - injection pressure, Pa

SMD – droplet mean Zauter diameter average along the whole cross section, m

T – temperature, K
U – velocity, m/s
α- coefficient of air excess
η - combustion efficiency
μ - dynamic viscosity, kg/(m·s)
ν - kinematic viscosity, m^2/s
θ_R - spray root angle, °
ρ - density, kg/m^3
σ - surface tension coefficient, N/m

Subscripts
a – air
C – combustion chamber
f - liquid fuel

7. References

Chuec S. G. (1993) Numerical Simulation of Nonswirling and Swirling Annular Liquid Jets. *AIAA Journal*. Vol.31. No.6. P.1022-1027.

Dityakin Yu. F., Klyachko L. A., Novikov B. V. and V. I. Yagodkin. (1977). *Spraying of Liquids* (in Russian), Mashinostroenie, Moscow.

Kulagin V.V. (2003). *The theory calculation and designing of aircraft engines and power plants.* (in Russian), Mashinostroenie, Moscow.

Lefebvre A.H. (1985). *Gas Turbine Combustion,* Hemisphere Publishing corporation, Washington, New York, London.

Lefebvre A.H. (1989). *Atomization and Sprays,* Hemisphere Publishing corporation, New York.

Anna Maiorova, Aleksandr Sviridenkov and Valentin Tretyakov (2010). The investigation of the mixture formation upon fuel injection into high-temperature gas flows. *Fuel Injection. Edited by D. Siano.* Published by Sciyo. P.121 - 142. ISBN 978-307-116-9.

Patankar S. (1980). *Numerical Heat Transfer and Fluid Flow,* Hemisphere Publishing, New York.

Vasil'ev A. Yu. (2007). Comparison of perfomance s of various types of the injectors working with use of an air stream. (in Russian). The bulletin of Samara State Aerospace University. No 2(13). P. 54-61.

A.Yu. Vasil'ev, A. I. Maiorova, A. A. Sviridenkov and V. I. Yagodkin (2009). Patent of Russian Federation No 86279.

A.Yu. Vasil'ev, A. I. Maiorova, A. A. Sviridenkov and V. I. Yagodkin (2010). Formation of Liquid Film Downstream of an Atomizer and Its Disintegration in Gaseous Medium. *Thermal Engineering.* Vol. 57, No. 2. P. 151-154. ISSN PRINT: 0040-6015. ISSN ONLINE: 1555-6301.

15

Acetals as Possible Diesel Additives

Ion Agirre, Victoria Laura Barrio, María Belén Güemez,
José Francisco Cambra and Pedro Luis Arias
Chemical and Environmental Engineering Department, Engineering School of Bilbao
Spain

1. Introduction

Nowadays, oil is the source of the vast majority of fuels used for transport, heating and of the hydrocarbons used in petrochemical industry. However, oil is a fossil fuel and some experts predict that its reserves will exhaust approximately in 20-30 years. Moreover, it seems that the demand of fossil fuels will increase at rates that can be estimated from "World Energy Outlook" elaborated by the International Energy Agency (IEA, 2007). Apart from all these data, there are evidences that the climate of the planet is changing due to the global warning. The temperature of the earth is increasing and the ice of the poles is beginning to melt; all these changes are attributed to the Greenhouse Effect. Besides, it has been estimated that 82 % of the anthropogenic CO_2 emissions are due to fossil fuel combustion so it is clear that alternative energy sources are needed (see Figure 1).

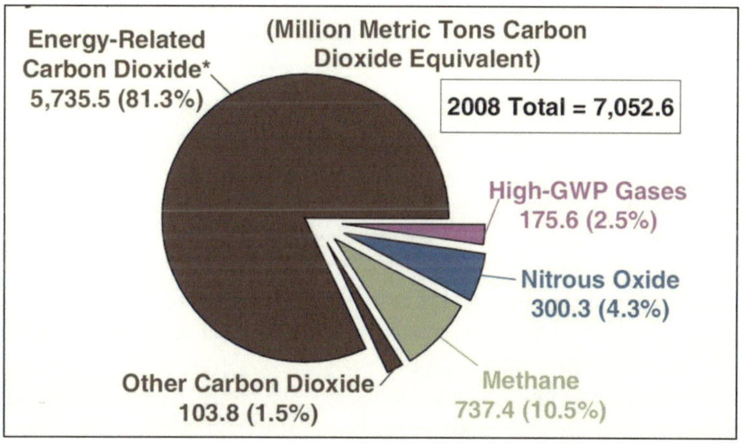

Fig. 1. U.S. Anthropogenic Green house Gas Emissions (IEA, 2008).

On the other hand, the use of other raw materials is being explored due to environmental concerns, economical fluctuations and the geopolitical instability in the producer countries. In order to overcome these difficulties, one possible alternative for transportation is the use of biofuels.

Biodiesel is an alternative fuel obtained from vegetable oils or animal fats and it has several technical advantages over petro-diesel such as the reduction of exhaust emissions, improved lubricity and biodegradability, higher flash point and lower toxicity. There are some other properties like cetane number, gross heat of combustion and viscosity that are very similar in biodiesels and in conventional diesels. Biodiesels show some worse properties when compared to conventional diesels: oxidation stability, nitrogen oxides emissions, energy content and cold weather operability (Moser et al., 2008). A possible solution to this problem is the use of additives.

A vast variety of fuel additives are added to diesel fuels to improve the engine efficiency and to reduce harmful emissions. An important group of diesel additives are metal-based ones that have been used as combustion catalysts for hydrocarbon fuels. These metals are manganese, iron, copper, barium, cerium, calcium and platinum which present catalytic activity in combustion processes (Burtscher et al., 1999; Keskin et al., 2008). The presence of this type of additives reduces diesel engine pollutant emissions and fuel consumption. The metallic function can react with water to produce hydroxyl radicals, enhancing soot oxidation, or can react directly with the carbon atoms in the soot lowering the emissions (Keskin et al., 2008). However, non-metallic, renewable ashless diesel combustion enhancer additives would be the best option, avoiding the emission of metallic compounds.

Nowadays ethers like MTBE (methyl ter-butyl ether) and ETBE (ethyl ter-butyl ether) are the most well known oxygenated additives for gasoline. ETBE is synthesized by reacting ethanol and isobutylene and it offers better characteristics than the ethanol being less volatile and more miscible with the gasoline.

Ethanol-diesel blend fuel has been studied because ethanol contains 34.3 % of oxygen (Kim & Choi, 2008) so it can reduce the emission of particulate matter (PM) in the diesel engine (Li et al., 2005). Ethanol is an appropriate additive for petrol engines due to its high octane number but due to its low cetane number and its high heat of vaporization it resists self-ignition in diesel engines (Kim & Choi, 2008). Besides that, ethanol – diesel blends are rather unstable even at low temperatures (Frusteri et al., 2007). An alternative to ethanol as oxygenated bio-additives for diesel fuel could be different dieters like acetals (1,1 diethoxy ethane and others). Mention that acetals can be produced from a completely renewable origin.

In the present chapter, a review on acetals will be presented. Its content will include:

- A presentation of different types of acetals (straight chain acetals or cyclic acetals) and their relevant properties.
- A discussion about their different production processes focusing on those processes that have a completely renewable origin.
- The different catalysts used in acetal synthesis, starting from homogenous catalysts to heterogeneous catalyst of various types like commercial or natural ones and also those under development at laboratory scale.
- Finally, an overview of some innovative reaction systems that are currently being tested and developed for this kind of reactions.

2. Acetals formation

Acetals are oxygenated compounds that can be produced following different types of reactions:

- Filley (Filley, 2005) studied the reaction between methyl 9,10 dihydroxystearate and long chain aldehydes to form the corresponding cyclic acetal in the presence of p-toluenesulfonic acid (PTSA).

- Reacting glyoxylic acid with aliphatic alcohols using cationic exchange resins as catalysts (Mahajani, 2000).
- From allylic ethers using as catalysts cobalt compounds (Chang, 1995).
- Reacting aldehydes and ketones with trimethyl/triethyl orthoformate at room temperature in the presence of copper(II) tetrafluoroborate as catalyst (Kumar et al., 2005).
- Reacting ethanol and acetaldehyde in the presence of an acid catalyst. The main reaction implies the production of 1,1 diethoxy ethane and water (Andrade et al., 1986; Capeletti et al., 2000; Chopade et al., 1997a; Chopade et al., 1997b; Kaufhold & El-Chahawi, 1996; Mahajani et al., 1995).

The last procedure is the most interesting one as it is quite a simple reaction and both reactants can have a renewable origin. Ethanol can be obtained via fermentation of sugar rich plants while the aldehyde can be obtained from its corresponding alcohol via partial oxidation or via dehydrogenation so the origin of the acetal can be totally bio/renewable. 1,1 diethoxy ethane has been used as a solvent, as an intermediate in chemical synthesis for the protection of the carbonyl group in ketones and aldehydes, in the fragrance industry as well as in alcoholic drinks like brandy or in several liquors (Capeletti et al. 2000). Moreover, da Silva Ferreira et al. (da Silva Ferreira et al., 2002) and Camara et al. (Camara et al., 2003) studied the presence heterocyclic acetals from glycerol and acetaldehyde in Port and Madeira wines in order to show the existence of a linear correlation between their amounts and the wine age (and its aroma).

One of the first manuscripts on acetalization reactions was published at the end of the 19th century and it is reported by Aksnes et al. (Aksnes et al., 1965). However, all these pioneering works are related to the isolation and identification of the different acetal isomers (Aksnes et al. 1965; Aksnes et al., 1966; Stefanovic et al., 1967).

It must be remarked the important role that glycerol can play in the field of bio-additives. Huge amounts of glycerol are being formed as a subproduct in the *transesterification* reactions of fatty acids (10 wt %). Currently, small amounts of this tri-alcohol are being used in pharmaceutical and personal care areas, so, in order to avoid its incineration, different alternatives are being investigated looking for high added value products from glycerol: hydrogen gas production, glycerin acetate production, citric acid production, cosmetic bonding agent for makeup including eye drops and conversion to propylene glycol, acrolein, ethanol and epichlorhydrin (Silva et al., 2010; Umbarkar et al., 2009). The obtention of glycerol additives would be another interesting use of the glycerol. Thus, it could be a good option to book out all the excedent in glycerol.

Acetal	**Flash point (K)**
Ethanol + acetaldehyde	252.15 ± 0.0
Ethanol + butyraldehyde	302.05 ± 18
Glycerol + formaldehyde	349.35 ± 4.2
Glycerol + butyraldehyde	371.55 ± 23.3
Diesel specifications (EN590:2004)	328.15

Table 2. Flash points of different acetals (Scifinder Scholar 2007 database) and their comparison with the flash point from the diesel specifications (EN590:2004).

However, not all the acetals can be used as diesel or biodiesel additive. Some acetals present low flash points and as a result they are not suitable to use as diesel additives. In Table 1 flash points of several acetals are reported. It can be observed how the glycerol acetals fulfill all the diesel specifications while acetals of ethanol require a big aldehyde in order to get an acceptable flash point. Lower molecular weight acetals are being used as surfactants, flavors, disinfectants (Capeletti et al. 2000; Silva et al. 2010), in cosmetics, food, pharmaceutical area or in fragrances (Umbarkar et al. 2009; Yang et al., 2006).

2.1 Acetals formation from alcohols and aldehydes

The reaction mechanism for straight chain acetals involves two reversible steps. In the first one alcohol molecule reacts with one aldehyde molecule leading to the formation of the corresponding hemiacetal liberating a significant amount of heat. This reaction takes place relatively quickly and in absence of catalyst also at room temperature. In the second step another alcohol molecule reacts with the hydroxyl group of the hemiacetal in order to form the corresponding acetal and water. This second step is also an exothermic reaction but it takes place in presence of an acid catalyst. (Agirre et al., 2010; Chopade & Sharma, 1997a) (see Figure 2).

Acetalization reactions involving polyalcohols (like glycerol) are also carried out in two reversible steps: the first one where the alcohol reacts with the aldehyde molecule leading to the formation of the corresponding hemiacetal and the second one, where two hydroxyl groups of the hemiacetal join to form the corresponding acetal, releasing a water molecule (Chopade & Sharma, 1997b). Silva et al. (da Silva Ferreira et al. 2002) explained in detail the reaction mechanism for the reaction between glycerol and acetaldehyde, indicating all the different steps of the reorganization of the hemiacetal molecule. However, in Figure 3 a simplified scheme of the reaction mechanism proposed by Sharma & Chopade (Chopade & Sharma, 1997b) is shown. It must be mentioned that in the case of reaction between glycerol and acetaldehyde acetals formation was observed in the absence of catalyst although the reaction rate was extremely low.

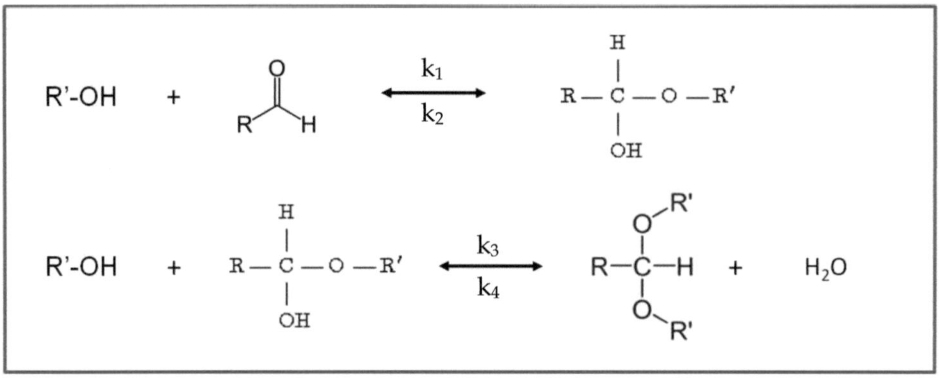

Fig. 2. Mechanism for the formation of straight chain acetals.

In the case of glycerol acetals two different acetals are formed: 1,3 dioxanes and 1,3, dioxalanes. Moreover, each one presents *cis* and *trans* stereoisomers (except when reacting glycerol with formaldehyde) being in the overall four different acetal isomers. Similar to the published results by Camara et al. and Da Silva Ferreira et al. (Camara et al. 2003; da Silva Ferreira et al. 2002), isomerization reaction were observed in our own experiments reacting formaldehyde and acetaldehyde with glycerol, respectively. In all the cases, isomerization of

1,3 dioxalanes to 1,3 dioxanes was observed. Moreover, they observed that 1,3 dioxalane isomers were formed faster but then the reaction mixture proceeded towards the isomerization equilibrium generating more 1,3 dioxanes. On the contrary, Ruiz et al. (Ruiz et al., 2010) observed the isomerization from dioxanes to dioxalanes.

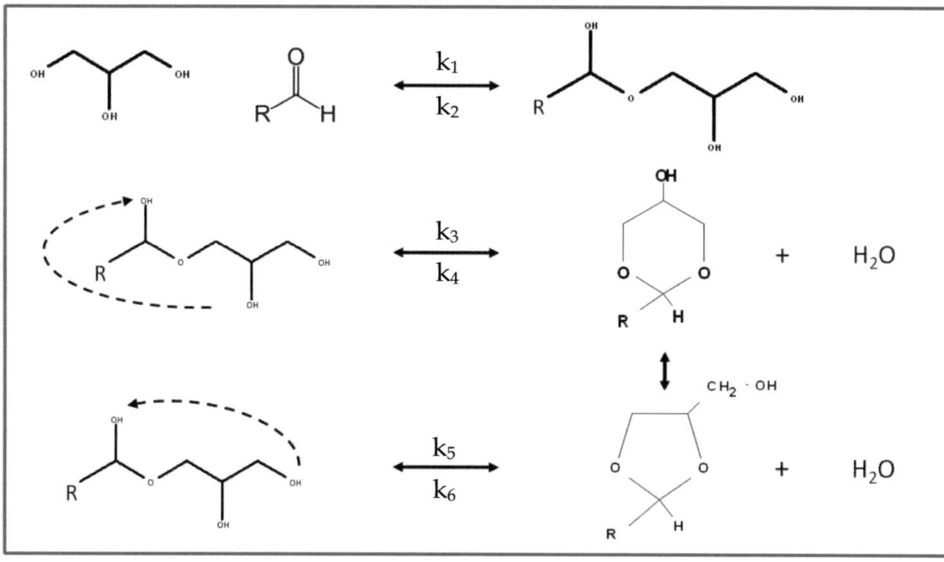

Fig. 3. Mechanism for the formation of cyclic acetals.

2.2 Kinetics of acetals formation reactions

All the information found in the literature shows that this type of reaction presents high thermodynamic limitations. Thus, Sharma and Chopade (Chopade & Sharma, 1997a; Chopade & Sharma, 1997b) achieved a maximum of 50% (at 363 K) of conversion reacting formaldehyde with ethylene glycol and less than 50% (at 343 K) reacting ethanol and formaldehyde. Agirre et al. (Agirre et al., 2010) studied the reaction between ethanol and butanal and the achieved maximum conversion was also around 50% (at 313 K). Other authors like Deutsch et al. (Deutsch et al., 2007) and Silva et al. (Silva et al., 2010) studied glycerol acetalization reactions with different aldehydes. Deutsch et al. (Deutsch et al., 2007) studied the acetalization reaction of glycerol with formaldehyde using different type of solvents and the maximum achieved conversions were between 58-77% (between 313 and 383 K). On the other hand, Silva et al. (Silva et al., 2010) studied the reaction of glycerol with butanal, pentanal, hexanal, octanal and decanal achieving a maximum of 80% (at 343 K) of conversion with butanal. In general, using heavier aldehydes they achieved lower conversions. In all cases thermodynamic limitations seems to be quite important

There are really few publications explaining the kinetics of acetalization reactions in the literature. Only Sharma & Chopade and Agirre et al. (Agirre et al. 2010; Chopade & Sharma, 1997a; Chopade & Sharma, 1997b) have published kinetic information. Thus, they were able to develop a pseudo-homogeneous kinetic model that is able to describe the reaction behaviour according to the reaction mechanism explained in Section 2.1.

It was explained that the hemiacetal formation and decomposition reaction rates are so high at the operating temperatures that the hemiacetal can be considered to be at equilibrium with the alcohol and the aldehyde.

$$[Hemiacetal] = K_1[RHO][R'OH] \quad (1)$$

The formation rate of acetal could be written as

$$\frac{d[Acetal]}{dt} = wk_3[Hemiacetal][R'OH] - wk_4[Aetal][Water] \quad (2)$$

Where w = (g cat)/(reaction volume)
Substituting [Hemiacetal] from Eq. 1,

$$\frac{d[Ac]}{dt} = wk[RHO][R'OH]^2 - wk_4[Acetal][Water] \quad (3)$$

Thus, theoretically it was concluded that the forward reaction is 2nd order with respect to alcohol and 1st order with respect to aldehyde and the reverse reaction is also 1st order with respect to acetal and water, i.e., elemental kinetics. This kinetic expression was confirmed experimentally by the authors in both cases.
Contrary to this behaviour, it was observed that the acetalization reaction between glycerol and acetaldehyde behaves as an irreversible reaction reaching 100% of conversion at different temperatures (Figure 4). In this case the kinetics is only function of the acetaldehyde concentration. The great advantage of this reaction is that, in principle, the industrial production of these acetals to be used as biodiesel additives or for other applications can be carried out in conventional reaction systems.

$$\frac{d[Ac]}{dt} = wk[AcHO] \quad (4)$$

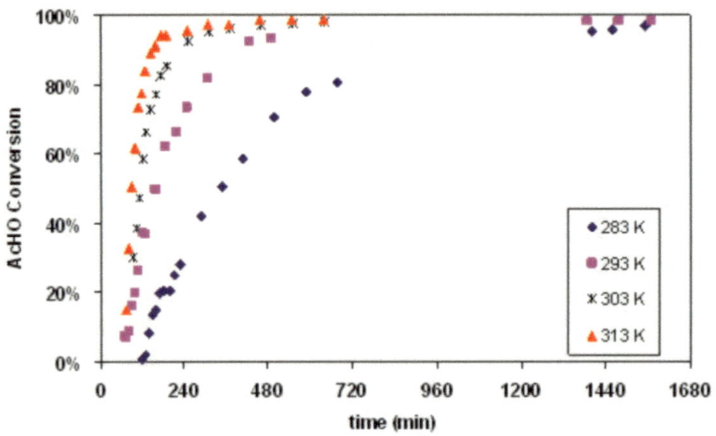

Fig. 4. Effect of the temperature on the acetalization reaction. Batch reactor, (700 rpm, feed ratio 3:1, 2 wt% Amberlyst 47) (Agirre et al. c)

2.3 Solid catalyst for acetal production

Acetals are produced via homogeneous catalytic processes using as catalyst strong mineral acids such as H_2SO_4, HF, HCl or p-toluensuphonic acid (Frusteri et al., 2007; Green, 1981;

Kaufhold & El-Chahawi, 1996). However, these processes entail corrosion problems, are uneconomical and they are not environmentally friendly. The use of a heterogeneous catalytic process would overcome all these problems so, nowadays, several solid acid catalysts are being tested.

One of the first heterogeneous catalytic process for acetal production was described by Andrade et al. (Andrade et al., 1986) in 1986. In this patent an acetal production process from saturated or unsaturated aldehydes and alcohols using strongly acidic ion exchange resins or zeolites is explained. The reaction takes place in the liquid phase and after removal of the catalyst the conversion mixture is extracted by means of water and by means of water insoluble organic solvents. This process is valid for certain alcohols and aldehydes:

An aldehyde of the formula:

$$R_1\text{-CHO}$$

Where R_1 is a straight chain alkyl group having 1 to 3 carbon atoms or alkenyl group with 2 or 3 carbon atoms is reacted with an alcohol of the formula:

$$R_2\text{-OH}$$

In which R_2 is an alkyl group of 1 or 2 carbon atoms.

The method of this invention serves particularly well for the preparation of acetals of the formula:

$$R_1\text{-CH-}(OR_2)_2$$

In terms of catalyst requirement it is recommended to use at least 1.0 g of ion exchange resin and 0.5 g of zeolites per mol of aldehyde used.

In order to find some new active, selective and stable solid acid catalysts for acetal production, Capeletti et al. (Capeletti et al., 2000) reported the performance of several solid acid catalysts of various types, from commercial, natural and laboratory sources shown in Table 2. After characterizing all these catalysts and determining their acidity, their catalytic performances were evaluated by means of experiments reacting ethanol and acetaldehyde. As a conclusion, Capeletti et al. proved that A15 ion exchange resin show better performance than other catalysts reaching equilibrium values much faster than with the others. They also observed that water, a reaction product, seems to have an inhibitory effect on the reaction rate (Capeletti et al., 2000; Mahajani, 2000).

In a review made by Sharma (Sharma, 1995) it is explained how ion-exchange resins, particularly the macroporous variety, are suitable catalysts for oligomerization of olefins, cross-dimerization of olefins, acetalization and ketalization reactions…offering high selectivity rates. Resin catalysts can be used in batch or semi-batch reactors as well as in continuous fixed, expanded or fluidized bed reactors. The heterogenized acidity can exceed the value of 100 % H_2SO_4. In Table 3 Hammett acidity function (H_0) for various acids used as catalysts is shown. More recently (Umbarkar et al., 2009) tested mesoporous MoO_3/SiO_2 solid acid catalyst for glycerol acetalization reaction with different aldehydes showing promising results. On the other hand (Ruiz et al., 2010) tested beta zeolites and Amberlyst resins and they compared them to soluble acids such as p-toluenesulfonic acid (PTSA). They found that in absence of water PTSA and Amberlyst resins are more active but in the presence of water hydrophobic Beta zeolites give better results.

However, acetalization reactions offer really low equilibrium conversions (around 50 % depending on the operating conditions) if they are carried out in a conventional batch

Catalyst	Surface area ($m^2\ g^{-1}$)	Pore vol. ($mL\ g^{-1}$)	Acidity ($meq\ g^{-1}$)
A15. Polystyrene-polydivinylbenzene sulphonic resin, Rohm & Haas	45	0.360	4.7
Acid-treated montmorillonite, Aldrich	345	0.564	0.273
Mordenite, Norton	436	0.210	0.649
Acid treated montmorillonite, natural	235	0.262	0.640
Zeolite FCC cat., Fresh BR1160, Engelhard, UCS: 24.72 Å	342	0.259	0.540
Zeolite FCC catalyst, Isoplus 1000, Engelhard, UCS: 24.40 Å	336	n.a.	0.474
Amorphous FCC catalyst, HA-HPV, Ketjen 25 % Al_2O_3	454	0.688	0.382
Amorphous FCC catalyst, LA-LPV, Ketjen 12 % Al_2O_3	559	0.642	0.350
Equilibrium zeolitic FCC catalyst, BR1160, Engelhard, UCS: 24.31 Å	175	0.213	0.065
Equilibrium zeolitic FCC catalyst, Octavision, FCC S.A., UCS: 24.24 Å	151	0.120	0.160

Table 2. Properties of different types of catalyst reported by (Capeletti M.R. et al. 2000).

reactor (Capeletti et al., 2000; Chopade & Sharma, 1997a; Chopade & Sharma, 1997b; Mahajani et al., 1995; Sharma, 1995). In order to enhance the performance of the acetalization reaction, innovative reaction systems are required. According to the literature reactive distillation processes as well as reactors integrating dehydration membranes seem to be the most promising systems (Benedict et al., 2006; Calvar et al., 2007; Chopade & Sharma, 1997a; Chopade & Sharma, 1997b; Domingues et al., 1999; Feng & Huang, 1996; Lim et al., 2002; Sanz et al., 2006a; Sanz et al., 2006b; Sharma 1995; Zhu et al., 1996).

Acid	**H_0**
p-Toluenesulfonic acid	+0.55
Montmorillonite	
Natural	1.5 to -3.0
Cation exchanged	-5.6 to -8.0
Amberlyst 15	-2.2
Sulfuric acid (40 %)	-2.4
Sulfuric acid (100 %)	-12.3
Nafion	-11 to -13
NY Zeolites	-13.6 to -12.7
$H_3PW_{12}O_4$ and $Cs_{2.5}H_{0.5}PW_{12}O_{40}$ (HPA)	-13.16
Lanthanum and cerium exchanged HY zeolites	<-14.5
Fluorosulfonic acid	-15.07
Sulfated zirconia	-16
H_3SO_3F-SbF_5	-20

Table 3. Hammett acidity function (H_0) values. (Sharma 1995)

3. Innovative reaction systems for acetalization reactions

As it is explained in sections 2.2 & 2.3 low equilibrium conversions are obtained for acetalization reactions using conventional batch reactors so innovative reaction systems like reactive distillation or reactors integrating dehydration membranes are required in order to achieve high conversions. In these both systems the reaction products, or at least one of the products, are being removed from the reaction shifting the reaction in the forward direction according to Le Chatelier's law.

3.1 Reactive distillation

Reactive distillation (RD) has become an interesting alternative to some conventional processes, especially for those that present high thermodynamic limitations like the acetalization reaction as well as etherification and esterification reactions. RD combines chemical reaction and thermal separation in the same unit. Thus, the reaction products are being removed from the reaction mixture and thermodynamic limitations can be overcome achieving high conversions. The scheme of a typical RD column is shown in Figure 5 Normally, the catalytic section is placed in the middle of the column having two different feed streams, one just above of the catalytic section and the second one just below the catalytic section. However, not always this configuration is the most optimum one.
Sharma and Chopade (Chopade & Sharma, 1997a; Chopade & Sharma, 1997b), Dhale et al. (Dhale et al., 2004) and Agirre et al. (Agirre et al., 2011a) used RD columns for acetalization reactions showing different results.

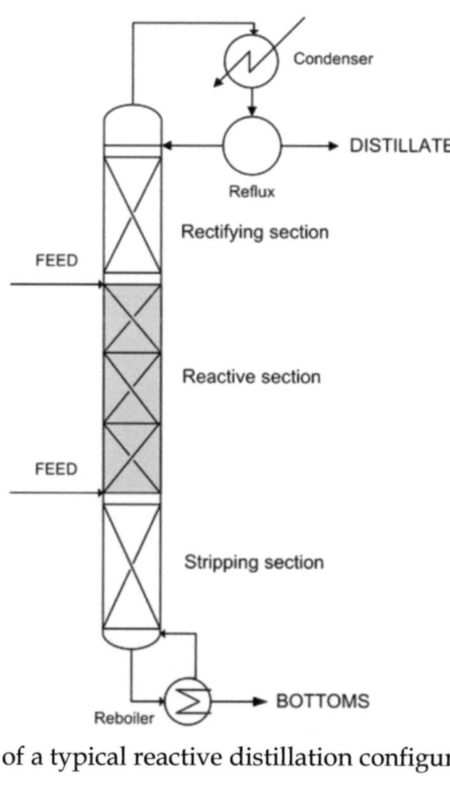

Fig. 5. Schematic diagram of a typical reactive distillation configuration.

Sharma and Chopade studied the reaction between ethanol and ethylen glycol with formaldehyde, respectively. They performed both, batch experiments and RD experiments following the proposed scheme in Figure 5 using two different types of macroporous cation-exchange resins, Indion 130 and Amberlyst 15. Conversions around 45% were achieved in the batch reactor for the first reaction carried out at 333K, 343K and 353K and they observed that the temperature did not have any significant effect on the equilibrium conversion. After that they studied the performance of the reaction in a RD distillation system. Conversions around 86-94% were achieved. At lower reflux ratios higher conversions than operating a high reflux ratios were achieved (at R=0.5 the conversion was 94.0% and at R=4, 86%). In terms of the catalyst the behaviour of Indion 130 and Amberlyst 15 was practically identical.

As indicated, Sharma and Chopade also studied the reaction between ethylene glycol with formaldehyde in a batch reactor and in a RD system. In this case, using an initial feed mol ratio of 1.5:1 ethylen glycol to formaldehyde the final conversion was increased from 42% to 74% using a RD system instead of a conventional batch reactor.

Also Calvar et al. (Calvar et al., 2007) and Klöker et al. (Klöker et al., 2004) showed the benefits of using reactive distillation systems in similar reactions like esterification of acetic acid with ethanol and in the synthesis of ethyl acetate.

In our case, the acetalization reaction between ethanol and butanal was studied. As well as Sharma and Chopade, a previous study was performed in a batch reactor and after that RD experiments were performed. Batch experiments allowed gathering kinetic data of the reaction and also observing the achievable conversions which were between 40% (at 333K) and 50% (293K) (Agirre et al., 2010). Reactive distillation experiments were not as satisfactory as in the previous cases. The maximum achieved conversion was 50% (Agirre et al., 2011a) while at this conditions in a batch reactor the equilibrium conversion would be 39%. Contrary to Sharma and Chopade, increasing the reflux ratio higher conversions were achieved as well as feeding a mixture of both reactants (ethanol and butanal) from the top part of the catalytic section.

All these experiments were performed using Amberlyst 47 ion exchange resin. This resin is equal to Amberlyst 15 resin, the most ussed catalyst in this kind of reactions, but it offers a better mechanical resistance. Another important issue is the way of placement of the resin. Usually the particle size is in the range of 0.5-3 mm so, a simple catalytic bed would offer unacceptable pressure drops. Taylor & Krishna (Taylor & Krishna, 2000) summarizes in a review the most common configurations for the reactive section of a reactive distillation process. The use of structured packing seems to be one of the most suitable options due to its low pressure drop and its high throughput (Klöker et al., 2004). However, when compared to conventional non-reactive structured packings, the specific surface area is moderate. This type of packing presents a really good radial distribution of the liquid phase. Besides, when the catalyst is spent and the columns must be shut down, the packing can be easily removed and replaced by another module. Within the structured packing KATAPAK modules are one of the most common ones.

It can be observed that thermodynamic limitations could be overcome somehow but not significantly. In this case it seems that there are some volatility constraints in order to achieve higher conversions. Taylor and Krishna (Taylor & Krishna, 2000) indicate in an extensive review article the advantages and also the disadvantages or constrains that a reactive distillation systems can offer being one of the constrains the volatility difference between reactants and products.

3.2 Membrane reactors

The use of dehydration membrane reactors is another promising alternative for acetal production. However, no publications were found in this topic apart from the one we published (Agirre et al., 2011b). The continuous water removal from the reaction mixture shifts the reaction to the product side obtaining higher conversions that the achievable ones in conventional reactors.

$$Alcohol\ +\ Aldehyde\quad \leftrightarrow\quad Acetal\ +\ Water$$

One of the first pervaporation processes studied using a dehydration membrane was the removal of water from ethanol-water mixtures. One of the advantages of doing this separation by using a pervaporation membrane is that complex distillation that is required to break the azeotrope can be avoided. Through the integration of distillation and a membrane step, high separation yields at relatively low capital and operational costs can be achieved (Steinigeweg et al., 2003). In the recent years several dehydration membranes as well as membrane processes for the production of ethers and esters were developed (Lee et al., 2006; Lee et al., 1997; Peters et al., 2007; Sanz & Gmehling, 2006a; Sanz & Gmehling, 2006b).

Sanchez Marcano and Tsotsis (Sanchez Marcano et al., 2002) were among the first to describe the advantages of a membrane reactor for the applications. The continuous removal of water from the reaction mixture through the application of a pervaporation membrane shifts the reaction to the product side and thus increases the yield (Benedict et al., 2003; Domingues et al., 1999; Feng & Huang, 1996; Lim et al., 2002; Sanz & Gmehling, 2006a; Sanz & Gmehling, 2006b; Zhu et al., 1996).

In pervaporation systems, the reaction and the separation can be carried out following different configurations:

1. Passive membrane in recycle loop: it is the most common option. The reaction takes place in a conventional reactor and then the desired or undesired product is separated in a membrane module (Benedict et al., 2003; Benedict et al., 2006; Domingues et al., 1999; Sanz & Gmehling, 2006a; Sanz & Gmehling, 2006b).
2. Passive membrane in reactor: the reaction and separation are carried out in the same unit using non-catalytic membranes and keeping catalyst particles as slurry in the reaction media (Agirre et al. 2011b).
3. Active membrane in reactor: when the reaction and separation takes place in the same unit using catalytically active membranes (Bagnell et al., 1993; Peters et al., 2005; Peters et al. 2007).

With regards to the level of complexity, the first configuration is the simplest one and the last one is the most complex one showing some limitations. On the one hand there is one degree of freedom less than using non-catalytic or inert membranes since the catalyst amount - membrane area ratio uses to be fixed. On the other hand, if the catalyst is deactivated the whole membrane must be replaced and vice versa, if the selective separation layer is damaged for a certain reason it must be replaced including the catalyst. These aspects could represent really big inconveniences at industrial scale. Moreover, due to this complexity, the time to market of catalytically active membranes will be longer than the non-catalytic membrane ones. In terms of the second configuration, passive membranes in reactor, the main issues are to have impact resistant membranes in case of slurry reactors and the design constraints that the presence of catalyst particles implies in case of multi-tube membranes.

In 1993 Bagnell et al. (Bagnell et al., 1993) studying the esterification of acetic acid with methanol and n-butanol, concluded that catalytically active membranes show higher permselectivities for water at the same or higher flux, compared to when no reaction was taking place within the membrane phase.

In 2005 Peters et al. (Peters et al., 2005) developed a zeolite-coated pervaporation membrane depositing zeolite H-USY layers on a silica membrane by dip-coating using TEOS and Ludox AS-40 as binder material. This membrane was tested in the esterification reaction between acetic acid and butanol. The catalytic activity of the membrane was comparable to the activity of the bulk zeolite catalyst. However, the performance of the system could be improved using a more active catalyst.

On the other hand, other authors have studied pervaporation processes apart from the reaction unit achieving also high efficiencies. Domingues et al. (Domingues et al., 1999) studied the esterification of benzyl alcohol with acetic acid achieving 96 % separation efficiency in water and a reaction conversion of 99 %. Benedict et al. (Benedict et al., 2006) studied the esterification of lactic and succinic acids with ethanol using a pervaporation unit. Removing water from the reaction mixture, they obtained reaction conversions very close to 1. High water fluxes through the membrane were obtained maintaining high recirculation rates and low permeate pressures. Sanz and Gmehling (Sanz & Gmehling, 2006a; Sanz & Gmehling, 2006b) studied the esterification of acetic acid with isopropanol removing the water formed using a pervaporation membrane. Also in this case, conversions above 90 % were achieved.

Another important classification concerns the membrane material, with the two main classes being ceramic and polymeric. In esterification processes, where the pervaporation unit is not integrated in the reaction unit, most of the articles report the use of commercial polymeric dehydration membranes (Benedict et al., 2006; Domingues et al., 1999; Sanz & Gmehling, 2006a; Sanz & Gmehling, 2006b), whereas in those processes where the reaction and the separation are combined in one single reactor polymeric, ceramic and polymeric/ceramic membranes are applied (Bagnell et al., 1993; Bernal et al., 2002; Peters et al., 2005; Peters et al., 2007; Zhu et al., 1996).

As an example, the acetalization of ethanol with butanal was carried out using the second configuration, i.e., placing a passive membrane in a reactor (Agirre et al., 2011b) showing a good behavior. In this work a hybrid silica membrane (HybSi®, developed at ECN) for dehydration was used. This HybSi® membrane has a very high hydrothermal stability and can be used in the dehydration of various organics (ECN, 2010; Castricum et al., 2008a; Castricum et al., 2008b; Castricum et al., 2008c; Kreiter et al., 2009). In the current research the high stability of this membrane under the presence of aggressive organic solvents, like the aldehyde butyraldehyde, and catalyst impacts was confirmed. Thus, by integrating the chemical reaction and the dehydration membrane in one single reactor the equilibrium conversion could be increased from 40% to 70% at 70 °C (see Figure 6) resulting dehydration membrane reactors more promising than RD systems for the acetalization reaction between ethanol and butanal.

There are not too many publications on the development of continuous membrane reactors. Zhu et al. (Zhu et al., 1996) performed continuous pervaporation experiments in a tubular

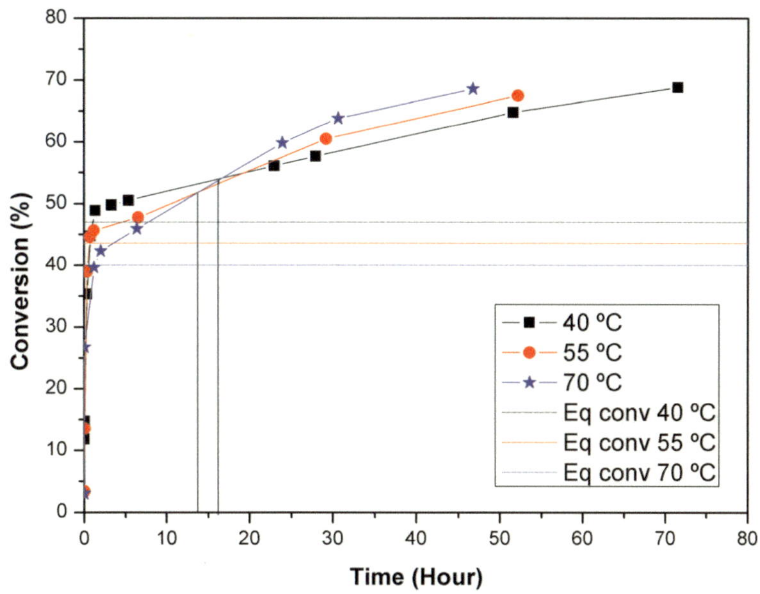

Fig. 6. Effect of the temperature and time on conversion. Conditions: ratio EtOH/Butyraldehyde 2:1 in moles, catalyst loading 0.5 wt%. (Agirre et al. 2011b)

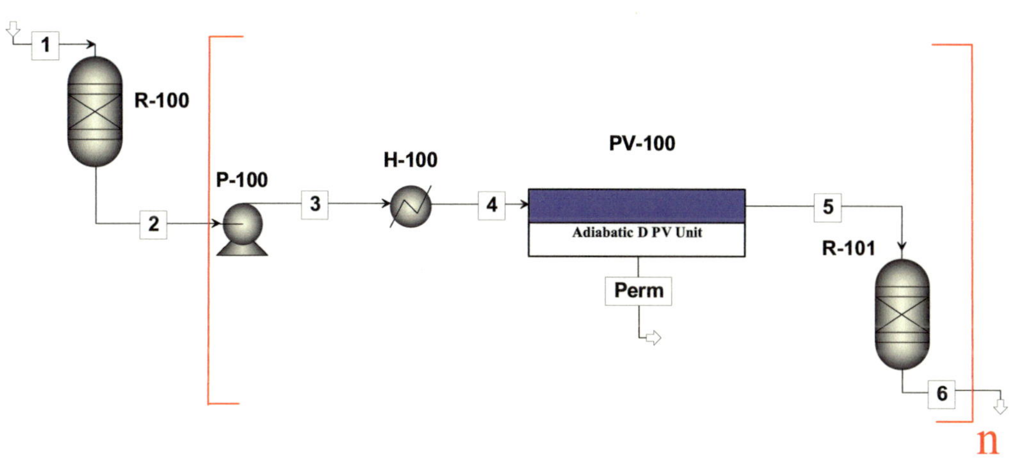

Fig. 7. Flow sheet diagram of PFR and PV modules placed in series.

pervaporation membrane reactor as well as modelling of esterification reactions using H_2SO_4 as homogenous catalyst. De la Iglesia et al. (de la Iglesia et al., 2007) also performed esterification reactions experiments in a continuous tubular reactor. In this case Amberlyst 15 was used as catalyst and it was placed inside the membrane. Lim et al. (Lim et al., 2002) studied different process configurations, and they concluded that tubular membrane reactors lead to a better performance than stirred tank membrane reactors. Nemec et al. (Nemec et al., 2005) analyzed multifunctional tubular reactors with the catalyst particles in the annular region between the membrane and the module shell; their results were not very satisfactory. In the case study showed in Figure 6 it can be observed that equilibrium conversion is achieved in few hours and then the pervaporation becomes the rate-determining step. Therefore, uncoupled reaction and pervaporation systems seem to be the most promising ones (Figure 7) in order to adjust both the reaction rate and the pervaporation rate.

4. Conclusions

An overview on acetals production has been presented showing different kind of acetals (straight chain and cyclic ones) and different studies in order to choose the most suitable catalyst. On the other hand, due to the high thermodynamic limitations that these reactions show, two different alternative innovative reaction systems have been presented. Depends on the reaction one alternative or the other one (or both) may be more suitable.

5. References

Agirre, I., Barrio, V.L., Güemez, B., Cambra, J.F., & Arias, P.L. (2010). The Development of a Reactive Distillation Process for the Production of 1,1 Diethoxy Butane from Bioalcohol: Kinetic Study and Simulation Model. *International Journal of Chemical Reactor Engineering*, 8, A86.

Agirre, I., Barrio, V.L., Güemez, B., Cambra, J.F., & Arias, P.L. (2011a). Catalytic reactive distillation process development for 1,1 diethoxy butane production from renewable sources. *Bioresource Technology*, 102, 1289-1297.

Agirre, I., Güemez, M.B., van Veen, H.M., Motelica, A., Vente, J.F., & Arias, P.L. (2011b). Acetalization reaction of ethanol with butyraldehyde coupled with pervaporation. Semi-batch pervaporation studies and resistance of HybSi® membranes to catalyst impacts. *Journal of Membrane Science*, 371, (1-2) 179-188.

Agirre, I., Güemez, M.B., Ugarte, A., Requies, J., Barrio, V.L., Cambra, J.F., Arias, P.L. c. Glycerol acetals as diesel additives: kinetic study of the reaction between glycerol and acetaldehyde. *Submitted to Biomass & Bioenergy.*

Aksnes, G. & Albriktsen, P. (1966). Intramolecular H bonding in cyclic acetal isomers of glycerol. *Acta Chem.Scand.*, 20, [5], 1330-1334.

Aksnes, G., Albriktsen, P., & Juvvik, P. (1965). Cyclic acetal and ketal isomers of glycerol. *Acta Chem.Scand.(1947-1973)*, 19, [4], 920-930.

Andrade J., Arntz D., Kraft M., & Prescher G. 1986, *Method for preparation of acetals*, 4579979, Degusa Aktiengesellschaft, (patent).

Bagnell, L., Cavell, K., Hodges, AM., Mau, A., & Seen, AJ. (1993). The use of catalytically active pervaporation membranes in esterification reactions to simultaneously increase product yield, membrane permselectivity and flux. *Journal of Membrane Science*, 85, (3) 291-299.

Benedict, D.J., Parulekar, S.J., & Tsai, S.P. (2003). Esterification of lactic acid and ethanol with/without pervaporation. *Ind.Eng.Chem.Res.*, 42, (11) 2282-2291.

Benedict, DJ., Parulekar, SJ., & Tsai, S. (2006). Pervaporation-assisted esterification of lactic and succinic acids with downstream ester recovery. *Journal of Membrane Science*, 281, (1-2) 435-445.

Bernal, M., Coronas, J., Menεndez, M., & Santamaria, J. (2002). Coupling of reaction and separation at the microscopic level: esterification processes in a H-ZSM-5 membrane reactor. *Chemical Engineering Science*, 57, (9) 1557-1562.

Burtscher, H., Matter, U., & Skillas, G. (1999). The effect of fuel additives on diesel engine particulate emissions. *Journal of Aerosol Science*, 30, (Supplement 1) S851-S852.

Calvar N., González B., & Dominguez A. (2007). Esterification of acetic acid with ethanol: Reaction kinetics and operation in a packed bed reactive distillation column. *Chemical Engineering and Processing*, 46, 1317-1323.

Camara, J. S., Marques, J. C., Alves, A., & Ferreira, A. C. S. (2003). Heterocyclic acetals in Madeira wines. *Anal Bioanal Chem*, 375, [8], 1221-1224.

Capeletti M.R., Balzano L., de la Puente G., Laborde M., & Sedran U (2000). Synthesis of acetal (1,1-diethoxyethane) from ethanol and acetaldehyde over acidic catalysts. *Applied Catalysis A: General*, 198, (1-2) L1-L4.

Castricum, H.L., Kreiter, R., van Veen, H.M., Blank, D.H.A., Vente, J.F., & ten Elshof, J.E. (2008a). High-performance hybrid pervaporation membranes with superior hydrothermal and acid stability. *Journal of Membrane Science*, 324, (1-2) 111-118.

Castricum, H.L., Sah, A., Kreiter, R., Blank, D.H.A., Vente, J.F., & ten Elshof, J.E. (2008b). Hybrid ceramic nanosieves: stabilizing nanopores with organic links. *Chem.Commun.(Cambridge, U.K.)* (9) 1103-1105.

Castricum, H.L., Sah, A., Kreiter, R., Blank, D.H.A., Vente, J.F., & ten Elshof, J.E. (2008c). Hydrothermally stable molecular separation membranes from organically linked silica. *J.Mater.Chem.*, 18, (18) 2150-2158.

Chang B.H. (1995). A facile synthesis of acetals and aldehydes from allylic ethers catalyzed by cobalt compounds. *Journal of Organometallic Chemistry*, 492, (1) 31-34.

Chopade, S.P. & Sharma M.M. (1997a). Reaction of ethanol and formaldehyde: use of versatile cation-exchange resins as catalyst in batch reactors and reactive distillation columns. *Reactive and Functional Polymers*, 32, (1) 53-65.

Chopade, S.P. & Sharma, M.M. (1997b). Acetalization of ethylene glycol with formaldehyde using cation-exchange resins as catalysts: batch versus reactive distillation. *Reactive and Functional Polymers*, 34, (1) 37-45.

da Silva Ferreira Antonio Cesar, Barbe, J., & Bertrand, A. (2002). Heterocyclic acetals from glycerol and acetaldehyde in Port wines: evolution with aging. *J Agric Food Chem*, 50, [9], 2560-2564.

de la Iglesia, O., Mallada, R., Menqndez, M., & Coronas, J. (2007). Continuous zeolite membrane reactor for esterification of ethanol and acetic acid. *Chemical Engineering Journal*, 131, (1-3) 35-39.

Deutsch, J., Martin, A., & Lieske, H. (2007). Investigations on heterogeneously catalysed condensations of glycerol to cyclic acetals. *Journal of Catalysis*, 245, (2) 428-435.

Dhale, A.D., Myrant, L.K., Chopade, S.P., Jackson, J.E., & Miller, D.J. (2004). Propylene glycol and ethylene glycol recovery from aqueous solution via reactive distillation. *Chemical Engineering Science*, 59, (14) 2881-2890.

Domingues, L., Recasens, F., & Larrayoz, M.A. (1999). Studies of a pervaporation reactor: Kinetics and equilibrium shift in benzyl alcohol acetylation. *Chemical Engineering Science*, 54, (10) 1461-1465.

ECN information available from URL: http://www.hybsi.com/

Feng, X. & Huang, R.Y.M. (1996). Studies of a membrane reactor: Esterification facilitated by pervaporation. *Chemical Engineering Science*, 51, (20) 4673-4679.

Filley J. (2005). New lubricants from vegetable oil: cyclic acetals of methyl 9,10-dihydroxystearate. *Bioresource Technology*, 96, (5) 551-555.

Frusteri, F., Spadaro, L., Beatrice, C., & Guido, C. (2007). Oxygenated additives production for diesel engine emission improvement. *Chem. Eng. J.*, 134, (1-3) 239-245.

Green T.W. 1981. *Protective Groups in Organic Synthesis.* Willey, ISBN: 0471160199, New York.

IEA 2007, *World Energy Outlook 2007.*

IEA 2008, *Emissions of Greenhouse Gases in the United States 2008.*

J.G.Sanchez Marcano & Th.T.Tsotsis 2002. *Catalytic Membrane and Membrane Reactors* Wiley, ISBN:3-527-30277-8, Weinheim.

Kaufhold M. & El-Chahawi M. 1996, *Process for preparing acetaldehyde diethyl acetal*, 5527969, Huels Aktiengesellschaft, (patent).

Keskin, A., Gürü, M., & AltIparmak, D. (2008). Influence of tall oil biodiesel with Mg and Mo based fuel additives on diesel engine performance and emission. *Bioresource Technology*, 99, (14) 6434-6438.

Kim, H. & Choi, B. (2008). Effect of ethanol-diesel blend fuels on emission and particle size distribution in a common-rail direct injection diesel engine with warm-up catalytic converter. *Renewable Energy*, 33, (10) 2222-2228.

Klöker M., Kenig E.Y., Górak A., Markusse A.P., Kwant G., & Moritz P. (2004). Investigation of different column configurations for the ethyl acetate synthesis vio reactive distillation. *Chemical Engineering and Processing*, 43, 791-801.

Kreiter, R., Rietkerk Marielle, D.A., Castricum, H.L., van Veen, H.M., ten Elshof, J.E., & Vente, J.F. (2009). Stable hybrid silica nanosieve membranes for the dehydration of lower alcohols. *ChemSusChem*, 2, (2) 158-160.

Kumar R. & Chakraborti A.K. (2005). Copper(II) tetrafluoroborate as a novel and highly efficient catalyst for acetal formation. *Tetrahedron Letters*, 46, (48) 8319-8323.

Lee, K.H., Youn, M.Y., & Sea, B. (2006). Preparation of hydrophilic ceramic membranes for a dehydration membrane reactor. *Desalination*, 191, (1-3) 296-302.

Lee, Y., Nam, S., & Woo, D. (1997). Pervaporation of ionically surface crosslinked chitosan composite membranes for water-alcohol mixtures. *Journal of Membrane Science*, 133, (1) 103-110.

Li, D.g., Zhen, H., Xingcai, L., Wu-gao, Z., & Jian-guang, Y. (2005). Physico-chemical properties of ethanol-diesel blend fuel and its effect on performance and emissions of diesel engines. *Renewable Energy*, 30, (6) 967-976.

Lim, Y., Park, B., Hung, F., Sahimi, M., & Tsotsis, T. (2002). Design issues of pervaporation membrane reactors for esterification. *Chemical Engineering Science*, 57, (22-23) 4933-4946.

Mahajani S.M. (2000). Reactions of glyoxylic acid with aliphatic alcohols using cationic exchange resins as catalysts. *Reactive and Functional Polymers*, 43, (3) 253-268.

Mahajani S.M., Kolah A.K., & Sharma M.M. (1995). Extractive reactions with cationic exchange resins as catalysts (acetalization of aldehydes with alcohols). *Reactive and Functional Polymers*, 28, (1) 29-38.

Moser, B.R. & Erhan, S.Z. (2008). Branched chain derivatives of alkyl oleates: Tribological, rheological, oxidation, and low temperature properties. *Fuel*, 87, (10-11) 2253-2257.

Nemec, D. & Van Gemert, R. (2005). Performing Esterification Reactions by Combining Heterogeneous Catalysis and Pervaporation in a Batch Process. *Ind.Eng.Chem.Res.*, 44, (25) 9718-9726.

Peters, T.A., van der Tuin, J., Houssin, C., Vorstman, M.A.G., Benes, N.E., Vroon, Z.A.E.P., Holmen, A., & Keurentjes, J.T.F. (2005). Preparation of zeolite-coated pervaporation membranes for the integration of reaction and separation. *Catalysis Today*, 104, (2-4) 288-295.

Peters, T.A., Benes, N.E., & Keurentjes, J.T.F. (2007). Preparation of Amberlyst-coated pervaporation membranes and their application in the esterification of acetic acid and butanol. *Applied Catalysis A: General*, 317, (1) 113-119.

Ruiz, V.R., Velty, A., Santos, L.L., Leyva-PΘrez, A., Sabater, M.J., Iborra, S., & Corma, A. (2010). Gold catalysts and solid catalysts for biomass transformations: Valorization of glycerol and glycerol-water mixtures through formation of cyclic acetals. *Journal of Catalysis*, 271, (2) 351-357.

Sanz, M.T. & Gmehling, J. (2006a). Esterification of acetic acid with isopropanol coupled with pervaporation: Part I: Kinetics and pervaporation studies. *Chemical Engineering Journal*, 123, (1-2) 1-8.

Sanz, M.T. & Gmehling, J. (2006b). Esterification of acetic acid with isopropanol coupled with pervaporation: Part II. Study of a pervaporation reactor. *Chemical Engineering Journal*, 123, (1-2) 9-14.

Sharma, M.M. (1995). Some novel aspects of cationic ion-exchange resins as catalysts. *Reactive and Functional Polymers*, 26, (1-3) 3-23.

Silva, P.H.R., Gonçalves, V.L.C., & Mota, C.J.A. (2010). Glycerol acetals as anti-freezing additives for biodiesel. *Bioresource Technology*, 101, (15) 6225-6229.

Stefanovic, D. & Petrovic, D. (1967). Structure of glycerol acetals. *Tetrahedron Lett.*, [33], 3153-3159.

Steinigeweg, S. & Gmehling, J. (2003). Transesterification processes by combination of reactive distillation and pervaporation. *Chem.Eng.Process.*, 43, [3], 447-456.

Taylor R. & Krishna R. (2000). Modelling reactive distillation. *Chemical Engineering Science*, 55, 5183-5229.

Umbarkar, S.B., Kotbagi, T.V., Biradar, A.V., Pasricha, R., Chanale, J., Dongare, M.K., Mamede, A.S., Lancelot, C., & Payen, E. (2009). Acetalization of glycerol using mesoporous MoO3/SiO2 solid acid catalyst. *Journal of Molecular Catalysis A: Chemical*, 310, (1-2) 150-158.

Yang, S., Wang, M., Liang, Y., & Sun, J. (2006). Eco-friendly one-pot synthesis of acetals and ketals by heterogeneously catalyzed liquid-solid phase reaction. *Rare Met.(Beijing, China)*, 25, [6], 625-629.

Zhu, Y., Minet, R.G., & Tsotsis, T.T. (1996). A continuous pervaporation membrane reactor for the study of esterification reactions using a composite polymeric/ceramic membrane. *Chemical Engineering Science*, 51, (17) 4103-4113.

16

Feedstocks for Second-Generation Biodiesel: Microalgae's Biology and Oil Composition

Patricia I. Leonardi, Cecilia A. Popovich and M. Cecilia Damiani
Laboratorio de Estudios Básicos y Biotecnológicos en Algas y Hongos, Centro de Recursos Naturales Renovables de la Zona Semiárida (CERZOS) - CONICET, Camino La Carrindanga, Bahía Blanca; Departamento de Biología, Bioquímica y Farmacia, Universidad Nacional del Sur, Bahía Blanca; Argentina

1. Introduction

The solar energy is an inexhaustible source, while other energy reserves, like fossil and nuclear fuels, are limited in quantity and are depleted as years go by. Renewable energy is necessary to replace petroleum-derived fuels. The first generation biofuels, which are produced from oil seeds and crops, are a possible alternative, but they are limited in their capacity to provide all the energy demanded in the world. Therefore, new sources for the sustainable production of renewable energy are being looked for. This concern has promoted the keen interest in developing second generation biofuels, which are produced from other feedstocks, such as microalgal oils (Schenk et al., 2008; Mata et al., 2010). Some microalgal species are capable of producing biomass yields containing high percentages of oils (Aaronson et. al., 1980). In addition, microalgal systems can use low value natural resources, such as arid lands and saline water, thus offering the potential for large biomass energy contributions without competing for prime agricultural or forest land. Most microalgae grow photoautotrophically by using solar energy and mainly carbon dioxide as carbon source. Alternatively, some species can grow heterotrophically or mixotrophically using organic compounds as energy and carbon sources (Kitano et al., 1997; Hu & Gao, 2003; Xu et al., 2006; Liang et al., 2009).

Some microalgae are called oleaginous because they synthesize and accumulate substantial amounts of neutral lipids, mainly as triacylglycerol (TAG), under diverse stress conditions (Bigogno et al., 2002; Hu et al., 2008; Gardner et al., 2010; Damiani et al., 2010). TAGs as storage lipids are the best substrate to produce biodiesel (Xu et al., 2006; Schenk et al., 2008). This biofuel is obtained by transesterification of oil or fat with a monohydric alcohol, yielding the corresponding mono-alkyl esters (Knothe, 2005). Since transesterification maintains the relative ratio of fatty acids present in the feedstock (Costa Neto et al., 2000), the profile of the fatty acid methyl esters is a reflection of the feedstock fatty-acid composition (Lang et al. 2001; Ferrari et al. 2005). Biodiesel production from microalgae is technically feasible (Xu et al., 2006; Patil et al., 2008; Francisco et al., 2010), but for an effective use of this renewable resource as biofuel, it is necessary to be able to modify microalgal growth conditions in order to obtain high biomass productivity and the desired lipid quantity and quality. Those interested in microalgal biomass production and lipid productivity are referred to recent reviews by Griffiths & Harrison (2009), Rodolfi et al. (2009) and Pruvost et al. (2011). In addition, it is important to have information about the various fatty-acid profiles of diverse microalga oils in order to evaluate their suitability as feedstocks for fuel-conversion processes (Gouveia & Oliveira, 2009; Damiani et al., 2010; Sobczuk & Chisti, 2010; Francisco et al., 2010; Popovich et al., 2011). Unlike land plants, oils of some microalga species have a significant amount of polyunsaturated fatty acids with four and more double bonds (Belarbi et al., 2000; Harwood & Guschina, 2009), which are valuable oils. This is a feature that limits the microalgal species that may be used for biodiesel production.

This chapter aims to provide an overview of the current status of research on microalgal feedstocks as regards biodiesel production. Since there are many species of microalgae with varied biological characteristics and lipid composition, a diversity of approaches for biodiesel production have been analysed. In this review the following relevant topics will be considered: 1) diagnostic characteristics of some microalgal main groups, such as Chlorophyceae, Eustigmatophyceae and Bacillariophyceae classes 2) triggering of lipid production, and 3) oil composition, i.e. lipid fractions, content of each lipid class and fatty acid composition of each fraction. In this context, how the latter might affect the biodiesel quality will be discussed. We hope this information provides a framework for future screening of oleaginous microalgae employed as feedstock for biofuel production.

2. Diagnostic characteristics of some main microalgal groups

In recent years plenty of research has been focused on promising microalgal species aiming at the development of sustainable, commercially feasible and economic processes for biodiesel production (Rodolfi et al., 2009; Mandal & Mallick, 2009; Mata et al., 2010). The first step for these studies includes the species selection, essential for a reliable analysis. This step requires knowledge of diagnostic characteristics of different microalgal groups to achieve correct species identification. Thus, contradictory or erroneous information about fatty acid profiles and other important features as reported by Zhukova & Aizdaicher (1995) and Goldberg & Boussiba (2011) can be avoided.

Algae are (with numerous exceptions) aquatic organisms that (with frequent exceptions) are photosynthetic and oxygenic autotrophs. They are typically smaller, except for the seaweeds, and less structurally complex than land plants (Graham & Wilcox, 2000). Microscopic algae are commonly named microalgae that live as solitary cells or as colonies. They vary a great deal with respect to their cell sizes, pigments, storage products, cell wall compositions and life cycles (van den Hoek, 1995). This highly specialised group of micro-organisms has the potential to adapt to diverse habitats as well as the ability to efficiently modify its lipid metabolism in response to changes in environmental conditions (Guschina & Harwood, 2006). Oleaginous microalgae can be found in diverse taxonomic groups and their total lipid content and fatty acid composition may vary noticeably among individual species or strains within and between taxonomic groups (Hu et al., 2008).

According to van den Hoek (1995) algae can be classified in ten major algal groups (Divisions). A number of characteristics has been traditionally used to distinguish these algal groups. The most prominent features are the types of photosynthetic pigments, storage reserves and the nature of the cell covering. One of the greatest groups is represented by Chlorophyta Division, commonly known as green algae because they look bright grass green. This colour is because the chlorophylls are usually unmasked by large amounts of accessory pigments. However, chlorophytes may not always have green colouring. Widely encountered examples include the flagellates *Haematococcus* and *Dunaliella,* whose deep red to purple colouring is due to astaxanthin and ß-carotene pigments, respectively (Borowitzka & Borowitzka 1988; Boussiba & Vonshak, 1991; Ben-Amotz, 1995). Features that are common to nearly all green algae include: chloroplasts enclosed by a double membrane, chlorophylls a and b, and starch storage (α-1,4-linked polyglucans) inside the chloroplasts. The light-harvesting systems of green algae resemble those of land plants. Hence, they are relatively well characterized (Larkum & Howe, 1997). In addition to chlorophylls and proteins, light-harvesting complexes also include carotenoids (Demming-Adams & Adams, 1992).

Ultrastructural and molecular evidences obtained within the past few decades have demonstrated the existence of several distinct green algal lineages (classes). Each lineage is characterized by specific differences in cellular features and primary habitat (Graham & Wilcox, 2000).

In Chlorophyta Division, Chlorophyceae Class represents the largest taxonomic group where oleaginous candidates have been identified (Hu et al., 2008). The Chlorophyceae Class includes some very familiar green algal genera. For example, *Chlamydomonas* is an important laboratory model system, while *Dunaliella, Haematococcus* and *Chlorella* can be valuable in production of industrially useful products (Borowitzka, 1992, 1995, 1997; Spolaore et al., 2006). Members of this class may occur as flagellate or non-flagellate unicells, either as individuals or colonies. Flagellate organisms inhabit fresh (or in a few cases, brackish or marine) waters. Non-flagellate forms occur in freshwaters or on soils. The genera differ in their types of asexual reproductive cells, i.e. formation of one or more zoospores, aplanospores or autospores within individual parental vegetative cells (Bold & Wynne, 1985). The chemical composition of cell walls varies greatly within the class. However, the cell wall of the most oleaginous species, like *Chlorella, Scenedesmus* and *Haematococcus'* cysts, consists of fibrillar polysaccharides and an outer coat of algaenan substance (Pickett-Heaps, 1975; Allard & Templier, 2000; Damiani et al., 2006). Algaenan walls are considered to be the single most decay-resistant biopolymer (Gelin et al., 1997), together with land plants' sporopollenin. It is noteworthy that this biopolymer's high resistance hinders oil extraction in these microalgae. Oleaginous green microalgae vary widely in cell sizes, ranging from 3 to 75 μm.

Heterokontophyta Division includes nine classes (van den Hoek, 1995). In this review some oleaginous species included in Eustigmatophyceae and Bacillariophyceae classes will be reviewed. The chloroplasts of all members of Heterokontophyta are enclosed by four membranes instead of two, as found in green algae and land plant chloroplasts (van den Hoek, 1995; Bozarth et al., 2009).

The Eustigmatophyceae Class includes small (2-32 μm) unicellular, coccoid microalgae. Cells have one or more yellow-green chloroplasts that only contain chlorophyll a. Violaxanthin is the major accessory pigment, which is also the main pigment involved in light harvesting (Whittle & Casselton, 1975). The storage product's chemical structure is unknown. Polysaccharide walls were indicated by van den Hoek (1995). Non-hydrolysable macromolecular constituents, i.e. algaenans, were also isolated from two species (Gelin et al., 1996, 1997). Asexual reproduction occurs by autospores production or in some cases by zoospores. Because of the similarities in morphology, reproduction, cell colour and chloroplast structure, eustigmatophyceans are commonly mistaken for coccoid green microalgae at the light-microscopy level. Its identification requires cell examination by transmission electron microscopy and/or pigment analysis by chromatography (Graham & Wilcox, 2000). There are about seven genera, most occurring in freshwater or in soil, but there are also some marine forms (van den Hoek, 1995). The oleaginous microalga *Nannochloropsis* is a marine coccoid form that resembles *Chlorella* (Santos & Leedale, 1995). This genus does not produce zoospores.

The Bacillariophyceae Class includes diatoms. They occur only as single cells or chains of cells. These microalgae are ubiquitous, occurring in marine and freshwaters, where they may be principally planktonic (they live suspended or growing in a fluid environment) or benthic (they live in the lowest level of a water body, often attached to the substrate bottom). There are diatoms in an immense variety of shapes. Circular, triangular, and modified square shapes are common. These diatoms are known as centric. Other diatoms, especially benthic forms, display varying types of bilateral symmetry and are termed pennate diatoms. Cell sizes range from less than 15 µm to 1 mm in length. Some species have been indicated as oleaginous microalgae (McGinnis, et al., 1997; Hu et al., 2008; Matsumoto et al., 2010; Yu et al., 2009; Popovich et al., 2011). The chloroplasts are usually golden-brown, because the chlorophylls a and c are masked by the accessory pigment fucoxanthin. The reserve polysaccharide is chrysolaninaran, a ß-1,3 linked glucan that is formed outside the chloroplast. They also store carbon in the form of natural oils (Bozarth et al., 2009). The cell wall is siliceous and is termed frustule (van den Hoek, 1995). Diatoms normally reproduce asexually by cell division. In terms of contributions to global primary productivity, diatoms are among the most important aquatic photosynthesizers. They dominate the phytoplankton of the oceans and recently circulated in lake waters.

3. Triggering of lipid production and oil composition

As it is usual in photosynthetic cells, microalgae contain polar and neutral lipids. Polar lipids include glycolipids and phospholipids. The glycolipids (monogalactosyl-, digalactosyl- and sulphoquinovosyldiacylglycerol) and phosphatidylglycerol have been attributed to chloroplast membranes, while the phospholipids (phosphatidylcholine and phosphatidylethanolamine) are considered more characteristic of extrachloroplastic membranes. Neutral lipid fraction is very diverse and compounds as different as sterols, free-fatty acids and acyl lipids (mono-, di- and triacylglycerols) can be found (Harwoord & Jones, 1989; Berge, et al., 1995). TAG accumulation specifically occurs in oil droplets distributed in the cytoplasm. Nile red fluorescence (Figs 1- 2) is a technique that has been used in some microalgae as a rapid screening method to detect TAG presence (Damiani et al. 2010; Popovich et al., 2011), as well as to determine the relative neutral lipid content (McGinnis et al., 1997; Yu et al., 2009).

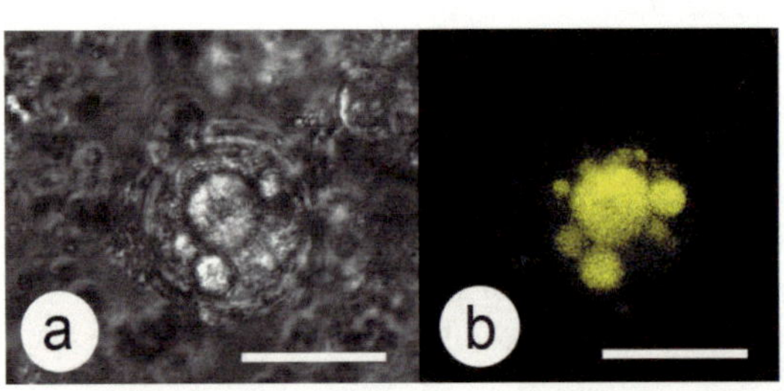

Fig. 1. Light micrographs of a *Haematococcus pluvialis* cyst. (a) Phase contrast microscopy. (b) Epifluorescent microscopy. Numerous yellow-gold neutral lipid droplets under stress condition (high light intensity) after 11-day growth are shown. Scale bars = 20 µm. Source: Damiani et al. (2010).

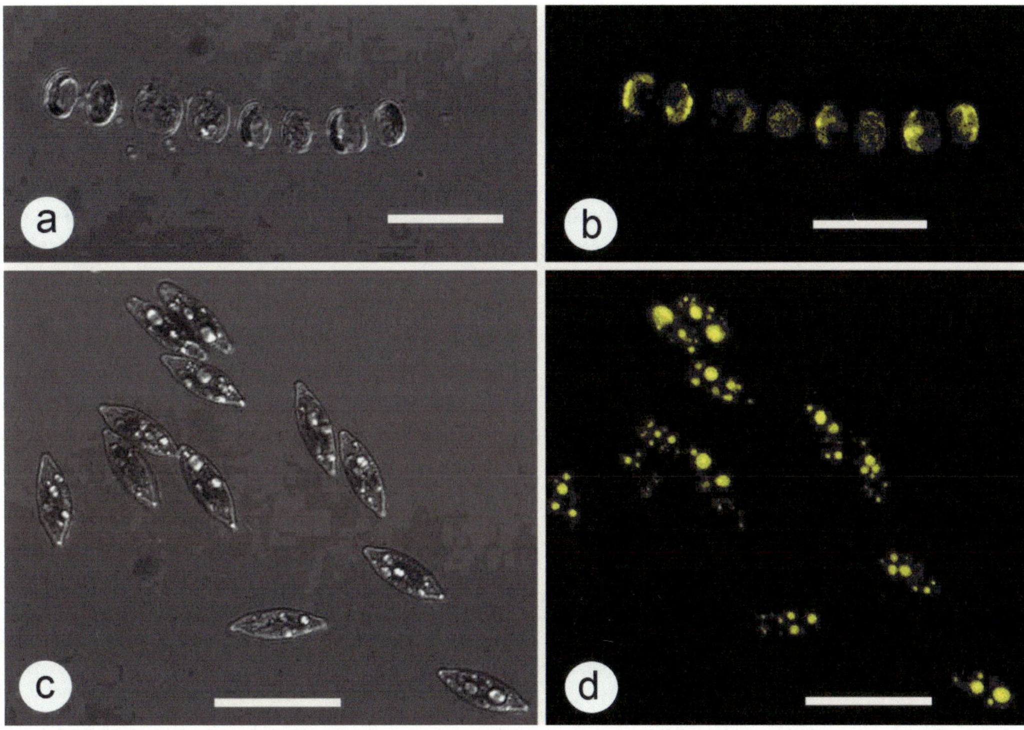

Fig. 2. Light micrographs of *Skeletonema costatum* (a, b) and *Navicula gregaria* (c, d). Phase contrast microscopy (a, c). Epifluorescent microscopy (b, d). Yellow-gold neutral lipid droplets after 15-day growth are shown. Lipid droplets are in a marginal position in *S. costatum*, while multiple lipid droplets homogenously distributed in the cytoplasm appear in *N. gregaria*. Scale bars = 25 µm. Source: Popovich et al. (2011).

Even though it is known that the intrinsic ability to produce large quantities of lipids is species- and strain specific (Hu et al., 2008), the energy storage capacity can be maximized by controlling the organism's metabolism. Tuning the microalgae's metabolism can lead to enhanced production of energy-rich compounds, such as fatty acids and glycerol. A single microalgal species may show remarkable variation in its metabolism, according to the conditions to which it is exposed, such as carbon dioxide supply, light intensity, temperature, nutrient concentrations and salinity (Shifrin & Chisholm, 1981; Roessler, 1990). Synthesis and accumulation of large amounts of TAGs accompanied by considerable alterations in lipid and fatty acid composition occur in the cell when oleaginous microalgae are placed under stress conditions imposed by chemical or physical environmental stimuli (Hu et al., 2008). Nutrient starvation (as nitrogen, phosphorus, and silicate), salinity and growth-medium pH are the major chemical stimuli employed, whereas light intensity and temperature are frequently used as physical stimuli (Hu et al., 2008). In addition, it is well known that certain levels of carbon dioxide supplementation in microalgal cultures can increase lipid content, especially TAG fraction (Gordillo et al., 1998; Huntley & Redalje, 2007; Tang et al., 2010; Francisco et al., 2010). Besides, the lipid content and fatty acid composition also depend on the age of culture and different life-cycle stages (Siron et al., 1989; López Alonso et al., 2000; Spolaore et al., 2006; Hu et al., 2008).

From a chemical point of view, oils from different sources have different fatty acid compositions. The fatty acids vary in their carbon chain length and in their number of unsaturated bonds. The fatty acids in land plant oils are very well studied and stearic acid (18:0), palmitic acid (16:0), oleic acid (18:1n9c), linoleic acid (18:2n6c) and linolenic acid (18:3n3) are commonly found (Durrett et al., 2008; Singh & Singh, 2010). As regards microalgae, there is a greater diversity of fatty-acid profiles of oils among different classes. Moreover, their information is very limited at present and most of the analyses of fatty acid composition have used total lipid content rather than the examination of individual lipid fractions (Hu et al., 2008). As previously indicated, the fatty acid composition can widely vary both quantitatively and qualitatively with the microalgae's physiological status and the environmental conditions (Hu et al., 2008; Rodolfi et al., 2009), making it difficult to compare microalgal species/strains across experimental conditions (Molina Grima et al., 1994).

Fatty-acid composition was used to predict the quality of fatty acid methyl esters of oils for use as biodiesel (Knothe, 2005). The most important characteristics include the ignition quality (i.e. cetane number), cold-flow properties and oxidative stability. For example, saturated oils produce a biodiesel with superior oxidative stability and a higher cetane number, but rather poor low-temperature properties. On the other hand, the biodiesel produced from feedstocks rich in polyunsaturated fatty acids (PUFAs) has good cold-flow properties. However, these fatty acids are particularly susceptible to oxidation (Knothe, 2005; Hu et al., 2008). Among PUFAs, some fatty acids should be taken into account. European standard EN 14214 limits linolenic acid's methyl ester for vehicle use to 12% and methyl esters with four and more double bonds to a maximum of 1%, (CEN EN-14214, 2003).

3.1 Chlorophyceae class

3.1.1 Triggering of lipid production

Nitrogen is the most commonly reported nutritional-limiting factor that triggers total lipid accumulation, mainly TAG in green microalgae (Hu et al., 2008; Pruvost et al., 2009, 2011). When nitrogen deprivation is imposed upon a culture exposed to suitable irradiances, photosynthesis continues, albeit at a reduced rate, and the flow of fixed carbon is diverted from protein to either lipid or carbohydrate synthesis (Shifrin & Chisholm, 1981). Some oleaginous microalgae seem to have the capacity for synthesizing *de novo* lipids when grown under nitrogen-deficiency, channelling the excess of carbon and energy into storage lipids, mainly TAGs (Shifrin & Chisholm, 1981; Rodolfi et al., 2009). The effect of nitrogen deficiency has been demonstrated a long time ago in numerous chlorophycean (Iwamoto et al., 1955; Fogg, 1959; Zhukova et al., 1969; Thompson, 1996). Moreover, a study of fifteen chlorophycean species by Shifrin & Chisholm (1981) showed that lipid content doubled or tripled when cells from exponential-phase growth were transferred to nitrogen-free conditions. In this way total lipid fractions of 30% to 50% dry weight (dw) were measured. In addition, at low nitrogen level, *Chlorella vulgaris* and *Scenedesmus obliquus* contained high percentage of total lipids (45% of the biomass); more than 70% of these were neutral lipids (Piorreck et al., 1984). Recently, an increase of total lipid content (from 12.7 to 43% dw) was obtained when stationary phase cultures of *S. obliquus* were transferred to media deficient in nitrate for 7 days (Mandal & Mallick, 2009). Higher lipid content values were reported under nitrogen-deficient conditions in *Chlorella emersonii* (63% dw) and *C. minutissima* (56% dw), but no changes were observed in *C. sorokiniana* under the same culture conditions

(Illman et al., 2000). Up to now, lipid contents as high as 85%, as reported earlier for *Chlorella pyrenoidosa* and *C. ellipsoidea* (Spoehr & Milner, 1949; Iwamoto et al., 1955), have been reported in no other chlorophycean species at all. It is important to remark that lipid content and lipid composition also vary according to the exposition time at nitrogen starvation. For example, a long time of nitrogen starvation (17 days) resulted in higher lipids and TAG accumulation (74%) than that obtained after 7 days of starvation in *Chlorella vulgaris*. In addition, lipid composition gradually changed from free fatty acid-rich lipid to TAG-rich lipids (Widjaja et al. 2009).

Chlorella protothecoides is a microalga that can grow photoautotrophically or heterotrophically under different culture conditions. Xu et al. (2006) reported that with the addition of organic carbon source (glucose) of the medium and the decrease of the inorganic nitrogen source, the heterotrophic *C. protothecoides* reached up to 55.2% dw lipid content, which was about four times higher than that in photoautotrophic cells (Miao & Wu, 2004).

In recent years special attention has also been given to *Neochloris oleoabundans* for its ability to accumulate high lipid content, especially TAGs. Nitrogen-starved cultures of this species showed 36-54% dw total lipid content, where more than 80% of these were TAGs (Tornabene et al., 1983). Similarly, 56% dw of lipid content after 6 days of nitrogen starvation was reported in the same species by Gouveia et al. (2009).

Moreover, formation of chloroplastic and extraplastidial lipid bodies containing both TAGs and carotenoids under nitrogen starvation was reported in *Dunaliella bardawil*, *Chlorella zofingiensis* and *Haematococcus pluvialis* (Thompson, 1996; Boussiba, 2000). In the latter species nitrogen starvation induced a sharp increase in TAG content (Zhekisheva et al., 2002).

There are few studies in green microalgae regarding phosphorus deficiency. According to Reitan et al. (1994), phosphorus deprivation results in decreased lipid content in *Nannochloris atomus*. However, phosphorus deprivation leads to a significant increase in the total fatty acid content of *Dunaliella tertiolecta* (Siron et al., 1989). In addition, a significant increase in lipid content (29.5% dw) was also obtained in *Scenedesmus obliquus* when stationary phase cultures were transferred to medium deficient in phosphate for 3 days (Mandal & Mallick, 2009).

On the other hand, Liu et al. (2008) reported that high iron concentration stimulated lipid storage in *Chlorella vulgaris*. Lipid content increased up to 56.6% dw when cells in late-exponential growth phase were re-inoculated into new medium containing 1.2 x 10^{-5} mol L^{-1} iron concentration.

Regarding physical stimuli, it is extensively known that low light intensity induces the formation of polar lipids, whereas high light intensity decreases total polar lipid content with a concomitant increase in the amount of neutral lipids (Hu et al., 2008 and cites therein). In *Haematococcus pluvialis* high light intensity (300 μmol photons m^{-2} s^{-1}) doubled the total lipid content (34.85% dw). In addition, the neutral lipid fraction also increased about 2-fold compared to the control (Table 1) (Damiani et al., 2010).

The effects of temperature on the total lipid content have only been reported for a few species in green microalgae; though a general trend cannot be established. A decrease in the growth temperature from 30 to 25°C led to an increase in the lipid content of *Chlorella vulgaris* from 5.9 to 14.7 % dw (Converti et al., 2009). However, no significant change in the lipid content was observed in *Chlorella sorokiniana* grown at various temperatures (14°, 22° and 38°C) (Patterson, 1970). On the other hand, maximum lipid content (56% dw) was

H. pluvialis	Total Lipids (% dw)	Neutral Lipids (% dw)	Glycolipids (% dw)	Phospholipids (% dw)
Control	15.61 ± 1.46 (b)	9.20 ± 0.67 (1)	3.70 ± 0.38(3)	1.87 ± 0.05(4)
High light intensity	34.85 ± 0.78 (c)	19.80 ± 0.14 (2)	7.85 ± 1.77(3)	9.50 ± 0.00 (3)

Table 1. Lipid content (percentage of dry weight biomass = % dw) and fractions (neutral, glycolipid and phospholipid) in *Haematococcus pluvialis* under control (90 μmol photons m^{-2} s^{-1}) and high light intensity (300 μmol photons m^{-2} s^{-1}) culture conditions. Values are means ± standard deviations of three replicates. Identical superscripts indicate non-significant (α = 0.05) differences. Adapted from Damiani et al. (2010).

reported in *Neochloris oleoabundans* grown under nitrogen starvation and 30ºC. Temperature has a major effect on the fatty acid composition of microalgae (Roessler, 1990; Hu et al., 2008). The general trend towards an increase in fatty acid unsaturation with decreasing temperature observed in land plants and other organisms (Raison, 1986) also occurs in many microalgae. Patterson (1970) reported a greater degree of fatty acid unsaturation in *Chlorella sorokiniana* cells grown at 22ºC relative to cells grown at 38ºC, although cells grown at 14ºC had an intermediate degree of unsaturation. On the other hand, Widjaja et al. (2009) have noticed that the drying temperature during lipid extraction from algal biomass not only affects lipid composition but also lipid content. For example, heating at higher than 60º C resulted in a decrease of total lipid content and TAG content in *Chlorella vulgaris*.

3.1.2 Lipid composition

Although variations have been reported in the fatty acid composition of some representatives of Chlorophyceae Class, in general the most abundant fatty acids saturated and mono-unsaturated are palmitic acid (C16:0) and oleic acid (C18:1n9c), respectively. In turn, the major polyunsaturated fatty acids found in green algae are linoleic acid (C18:2n6c) and linolenic acid (C18:3n3). PUFAs above C18 are not usually present as majority fatty acids (Hu et al., 2008; Gouveia & Oliveira 2009; Gouveia et al., 2009; Damiani et al., 2010; Ho et al., 2010). Some examples of fatty acids profiles of green microalgae are shown in Table 2. As to saturated fatty acids (SFAs), a significant percentage (25-17%) of palmitic acid was indicated in *Chlorella vulgaris* INETI 58, *Scenedesmus obliquus* FCTU Coimbra, *Neochloris oleoabundans* UTEX # 1185 and *Dunaliella tertiolecta* (IPIMAR) by Gouveia & Oliveira (2009). Similarly, high palmitic acid contents were reported in *Haematococcus pluvialis* (19%) (Leonardi et al., 2008) and *S. obliquus* CNW-N (15%) (Ho et al., 2010). Two strains of *S. obliquus* are shown in Table 2; however, high content of stearic acid (15 %) has only been found in the strain CNW-N. This SFA is commonly found in land plant oils (Ramos et al., 2009; Singh & Singh, 2010). The monounsaturated oleic acid was well represented in all species studied, showing *D. tertiolecta* the lowest percentage. Among PUFAs, linoleic acid was also well represented in all species. Regarding linolenic acid, except for the oils extracted from *S. obliquus* FCTU Coimbra and *S. obliquus* CNW-N, the other oils showed contents above 12%. The length of dominant fatty acids in all species was intermediate with a maximum of 18 carbons and the maximum degree of chain unsaturation was three (Table 2).
As to stress conditions, a marked increase in the level of SFAs and MUFAs with a concomitant decrease in PUFAs is usually associated with nitrogen deficiency (Piorreck et al., 1984). For example, *Neochloris oleoabundans* UTEX # 1185 grown under nitrogen starvation showed oleic acid as the main fatty acid present, followed by palmitic and stearic

Fatty acids (% of total fatty acids)	*Chlorella vulgaris* (1)	*Scenedesmus obliquus* FCTU (1)	*Scenedesmus obliquus* CNW-N (2)	*Dunaliella tertiolecta* (1)	*Neochloris oleoabundans* (1)	*Haematococcus pluvialis* (3)
C12:0	n.d.	n.d.	0.99	n.d.	n.d.	0.07
C13:0	n.d.	n.d.	1.43	n.d.	n.d.	n.d.
C14:0	3.07	1.48	0.91	0.47	0.43	0.60
C15:0	n.d.	n.d.		n.d.	n.d.	0,17
C16:0	25.7	21.78	15.05	17.70	19.35	18,88
C17:0	n.d.	n.d.	2.36	n.d.	n.d.	0,25
C18:0	0.63	0.45	17.16	n.d.	0.98	2,93
C20:0	0.09	n.d.		n.d.	n.d.	0,35
C22:0	n.d.	n.d.	0.33	n.d.	n.d.	0.18
C14:1	n.d.	n.d.	0.36	n.d.	n.d.	n.d.
C16:1	5.25	5.95	2.30	0.88	1.85	0,52
C17:1	n.d.	n.d.	0.37	n.d.	n.d.	n.d.
C18:1	12.64	17.93	15.55	4.87	20.29	
C18:1 n9c						15,98
C18:1 n9t						0,17
C20:1	0.93	n.d.	0.75	n.d.	n.d	0,09
C16:2	n.d.	3.96		3.03	1.74	n.d.
C16:3	1.27	0.68		1.24	0.96	n.d.
C16:4	4.06	0.43		10.56	7.24	n.d.
C18:2	7.19	21.74	13.39	12.37	12.99	
C18:2 n6c						24,82
C18:2 n6t						7,25
C18:3	19.05	3.76	3.97	30.19	17.43	
C18:3 n3						16,06
C18:3 n6						1.40
C18:4	n.d.	0.21		n.d.	2.10	n.d.
C20:2	n.d.	n.d.		n.d.	n.d.	0.25
C20:3	0.83	n.d.		n.d.	n.d.	n.d.
C20:4	0.23	n.d.		n.d.	n.d.	1.25
C20:5	0.46	n.d.		n.d.	n.d	0.50
SFA (%)	28.56	23.71	37.9*	18.17	20.76	23,59
MUFA (%)	18.82*	23.88*	19.33*	5.75*	22.14*	16,76
PUFA (%)	33.09*	30.78*	17.36*	57.39*	42.46*	51,73
Unsaturated	51.91	54.66		63.14	64.60	68.49
Source	Gouveia & Oliveira 2009	Gouveia & Oliveira 2009	Ho et al. 2010	Gouveia & Oliveira 2009	Gouveia & Oliveira 2009	Leonardi et al. 2008

Table 2. Fatty acid profiles of green microalgae. (1) *Chlorella vulgaris* INETI 58, *Scenedesmus obliquus* FCTU Coimbra, *Neochloris oleoabundans* UTEX # 1185 and *Dunaliella tertiolecta* IPIMAR cultivated in an appropriate growth medium with bubbling air, under 150 µmol E $m^{-2} s^{-1}$ of light intensity and finally, outdoors during 4 months. (2) *Scenedesmus obliquus* CNW-N cultivated in a nutrient-rich medium with 10% CO_2, under 60 µmol E $m^{-2} s^{-1}$ of light intensity. (3) *Haematococcus pluvialis* cultivated in Basal Bold medium with bubbling air, under 150 µmol E $m^{-2} s^{-1}$ of light intensity. *In order to compare total SFAs, MUFAs and PUFAs, the sums were performed according to published data.

acids (Gouveia et al., 2009). In contrast, linolenic acid proportion was below 12%, which was lower than the content (17.43%) reported for the same strain grown under sufficient nitrogen (Gouveia et al., 2009; Gouveia & Oliveira, 2009). In a similar way, Mendoza Guzmán et al. (2011) reported in *Dunaliella salina* ITC-5.003 a significant decrease in relative PUFA content and unsaturation index in cultures exposed to nitrogen starvation when compared to control conditions (Table 3). This variation occurred primarily related to variations in MUFA and PUFA contents. The MUFA content increased in the nitrogen-starved cultures, while PUFA had a higher relative content in control cultures. Conversely, in *Chlorella pyrenoidosa* BNA-10-013 no significant variation was observed in the content of fatty acids in control and nitrogen-starvation conditions (Table 3). This species' culture probably required a more prolonged exposure to nitrogen deficiency in order to be able to observe significant variations in the fatty acid composition (Mendoza Guzmán et al., 2011).

Fatty acids (% of total fatty acids)	*Dunaliella salina* Control exponential growth phase	*Dunaliella salina* Nitrogen-starved	*Chlorella pyramimosa* Control exponential growth phase	*Chlorella pyramimosa* Nitrogen-starved
C13:0	0.68	0.68		2.95
C14:0	0.92	1.08	0.85	0.85
C16:0	24.99	34.33	20.54	21.13
C18:0	15.92	12.60	0.24	0.38
C14:1	0.06	0.06	0.05	0.13
C16:1	3.03	2.62	3.29	2.48
C18:1 n9c	3.59	15.52	2.30	3.45
C18:1 n9t	3.22	1.95	0.90	0.80
C16:2	0.08	1.20	7.59	10.39
C16:3	0.52	0.62		
C16:4	2.17	1.21	19.36	13.00
C18:2 n6	8.59	6.64	17.75	18.10
C18:3 n3	34.86	20.81	26.36	24.91
C18:3 n6			1.37	0.68
C22:2			0.27	1.43
SFA (%)	42.51*	48.69*	21.63*	25.31*
MUFA (%)	9.9*	20.15*	6.54*	6.86*
PUFA (%)	46.22*	30.48*	72.70*	68.51*

Table 3. Fatty-acid profiles of *Dunaliella salina* ITC-5.003 and *Chlorella pyramimosa* BNA-10-013 cultivated under two experimental conditions. Control at exponential growth-phase and nitrogen starved. Adapted from Mendoza Guzmán et al. (2011). *In order to compare total SFAs, MUFAs and PUFAs, the sums were performed according to published data.

High content of TAGs (more than 70% of the total lipids) containing mainly palmitic and oleic acids was reported in *Chlorella vulgaris* and *Scenedesmus obliquus* grown under low nitrogen (Piorreck et al., 1984). A detailed TAG-composition study under nitrogen starvation was performed by Zhekisheva et al. (2002) in a German strain of *Haematococcus pluvialis*. After 1 day of nitrogen starvation, the proportion of oleic acid increased sharply to 24.1%, compared with 5% in the control cells, while PUFA content decreased. In the following days, there was a further decrease in the unsaturation level that was expressed by an increase in the proportion of linoleic acid at the expense of PUFAs (C16:4 and C18:3). On the other hand, Zhekisheva et al. (2002) found in *H. pluvialis* a response of the TAG's fatty acid composition to high light intensity (350 μmol photons $m^{-2} s^{-1}$), which was similar to the behaviour observed under nitrogen starvation. The most outstanding change was noted in the proportion of oleic acid that increased from 5.2 % in the control to 19.8% after 1.5 days and thereafter decreased. Recently, Damiani et al. (2010) studied the fatty acid composition of neutral fraction in an Argentinian strain of *Haematococcus pluvialis* under control and high

	Neutral lipids	
Fatty acids (%)	**Control**	**High light intensity**
C6:0	nd	1.33
C8:0	nd	0.27
C10:0	tr	0.23
C12:0	0.21	0.30
C13:0	nd	nd
C14:0	1.25	1.35
C14:1	tr	nd
C15:0	0.19	0.27
C15:1	nd	nd
C16:0	22.49	18.87
C16:1	0.64	0.58
C17:0	0.19	0.32
C17:1	tr	tr
C18:0	3.15	7.07
C18:1n9t	tr	0.67
C18:1n9c	19.36	18.25
C18:2n6t	6.67	5.37
C18:2n6c	20.23	22.06
C20:0	0.20	0.32
C18:3n6	0.86	1.02
C20:1	0.13	0.23
C18:3n3	16.18	12.01
C21:0	tr	tr
C20:2	0.32	1.15
C22:0	0.18	0.31
C22:1n9	tr	0.17
C20:4n6	0.89	1.21
C24:0	tr	0.20
C20:5n3	0.57	0.48
C22:5n3	nd	nd
SFA %	27.81 ± 0.42 (a)	30.36 ± 1.19 (b)
MUFA %	20.07 ± 0.06 (g)	19.91 ± 0.12 (g)
PUFA %	45.80 ± 0.18 (k)	43.15 ± 0.68 (l)

Table 4. Neutral fraction fatty-acid profiles (percentage of total fatty acids) of *Haematococcus pluvialis* under control and high light intensity culture conditions. Identical superscripts indicate non significant differences in the values (α = 0.05). Values are the means of four replicates. tr: trace, nd: no detected. Source: Damiani et al. (2010).

light intensity (300 μmol photons $m^{-2} s^{-1}$) conditions during 14 days of growth. The fatty acid profile was similar under both culture conditions, and the major components were palmitic, stearic, oleic, linoleic, linolenic and linolelaidic acids (Table 4). The percentage of SFAs was significantly higher in cultures grown under high light intensity, when compared to the control. The palmitic acid content slightly declined under stress condition. In contrast, the stearic acid's relative content increased. The MUFAs showed no significant differences between control and stress conditions. In addition, PUFAs presented a significant decrease in the stress condition compared to the control, with a concomitant decrease in linolenic acid.

Even though the concept of using microalgae as a source of biofuel is old (Sheehan et al., 1998), at present there are few studies related to lipid valorisation as biodiesel using chlorophycean and most of the microalgae tested are species currently cultivated for aquaculture or for human nutritional products. Some studies concerning to the potential use of green algae for biodiesel include the analysis of the total fatty acid quality- for example, *Chlorella vulgaris* INETI 58, *Scenedesmus obliquus* FCTU Coimbra, *Dunaliella tertiolecta* IPIMAR (Gouveia & Oliveira, 2009), *Neochloris oleoabundans* UTEX # 118 (Gouveia & Oliveira, 2009; Gouveia et al., 2009) and *Scenedesmus obliquus* CNW-N (Ho et al., 2010), while only one analyses the neutral lipid fraction: *Haematococcus pluvialis* (Damiani et al., 2010). On the other hand, some studies are focused on the analysis of fatty acid methyl esters (FAMEs)- for example, *Chorella protothecoides* (Xu et al. 2006); *Chorella vulgaris* CCAP 211 (Converti et al. 2009), *Scenedesmus obliquus* SAG 276-3a (Mandal & Mallick, 2009); *Chlorella vulgaris* UTCC90, *Scenedesmus obliquus* UTCC 5, *Dunaliella tertiolecta* UTCC 420 (Francisco et al., 2010) and *Chlorella* sp. (Rasoul-Amini et al., 2011).

Recently, Chinnasamy et al. (2010) evaluated the feasibility of producing biodiesel from microalgal consortium of fifteen isolates (consisting of chlorophycean and cyanobacterial species) grown in treated wastewater.

3.2 Eustigmatophyceae class

3.2.1 Oil composition and triggering of lipid production

The Eustigmatophyceae Class is characterized by a typical fatty acid composition that includes four abundant fatty acids: palmitic acid, palmitoleic acid (16:1n7), arachidonic acid (ARA, 20:4n6) and eicosapentaenoic acid (EPA, 20:5n3) (Volkman et al., 1993; Goldberg & Boussiba, 2011). In contrast with green algae, fatty acids with C18 chain length are present as relatively minor components. Regarding lipid fractions, there are only few studies related to TAG composition in *Nannochloropsis* (Sukenik et al., 1989; Hodgson et al., 1991). *Nannochloropsis* spp. are widely used in aquaculture and have been investigated as a potential EPA source (Sukenik, 1999). EPA is an essential fatty acid currently sourced from fish oil. No other sources are commercially available. In recent years special interest has developed in this microalga as biodiesel feedstock due to its high lipid content (Rodolfi et al., 2009) and its composition (Gouveia & Oliveira, 2009). Biodiesel was specifically obtained from two *N. oculata* strains by Umdu et al. (2009) and Converti et al. (2009) and *Nannochloropsis* sp. by Koberg et al. (2011). However, for biodiesel production approaches more related to culture conditions tending to a PUFA decrease in *Nannochloropsis* spp. should be considered. Nitrogen deprivation is a triggering factor in *Nannochloropsis* spp. The strain *Nannochloropsis* sp. Q2 grown under nitrogen deficiency was characterized by high lipid content (from 24% to 55%) and the principal lipid fraction synthesized was TAG (79%) rich in SFAs (Suen et al., 1987). In addition, lipid content of *Nannochloropsis* sp. PP983 that was grown in low nitrogen level increased up to 62% dw (Hu & Gao, 2006). Similarly, *Nannochloropsis* sp. F&M-M24 growing in photobioreactors reached 60% lipid content after nitrogen starvation (Rodolfi et al., 2009). The effect of irradiance on lipid content has been reported for many strains of *Nannochloropsis* growing under different culture conditions and modes. For example, Greek

Fatty acids (% of total fatty acids)	***Nannochloropsis* sp. Irradiance µmol quanta m-2 s-1**		***Nannochloropsis* sp. MFD-2 Temperature °C**			
	40	480	20	25	30	35
C14:0	2.14	6.35	5.60	6.45	7.75	10.85
C15:0			0.40	0.40	0.55	0.50
C16:0	25.88	40.44	26.70	20.85	24.75	42.55
C17:0			0.40	0.40	0.35	0.20
C18:0	4.00	0.84	5.50	0.40	0.55	1.15
C19:0			0.30	0.30	0.15	n.d.
C20:0			n.d.	n.d.	n.d.	0.20
C14:1n9			n.d.	n.d.	n.d.	0.25
C16:1 n7	18.25	36.15	30.05	30.35	27.20	24.90
C16:1n9	1.06	0.46				
C18:1 n9	3.05	3.33	n.d.	3.75	2.65	3.80
C18:1n7	9.92	0.35				
C20:1 n9			n.d.	n.d.	0.55	n.d.
C22:1 n9			n.d.	0.35	n.d.	n.d.
C18:2 n6	1.88	0.65	1.40	3.05	3.35	1.60
C18:3 n6	1.76	0.42	n.d.	n.d.	n.d.	0.30
C18:3 n3			1.35	0.90	n.d.	n.d.
C20:3n3			5.50	5.15	3.80	3.00
C20:3n6	n.d.	0.37				
C20:4 n6	3.19	2.31				
C20:5 n3	28.90	8.02	18.90	23.70	19.50	7.40
C22:1 n9			n.d.	0.35	n.d.	n.d.
C22:3 n3			n.d.	0.30	n.d.	n.d.
C22:6n3			n.d.	0.40	n.d.	n.d.
SFA (%)	32.57	48.78	38.90 *	28.80 *	34.10 *	55.45 *
MUFA (%)	32.30	40.06	30.05 *	34.45 *	30.40 *	28.95 *
PUFA (%)	35.13	11.16	27.15 *	33.85 *	26.65 *	12.30 *
Total lipids (%dw)			29.09	20.20	17.65	32.10
Source	Fábregas et al. 2004		James et al. 1989			

Table 5. Fatty acid profiles of *Nannochloropsis* spp. cultures. *Nannochloropsis* sp. grown in semicontinuous culture under light-limited (40 µmol quanta $m^{-2} s^{-1}$) and light-saturated (480 µmol quanta $m^{-2} s^{-1}$) conditions (Fábregas et al. 2004). Massive cultures of *Nannochloropsis* MFD-2 at different temperatures (Adapted from James et al. 1989). * In order to compare total SFAs, MUFAs and PUFAs, the sums were performed according to published data.

Nannochloropsis sp. strain growing in continuous cultures under saturating light (290 µmol quanta m^{-2} s^{-1}) was characterized by high lipid content with a TAG increase. This fraction contained mainly palmitic and palmitoleic acids with myristic (C:14) and stearic acids as minor fatty acids (Sukenik et al., 1989). *Nannochloropsis* sp. growing semicontinuously showed an increase of SFAs with increasing irradiance to the detriment of unsaturated fatty acids (Table 5) (Fábregas et al., 2004). A similar trend was observed in *N. oculata* CS179 growing under high irradiance (1100 µE m^{-2} s^{-1}) in outdoor cultures (Renaud et al., 1991). The strain showed a decrease in the ratio of total C16 unsaturated fatty acids to saturated C16:0, and a decrease in the ratio of total C18 unsaturated fatty acids to saturated C18:0.

The effect of temperature on lipid content and fatty acid composition of *Nannochloropsis* seems to be quite complex, so a general trend cannot be established. An increase in temperature from 20 to 25°C practically doubled the lipid content (from 7.90 to 14.92% dw) of *N. oculata* (Converti et al., 2009). However, in *Nannochloropsis* sp. MFD-2 an increase in temperature from 15° to 20°C increased the total lipid content (from 21.60 to 29.05% dw), while an increase from 20° to 30°C brought about a decrease of the total lipid content (from 29.05 to 17.65% dw). Maximum lipid content (32.10%) was observed at 35°C (Table 5) (James et al., 1989). Among PUFAs, EPA constituted the major component. This fatty acid showed an increasing trend with increasing temperature up to 25°C, while a declining trend in EPA was observed from 25° to 35°C (Table 5). In a similar way, PUFAs and EPA were accumulated at lower temperatures in *Nannochloropsis* sp. PP983 (Hu & Gao, 2006).

3.3 Bacillariophyceae class

3.3.1 Oil composition and triggering of lipid production

The diatoms' fatty acids have been studied more extensively than other microalgal classes. This interest is related to the wide use of diatoms in aquaculture (Zhukova & Aizdaicher, 1995; Lebeau & Robert, 2003a, 2003b; Bozarth et al., 2009). Diatoms are characterized by an unusual distribution of fatty acids compared to green algae and land plants (Darley, 1977). The C14, C16 and C20 acids comprise the bulk of thc diatom fatty acids, while unsaturated C18 acids, particularly linolenic acid, are either absent or present at very low levels. In general, SFAs and MUFAs in diatoms are myristic acid, palmitic acid and palmitoleic acid (Reitan et al., 1994; Hu et al., 2008 and cites therein). Although this composition is relatively constant, some differences have been observed in the number of fatty acids detected among species (Dunstan et al., 1994; Rousch et al., 2003), and among individual microalgal strains within a species (Johansen et al., 1990). For example, the fatty-acid numbers detected in *Skeletonema costatum* were 19 (Berge et al., 1995), 24 (Zhukova & Aizdaicher, 1995), 12 (Nahon et al., 2010) and 28 (Popovich et al., 2011). However, it is noticeable that different strains, conditions and growth phases were analysed in these studies. High ratios of C16:1n7/ C16:0 and $\sum$C16/$\sum$C18 are also characteristic of diatoms (Budge & Parrish, 1999). Furthermore, this prevalence and the high level of C14:0 have been used as markers of diatoms (Zhukova & Aizdaicher, 1995; Napolitano et al., 1997). Another characteristic feature of the composition of diatoms' fatty acid is the PUFA predominance, such as EPA and ARA (Reitan et al., 1994; Belarbi et al., 2000; Molina Grima et al., 2005). Like EPA, ARA is another essential fatty acid. Chu et al. (1994) reported ARA production by *Nitzschia conspicua* in a range of 0.6-4.7% total fatty acids.

The lipid content of oleaginous diatoms of freshwater and marine origin growing under normal and stress culture conditions was summarized by Hu et al. (2008). Statistical analysis

indicated that the average lipid content of oleaginous diatoms was 22.7% dw when maintained under normal growth conditions, whereas a total lipid content of 44.6% dw was achievable under stress conditions. However, lipid content and lipid fraction composition in diatoms are subjected to variability during the growth cycle and nutrient-deficiency (Table 6). Specifically, TAG synthesis and accumulation take place naturally in the stationary growth-phase (Siron et al., 1989; Sicko-Goad & Andresen, 1991; Lombardi & Wangersky, 1995; López Alonso et al. 1998, 2000). This fact occurs when photosynthetic assimilation is carried out while cell division is blocked due to a nutritional deficiency (Siron et al., 1989; Dunahay et al. 1996). In addition, a marked increase in the level of SFAs and MUFAs (e.g. 16:0, 16:1n7 and 18:1n9), with a concomitant decrease in the levels of PUFAs (e.g. 16:3n4 and 20:5n3) with increasing culture age has been observed in *Phaeodactylum tricornutum* and *Chaetoceros muelleri* (Liang et al., 2006). Table 6 gives evidence that culture age led to a significant increase of SFAs and MUFAs in detriment of PUFAs in both the total fatty acid and the TAG profiles in different diatoms. For example, Siron et al. (1989) reported that in *P. tricornutum* during exponential-growth phase the fatty acid pattern was characterized by a large fraction of EPA, palmitic and palmitoleic acids (Table 6). Conversely, this species in stationary-growth phase showed an increase of palmitic and palmitoleic acids in detriment of EPA (Table 6). The latter pattern was related to lipid storage. López Alonso et al. (2000) showed that for the same species, culture age had almost no influence on the fatty acid content. Conversely, age had a high impact on lipid fractions, producing a TAG increase (68%) and a polar lipids' decrease. Berge et al. (1995) studied the fatty acid composition of each lipid fraction of *Skeletonema costatum* during exponential growth-phase. They showed that polar lipids were mostly represented and TAGs were less abundant (ca. 10%). The authors reported that EPA, C16:3n4 and C16:4n1 accounted for about 61% of the total fatty acids (Table 6). Popovich et al. (2011) studied the fatty acid composition of each lipid fraction of *Skeletonema costatum* and *Navicula gregaria* during stationary growth-phase under similar culture conditions to Berge et al. (1995). TAGs were the main fraction of total lipids in both species, respectively accounting for ca. 65 and 76%, at 15 days of culture. *Skeletonema costatum* predominated in SFAs and MUFAs (Table 6), while *N. gregaria* was predominant in MUFAs, followed by SFAs. In *S. costatum*, the main fatty acids in the neutral lipid fraction were myristic, palmitic, palmitoleic and oleic acids (Table 6), while the main ones in *N. gregaria* were palmitic (28.74%) and palmitoleic (50.28%) acids (Popovich et al., 2011).
Regarding nutrient availability, high TAG amounts in diatoms have been related to silicon (Hu et al., 2008) and phosphorus deficiency (Siron et al., 1989; Reitan et al., 1994). Silicate is the essential compound of a diatom's cell wall. This nutrient has a positive effect on growth (Turpin et al., 1999) and affects cellular lipid metabolism (Hu et al., 2008). The response to

Fatty acids (%)	***P. tricornutum* (1)**	***P. tricornutum* (2)**	***S. costatum* (3)**	***S. costatum* (4)**	***P. tricornutum* (5)**	***P. tricornutum* (6)**
C6:0				0.11		
C10:0				0.08		
C12:0	1.0	1.8		0.57	0.4	
C13:0				0.16		
C14:0	5.3	6.3	5.4	20.19	5.6	9.4
C15:0				0.95		
C16:0	15.5	31.5	4.2	15.77	32.5	24.3

Fatty acids (%)	***P. tricornutum* (1)**	***P. tricornutum* (2)**	***S. costatum* (3)**	***S. costatum* (4)**	***P. tricornutum* (5)**	***P. tricornutum* (6)**
C17:0				0.38		
C18:0	0.4	0.4	0.3	5.21	0.8	
C20:0				tr.		
C22:0	0.1	tr.			0.3	
C14:1				0.38		
C16:1 n7	24.2	43.0	9.6	19.47	40.4	30.0
C16:1 n5			1.6			
C17:1				8.45		
C18:1 n9	0.6		1.1		0.6	3.3
C18:1 n9c				11.06		
C18:1 n7		2.1	0.2			0.3
C20:1 n9			0.2	1.6		
C22:1 n9			0.3			
C16:2	11.7	4.5			1.1	
C16:2 n4			4.5			1.9
C16:3	1.8	0.4			0.4	
C16:3 n4			13.1			1.9
C16:4 n4			0.1			
C16:4 n1			12.5			0.1
C18:2 n6			1.0		1.1	1.0
C18:2 n3			0.5			
C18:2 n6c	1.0	0.5		3.38		
C18:2 n6t				0.12		
C18:3 n3	0.3	tr.		0.19	3.4	
C18:3 n6				0.25		
C18:4 n3	0.4	1.3	4.4		0.5	
C18:n[a]	2.2	tr.			3.7	
C20:2 n6				2.37		
C20:3 n6			0.1			
C20:4 n6				0.22		0.1
C20:5 n3	21.3	4.0	35.3	6.63	6.9	18.5
22:1 n9			3.3			
C22:6 n3	0.9	tr	5.6	0.58	0.3	0.6
others						8.9
SFA (%)	22.3*	40*	9.9	43.48	39.6*	33.7*
MUFA (%)	24.8*	45.1*	13.0	40.11	41*	33.6*
PUFA (%)	39.6*	10.7*	77.1	13.74	17.4*	23.1*
Source	Siron et al., 1989	Siron et al., 1989	Berge et al., 1995	Popovich et al., 2011	Siron et al., 1989	López Alonso et al., 2000

Table 6. Fatty acid profiles of diatoms. Total fatty acid profiles (TFA) of *Phaeodactylum tricornutum* growing in appropriate medium at (1) exponential growth-phase and (2) late stationary growth-phase, (3) TFA of *Skeletonema costatum* at exponential growth-phase growing in appropriate medium, and (4) Neutral lipid profile (TAGs) of Argentinian strain *S. costatum* at late stationary growth-phase growing in appropriate medium. (5) TFA of *P. tricornutum* under phosphorus-deficient condition. (6) TAGs of *P. tricornutum* under nitrogen deprivation. * In order to compare total SFAs, MUFAs and PUFAs, the sums were performed according to published data. [a] The number of double bonds ($n \geq 3$) was not determined.

stress induced by silicate starvation indicates a rapid increase in neutral lipids. For example, after 6h of silicon deprivation, the total lipid fraction in *Cyclotella cryptica* increased from 30 to 42% dw (Shifrin & Chisholm, 1981). In addition, silicon-deficient *C. cryptica* cells had higher TAG levels and higher proportions of saturated and monounsaturated fatty acids than silicon-replete cells (Roessler, 1988). The lipid content in several species of *Navicula* reached up to 49% dw under silicon deficiency conditions (Griffiths & Harrison, 2008 and cites therein). On the other hand, phosphorus deficiency resulted in increased lipid content, mainly neutral lipids, in *Phaeodactylum tricornutum* (Siron et al., 1989), *P. tricornutum* and *Chaetoceros* sp. (Reitan et al., 1994) and *C. gracilis* (Lombardi & Wangersky, 1995). In addition, the PUFA percentage decreased and the percentage of MUFAs and SFAs increased with phosphorus limitation in *Phaeodactylum tricornutum* (Table 6) and *P. tricornutum* and *Chaetoceros* sp. (Reitan et al., 1994). *Stephanodiscus minutulus* showed an increase in TAGs and a decrease of polar lipids under silicon, phosphorus or nitrogen limitation (Lynn et al., 2000). However, large variability exists in the response of diatoms to nitrogen deficiency, e.g. under such condition some strains increase in their lipid content, particularly neutral lipids (Shifrin & Chisholm, 1981; Parrish & Wangersky, 1987; McGinnis et al., 1997) and others decrease (Shifrin & Chisholm, 1981 and cites therein). TAG was the lipid fraction with the highest increase under nitrogen limitation in *Chaetoceros muelleri*, *Navicula saprophila* (Chelf, 1990) and *P. tricornutum* (Table 6) (López Alonso et al., 2000) cultures. The latter authors indicated that saturated and monounsaturated fatty acids accumulated when nitrogen was decreased (Table 6). Under nitrogen deficiency conditions the lipid content in several species of *Navicula* also reached up to 51% dw, while in *Skeletonema costatum* it increased from 16 to 25% dw (Griffiths & Harrison, 2008 and cites therein). However, other strains remain unchanged in lipid content (Roessler, 1990; Sheehan et al., 1998; Rodolfi et al., 2009). This phenomenon might be due to the fact that diatoms have relatively high lipid content during the exponential growth-phase, and they do not increase their lipid content by nutrient starvation (Shifrin & Chisholm, 1981). In addition, the response of a microalgal culture to a factor (cell age or nutrient level) is also affected by the type of culture, either batch or continuous (López Alonso et al., 2000), and by the culture step during the batch scale-up process and the semi-continuous mode (Pernet et al., 2003).

Light intensity is another factor that affects lipid composition in diatoms (Orcutt & Patterson, 1974; Roessler, 1990). For example, increased TAG levels were observed in *Cylindrotheca fusiformis* under high intensity (Orcutt & Patterson, 1974). Temperature has a major effect on the types of fatty acids produced by microalgae (Thompson et al., 1992). Many microalgal species respond to decreased growth temperature by increasing the ratio of unsaturated to saturated fatty acids (Mortensen et al., 1988; James et al., 1989; Thompson et al., 1992; Renaud et al., 1995; Oliveira et al., 1999). However, the response to growth temperature varies from species to species, with no overall consistent relationship between temperature and fatty acid unsaturation (James et al., 1989; Thompson et al., 1992; Renaud et al., 1995). Rousch et al. (2003) investigated the effect of heat stress on the fatty acid composition of thermo-intolerant (*Phaeodactylum tricornutum*) and thermo-tolerant (*Chaetoceros muelleri*) marine diatoms under laboratory conditions. They found a production of fatty acids with greater saturation during heat stress. They also observed that changes in both fatty acid composition and fatty acid saturation degree occur more quickly in diatoms in response to increased temperature than under nutrient starvation. This implies that mechanisms associated with lipid changes in response to temperature may differ from those associated with other stresses (Rousch et al., 2003) Furthermore, the opposed trend to

produce unsaturated fatty acids during cooler temperatures has also been observed in several diatoms (Blanchemain & Grizeau, 1999; Rousch et al., 2003).
The number of genera and species of diatoms is in the order of 250 and 100,000, respectively (Norton et al., 1996; van den Hoek et al., 1995). In spite of their abundance and diversity in nature, cultures of diatoms of biotechnological interest are still at the early stage of development, except for aquaculture. This is most likely due to difficulties in their cultivation. In addition to EPA, which is usually extracted on a semi-industrial scale, and biomass for feeding in aquaculture, silicon production from diatoms' frustules is the most promising application, particularly in the field of nanotechnology (Lebeau & Robert, 2003a, 2003b; Bozarth et al., 2009). Only a few authors have reported lipid valorisation as biodiesel using diatoms, for example in *Hantzschia* DI-60 (Sriharan et al., 1990), *Chaetoceros muelleri* (McGinnis et al., 1997), *Skeletonema costatum* and *Navicula gregaria* (Popovich, et al., 2011). Among these studies, only Popovich et al. (2011) analyzed TAG profiles. Others studies have focused on the analysis of FAMEs. For example, there are analyses in *Navicula* JPCC DA0580 (Matsumoto et al., 2010) and *Phaeodactylum tricornutum* (Francisco et al., 2010). On the other hand, aiming at the production of biodiesel from microalgae, *Cyclotella cryptica* and *Navicula saprophila* were genetically manipulated by Dunahay et al. (1996) to optimise lipid production. Yu et al. (2009) identified and compared the molecular species of TAGs in *P. tricornutum* CCAP1055/1 (CCMP632) and *Thalassiosira pseudonana* 3H (CCMP1335) under control and nutrient-limitation. This study represents an advance in the development of molecular genetic tools for the manipulation of these strains.

4. Influence of oil composition on biodiesel quality

As was previously indicated, the most important properties of biofuel- i.e. cetane number (ignition quality), cold-flow properties, oxidative stability, and iodine value- are determined by the structure of fatty esters, which form essential part of the biodiesel (Knothe, 2005; Chisti, 2007). In turn, the properties of fatty esters are determined by the characteristics of fatty acid's oil- i.e. carbon chain length, its unsaturation degree, and the alcohol moieties that comprise a fatty ester (Knothe, 2005). Thus, the fatty-acid composition of different oils has a significant effect on the characteristics of the produced biodiesel. For instance, palmitic, stearic, oleic, linoleic and linolenic acids were recognized as the most common fatty acids contained in biodiesel from land plants (Miao & Wu, 2006; Knothe, 2008). These fatty acids are well represented in some green microalgae recently examined.
The selection of appropriate microalgal strains is an important factor for the overall success of biofuel production from microalgae. Rigorous selection is challenging owing to the large number of microalgal species available, the limited characterization of these organisms and their varying sets of characteristics. At present no ideal species have been found for this purpose; however, some examples of promising microalgal species are the following. Regarding chlorophycean, some research has been performed with the aim to analyze fatty acid composition so as to infer biodiesel quality. Total fatty acid profiles of *Chlorella vulgaris* INETI 58, *Scenedesmus obliquus* FCTU, *Neochloris oleoabundans* UTEX # 1185 and *Dunaliella tertiolecta* (IPIMAR) (Table 2) were studied by Gouveia & Oliveira (2009). They indicated that only the oils extracted from *S. obliquus* FCTU presented linolenic acid contents within European standard EN 14214 specifications (≤ 12%)(CEN EN-14214, 2003). Another aspect reported by these authors was the iodine value. This is a parameter that only depends on the oil origin (Mittelbach, 1996) and it is a measurement of oil unsaturation (Knothe, 2002).

Senedesmus obliquus and *N. oleoabundans* were the only species that presented iodine values (69 g I_2/100 g and 102 g I_2/100 g, respectively) below of the European standard EN 14214 allowed (maximum value of 120 g I_2/100 g). They concluded that if the purpose is to produce biodiesel only from one species, *S. obliquus* presented the most adequate fatty acid profile. However, *N. oleoabundans* and *D. tertiolecta* oils may be used for good quality biodiesel if associated with other oils. Under nitrogen-starvation conditions Gouveia et al. (2009) indicated a better oil quality in *N. oleoabundans* UTEX # 1185, with linolenic acid bellow 12% and iodine value of 72 I_2/100 g.

As was reported, lipids produced from green microalgal species usually contain fatty acid profiles of mainly C16 and C18. In *Senedesmus obliquus* CNW-N, C16 and C18 groups accounted for about 70% and 89% of total fatty acids. This fact was observed in cells growing under both nutrient-rich and nutrient-deficient media respectively (Ho et al., 2010). In both conditions the authors reported linolenic acid values below 12%, indicating that the lipid produced from *S. obliquus* CNW-N would be suitable for biodiesel production. The eustigmatophycean *Nannochloropsis* sp. also presented linolenic acid and iodine value (52 g I_2/100 g) within European standard (EN 14214) specifications (Gouveia & Oliveira, 2009). However, due to the high content of EPA and ARA, the authors remarked that the oil from this species may be used for good quality biodiesel, if associated with other oils.

Even though it is important to have information about total fatty acid profiles to screen microalgae for biodiesel production, transesterification includes TAG conversion to diglycerides, monoglycerides and then esters and glycerol (Mata et al., 2010). In this context, the utility of microalgal oils as biodiesel will depend on fatty acids' quantity and composition in TAG fraction. The analysis of TAG profiles for Argentinian *Haematococcus pluvialis* strain (Table 4) showed that the fatty acid composition was in agreement with the European standard: linolenic acid was below 12 % and oil's iodine values were 110.95 g I_2/100 g and 99.64 g I_2/100 g under control and high light intensity culture conditions, respectively. In addition, the maximum unsaturation degree of the PUFA chains was three and the length of the main fatty acid was intermediate with a maximum of 18 carbons. These features allowed them to infer that quality *H. pluvialis*' oil would be suitable for biodiesel production (Damiani et al., 2010).

In a similar way, the TAG profiles were analyzed in the diatoms *Skeletonema costatum* (Table 6) and *Navicula gregaria* under stationary growth phase (Popovich et al., 2011). The oils extracted from these species presented linolenic-acid contents within European biodiesel's quality specifications. However, both species showed EPA levels higher than the required limit. The iodine values for *S. costatum* (35.87 g I_2/100 g) and *N. gregaria* (51.93 g I_2/100 g) oils were well below the allowed European biodiesel standards. On the other hand, Knothe (2010) demonstrated that fatty-acid profiles enriched in palmitoleic acid may impart overall favourable properties to a biodiesel fuel, giving especially improved cold-flow properties. This fatty acid occurs in small amounts in land plant oils; however, it is noteworthy that *S. costatum* and *N. gregaria* respectively presented 19.45% and 50.28% of palmitoleic acid. In addition, cetane number and cold-flow properties were inferred in both diatoms. Cetane number (CN) is the dimensionless descriptor of a diesel fuel's ignition quality, and it is a prime indicator of biodiesel quality. It augments with both increasing saturation and straight-chain length in the fatty acids (Knothe, 2008). The minimum limits allowed in biodiesel standards are equal to 51 in European EN 14214, and 47 in American ASTM D675 (ASTM D675, 2002). According to SFA composition in both diatoms, biodiesel derived from these oils may present an acceptable CN. On the other hand, biodiesel coming from oils with high amounts of saturated long-chain fatty acids tends to have relatively poor cold-flow properties (Chiu et al., 2004). Thus, high SFA content found in both diatoms would likely indicate poor cold flow properties (Popovich et al., 2011). The authors concluded that lipid quality in *S. costatum* and *N. gregaria* indicated the microalgae's potential as biodiesel feedstocks.

As to FAME analysis, there are few studies focused on biodiesel production from different microalgal species. It is well known that certain methyl esters allow us to improve biodiesel quality. For instance, methyl oleate has been suggested as a possible candidate since it exhibits a combination of improved fuel properties (Knothe, 2005). Mandal & Mallick (2009) reported a high amount of methyl palmitate (38.8%) and methyl oleate (35.4%) fatty acids in biodiesel from *Scenedesmus obliquus* SAG 276-3a, which gives its high oxidative stability. In a similar way, *Chlorella protothecoides* growing under heterotrophic conditions showed oleic acid methyl ester as the most abundant (60.84%). In addition, in this species the total content of oleic-, linoleic- and stearic-acid methyl esters was over 80%. These features resulted in the biodiesel's high quality, according to Xu et al. (2006). Moreover, it is important to remark that biodiesel properties, such as density (0.864 kg L^{-1}), viscosity (5.2 x 10^{-4} Pas at 40°C), flash point (115°C), cold filter plugging point (-11°C), solidifying point (-12°C) and heating value (41 MJ kg^{-1}) were determined only for this heterotrophic species (Xu et al. 2006). All these physical and fuel properties studied were in agreement with the US Standard (ASTM 6751), and, specifically, its cold filter plugging point (-11°C) was much lower in comparison with the diesel fuel's. The authors concluded that the biodiesel from heterotrophic microalgal oil might be a competitive alternative to conventional diesel fuel.

Francisco et al. (2010) analyzed feedstocks of six microalgal strains for biodiesel production, taking into account important properties like ester content, cetane number, iodine value and unsaturation degree. In this study they found *Chlorella vulgaris* UTCC 90 proves to be the best strain for use as a feedstock for biodiesel production. Qualitative analysis of FAMEs demonstrated the predominance of saturated (43.5%) and monounsaturated (41.9%) fatty acids. The biodiesel's quality properties were an ester content of 99.8%, a cetane number of 56.7%, an iodine value of 65 g I_2/100 g, an unsaturation degree of 74.1 % and a cold filter plugging point of 4.5°C. All these parameters agreed with the limits established by the US Standard (ASTM 6751), the European Standard (EN 14214), the Brazilian National Petroleum Agency Standard (ANP 255) (ANP 255, 2003) and the Australian Standard for biodiesel quality (Fuel Standard, 2003).

As regards Eustigmatophyceae, Koberg et al. (2011) reported that the major composition of biodiesel produced from *Nannochloropsis* sp. consisted of methyl esters of palmitic and palmitoleic acids. This fact agrees with the concept indicated above, where fatty-acid profiles enriched in palmitoleic acid may impart overall favourable properties to a biodiesel fuel (Knothe, 2010). Moreover, the low percentage of methyl esters with a carbon chain of > 18 carbons observed in this species would guarantee a low viscosity for the biodiesel. Similarly, the FAME profile of the marine diatom *Navicula* sp. JPCC DA 0580 mainly contained methyl palmitate and methyl palmitoleate. This oil might be suitable for biodiesel production according to Matsumoto et al. (2010).

5. Conclusions

The production of biodiesel from microalgae is an emergent area greatly promising for the gradual replacement of diesel fuel. This review covers only a small part of various aspects that should be taken into account when choosing a microalgal species to produce biodiesel. The results clearly indicate the need for more research on microalgal lipids, especially TAG and fatty-acid profiles. Besides, rigorous comparisons across experiments under different conditions are impossible to carry out. As has been shown, the lipid content and fatty acid composition can vary with: a) strain/species, b) environmental conditions, c) type of stress condition, d) duration of stress condition, e) nutrient's concentration or irradiance's intensity during cultivation, f) life cycle (exponential or stationary phase), g) culture age, h) culture strategies (one- or two-stage cultivation, indoor or outdoor), among others. Even though no generalization can be strictly made, stress conditions seem to be effective in promoting lipid accumulation and improving TAG amount in many microalgae. These features are often associated with low productivity of biomass and lipids. Thus, the major challenge in process development for microalgal biodiesel production consists in choosing the best strains and defining cultivation strategies in order to simultaneously achieve three objectives: to obtain high biomass yields, high lipid contents and lipids with adequate fatty acid profiles.

6. Acknowledgement

Thanks to Secretaría de Ciencia, Tecnología e Innovación Productiva (SeCyT) de la República Argentina, Estudio Exploratorio N° E658/07PET29, Consejo Nacional de Investigaciones Científicas y Técnicas de la República Argentina (CONICET) PIP 112-200801-00234, Secretaría de Ciencia y Tecnología de la Universidad Nacional del Sur, PGI TIR and Agencia Nacional de Promoción Científica y Tecnológica, PICT-2010-0959. P.I.L. is Research Member of CONICET.

7. References

Aaronson, S., Berner, T. & Dubinsky, Z. (1980). Microalgae as a source of chemicals and natural products. In Shelef, G. & Soeder, C.J. eds. Algae Biomass. Elsevier/North Holland Biomedical Press. pp. 575-601. ISBN 0-4444802428.

Allard, B. & Templier, J. (2000). Comparison of neutral lipid profile of various trilaminar outer cell wall (TLS)-containing microalgae with emphasis on algaenan occurrence. *Phytochemistry* 54: 369-380.

ANP 255. (2003). Provisional Brazilian Biodiesel Standard ANP (Agencia Nacional do Petróleo).

ASTM 6751. (2002). Standard Specification for Biodiesel Fuel (B100) Blend Stock for Distillate Fuels.

Belarbi, E.H., Molina Grima, E. & Chisti, Y.A. (2000). A process for high yield and scaleable recovery of high purity eicosapentaenoic acid esters from microalgae and fish oil. *Enzyme and Microbial Technology* 26: 516-29.

Ben-Amotz, A. (1995). New mode of *Dunaliella* biotechnology: two-phase growth for ß-carotene production. *Journal of Applied Phycology* 7: 65-68.

Ben-Amotz, A., Tornabe, T.G. & Thomas, W.H. (1985). Chemical profile of selected species of microalgae with emphasis on lipids. *Journal of Phycology* 21: 72-81.

Berge, J.P., Gouygou, J.P., Dubacq, J.P. & Durand, P. (1995). Reassessment of lipid composition of the diatom *Skeletonema costatum*. *Phytochemistry*. 39: 1017-1021.

Blanchemain, A. & Grizeau, D. (1999). Increased production of eicosapentaenoic acid by *Skeletonema costatum* cells after decantation at low temperature. *Biotechnology Techniques* 13: 497-501.

Bigogno, C., Khozin-Goldberg, I., Boussiba, S., Vonshk, A. & Cohen, Z. (2002). Lipid and fatty acid composition of the green oleaginous alga *Parietochloris incisa*, the richest plant source of arachidonic acid. *Phytochemistry* 60: 497-503.

Bold, H.C. & Wynne, M.J. (1985). *Introduction to the Algae. Structure and Reproduction.* Second Edition. Prentice Hall, Inc. Englewood Cliffs, N.J. 07632. ISBN 0-13-477746-8.

Borowitzka, M.A. (1988). Fats, oils and hydrocarbons. In: Borowitzka, M.A. & Borowitzka, L.J. eds. Micro-algal biotechnology. Cambridge University Press, Cambridge, pp. 257-287. ISBN 0-521-323495.

Borowitzka, M.A. (1992). Algal biotechnology products and processes: matching science and economics. *Journal of Applied Phycology* 4: 267-279.

Borowitzka, M.A. (1995). Microalgae as sources of pharmaceuticals and other biologically active compounds. *Journal of Applied Phycology* 7: 13-15.

Borowitzka, M.A. (1997). Algae for aquaculture: Opportunities and constrains. *Journal of Applied Phycology* 9: 393-401.

Borowitzka, M.A. & Borowitzka, L.J. (1988). *Dunaliella.* In: Borowitzka, M.A. & Borowitzka, L.J. eds. Micro-algal biotechnology. Cambridge University Press, Cambridge, pp. 27-58. ISBN 0-521-323495.

Boussiba, S. (2000). Carotenogenesis in the green alga *Haematococcus pluvialis.* Cellular physiology and stress response. *Physiologia Plantarum* 108: 111-117.

Boussiba, S. & Vonshak, A. (1991). Astaxanthin accumulation in the green alga *Haematococcus pluvialis. Plant Physiology* 3: 1077-82.

Boussiba, S., Vonshak, A. Cohen, Z., Avissar, Y. & Richmond, A. (1987). Lipid and biomass production by halotolerant microalga *Nannochloropsis salina. Biomass* 12: 37-47.

Bozarth, A., Maier, U.G. & Zauner, S. (2009). Diatoms in biotechnology: modern tools and applications. *Applied Microbiology Biotechnology* 82: 195-201.

Budge, S.M. & Parrish, C.C. (1999). Lipid class and fatty acid composition of *Pseudo-nitzschia multiseries* and *Pseudo-nitzschia pungens* and effects of lipolytic enzyme deactivation. *Phytochemistry* 52: 561-566.

CEN EN-14214. (2003) Automotive fuels. Fatty acid methyl esters (FAME) for diesel engines. Requirements and test methods.

Chelf, P. (1990). Environmental control of lipid and biomass production in two diatom species. *Journal of Applied Phycology* 2: 121-129.

Chinnasamy, S., Bhatnagar, A., Hunt, R.W. & Das, K.C. (2010). Microalgae cultivation in a wastewater dominated by carpet mill effluents for biofuel applications. *Bioresource Technology* 101: 3097-3105.

Chisti, Y. (2007). Biodiesel from microalgae. *Biotechnology Advances* 25: 294–306.

Chiu, C., Schumacher, L.G. & Suppes, G.J. (2004). Impact of cold flow improvers on soybean biodiesel blend. *Biomass Bioenergy* 27: 485-491.

Chu, W.L., Phang, S. M. & Goh, S.H. (1994). Studies on the production of useful chemicals, especially fatty acids in the marine diatom *Nizschia conspicua* Grunow. *Hydrobiologia* 285: 33-40.

Costa Neto, P.R., Rossi, L.F.S., Zagonel, G.F., Ramos, L.P. (2000). Produção de biocombustivel alternativo ao óleo diesel a través da transesterificação de óleo de soja usado em frituras. *Química Nova* 23: 531-537.

Converti, A., Casazza, A.A., Ortiz, E.Y., Perego, P. & Borghi, M.D. (2009). Effect of temperature and nitrogen concentration on the growth and lipid content f *Nannochloropsis oculata* and *Chlorella vulgaris* for biodiesel production. *Chemical Engineering and Processing: Process Intensification* 48: 1146-1151.

Damiani, M.C., Leonardi, P.I., Pieroni, O. & Cáceres, E.J. (2006). Ultrastructure of the cyst wall of *Haematococcus pluvialis* (Chlorophyceae): wall development and behaviour during cyst germination. *Phycologia* 45: 616-623.

Damiani, M.C., Popovich, C.A., Constenla, D. & Leonardi, P.I. (2010). Lipid analysis in *Haematococcus pluvialis* to assess its potential use as biodiesel feedstock. *Bioresource Technology* 101: 3801-3807.

Darley, W.M. (1977). Biochemical composition. In: *The Biology of Diatoms.* Werner, D. ed. University of California Press. Berkeley and Los Angeles, California, pp. 198-223. ISBN: 0-5200-3400-7.

Demming-Adams, B. & Adams, W.W. (1992). Photoprotection and other responses of plants of high light stress. *Annual review of Plant Physiology and Plant Molecular Biology* 43: 599-626.

Dunahay, T.G., Jarvis, E.E., Dais, S.S. & Roessler, P.G. (1996). Manipulation of microalgal lipid production using genetic engineering. *Journal of Applied Phycology* 57/58:223-231.

Dunstan, G.A., Volkman, J.K., Barrett, S.M., Leroi, J.M. & Jeffrey, S.W. (1994). Essential polyunsaturated fatty acids from 14 species of diatom (Bacillariophyceae). *Phytochemistry* 35: 155-161.

Durrett, T., Benning, C., Ohlrogge, J. (2008). Plant triacylglycerols as feedstocks for the production of biofuels. *Plant Journal* 54: 593–607.

Fábregas, J., Maseda, A., Domínguez, A. & Otero, A. (2004). The cell composition of *Nannochloropsis* sp. changes under different irradiances in semicontinuous culture. *World Journal of Microbiology & Biotechnology* 20: 31-35.

Ferrari, R.A., Oliveira, V.S. & Scabio, A. (2005). Oxidative stability of biodiesel from soybean oil fatty acid ethyl esters. *Scientia Agricola* 62: 291-295.

Fogg, G.E. (1959). Nitrogen nutrition and metabolic patterns in algae. *Symposia of the Society for Experimental Biology* 13: 106-125.

Francisco, E.C., Neves, D.B., Jacob-Lopes, E. & Franco, T.T. (2010). Microalgae as feedstock for biodiesel production: Carbon dioxide sequestration, lipid production and biofuel quality. *Journal of Chemical Technology and Biotechnology* 85: 395-403.

Fuel Standard (2003) Biodiesel Determination, Approved Under the Fuel Quality Standard Act. 2000 by the Australian Minister for the Environmental and Heritage.

Gardner, R., Peters, P., Peyton, B. & Cooksey, K.E. (2010). Medium pH and nitrate concentration effects on accumulation of triacylglycerol in two members of the chlorophyta. *Journal of Applied Phycology*. DOI 10.1007/s10811-010-9633-4.

Gelin, F., Boogers, I., Noordelos, A.A.M., Sinnighe Damsté, J.S., Hatcher, P.G. & de Leeuw, J.W. (1996). Novel, resistant microalgal polyethers: An important sink of organic carbon in the marine environment? *Geochimica et Cosmochimica Acta* 60: 1275-1280.

Gelin, F., Boogers, I., Noordelos, A.A.M., Sinnighe Damsté, J.S. & de Leeuw, J.W. (1997). Resistant biomacromolecules in marine microalgae of the class eustigmatophyceae and Chlorophyceae: Geochemical implications. *Organic Geochemistry* 26: 659-675.

Goldberg, I.K. & Boussiba, S. (2011). Concerns over reporting of inconsistent data on fatty acid composition for microalgae of the genus *Nannochloropsis* (Eustigmatophyceae). *Journal of Applied Phycology* DOI 10.1007/s10811-011-9659-2.

Gordillo, F.J., Goutx, M., Figueroa, F.L. & Niell, F.X. (1998). Effects of light intensity, CO_2 and nitrogen supply on lipid class composition of *Dunaliella viridis*. *Journal of Applied Phycology* 10: 135-144.

Gouveia, L., Marques, A.E., da Silva, T.L. & Reis, A. (2009). *Neochloris oleabundans* UTEX # 1185: a suitable renewable lipid source fro biodiesel production. *Journal of Industrial Microbiology and Biotechnolology* 36: 821-826.

Gouveia, L. & Oliveira, A.C. (2009). Microalgae as a raw material for biofuels production. *Journal of Industrial Microbiology and Biotechnolology* 36: 269-274.

Graham, L.E. & Wilcox, L.W. (2000). *Algae*. Prentice Hall. Upper Saddle River, NJ 07458. ISBN 0-13-660333-5.

Griffiths, M.J. & Harrison, S.T.L. (2009). Lipid productivity as a key characteristic for choosing algal species for biodiesel production. *Journal of Applied Phycology* 276: 23-25.

Guschina, I.A. & Harwood, J.L. (2006). Lipids and lipid metabolism in eukaryotic algae. *Progress in Lipid Research* 45: 160-186.

Harwood, J.L. & Jones, A.L. (1989). Lipid metabolism in algae. *Advances in Botanical Research* 16:1-53.

Harwood, J.L. & Guschina, I.A. (2009). The versatility of algae and their lipid metabolism. *Biochimie* 91: 679-684.

Ho, S., Chen, W. & Chang, J. (2010). *Scenedesmus obliquus* CNW-N as a potential candidate for CO_2 mitigation and biodiesel production. *Bioresource Technology* 101: 8725-8730.

Hodgson, P.A., Henderson, R.J., Sargent, J.R. & Leftley, J.W. (1991). Patterns of variation in the lipid class and fatty acid composition of *Nannochloropsis oculata* (Eustigmatophyceae) during batch culture. I. The growth cycle. *Journal of Applied Phycology* 3: 169-181.

Hu, Q. (2004). Environmental effects on cell composition. In: Richmond A. ed. Handbook of Microalgal Culture: *Biotechnology and Applied Phycology*. Oxford: Blackwell Science Ltd. ISBN 0-632-05953-2.

Hu, H. & Gao, K. (2003). Optimization of growth and fatty acid composition of an unicellular marine picoplankton, *Nannochloropsis* sp., with enriched carbon sources. *Biotechnology Letters* 25: 421-425.

Hu, H. & Gao, K. (2006). Response of growth and fatty acid compositions of *Nannochloropsis* sp. to environmental factors under elevated CO_2 concentration. *Biotechnology Letter* 28: 987-992.

Hu, Q., Sommerfeld, M., Jarvis, E., Ghirardi, M., Posewitz, M., Seibert, M. & Darzins, A. (2008). Microalgal triacylglycerols as feedstocks for biofuel production: perspectives and advances. *The Plant Journal* 54: 621-639.

Huntley, M.E. & Redalje, D.G. (2007). CO_2 mitigation and renewable oil from photosynthetic microbes: a new appraisal. *Mitigation and Adaptation Strategies for Global Change* 12: 573-608.

Illman, A.M., Scragg, A.H. & Shales, S.W. (2000). Increase in *Chlorella* strains caloric values when grown in low nitrogen medium. *Enzyme and Microbial Technology* 27: 631-635.

Iwamoto, H., Yonekawa, G. & Asai, T. (1955). Fat synthesis in unicellular algae. I. Culture conditions for fat accumulation in *Chlorella* cells. *Bulletin of Agricultural Chemistry Society of Japan* 19: 240-252.

James, C.M., Al-Hinty, A. & Salman, A.E. (1989). Growth and ω3 fatty Acid and Amino Acid Composition of Microalgae Under Different Temperature Regimes. *Aquaculture* 77: 337-351.

Johansen, J.R., Barclay, W.R. & Nagle, N. (1990). Physiological variability within ten strains of *Chaetoceros muelleri* (Bacillariophyceae). *Journal of Phycology* 26: 271- 278.

Kitano, M., Matsukawa, R. & Karube, I. (1997). Changes in eicosapentaenoic acid content of *Navicula saprophila, Rhodomonas salina* and *Nitzschia* sp. under mixotrophic conditions. *Journal of Applied Phycology* 9: 559-563.

Koberg, M., Cohen, M., Ben-Amotz, A. & Gedanken, A. (2011). Bio-diesel production directly from the microalgae biomass of *Nannochloropsis* by microwave and ultrasound radiation. *Bioresource Technology* 102:4265-4269.

Knothe, G. (2005). Dependence of biodiesel fuel properties on the structure of fatty acid alkyl esters. *Fuel Processing Technology* 86: 1059-1070.

Knothe, G. (2008). "Designer" biodiesel: optimizing fatty ester composition to improve fuel properties. *Energy Fuel* 22:1358-1364.

Knothe, G. (2010). Biodiesel Derived from Model Oil Enriched in Palmitoleic Acid, Macadamia Nut Oil. *Energy Fuels* 24: 2098-2103.

Lang, X., Dalai, A.K., Bakhshi, N.N., Reaney, M.J. & Hertz, P.B. (2001). Preparation and characterization of bio-diesels from various bio-oils. *Bioresource Technology* 80: 53-62.

Larkum, T. & Howe, J. (1997). Molecular aspects of light-harvesting processes in algae. *Advances in Botanical Research* 27: 258-330.

Lebeau, T. & Robert, J.M. (2003a). Diatom cultivation and biotechnologically relevant products. Part I: Cultivation at various scales. *Applied Microbiology and Biotechnology* 60: 612-623.

Lebeau, T. & Robert, J.M. (2003b). Diatom cultivation and biotechnologically relevant products. Part II: Current and putative products. *Applied Microbiology and Biotechnology* 60: 624-632.

Leonardi, P.I., Damiani, M.C. & Popovich, C.A. (2008). Obtención de aceites para la producción de biodiesel a partir de microalgas. Informe Técnico. Secretaría de Ciencia, Tecnología e Innovación Productiva. Buenos Aires, Argentina. pp.1-18.

Liang, Y., Beardall, J. & Heraud, P. (2006). Changes in growth, chlorophyll fluorescence and fatty acid composition with culture age in batch cultures of *Phaeodactylum tricornutum* and *Chaetoceros muelleri* (Bacillariophyceae). *Botanic Marina* 49: 165-173.

Liang, Y., Sarkany, N. & Cui, Y. (2009). Biomass and lipid productivities of *Chlorella vulgaris* under autotrophic, heterotrophic and mixotrophic growth conditions. *Biotechnology Letters* 31: 1043-1049.

Liu, Z., Wang, G. & Zhou, B. (2008). Effect of iron on growth and lipid accumulation in *Chlorella vulgaris. Bioresource Technology* 99: 4717-4722.

Lombardi, A.T. & Wangersky, P.J. (1995). Particulate lipid class composition of three marine phytoplankters *Chaetoceros gracilis, Isochrysis galbana* (Tahiti) and *Dunaliella tertiolecta* grown in batch culture. *Hydrobiologia* 306: 1-6.

López Alonso, D., Belarbi, E.H., Rodríguez-Ruiz, J., Segura, C.I. & Jiménez, A. (1998). Acyl lipids of three microalgae. *Phytochemistry* 47: 1473-1481.

López Alonso, D., Belarbi, E-H., Fernández-Sevilla, J.M., Rodríguez-Ruiz, J. & Molina Grima, E. (2000). Acyl lipid composition variation related to culture age and nitrogen concentration in continuous culture of the microalga *Phaeodactylum tricornutum*. *Phytochemistry* 54: 461-471.

Lynn, S.G., Kilham, S.S., Kreeger, D. A. & Interlandi, S.J. (2000). Effect of nutrient availability on the biochemical and elemental stoichiometry in the freshwater diatom *Stephanodiscus minutulus* (Bacillariophyceae). *Journal of Phycology* 36: 510-522.

Mandal, S. & Mallick, N. (2009). Microalga *Scenedesmus obliquus* as a potential source for biodiesel production. *Biotechnological Products and Process Engineering* 84: 281-291.

Mata, T.M., Martins, A & Caetano, N.S. (2010). Microalgae for biodiesel production and other applications: A review. *Renewable and Sustainable Energy Reviews* 14: 217-232.

Matsumoto, M., Sugiyama, H., Macda, Y., Sato, R., Tanaka, T. & Matsunaga, T. (2010). Marine Diatom *Navicula* sp. Strain JPCC DA0580 and Marine Green Alga, *Chlorella* sp. Strain NKG400014 as Potential Sources for Biodiesel Production. *Applied Biochemistry and Biotechnology* 161: 483-490

McGinnis, K.M., Dempster, T.A. & Sommerfeld, M.R. (1997). Characterization of the growth and lipid content of the diatom *Chaetoceros muelleri*. *Journal of Applied Phycology* 9: 19-24.

Mendoza Guzmán, H.M., Jara Valido, A., Duarte, L.C. & Presmanes, K.F. (2011). Analysis of interspecific variation in relative fatty acid composition: use a flow cytometry to estimate unsaturation index and relative polyunsaturated fatty acid content in microalgae. *Journal of Applied Phycology* 23: 7-15.

Miao, X.L. & Wu, Q.Y. (2004). High yield bio-oil production from fast pyrolisis by metabolic controlling of *Chlorella protothecoides*. *Journal of Biotechnology* 110: 85-93.

Mittelbach, M. (1996). Diesel fuel derived from vegetable oils, VI: specifications and quality control of biodiesel. *Bioresource Technology* 56: 7–11.

Molina Grima, E., García Camacho, F. & Acién Fernández, F.G. (2005). Production of EPA from *Phaeodactylum tricornutum*. In: *Chemicals from microalgae.* Cohen, Z. (ed.) Taylor & Francis Ltd, Philadelphia, pp. 57-92. ISBN: 0-7484-0515-1.

Molina Grima, E., Robles Medina, A., Giménez Giménez A., Sánchez Pérez, J.A., García Camacho, F. & García Sánchez, J.L. (1994). Comparison between extraction of lipids and fatty acids from microalgal biomass. *Journal of the American Oil Chemist's Society* 71: 955-959.

Mortensen, S.H., Borsheim, K.Y., Rainuzzo, J.K. & Knutsen, G. (1988). Fatty acid and elemental composition of the marine diatom *Chaetoceros gracilis* Schutt. Effects of silicate deprivation, temperature and light intensity. *Journal of Experimental Marine Biology and Ecology* 122: 173-185.

Nahon, S., Charles, F., Lantoine, F., Vétion, G., Escoubeyrou, K., Desmalades, M. & Pruski, A.M. (2010). Ultraviolet radiation negatively affects growth and food quality of the pelagic diatom *Skeletonema costatum*. *Journal of Experimental Marine Biology and Ecology* 383: 164-170.

Napolitano, G.E., Pollero, R.J., Gayoso, A.M., Macdonald B.A. & Thompson, R.J. (1997). Fatty Acids as Trophic Markers of Phytoplankton Blooms in the Bahía Blanca Estuary (Buenos Aires, Argentina) and in Trinity Bay (Newfoundland, Canada). *Biochemical Systematics and Ecology* 25: 739-755.

Norton, T.A., Melkonian, M. & Andersen, R.A. (1996). Algal biodiversity. *Phycologia* 35: 308-326.

Oliveira, M.A.S., Monteiro, M.P., Robbs, P.G. & Leite, S.G. (1999). Growth and chemical composition of *Spirulina maxima* and *Spirulina platensis* biomass at different temperatures. *Aquaculture International* 7: 261-275.

Orcutt, D.M. & Patterson, G.W. (1974). Effect of light intensity upon lipid composition of *Nitzschia closterium* (*Cylindrotheca fusiformis*). *Lipids* 9: 1000-1003.

Patil, V., Tran, K. & Giselrod, H.R. (2008). Towards Sustainable Production of Biofuels from Microalgae. *International Journal of Molecular Science* 9: 1188-1195.

Patterson, G. (1970). Effect of temperature of fatty acid composition of *Chlorella sorokiniana*. *Lipids* 5: 597-600.

Parrish, C.C. & Wangersky P.J. (1987). Particulate and dissolved lipid classes in cultures of *Phaeodactylum tricornutum* grown in cage culture turbidostats with a range of nitrogen supply rates. *Marine Ecolocogy Progress Series* 35: 119-128.

Pernet, F., Tremblay, R., Demers, E. & Roussy, M. (2003). Variation of lipid class and fatty acid compostion of *Chaetoceros muelleri* an *Isochrysis* sp. grown in a semicontinuous system. *Aquaculture* 221: 393-406.

Pickett-Heaps, J.D. (1975). *Green Algae. Structure, reproduction and Evolution in Selected Genera*. ISBN 0-87893-652-1. Sinauer Associates, Inc., Publishers Sunderland, Massachusetts.

Piorreck, C.C., Baasck, K.H. & Pohl, P. (1984). Biomass production, total protein, chlorophylls, lipids and fatty acids of freshwater green and blue-green algae under different nitrogen regimes. *Phytochemistry* 23: 207-216.

Popovich, C.A., Damiani, M.C., Constenla, D. & Leonardi, P.I. (2011). Lipid quality of *Skeletonema costatum* and *Navicula gregaria* from South Atlantic Coast (Argentina): evaluation of its suitability as biodiesel feedstock. *Journal of Applied Phycology* DOI 10.1007/s10811-010-9639-y.

Pruvost, J., Van Vooren, G., Cogne, G. & Legrand, J. (2009). Investigation of biomasa and lipid production with *Neochloris oleoabundans* in photobioreactor. *Bioresource Technology* 100: 5988-5995.

Pruvost, J., Van Vooren, G., Le Gouic, B., Couzinet-Mossion, A. & Legrand, J. (2011). Systematic investigation of biomass and lipid productivity by microalgae in photobioreactors for biodiesel application. *Bioresource Technology* 102: 150-158.

Raison, J.K. (1986). Alterations in the physiological properties and thermal responses of membrane lipids: correlations with acclimation to chilling and high temperature. In: *Frontiers of Membrane Research in Agriculture*. St John, J.B., Berlin, E. & Jackson, P.G. eds. Totowa, N.J.: Rowman & Allanheld, pp. 383-401. ISBN 0-847674266.

Ramos, M.J., Fernández, C.M., Casas, A., Rodríguez, L. & Pérez, A. (2009). Influence of fatty acid composition of raw materials on biodiesel properties. *Bioresource Technology* 100: 261-268.

Rasoul-Amini, S., Montazeri-Najafabady, N., Mobasher, M.A., Hoseini-Alhashemi, S. & Ghasemi, Y. (2011). *Chlorella* sp.: A new strain with highly saturated fatty acids for biodiesel production in bubble-column photobioreactor. *Applied Energy* DOI: 10.1016/j.apenergy.2010.12.040.

Reitan, K.I., Rainuzzo, J.R. & Olsen Y. (1994). Effect of nutrient limitation on fatty acid and lipid content of marine microalgae. *Journal of Phycology* 30: 972-979.

Renaud, S.M., Parry, D.I., Thinh, L.V., Kuo, C., Padovan, A. & Sammy, N. (1991). Effect of light intensity on the proximate biochemical and fatty acid composition of *Isochrysis* sp. and *Nannochloropsis oculata* for use in tropical aquaculture. *Journal of Applied Phycology* 3: 43-53.

Rodolfi, L., Chini Zittelli, G., Bassi, N., Padovani, G., Biondi, N., Bonini, G. & Tredici, M.R. (2009). Microalgae for Oil: Strain Selection, Induction of Lipid Synthesis and Outdoor Mass Cultivation in a Low-Cost Photobioreactor. *Biotechnology and Bioengineering* 102: 100-112.

Roessler, P.G. (1988). Changes in the activities of various lipid and carbohydrate biosynthetic enzymes in the diatom *Cyclotella cryptica* in response to silicon deficiency. *Archives Biochemistry and Biophysics* 267: 521-528.

Roessler, P.G. (1990). Environmental control of glycerolipid metabolism in microalgae: commercial implications and future research directions. *Journal of Phycology* 26: 393-399.

Rousch, J.M., Bingham, S.E. & Sommerfeld, M. R. (2003). Changes in fatty acid profiles of thermo-intolerant and thermo-tolerant marine diatoms during temperature stress. *Journal of Experimental Marine Biology and Ecology* 295: 145– 156.

Santos L.M.A. & Leedale, G.F. (1995). Some notes on the ultrastructure of some azoosporic members of the algal class Eustigmatophyceae. *Nova Hedwigia* 60: 219-225.

Schenk, P.M., Thomas-Hall, S.R., Stephens, E., Markx, U.C., Mussgnug, J.H., Posten, C., Kruse, O. & Hankamer, B. (2008). Second Generation Biofuels: High-Efficiency Microalgae for biodiesel Production. *Bioenergy Research* 1: 20-43.

Sheehan, J., Dunahay, T., Benemann, J., Roessler, P. (1998). A Look Back at the U.S. Department of Energy's Aquatic Species Program - Biodiesel from Algae, National Renewable Energy Laboratory, Golden, Colorado. 294 pp.

Shifrin, N.S. & Chisholm, S.W. (1981). Phytoplankton lipids: Interspecific differences and effects of nitrate, silicate and light-dark cycles. *Journal of Phycology* 17: 374-384.

Sicko-Goad, L. & Andresen, N.A. (1991). Effect of growth and light/dark cycles on diatom content and composition. *Journal of Phycology* 27: 710-718.

Singh, S.P. & Singh, D. (2010). Biodiesel production through the use of different sources and characterization of oils and their esters as the substitute of diesel: A review. *Renewable and Sustainable Energy Reviews* 14: 200-216.

Siron, R., Giusti, G. & Berland, B. (1989). Changes in the fatty acid composition of *Phaeodactylum tricornutum* and *Dunaliella tertiolecta* during growth and under phosphorus deficiency. *Marine Ecology Progress Series* 55: 95-100.

Sobczuk, T.M. & Chisti, Y. (2010). Potential fuel oils from the microalga *Choricistis minor*. *Journal of Chemical Technology and Biotechnology* 85: 100-108.

Spoehr, H.A. & Milner, H.W. (1949). The chemical composition of *Chlorella*. Effect of environmental conditions. *Plant Physiology* 24: 120-149.

Spolaore, P., Joannis-Cassan, C., Duran, E. & Isambert, A. (2006). Commercial Applications of Microalgae. *Journal of Bioscience and Bioengineering* 101: 87-96.

Sriharan, S., Bagga, D. & Sriharan, T.P. (1990). Effects of nutrients and temperature on lipid and fatty acid production in the diatom *Hantzschia* DI-60. *Applied Biochemistry Biotechnology* 24/25: 309-316.

Suen, Y., Hubbard, J.S., Holzer, G. & Tornabene, T.G. (1987). Total lipid production of the green alga *Nannochloropsis* sp. QII under different nitrogen regimes. *Journal of Phycology* 23: 289-296.

Sukenik, A. (1999). Production of eicosapentaenoic acid by the marine eustigmatophyte *Nannochloropsis*. In: Cohen, Z. (Ed.). Chemical from Microalgae. Taylor and Francis, London pp. 41-56. ISBN: 0-7484-0515-1

Sukenik, A., Carmeli, Y. & Berner, T. (1989). Regulation of fatty acid composition by irradiance level in the Eustigmatophyte *Nannochloropsis* sp. *Journal of Phycology* 25: 666-692.

Tang, D., Han, W., Li, P., Miao X. & Zhong J. (2010). CO_2 biofixation and fatty acid composition of *Scenedesmus obliquus* and *Chlorella pyrenoidosa* in response to different CO_2 levels. *Bioresource Technology* 102: 3071-3076.

Thompson, G.A. (1996). Lipids and membrane function in green algae. *Biochimica et Biophysica Acta* 1302: 17-45.

Thompson, P.A., Guo, M-X., Harrison, P.J. & Whyte, J.N.C. (1992). Effects of variation in temperature. II. On the fatty acid composition of eight species of Marine phytoplankton. *Journal of Phycology* 28: 488-497.

Tornabene, T.G., Holzer, G., Lien, S. & Burris, N. (1983). Lipid composition of the nitrogen starved green alga *Neochloris oleoabundans*. *Enzyme Microbiology Technology* 5: 435-440.

Turpin, V., Robert, J.M. & Goulletquer, P. (1999). Limiting nutrients of oyster pond seawaters in the Marennes-Oleron region for *Haslea ostrearia*: applications to the mass production of the diatom in mesocosm experiments. *Aquatic Living Resources* 12: 335-342.

Umdu, E.S., Tuncer, M. & Seker, M. (2009). Transesterification of *Nannochloropsis oculata* microalga's lipid to biodiesel on Al_2O_3 supported CaO and MgO catalysis. *Bioresource Technology* 100: 2828-2831.

van den Hoek, C., Mann, D.G., Jahns & H.M. (1995). *Algae. An introduction to Phycology*. ISBN 0-521- 30419-9. Cambridge University Press.

Volkman, J.K., Brown, M.R., Dunstan, G.A. & Jeffrey, S.W. (1993). Biochemical composition of marine microalgae from the class Eustigmatophyceae. *Journal of Phycology* 29: 69-78.

Whittle, S.J. & Casselton, P.J. (1975). The chloroplast pigments of the algal classes Eustigmatophyceae and Xanthophyceae. I. Eustigmatophyceae. *British Phycological Journal* 10: 179-181.

Widjaja, A., Chien, C. & Ju, Y. (2009). Study of increasing lipid production from fresh water microalga *Chlorella vulgaris*. *Journal of the Taiwan Institute of Chemical Engineers* 40: 13-20.

Xu, H., Miao, X. & Wu, Q. (2006). High quality biodiesel production from a microalga *Chlorella protothecoides* by heterotrophic growth in fermenters. *Journal of Biotechnology* 126: 499-507.

Yu, E.T., Zendejas F.J., Lane, P.D., Gaucher, S., Simmons, B.A. & Lane, T.W. (2009). Triacylglycerol accumulation and profiling in the model diatoms *Thalassiosira pseudonana* and *Phaeodactylum tricornutum* (Bacillariophyceae) during starvation. *Journal Applied of Phycology* 21: 669-681.

Zhekisheva, M., Boussiba, S., Khozin-Goldberg, I., Zarka, A. & Cohen, Z. (2002). Accumulation of oleic acid in *Haematococcus pluvialis* (Chlorophyceae) under nitrogen starvation or high Light is correlated with that of astaxanthin esters. *Journal of Phycology* 38: 325-331.

Zhukova, N.V. & Aizdaicher, N.A. (1995). Fatty acid composition of 15 species of marine microalgae. *Phytochemistry* 39: 351-356.

Zhukova, T.S., Klaychko-Gurvich, T.L., Vladimirova, M.T. & Kurnosova, T.A. (1969). Comparative characteristics of the growth and direction of biosynthesis of various strains of *Chlorella* under conditions of nitrogen starvation II. Formation of carbohydrates and lipids. *Soviet Plant Physiology* 16: 96-101.

Biodiesel Features in the Railway Transport

Gediminas Vaiciunas and Leonas Povilas Lingaitis
Vilnius Gediminas Technical University, Faculty of Transport Engineering, Department of railway transport, Vilnius, Lithuania

1. Introduction

To calculate enterprise expenses on fuel, the above tax should be taken into account. Consumption and the amount of RME, compared to pure diesel oil consumption when 20 % of RME is added, fuel consumption grows by about 4 % (Lingaitis and Pukalskas, 2008), compared to pure diesel oil consumption while the use of 30 % of RME in the mixture with diesel oil increases fuel consumption by about 2.5 %. A further increase in the percentage of RME in the mixture leads to a further increase in fuel consumption. When smaller RME amounts are used, the expenses are increased insignificantly.
A mathematical model allowing fuel costs to be determined for a locomotive engine using a certain amount of biological diesel oil was developed referring to the methods of economic effect calculation described in the present paper. The relationship between the proportion of hydrocarbons (CxHy) in the exhaust emissions and the engine load is expressed by regression equations for all types of mixtures. CxHy emission is growing for all the mixtures except those containing 20 and 30 % of RME (Lingaitis and Pukalskas, 2008). The maximum is reached at 50 % P_{max}, with further decrease of emission. The smallest effect of the engine load on CxHy emission can be observed, when pure diesel oil is used. For the mixture of 10 % RME and diesel oil, CxHy emission is the lowest, being comparable with that of pure diesel oil. When the total amount of contaminants is considered without PM emission, the use of biodiesel with 30-40 % RME added is most rational from the ecological perspective.
According to the law of the Republic of Lithuania on environment pollution taxes (Lingaitis and Pukalskas, 2007), all physical and juridical persons engaged in commercial activities and using mobile equipment causing environment pollution should pay the environmental tax depending on the type of transport facilities used. The annual fixed tax of 7.53 EUR per one ton of burnt fuel should be used for rail transport in the period from 2005 to 2009.

2. Influence of biodiesel to energetic and ecological indicators of a diesel engine

Research carried out by the Austrian company "AVL LIST" and the Swedish company "MTC" (Makarevičienė et al., 2001) show that the amount of SP (solid particles) in oxides is proportional to the amount of sulphur in the fuel. Having increased amount of sulphur by 0.1 % of the mass, the SP emission increases by 0.027 g/(kW·h). The SP emission may be reduced significantly by using alternative fuel (Xiao et al., 2000). Research carried out by Swedish scientists show that, when using ethanol, gas or dimethyl ether (DME) in a diesel engine, the SP emission is reduced 7 times than when the engine operates using the Ec1 diesel or RME (rapeseed methyl ester) complying with the strictest standard. Besides, the results of these research state that, when using a mixture of 15 % of ethanol and 85 % of diesel, the SP emission is reduced by 30…50 %, the CO emission is reduced slightly, the amount of HC increases, whereas the amount of NOx is the same as when operating using pure diesel. The results of the research are shown in fig. 1 (Lingaitis and Pukalskas, 2008).

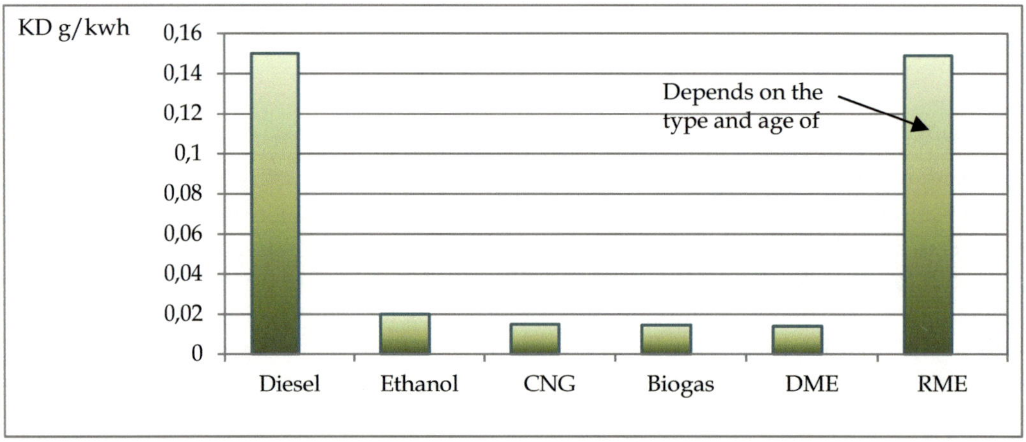

Fig. 1. The SP emission of diesel engines when using various fuel and operating at the ECE R49 cycle: CNG - compressed natural gas; DME - dimethyl ether; RME - rapeseed oil fatty acid methyl ester

It is also known (Wang et al., 1997; Zelenka et al., 1990) that approximately 90 % of oxides of nitrogen oxides contain NO. The reaction takes place at a very high temperature exceeding 1800°C where there is enough oxygen. Thus the origin of NO is thermal. When the combustion process is perfect, the temperature of the cycle is higher (the COP of the cycle is also higher) and the oxides contain more NO. It was determined that the chemical composition of the fuel does not influence the amount of NO in oxides.

A thorough research generalizing works carried out by approximately 80 researchers was carried out in the USA. This research was carried out by the Environmental Protection Agency, an agency famous in all states that carries out environment protection research. Data from various works analyzing different diesel vehicles from heavy goods trucks to cars was collected and analyzed for this generalizing research. All these vehicles were designed for using the ordinary diesel but they used pure biodiesel or various mixtures containing diesel during the experiments.

Both vegetable- and animal-based biodiesel was used in the research; therefore, calorific value thereof fluctuated from 32.25 to 33.230 MJ/l, whereas average calorific value of diesel amounted to 36.094 MJ/l (Lingaitis and Pukalskas, 2008). Having analyzed the data available, it turned out that increase of fuel consumption was only 4.6 % when using diesel. It was theoretically reasoned that this difference should be higher, as calorific value of vegetable-based biodiesel is 7.9 % and animal-based biodiesel is 10.6 % lower than that of mineral diesel. Comparative data about emission of pollutants is presented in fig. 2.

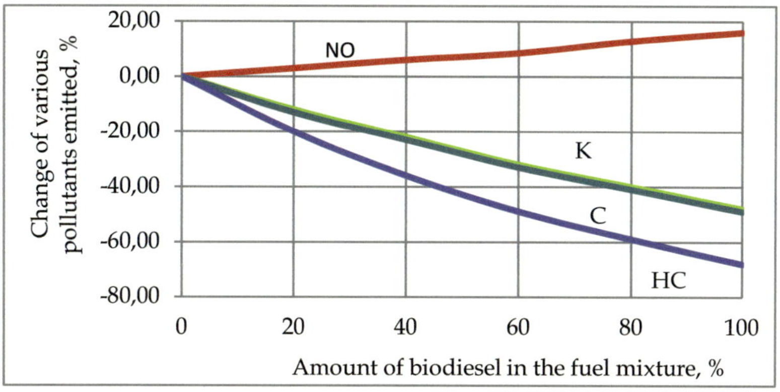

Fig. 2. Dependence of the change of various pollutants on the amount of biodiesel in the fuel mixture

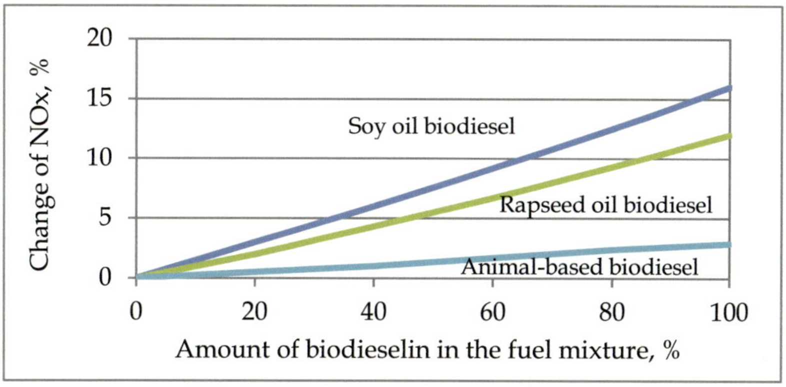

Fig. 3. Dependence of the amount of nitrogen oxides NOx on the type of biodiesel and the amount thereof in the fuel mixture

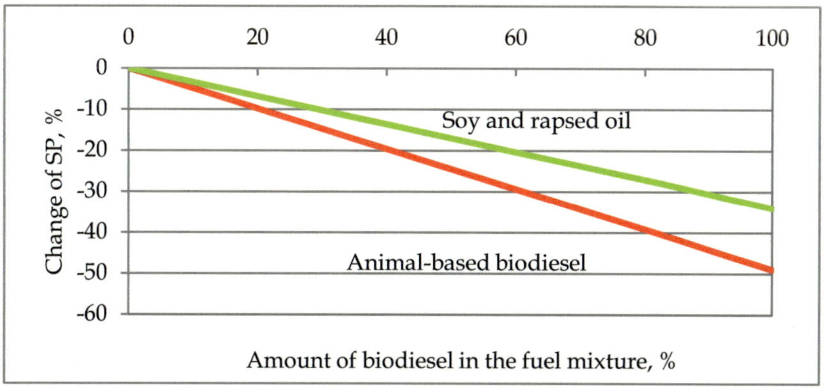

Fig. 4. Dependence of solid particles on the type of biodiesel and the amount thereof in the fuel mixture

It is obvious that biodiesel increases the NOx emission and reduces the emissions of solid particles, CO and CxHy. The research states that an addition of 10 % of biodiesel increases the NOx emission by 1 %. However, it is noteworthy that biodiesel of different origin affects the composition of the pollutants emitted differently (fig. 3, 4 and 5 (Lingaitis and Pukalskas, 2008)).

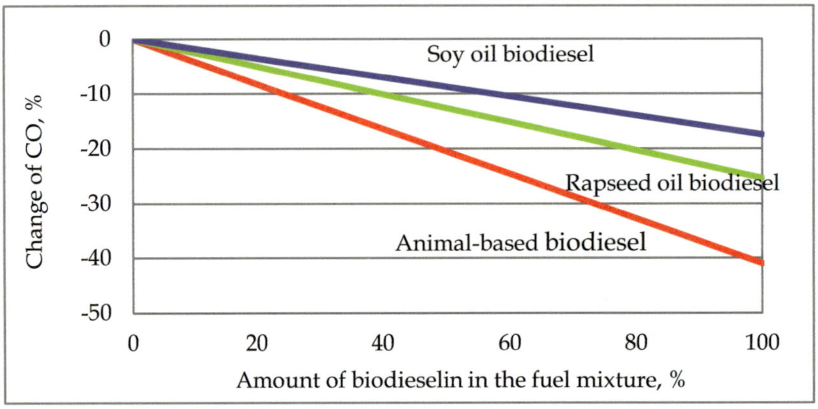

Fig. 5. Dependence of the amount of carbon monoxide CO on the type of biodiesel and the amount thereof in the fuel mixture

It can be seen from the examples presented that soy oil biodiesel increases the amount of NOx pollutants the most (fig. 3) and reduces the amounts of solid particles (fig. 4) and CO (fig. 5) the least, whereas animal-based biodiesel is the best one in ecological respect, although having the lowest calorific value, as it increases the amount of NOx the least (fig. 4) and reduces the amounts of solid particles (fig. 5) and CO (fig. 5) the most.

2. Experimental biodiesel tests

The experimental tests were carried out in the production premises designed for testing engines of UAB (Private Limited Liability Company) "Vilniaus lokomotyvų remonto depas". The tests were carried out in a closed room. The ambient air temperature fluctuated from 5 to 10°C.

The engine of a Hungarian-made diesel train D1 was used for the tests. The characteristics thereof are presented in table 1.

During the course of the experiments, the engine being tested was secured in the engine testing stand. The stand consists of an AC generator, a liquid rheostat and a control panel. The generator is connected to the engine by a cardan shaft. The engine produces electrical current by rotating the generator; the current is supplied to the liquid rheostat. Here electrical current is transformed into heat that disperses in the environment. The engine load is changed by regulating the depth of submersion of the electrodes of the liquid rheostat in electrolyte.

The cooling system of the engine does not contain a cooling radiator. A water tank of approximately 2 m^3 volume is used for cooling. When water in it reaches maximum permissible temperature, a tap is opened and the tank is cooled by water from water-supply.

No	Indicator	Value
1.	Make	12VFE17/24
2.	Type	four strokes, pre-chamber, V-shaped
3.	Number of cylinders	12
4.	Cylinder diameter, mm	170
5.	Piston stroke, mm	240
6.	Engine revolutions, min^{-1}: minimum maximum	 550 1250
7.	Compression degree	13.6
8.	Average effective pressure, MPa	0.551
9.	Nominal power, kW	538
10.	Operating volume of the engine, l	65.3
11.	Average piston speed, m/s	10
12.	Comparative fuel consumption at nominal power, g/(kW h): fuel engine oil	225+8 % 4
13.	Maximum permissible temperature, °C: water engine oil (recommended)	 95 95 (80)
14.	Type of turbo compressor	centrifugal, single-step
15.	Rotor revolutions of turbo compressor, min^{-1}	21 000
16.	Inflation pressure, MPa	0.09±0.015
17.	Fuel injection angle, °	21±1.5
18.	Engine dimensions, mm: length × width × height	 3210 × 1300 × 1700
19.	Engine mass without oil and water, kg	4600

Table 1. Technical characteristics of the engine of the diesel train D1

An additional pump was used for supplying fuel from the fuel tank to the high-pressure fuel pump of the engine.

The control panel of the stand regulates the engine load (by changing the value of the current produced by the generator) and revolutions (by regulating the fuel pump).

Fuel of two types was used for the tests: diesel produced by UAB "Mažeikių nafta" conforming with the requirements of the standard LST EN 590 and biodiesel produced by UAB "Rapsoila", i.e. fatty acid methyl ester (RME), conforming with the requirements of the standard LST EN 14214:2003.

Three different methods and units were used for analyzing various components contained in exhausted gas:

1. the weight method was used for determining the amount of solid particles;
2. the electrochemical method and the unit "Testo 350-M/XL" was used for determining the amounts of nitrogen oxides (NOx), carbon monoxide (CO) and carbon dioxide (CO_2);

3. the amount of carbohydrates (CxHy) was determined by the chromatographic method and the chromatograph SRI 8610.

The concentration of solid particles in the oxides of the engine was determined by the weight method. Perchlorvinyl AFA-VP-20 type filters permeable to air, on which solid particles contained in the oxides attracted by the aspirator deposit, were used for that purpose. The difference of the weight of the filters before and after sampling divided by the volume of the air pumped shows the concentration of solid particles. The filters are weighted by the analytical scale VLR-200. The filters are kept in the exicator for 3...4 hours before weighting in order to remove moisture from them. Error of the method: ±25 %.

The characteristics of the parameters measured by the unit "Testo 350-M/XL" are presented in table 2.

No	Parameter measured	Measurement limits	Measurement accuracy
1.	Oxygen (O_2), %	0...25	< 0,8 %
2.	Carbon monoxide (CO), ppm	0...10,000	<5 ppm at 0...99 ppm <5 % mv at 100...2,000 ppm <10 % mv at 2,100...10,000 ppm
3.	Nitrogen oxides, ppm: NO	0...3,000	<5 ppm at 0...99 ppm <5 % mv at 100...2,000 ppm <10 % mv at 2,100...3,000 ppm
	NO_2	0...500	<5 ppm at 0...99 ppm <5 % mv at 100...500 ppm
	NO_x	0...3,000	<5 ppm at 0...99 ppm <5 % mv at 100...2,000 ppm <10 % mv at 2,100...3,000 ppm
4.	Sulphur dioxide (SO_2), ppm	0...5,000	<5 ppm at 0...99 ppm <5 % mv at 100...2,000 ppm <10 % mv at 2,100...5,000 ppm
5.	Lambda λ	0.5...2.0	0.1
6.	Carbon dioxide (CO_2), %	0...50 %	<0.3 % at 0...25 % <0.5 % at 25..50 %
7.	Gas temperature, °C	0...1,000	<0.1°C at 100°C <1°C at 100...1,000°C

Table 2. Technical characteristics of the unit "Testo 350-M/XL" measuring pollutants

Measurements of the speed of gas flow were also carried out. Pneumatic pipes allowing measuring dynamic pressure of gas by a micromanometer are used for determining the speed of gas flow in the air duct. Dynamic pressure PD is equal to the difference of total pressure PB and pressure of the building PS. As dynamic pressure is proportionate to the speed of gas square, the speed of gas flow in the air duct is measured indirectly by recording fluctuations of dynamic pressure.

A short description of the course of experimental tests is presented. The following was measured during the tests:

1. amount of fuel used by the engine (by the weight method);

2. time of tests;
3. revolutions of the engine;
4. value of electrical current and voltage produced by the load generator;
5. components of exhausted gas: NOx, CO, CO_2, CxHy and SP;
6. pressure generated by the turbo compressor in the air intake collector;
7. temperature of the air sucked;
8. temperature of engine oil and coolant.

The tests were carried out using the engine heated up to the operating temperature. Temperature of water cooling the engine fluctuated between 70...80°C during the tests. It was not possible to keep a more stable temperature mode of the engine because of the old, non-automated oil and engine cooling units of the stand.

The plan of the tests containing the course of the experiment is presented in table 3. In the beginning, the tests were carried out using pure diesel at 4 different engine modes. Fuel consumption was measured, exhausted gas was analyzed and the temperature state of the engine was observed during them. After that, 10 % of RME was added to diesel and tests were carried out at the same engine operation modes as when testing pure diesel. In this way, tests were repeated with the engine operating with an increasing load and, after that, by increasing the amount of RME in the fuel mixture by 10 %. The results obtained were recorded in the table.

No.	Fuel	Engine mode		
		revolutions, min^{-1}	generator current, A	generator voltage, V
1.	diesel	540	0	0
2.	diesel	1,050	600	300
3.	diesel	1,250	700	350
4.	diesel	1,325	800	400
5.	diesel + 10 % of RME	540	0	0
6.	diesel + 10 % of RME	1,050	600	300
7.	diesel + 10 % of RME	1,250	700	350
8.	diesel + 10 % of RME	1,325	800	400
9.	diesel + 20 % of RME	540	0	0
10.	diesel + 20 % of RME	1,050	600	300
11.	diesel + 20 % of RME	1,250	700	350
12.	diesel + 20 % of RME	1,325	800	400
13.	diesel + 30 % of RME	540	0	0
14.	diesel + 30 % of RME	1,050	600	300
15.	diesel + 30 % of RME	1,250	700	350
16.	diesel + 30 % of RME	1,325	800	400
17.	diesel + 40 % of RME	540	0	0
18.	diesel + 40 % of RME	1,050	600	300
19.	diesel + 40 % of RME	1,250	700	350
20.	diesel + 40 % of RME	1,325	800	400

Table 3. Plan of experimental tests

An electronic scale with a tank with the fuel being tested placed on it was used for measuring fuel consumption. Each test lasted for about 10 min. The amount of fuel used was weighted and measurements of pollutants were carried out during this time. Each test was repeated three times in order to ensure accuracy.
The results of the tests were processed on the basis of the least squares method and are presented in fig. 6–16.
When analyzing the charts presented, we can see that comparative fuel consumption (g/(kW·h)) increases slightly when the engine operates at low load mode up to 0.30...0.35 maximum power (P_{max}) and having mixed rapeseed oil fatty acid methyl ester (RME) compared to consumption of pure diesel. It may be maintained that this increase is directly proportionate to the amount of RME in diesel. However, consumption of the mixtures approaches consumption of pure diesel when increasing the engine load and, having reached 0,6 P_{max}, this consumption becomes similar to consumption when using pure diesel.

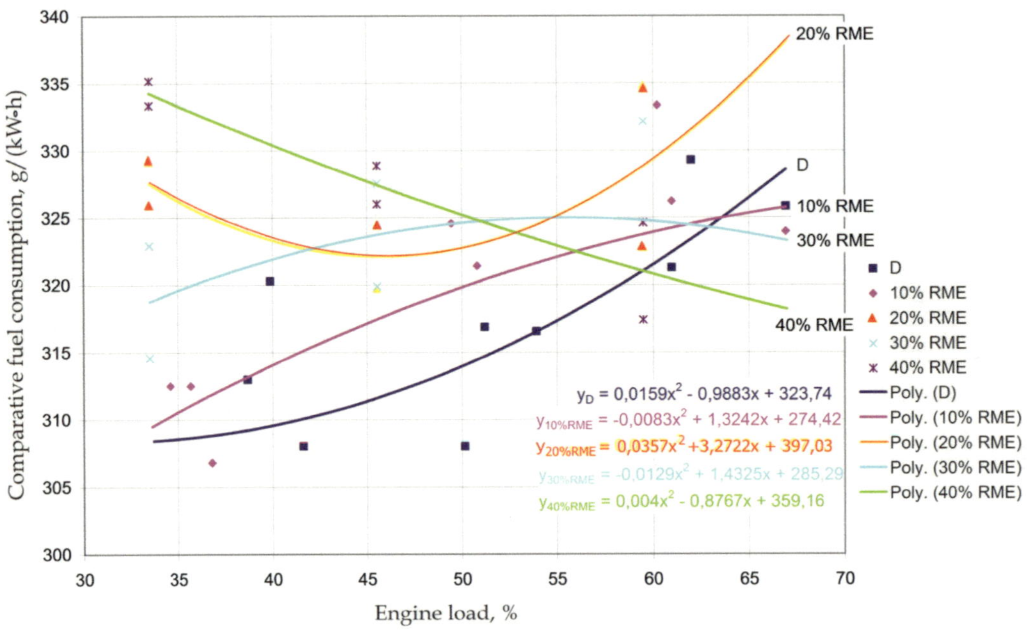

Fig. 6. Dependence of comparative fuel consumption on the engine power (% P_{max}) when using various fuel mixtures

We believe that oxygen contained in RME does not significantly influence improvement of effectiveness of the combustion process and fuel consumption increases slightly due to lower calorific value of RME (approximately 8…9 % compared to diesel) when the engine operates at low loads with a high air surplus coefficient. However, as mentioned before, having increased the engine load more than 0.5 P_{max} and reduced the air surplus coefficient in the engine cylinders, this additional oxygen improves the combustion process of the mixture and reduces fuel consumption.
This can be obviously seen in the chart of percentage comparison of comparative fuel consumption of various fuel mixtures and pure diesel presented in fig. 7. When the engine power amounts to $P \leq 0.5 P_{max}$, comparative fuel consumption increases in proportion to the

increase of the amount of RME in the mixture; however, when the engine power reaches 0.65 P_{max}, comparative fuel consumption of the mixtures starts decreasing (excluding the mixture of 10 % of RME and diesel). Here we can see a reverse result: the higher the amount of RME in the mixture, the lower the comparative fuel consumption.

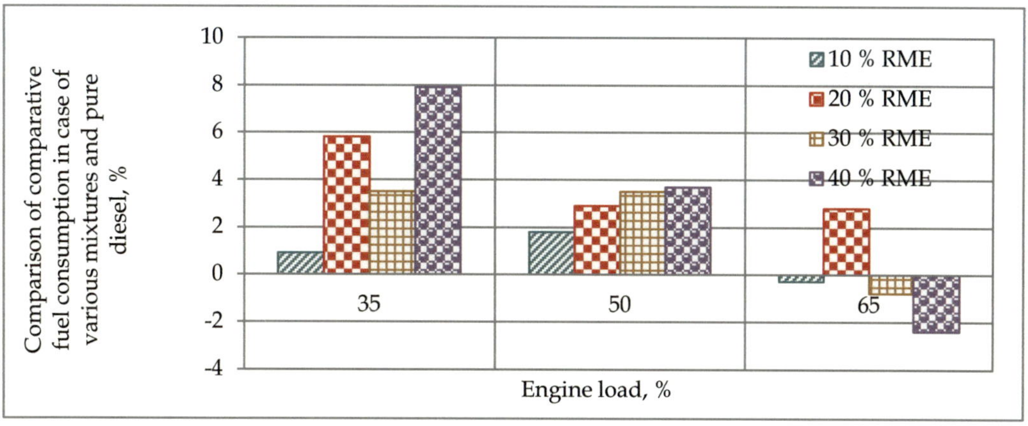

Fig. 7. Percentage comparison of comparative fuel consumption in case of various fuel mixtures and pure diesel

When analyzing emission of solid particles (exhaust smoke) presented in fig. 8, we can also see the same situation as in case of fuel consumption. When the engine power reaches 0.5 P_{max}, exhaust smoke is the lowest and when using the mixture containing 10 % and 20 % of RME and diesel, it is even lower than that of pure diesel.

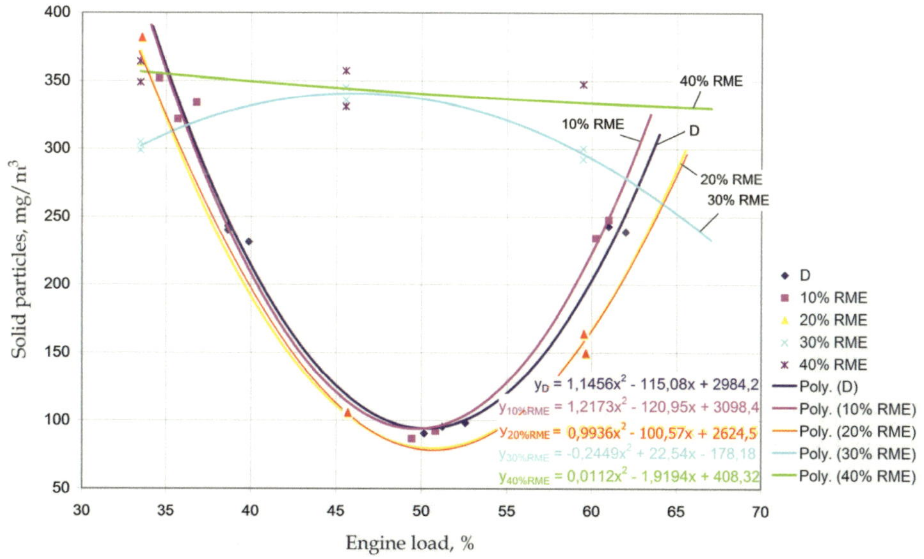

Fig. 8. Dependence of emission of solid particles on the engine power (% P_{max}) when using various fuel mixtures

However, when the amount of RME in diesel amounts to 30 % or 40 %, regularity of exhaust smoke changes radically. Trying to find out the reasons of this change was unsuccessful. The curves of emission of solid particles of the mixtures containing 30 % and 40 % of RME and diesel show that exhaust smoke is reduced when the engine load increases; however, exhaust smoke is higher when using the mixture containing 40 % of RME and diesel than when using the mixture containing 30 % of RME and diesel.

Higher exhaust smoke means poorer combustion of the mixture in the engine cylinders, although it should be vice versa due to a higher amount of oxygen in biodiesel. In order to clarify contradiction once and for all, additional tests with a few different engines should be carried out.

It can be seen from the chart presented in fig. 9 that SP emission is lower than when using diesel in almost all cases. Its values exceed SP emission of diesel even 2.5 times at two points, reliability of which is the most doubtful, only.

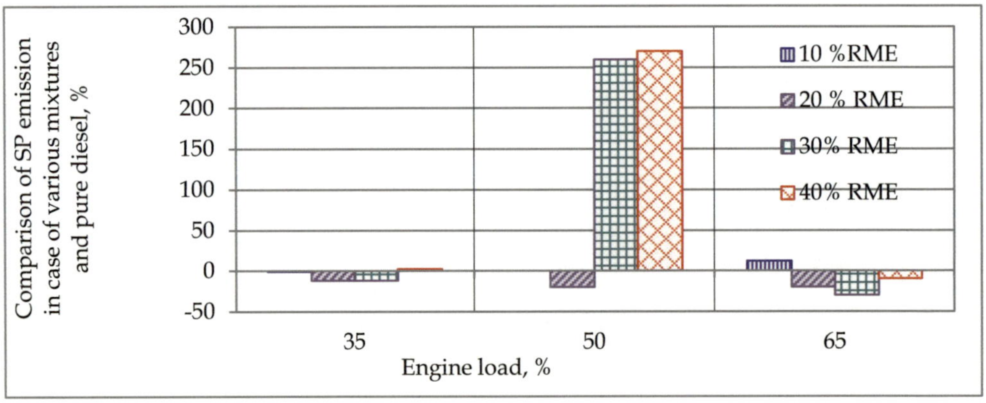

Fig. 9. Percentage comparison of emission of solid particles in case of various mixtures to emission when using pure diesel

Emissions of nitrogen oxides (NOx) are presented in fig. 10. The charts clearly show that the amount of NOx in oxides decreases when the amount of RME in diesel increases (although a reverse effect is known from the literature). Besides, it also depends on the engine power achieved: the higher the engine power, the lower the amount of NOx.

Knowing that intensity of formation of NOx is proportionate to the combustion temperature (the higher the temperature, the higher the amount of NOx), reduction of pollutants when the engine power increases can be easily explained: as the load increases and the air surplus coefficient decreases, the mixture becomes richer; therefore, combustion temperature decreases and conditions for formation of NOx worsen.

The values of carbon dioxide CO_2 are very important. This gas causes the "greenhouse effect". Dependence of the change of emission thereof on the amount of RME in diesel and on the engine load is shown in fig. 11. Compared to pure diesel when the engine load does not exceed 0,5 P_{max} at various amounts of RME (up to 30 %) in diesel, the amount of CO_2 gas emitted is always lower, excluding the case when the amount of RME in diesel is 40 %. As the engine load increases, emissions of CO_2 gas of all mixtures and pure diesel increase proportionately and regularities thereof are similar. Only CO_2 values of the mixtures containing 20 % and 30 % of RME may be distinguished; they increase slightly more sharply than those of other mixtures or pure diesel as the engine load increases. However, they increase insignificantly and, having assessed errors, it may be maintained that CO_2 values of all mixtures are similar.

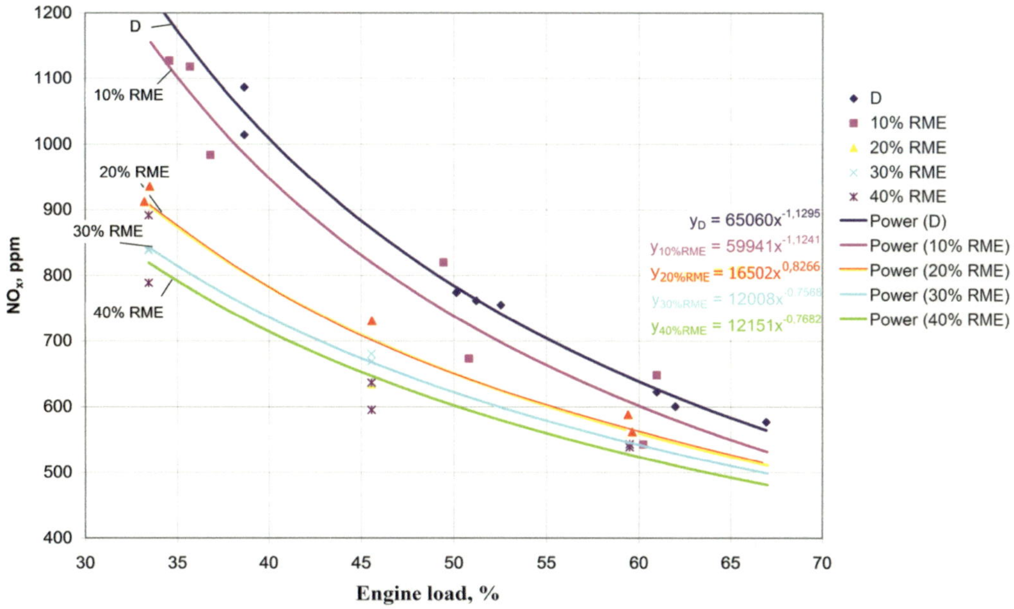

Fig. 10. Dependence of emission of nitrogen oxides on the engine power (% P_{max}) when using various fuel mixtures

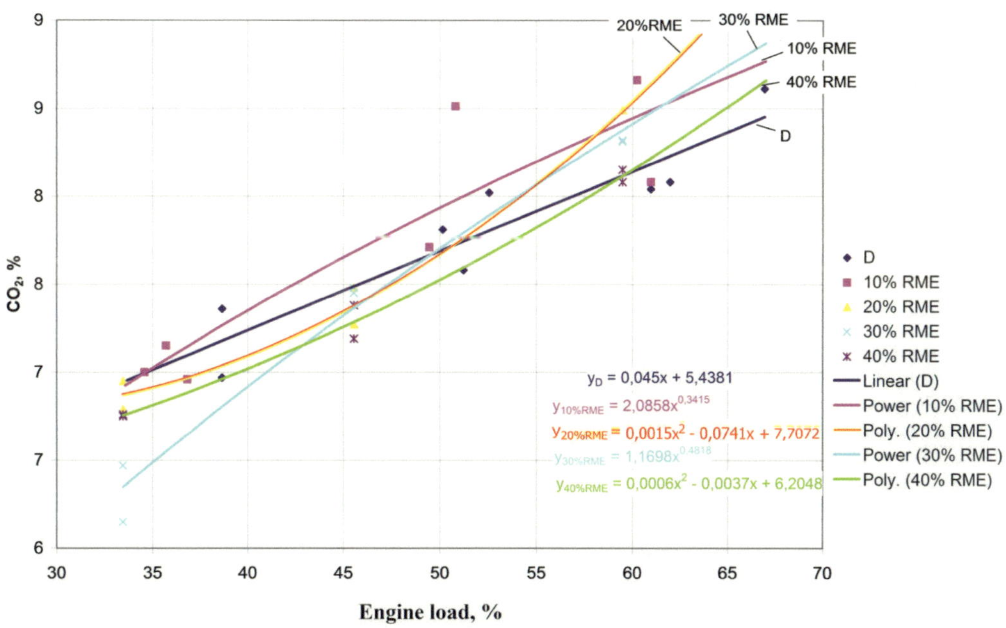

Fig. 11. Dependence of emission of CO_2 on the engine power (% P_{max}) when using various fuel mixtures

On the one hand, lower formation of CO_2 seemingly reduces formation of the "greenhouse effect"; on the other hand, higher emission of CO_2 gas shows better combustion of the combustible mixture and, consequently, lower fuel consumption.

Scientific literature upholds the opinion that burning biofuel does not increase the concentration of CO_2 in the atmospheric air, as growing plants, from which this fuel is produced, absorb the same amount of CO_2 that was emitted during combustion.

When analyzing emissions of charcoal fumes, i.e. carbon monoxide (CO) (fig. 12), it can be seen that, as the engine load increases, when $P \geq 0,5\ P_{max}$, CO increases at all compositions of all the mixtures tried. It increases slightly slowly when the amount of RME in diesel is 30...40 %.

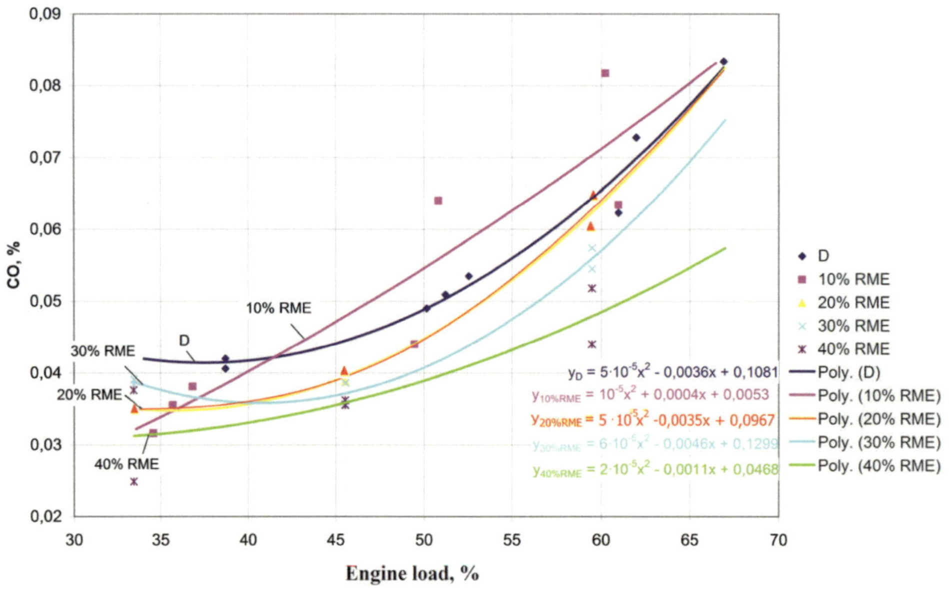

Fig. 12. Dependence of emission of CO on the engine power (% P_{max}) when using various fuel mixtures

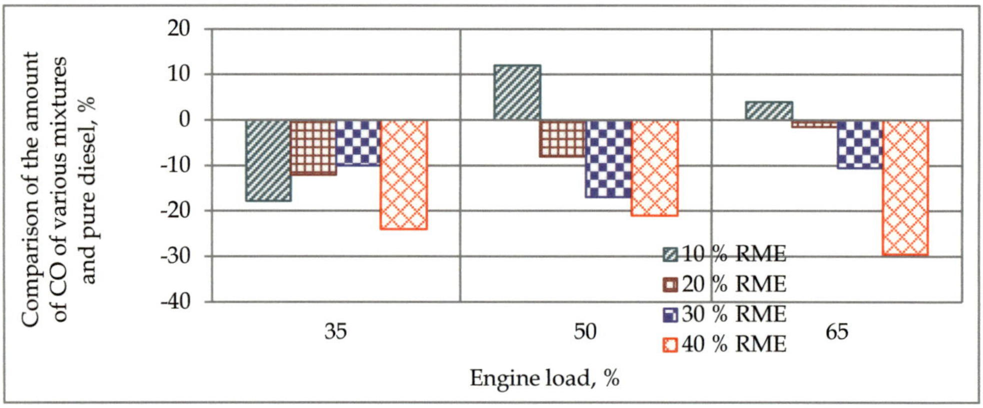

Fig. 13. Percentage comparison of the amount of CO to CO emission when using pure diesel

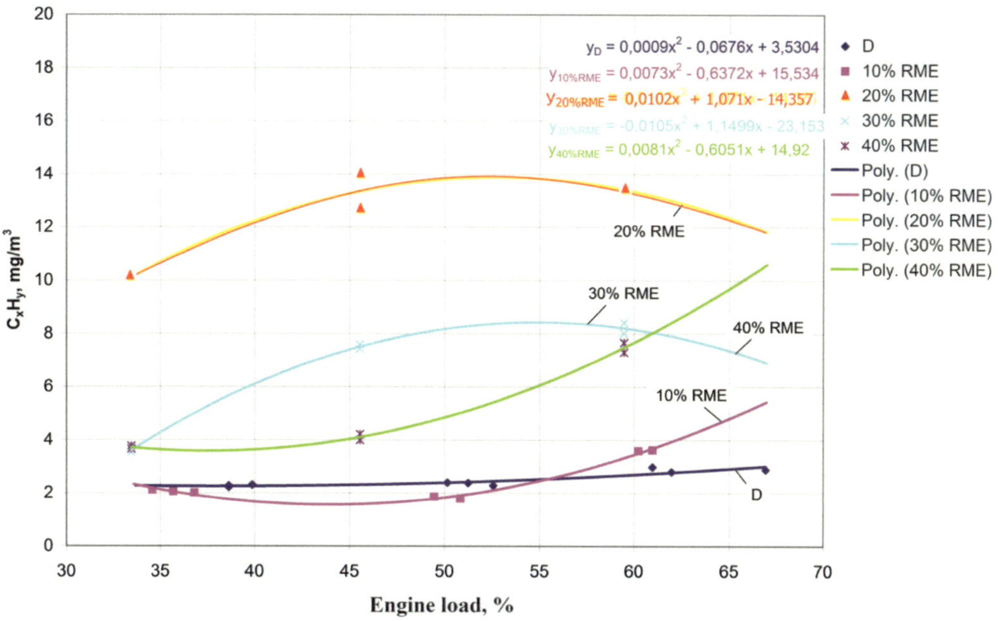

Fig. 14. Dependence of emission of CxHy on the engine power (% P_{max}) when using various fuel mixtures

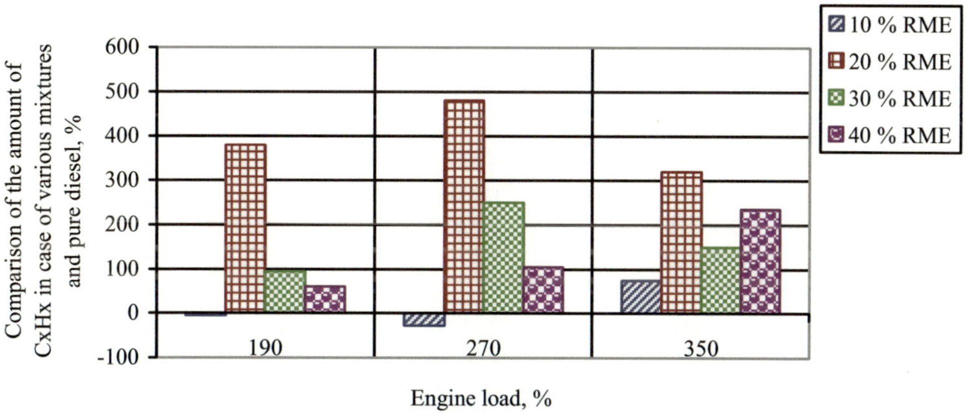

Fig. 15. Percentage comparison of CxHy emission to CxHy emission when using pure diesel

It can be seen from these charts that, as the engine load increases, these emissions increase too, excluding the mixtures containing 20 % and 30 % of RME. Emissions thereof reach maximum at the value 50 % P_{max} and keep decreasing afterwards. The load has the lowest impact on the CxHy emission when the engine operates using pure diesel. The values of CxHy of the mixture containing 10 % RME and diesel are ones of the lowest and close to the values of CxHy pollutants of diesel.

As the engines of diesel trains operate at idle run a lot of time, i.e. they are not turned off in railway stations, when waiting for the permissible traffic lights signal etc, it is rational to perform tests on the unloaded engine using mixtures of various compositions.

Percentage comparison of emission of solid particles when mixtures of various compositions (10 %, 20 %, 30 % and 40 % of RME) are combusted and the engine is unloaded to emissions when pure diesel is combusted is shown in fig. 16. We can see that exhaust smoke increases when the mixture contains 30 % and 40 % of RME; in all other cases, exhaust smoke is lower than exhaust smoke of pure diesel.

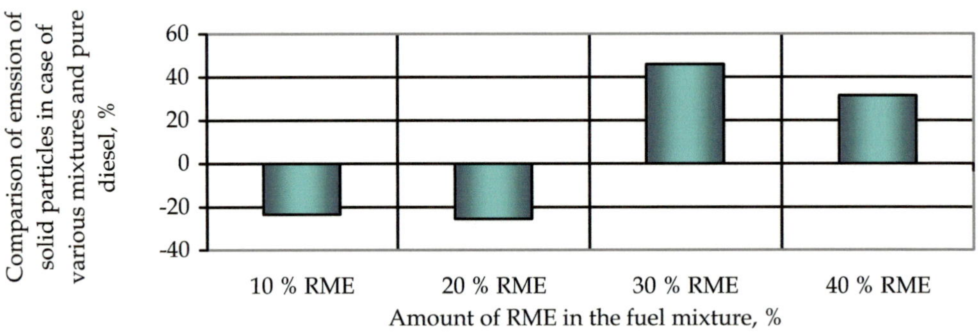

Fig. 16. Comparison of emission of solid particles in case of various mixtures to emission when using pure diesel

Naturally, it impossible to choose a different composition of the mixtures for all the engine loads; these values suggest that the most rational option is using biodiesel containing 20 % of RME, as, in this case, exhaust smoke is reduced by 25 % compared to exhaust smoke of pure diesel.

We can see from fig. 17 that the amount of nitrogen oxides in oxides increases significantly (up to 27 % compared to pure diesel) when the mixture contains 20 % of RME; in all other cases, emissions of these pollutants are significantly lower. With regard to these amounts of pollutants, the amounts of NOx are the lowest in case of the mixture containing 10 % of RME. They are lower by 15 % compared to the amount of NOx in case of pure diesel.

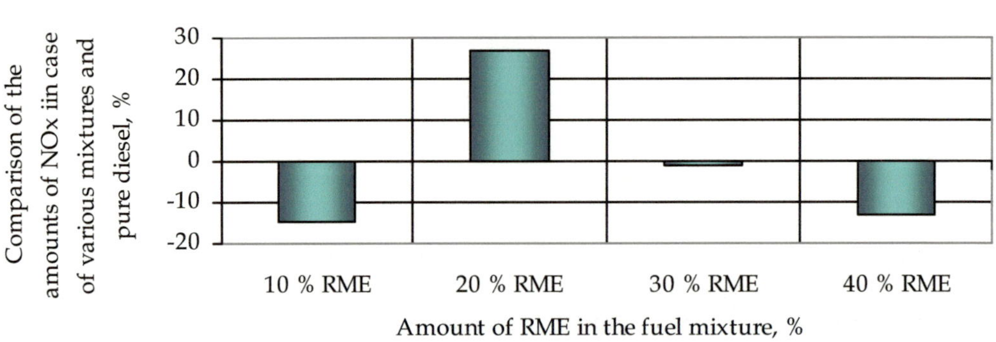

Fig. 17. Comparison of NOx emission in case of various mixtures to emission when using pure diesel

We can see CO_2 emissions to the environment at idle run and different composition of fuel in fig. 18. It can be clearly seen here that CO_2 emissions to the environment increase by up to 3 % when the mixture contains 20 % of RME compared to pure diesel. In all other cases, CO_2 decreases and, when the mixture contains 40 % of RME, CO_2 decreases by up to 10 % compared to emissions when using pure diesel.

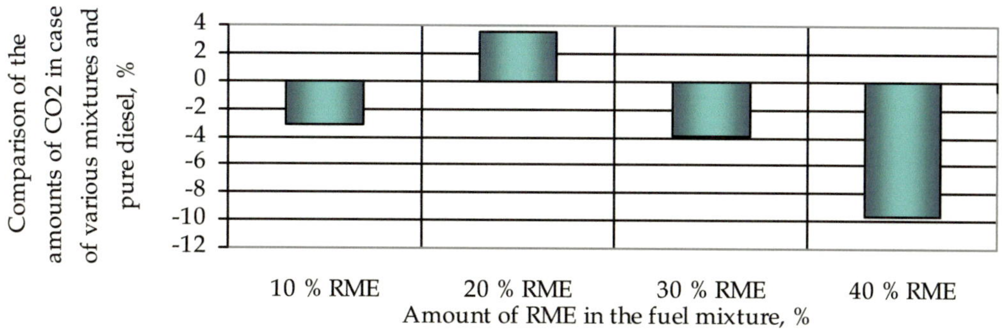

Fig. 18. Comparison of CO_2 emission in case of various mixtures to emission when using pure diesel

The amount of charcoal fumes (CO) emitted to the environment decreased from 5 to 26 percent in all cases compared to emissions when using pure diesel (fig. 19). The amount of CO decreased in the oxides the most when the engine operated using a mixture containing 20 % of RME and diesel: even up to 26 % compared to CO emission when using pure diesel.

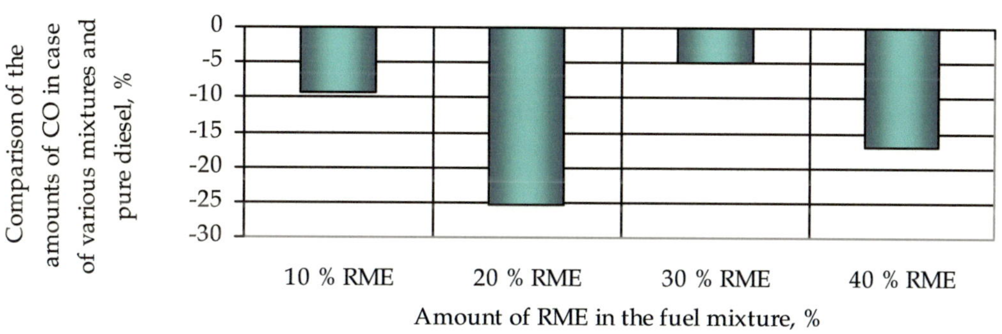

Fig. 19. Comparison of CO emission in case of various mixtures to emission when using pure diesel

When adding RME to diesel, the amount of carbohydrates (CxHy) in the oxides increases from 151 % to 290 % compared to emissions when using pure diesel (fig. 20), excluding the mixture where the amount of RME is 10 %. In this case, the amount of CxHy emitted to the environment is reduced by approximately 30 %.

Generalizing the results obtained, an optimal composition of the mixture should be chosen in accordance with all the criteria analyzed and not only the pollutants emitted to the environment together with oxides should be taken into account but also the level of toxicity of these pollutants and fuel consumption.

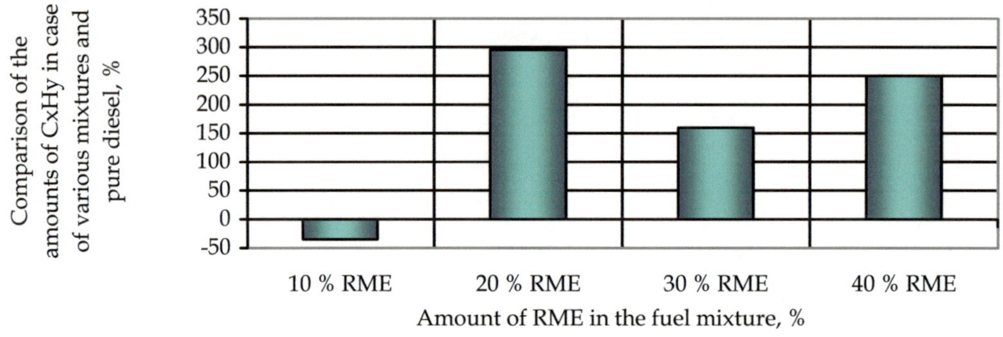

Fig. 20. Comparison of CxHy emission in case of various mixtures to emission when using pure diesel

Mixtures containing 5 % of fatty acid methyl esters with mineral diesel are used in a lot of EU countries. They conform to the requirements of the standard LST EN 590 and do not cause any operational issues. Biodiesel is used in such mixtures as an addition to mineral diesel, such mixtures must not be specially market at the selling spots and are sold as ordinary diesel.
Higher-concentration biodiesel mixtures containing mineral diesel are used in Italy (25 % of RME) and the Czech Republic (30 % of RME). It was determined that a mixture containing 30 % of biodiesel and mineral diesel is optimal with regard to the operating and environment protection features. When using such a mixture, fuel consumption increases slightly and engines do not need special preparation (this is a must when using pure biodiesel). Besides, less nitrogen oxides that are characterized by features causing the greenhouse effect are emitted to the environment and biologic decomposition of the mixture in the nature conforms to the requirements established for biofuel. The rules on trade in oil products, biofuel, bio oil and other combustible liquid products in the Republic of Lithuania stipulate that pure RME and mixtures containing 5 % and 30 % (volume) of RME with mineral diesel may be sold in our Republic.

3. Generalization of stand tests

Having processed the results obtained during these tests, mathematic dependence of fuel consumption of the engine power and the amount of RME was designed:

$$B = 2.92481 + 0.27641 \cdot P + 0.10377 \cdot$$

$$10^{-3} \cdot P^2 + 0.25431 \cdot B_{RME\%} - 5.07103 \cdot 10^{-3} \cdot B^2_{RME\%}\,; \quad (1)$$

here: P – engine power, kW; $B_{RME\%}$ – amount of RME in total volume of fuel, %.
The graphic form of this mathematic expression is presented in fig. 21.
We can see in the chart presented that, as the engine power increases, fuel consumption increases rapidly, as more fuel is needed for obtaining energy and the amount of biofuel influences fuel consumption slightly.

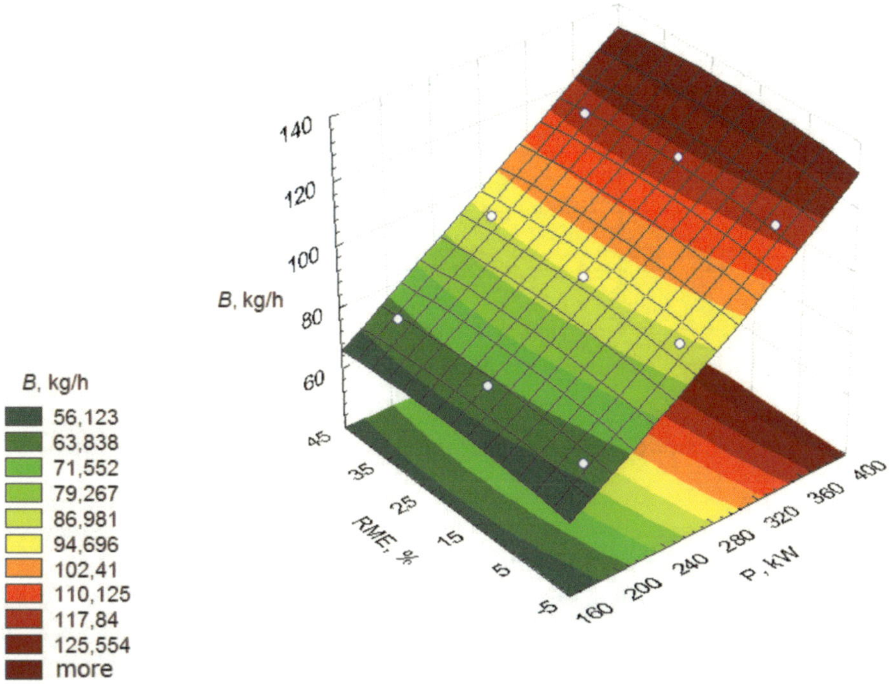

Fig. 21. Dependence of total fuel consumption B (when using a mixture of diesel and RME) on the amount of RME and the engine power P

4. Economic calculations

We calculate annual expenses for fuel of diesel passenger trains. It is complicated to calculate accurately the economic effect that may be achieved when using RME, as fuel consumption varies and prices fluctuate constantly.

Annual (year 2004...2006) fuel consumption of passenger trains of AB (Public Limited Liability Company) "Lietuvos gelezinkeliai" is presented in table 4.

Year	Fuel consumption, t
2004	1301.5
2005	1632.1
2006	3251.5

Table 4. Annual fuel consumption of diesel passenger trains, year 2004...2006

Having assessed annual increase of fuel consumption, it may be forecasted that it shall amount to 3251.5 t/m in 2007 (fig. 22).

Knowing the amount of fuel consumed Bdyz and prices of fuel Kdyz, we can calculate annual expenses for fuel Z, (EUR/y.):

$$Z = B_{dyz} \cdot K_{dyz}; \qquad (2)$$

here: B_{dyz} – annual fuel consumption of diesel trains, t/m.; K_{dyz} – price of diesel paid by the company, EUR/t.

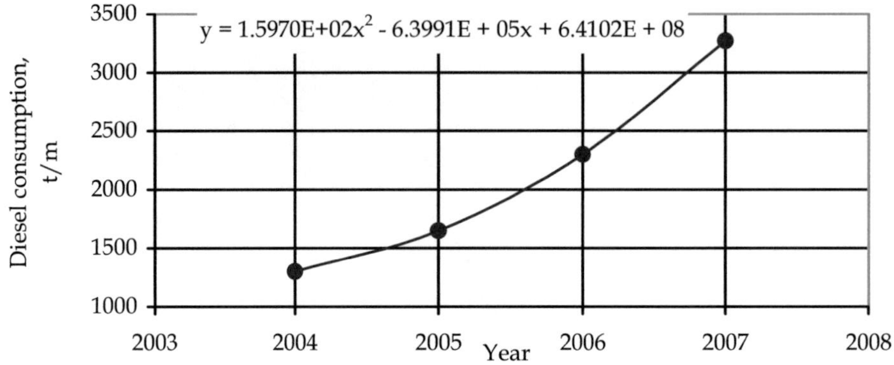

Fig. 22. Annual fuel consumption of diesel passenger trains (the fact and prognosis for the year 2007)

Having performed experimental tests, it was determined that fuel consumption increased by 2.95 % at an average when using RME compared to consumption of pure diesel; therefore, this increase must be assessed when calculating annual expenses for fuel. When using RME, total fuel consumption changes not in proportion to the amount thereof but in accordance with a certain law which is presented in fig. 23.

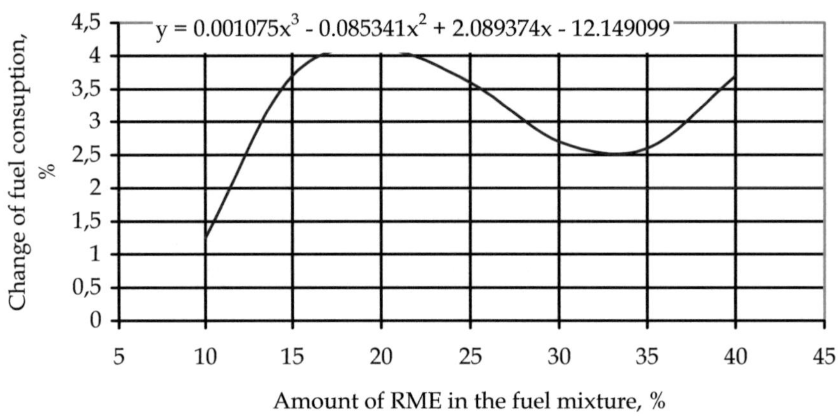

Fig. 23. Dependence of the change of fuel consumption on the amount of RME compared to consumption of pure diesel

The chart presented in fig. 23 shows that fuel consumption increases by up to 4 % when using a mixture containing 20 % of RME compared to consumption of pure diesel, it increases by up to 2.5 % using a mixture containing 30 % of RME and consumption keeps increasing when the amount of RME in the mixture is increased.
Knowing that total fuel consumption is:

$$B = B'_{dyz} + B_{RME}; \tag{3}$$

here: B'_{dyz} – amount of diesel when using biodiesel and having assessed total increase of fuel consumption due to the difference in calorific value.
The chart of the change of expenses for fuel is presented in fig. 24. The calculations were carried out at the prices of fuel and taxes in the 4th quarter of 2006 and having assessed the change of fuel consumption due to usage of RME.

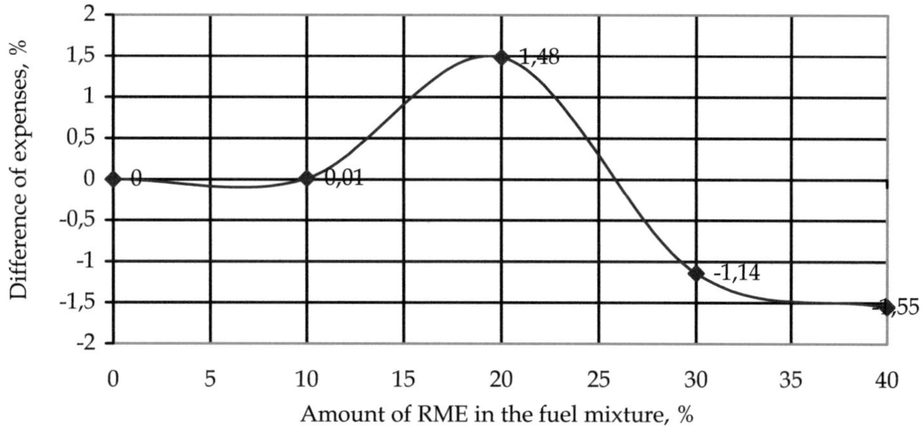

*The prices of fuel valid in the end of 2006 were used for the calculations

Fig. 24. Difference of expenses for fuel when using a different amount of RME and having assessed the change of fuel consumption due to usage of RME compared to expenses when using pure diesel

The figure (fig. 24) shows that expenses for fuel (and environmental taxes) shall change insignificantly when using a fuel mixture containing 10 % of RME compared to expenses when using pure diesel; expenses shall increase by 1.48 % when using a fuel mixture containing 20 % of RME; expenses shall decrease by 1.14 % when using a fuel mixture containing 30 % of RME; expenses shall decrease by 1.5 % when using a fuel mixture containing 40 % of RME.

5. Environment protection calculations

When analyzing data concerning pollutants obtained during the experiment (fig. 6, 21 and 23), a calculation was carried out in accordance with the "Methodology of assessment of pollutants emitted to the atmosphere by locomotives and diesel trains" presented in the normative instrument LAND 18-2003/M-03 for environment protection.
This instrument is a corrected methodology of calculation of pollutants emitted by the railway transport presented in the normative instrument LAND 18-96/M-03 compiled having assessed the recommendations of the program CORINAIR of the European Union (EU) and other methodologies of calculation of pollutants emitted to the atmosphere by traction rolling-stock.
The calculation methodology was corrected having additionally assessed the fact that the amount of pollutants emitted to the atmosphere depends on the currently changed operation conditions of traction rolling-stock with diesel engines in Lithuania, comparatively high average mass of freight trains, change of the engines of old traction

rolling-stock by new engines conforming to the standards of emission of pollutants (UIC I and UIC II) established by UIC (International Union of Railways).
The purpose of this calculation is to present comparable estimates of various pollutants that would enable determining the most ecological ration of diesel and RME.
The difference of the averages of all the amounts of pollutants analyzed during the experiment is calculated based on the available results of the tests compared to pollutants emitted by an engine using pure diesel. The data about the differences calculated is presented in table 5.

	Engine load, %	Amount of RME in the fuel mixture, %			
		10	20	30	40
Difference CO, %	0[1]	-9.1	-22.5	-5.0	-16.9
	25	-53.4	-17.4	2.0	-36.1
	50	11.8	-8.3	-16.6	-20.5
	75	-7.3	1.7	-5.0	-35.6
Difference CO_2, %	0	-3.1	-4.6	-3.9	-9.6
	25	-5.3	2.5	-16.6	-1.8
	50	3.3	-0.1	0.3	-2.0
	75	3.0	17.7	5.9	5.3
Difference C_xH_y, %	0	-31.6	283.7	159.8	251.8
	25	73.7	150.8	-58.098	102.2
	50	-22.6	478.5	242.9	104.1
	75	147.1	152.1	19.2	328.7
Difference NO_x, %	0	-14.5	5.3	-0.5	-12.7
	25	-6.3	-32.8	-38.8	-40.3
	50	-5.9	-16.9	-20.6	-23.1
	75	-5.7	-6.2	-7.7	-11.1
Difference SP, %	0	-23.4	-26.9	46.0	31.5
	25	1.5	-11.2	-72.1	-55.7
	50	0.1	-15.4	256.8	261.0
	75	9.7	-15.9	-83.4	-59.2

Table 5. Comparison of various pollutants when using different amounts of RME in diesel to pollutants emitted by an engine using pure diesel (the values are presented in %)

The amount of fuel Q (t/h) consumed at different engine load modes is presented in table 6.

Fuel	Engine load, %			
	0	25	50	75
diesel	0.0031	0.0095	0.0180	0.0146
10 % of RME	0.0030	0.0100	0.0182	0.0146
20 % of RME	0.0032	0.0116	0.0183	0.0153
30 % of RME	0.0031	0.0105	0.0184	0.0145
40 % of RME	0.0031	0.0116	0.0184	0.0141

Table 6. The amount of fuel Q (t/h) consumed at different engine load modes

[1]The engine was operating at idle run.

The values of comparative pollution by trains D1 depending on the amount of fuel consumed when the engine operates at different modes is presented in table 7.

Pollutant	Engine load, %			
	0	25	50	75
CO	3.5	5.0	5.7	6.0
NOx	12.0	25.0	35.0	86.0
CxHy	0.7	0.8	0.9	1.1
SP	1.6	2.2	2.5	3.2

Table 7. Comparative values of various pollutants

Amounts of various pollutants $W(t)$ are calculated then:

$$W_{k,j,i} = l_{k,j,i} \cdot Q_j; \tag{4}$$

here: $l_{k,j,i}$ – comparative portion of the pollutant "k" at the load mode "j" in a locomotive, engine unit of a diesel train or automotrice of series "i" in kilograms per one ton of fuel consumed (table 8).
The W values calculated are presented in table 8.

Fuel	CO, t/h				C_xH_y, t/h				NO_x, t/h				SP, t/h			
	Engine power, %				Engine power, %				Engine power, %				Engine power, %			
	0	25	50	75	0	25	50	75	0	25	50	75	0	25	50	75
diesel	0.011	0.047	0.102	0.087	0.002	0.008	0.016	0.016	0.037	0.237	0.629	1.254	0.005	0.021	0.045	0.047
10% of RME	0.010	0.022	0.115	0.081	0.001	0.013	0.013	0.040	0.031	0.222	0.592	1.182	0.004	0.021	0.045	0.051
20% of RME	0.008	0.039	0.094	0.089	0.008	0.019	0.094	0.040	0.039	0.159	0.522	1.177	0.004	0.019	0.038	0.039
30% of RME	0.010	0.048	0.085	0.083	0.006	0.003	0.055	0.019	0.037	0.145	0.500	1.157	0.007	0.006	0.160	0.008
40% of RME	0.009	0.030	0.081	0.056	0.008	0.015	0.033	0.069	0.032	0.141	0.483	1.115	0.006	0.009	0.162	0.019

Table 8. Amounts of various pollutants W

Having performed these calculations, different harmful effect of pollutants to the environment is assessed. It is assessed in accordance with the aggressiveness indicator that is presented in table 9.

Pollutant	Aggressiveness indicator A
CO	1.0
NOx	41.1
CxHy	3.16
SP	300

Table 9. Aggressiveness indicators of pollutant

Conditional amounts of pollutants M assessing the different effect of each pollutant to the environment are calculated (table 10).

Fuel	CO, t/h				C_xH_y, t/h				NO_x, t/h				SP, t/h			
	Engine power, %				Engine power, %				Engine power, %				Engine power, %			
	0	25	50	75	0	25	50	75	0	25	50	75	0	25	50	75
diesel	0.011	0.047	0.102	0.087	0.007	0.024	0.051	0.051	1.511	9.730	25.847	51.534	1.471	6.250	13.476	13.997
10% of RME	0.010	0.022	0.115	0.081	0.005	0.042	0.040	0.125	1.292	9.121	24.321	48.596	1.127	6.344	13.487	15.349
20% of RME	0.008	0.039	0.094	0.089	0.026	0.060	0.296	0.128	1.591	6.536	21.467	48.357	1.074	5.551	11.396	11.774
30% of RME	0.010	0.048	0.085	0.083	0.018	0.010	0.175	0.060	1.503	5.953	20.531	47.568	2.146	1.745	48.087	2.330
40% of RME	0.009	0.030	0.081	0.056	0.024	0.048	0.104	0.217	1.320	5.806	19.866	45.818	1.934	2.772	48.644	5.708

Table 10. Conditional amounts of pollutants *M* having assessed aggressiveness

By adding these conditional amounts of pollutants *M*, it is possible to determine the most ecological composition of fuel: the lower the value *M*, the more ecological the fuel (table 11).

Fuel	Total amounts of pollutants
diesel	124.197
10 % of RME	120.075
20 % of RME	108.485
30 % of RME	130.333
40 % of RME	132.438

Table 11. Total amounts of pollutants M having assessed aggressiveness

Dependence of the total amount of pollutants on the percent of RME may be determined in accordance with the data of the table (fig. 25).

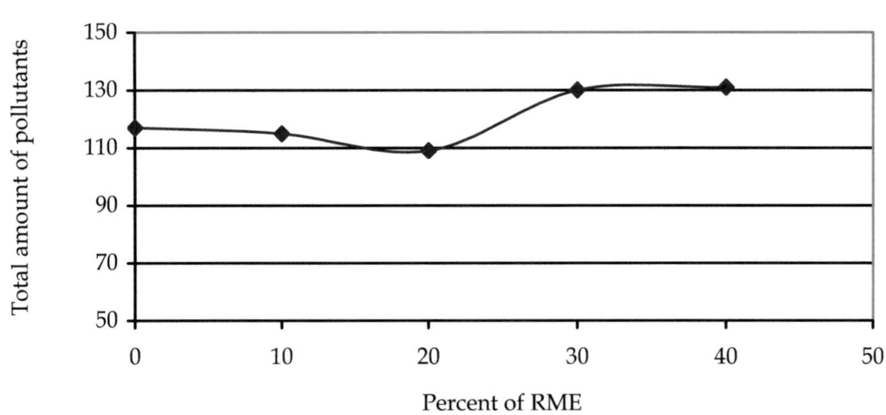

Fig. 25. Dependence of the total amount of pollutants on the percent of RME

It is obvious from table 11 and figure that the lowest total amount of pollutants is obtained when RME amounts to 20 %.

6. Conclusions

Having analyzed the results of the tests obtained, the following conclusions and recommendations may be presented:

1. Technically, there are no obstacles for using biodiesel in an engine that is not adapted to it, as physical characteristics of diesel and biodiesel (RME) are very similar. The only problem that may occur is shorter useful life of the rubber parts of the engine (hoses, sealing parts etc), which are not adapted specially, that have direct contact with fuel due to the acids contained in biodiesel.
2. Technologically, the ratio of RME and diesel is not important at all, as this fuel mixes well, does not form layers and stable mixtures are obtained.
3. Economic aspect.
   a. Under a low (up to 50 % P_{max}) load, fuel consumption increases in proportion to the amount of RME in the mixture; however, when the load increases more, fuel consumption approaches consumption of pure diesel, whereas it decreases by over 2 % when using the mixture containing 40 % of RME and diesel. Only the mixture containing 20 % of RME and diesel may be distinguished, as fuel consumption using it is 4 % higher than when using pure diesel at an average.
   b. An economic calculation assessing the pollution tax and the change of fuel consumption due to lower calorific value of biodiesel was carried out in order to determine expenses for fuel and showed that the highest expenses shall be incurred when using the mixture containing 20 % of RME (they shall increase by 1.48 % compared to diesel at an average), whereas the lowest shall be incurred when using the mixture containing 40 % of RME and diesel (they shall decrease by 1.55 % compared to diesel at an average). Expenses for fuel shall decrease by 1.14 % compared to diesel at an average when using the mixture containing 30 % of RME and diesel.
4. Environment protection aspect. Having measured the composition of oxides, calculated the values of pollutants and assessed different aggressiveness of pollutants, a total amount of pollutants was determined which showed that the most ecological fuel, having assessed all components of oxides, is the mixture containing 20 % of RME and diesel (value: 108.485), whereas the most polluting one is the mixture containing 40 % of RME and diesel.

With regard to fuel consumption measured, expenses calculated and having assessed harmfulness of pollutants, it may be maintained that the most rational option is to use the mixture containing 30 % of RME and diesel. Although expenses incurred when using this mixture are not the lowest and this mixture is not the most ecological one, it differs from the mixture containing 40 % of RME and diesel (expenses are the lowest for this mixture) only by 0.5 % with regard to expenses, whereas it is in the middle with regard to the estimates of harmfulness of pollutants: between the most ecological and the most polluting mixtures when calculating in one manner and is only by approximately 3.5 % more polluting than then most ecological mixture when calculating in another manner.

7. References

A Comprehensive Analysis of Biodiesel Impact on Exhaust Emissions. Draft Technical Report. Environmental Protection Agency, EPA420-P-02-001. October 2002

ADEME (1999). Referentiel pour le calcul des bilans energetiques. Production Agricole. Project de brochure. Paris, 1999

Available from http://biodiesel.ag.uidaho.edu/main/information.html

Available from http://www.greenfuels.org

Available from http://www.vegievan.org/biodiesel.html

Available from http://www.caddet–ee.org

Available from http://www.oceanairenvironmental.com

Council Directive 2003/96/EC of 27 October 2003 on restructuring the Community framework for the taxation of energy products and electricity. OJ L283, 2003, p. 51–70

Council Regulation (EC) No. 1782/2003 of 29 September 2003 on establishing common rules for direct support schemes under the common agricultural policy and establishing certain support schemes for farmers and amending Regulations (EEC) No. 2019/93, (EC) No. 1452/2001, (EC) No. 1453/2001, (EC) No. 1454/2001, (EC) No. 1868/94, (EC) No. 1251/1999, (EC) No. 1254/1999, (EC) No. 1673/2000, (EEC) No. 2358/71 and (EC) No. 2529/2001. OJ L203, 2003, p. 1

Directive 2003/03/EC of the European Parliament and the Council of 8 May 2003 on the promotion of the use of biofuels or other renewable fuels for transport. OJ L 123, 2003, p. 42–46

Energy for the future: Renewable sources of energy. White paper for a Community strategy and action plan. Communication from the Commission (COM (97) 0599-C4-0047/98)

Engler, C. R.; Lepori, W. A.; Johnson, L. A.; Yarbrough, C. M. (1992). Processing requirements form plant oils as alternative diesel fuels // *Proc. of an Alternative Energy Conference. Liquid fuels from renewable resources.* St. Joseph, MI, p. 79–88

European Biodiesel Board. Avenue de Tervuren, 363 - 1150 Bruxelles – Belgium. Available from http://www.ebb-eu.org/ebbstyle.css

Industrial use of rape for biodiesel production in the Czech Republic. Prague, 1997

Janulis, P.; Makarevičienė,V. (2004). Biodegalų ir bioalyvų naudojimas Lietuvoje (Use of biofuel and bio oils in Lithuania). Vilnius. p. 72

Kerstin, G. (1999). H–D Vehicle Exhaust Emissions and Alternative Fuels // MTC AB

Krahl, J.; Munack, A.; Bahadir, M.; Schumacher, L. (1996). Survey about biodiesel exhaust emissions and their environmental effects // Liquid fuels and industrial products from renewable resources: *proceedings of the Third fuel conference.* Michigan, p. 136–148

Kyoto Protocol to the United Nations Framework Convention on Climate Change. UNFCCC, 1997 (FCCC/CP/1997/L.7/Add.1, December 1997)

Labeckas, G.; Slavinskas, S. (1998). Biodegalų naudojimo dyzeliniame variklyje tyrimas // Transporto technologijų tyrimai, kūrimas, energetinių resursų taupymas ir aplinkosauga (Research on use of biogas in a diesel engine // Research and development of transport technologies, saving energetic resources and environment protection). *Material for the national conference.* Vilnius, p. 409–415

Lingaitis, L. P.; Pukalskas, S. (2008). Ecological aspects of using biological diesel oil in railway transport / Leonas Povilas Lingaitis, Saugirdas Pukalskas // *Transport*. Vilnius: Technika. ISSN 1648-4142. 23(2): 138–143

Lingaitis, L. P.; Pukalskas, S. (2008). The economic effect of using biological diesel oil on railway transport / Leonas Povilas Lingaitis, Saugirdas Pukalskas // *Transport*. Vilnius: Technika. ISSN 1648-4142. 23(4): 287–290

Lingaitis, L. P.; Pukalskas, S. (2007). Determining the consumption of biodiesel by locomotive engines / L. P. Lingaitis, S. Pukalskas // *Transport Means - 2007: proceedings of the 11th international conference*, 18–19 October 2007. Kaunas University of Technology, Lithuania / Kaunas University of Technology, IFTOMM National Committee of Lithuania, SAE Lithuanian Branch, The Division of Technical Sciences of Lithuanian Academy of Sciences, Klaipėda University, Vilnius Gediminas Technical University. Kaunas: Technologija. ISSN 1822-296X. 2007, p. 194–197

LST EN 14214 Automotive fuels. Fatty acid methyl esters (FAME) for diesel engines. Requirements and test methods.

LST EN 590:2004 Automotive fuels. Diesel. Requirements and test methods

Makarevičienė, V. (2001). Rapsų aliejaus riebalų rūgščių etilesterio gavimas, vartojimas degalais ir poveikis aplinkai (Production of rapeseed oil fatty acid methyl ester, use as a fuel and impact on environment): *doctoral thesis*. Kaunas: LŽŪU. p. 113

Makarevičienė, V.; Pukalskas, S.; Janulis P. (2001). Palyginamieji rapsų aliejaus etil- ir metilesterių deginių emisijos tyrimai // *Aplinkos inžinerija* (Comparative research on emission of oxides of rapeseed oil ethyl- and methyl esters // *Environmental engineering*). Vilnius: Technika, 9(3): 158–163

Makarevičienė, V.; Sendzikiene, E.; Janulis, P. (2005). Solubility in multicomponent biodiesel fuel system // *Bioresource technology*. 96(5): 611–616

Peterson, C. L.; Reece, D. I.; Cruz, R.; Thompson, J. (1992). A comparison of ethyl and methyl esters of vegetable oils as diesel fuel substitutes // *Proc. of an Alternative Energy Conference*. Liquid fuels from renewable resources. St. Joseph, MI, p. 99–110

Rules on trade in oil products, biofuel, bio oil and other combustible liquid products in the Republic of Lithuania. "*Žin.*", 2001, No 37-1269, 2003 No 36-1596, No 121-5506

Wang, W. G.; Clark, N. N.; Lyons, D. W. (1997). Emissions comparisons from alternative fuel buses and diesel buses with a chassis dynamometer testing facility // *Environmental Science & Technology*. ISSN: 0013–936X., 31: 3132–3137

Worldwide Review on Biodiesel Production. (2003). IEA Bioenergy Task 39 Subtask "Biodiesel". Austrian Biofuels Institute. p. 92

Xiao, Z.; Ladommatos, N.; Zhao, H. (2000). The effect of aromatic hydrocarbons and oxygenates on diesel engine emissions // Proceedings of the institution of mechanical engineers part D-*Journal of automobile engineering*. ISSN: 0954-4070., 214(D3): 307–332

Zelenka, P.; Kriegler, W.; Herzog, P. L.; Cartellieri, W. P. (1990). Ways toward the clean heavy-duty diesel // SAE Tech. Pap. Ser. Nr. 900602. p. 13

Zubr, J.; Matzen, R. (1996). Mixed vegetable and diesel oil as fuel // 9th European Bioenergy Conference, 1st European Technology Exhibition on Biomass for Energy, Copenhagen, 24–27 June. p. 10

18

Understanding the Developing Cellulosic Biofuels Industry through Dynamic Modeling

Emily Newes, Daniel Inman and Brian Bush
National Renewable Energy Laboratory
United States of America

1. Introduction

Biofuels are promoted in the United States through aggressive legislation, as one part of an overall strategy to lessen dependence on imported energy as well as to reduce the emissions of greenhouse gases (Office of the Biomass Program and Energy Efficiency and Renewable Energy, 2008). For example, the Energy Independence and Security Act of 2007 (EISA) mandates 36 billion gallons of renewable liquid transportation fuel in the U.S. marketplace by the year 2022 (U.S. Government, 2007). Meeting such large volumetric targets has prompted an unprecedented increase in funding for biofuels research. Language in the EISA legislation limits the amount of renewable fuel derived from starch-based feedstocks (which are already established and feed the commercially viable ethanol industry in the United States); therefore, much of the current research is focused on producing ethanol—but from cellulosic feedstocks. These feedstocks, such as agricultural and forestry residues, perennial grasses, woody crops, and municipal solid wastes, are advantageous because they do not necessarily compete directly with food, feed, and fiber production and are envisaged to require fewer inputs (e.g., water, nutrients, and land) as compared to corn and other commodity crops. In order to help propel the biofuels industry in general and the cellulosic ethanol industry in particular, the U.S. government has enacted subsidies, fixed capital investment grants, loan guarantees, vehicle choice credits, and aggressive corporate average fuel economy standards as incentives. However, the effect of these policies on the cellulosic ethanol industry over time is not well understood. Policies such as those enacted in the United States, that are intended to incentivize the industry and promote industrial expansion, can have profound long-term effects on growth and industry takeoff as well as interact with other policies in unforeseen ways (both negative and positive). Qualifying the relative efficacies of incentive strategies could potentially lead to faster industry growth as well as optimize the government's investment in policies to promote renewable fuels.

The purpose of this chapter is to discuss a system dynamics model called the Biomass Scenario Model (BSM), which is being developed by the U.S. Department of Energy as a tool to better understand the interaction of complex policies and their potential effects on the burgeoning cellulosic biofuels industry in the United States. The model has also recently been expanded to include advanced conversion technologies and biofuels (i.e., conversion pathways that yield biomass-based gasoline, diesel, jet fuel, and butanol), but we focus on cellulosic ethanol conversion pathways here. The BSM uses a system dynamics modeling approach (Bush et al., 2008) built on the STELLA software platform (isee systems, 2010) to

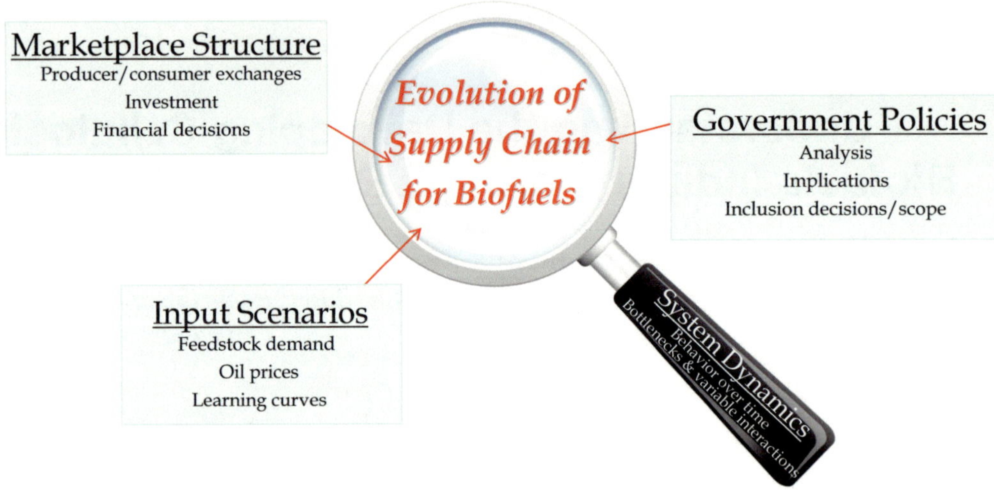

Fig. 1. Key components of the Biomass Scenario Model

model the entire biomass-to-biofuels supply chain. Key components of the BSM are shown in Figure 1. In addition to describing the underpinnings of this model, we will share insights that have been gleaned from a myriad of scenario- and policy-driven model runs. These insights will focus on how roadblocks, bottlenecks, and incentives all work in concert to have profound effects on the future of the industry.

2. Model background

The major sectors of the ethanol supply chain are shown in Figure 2. Each sector (feedstock production, feedstock logistics, biofuels production, biofuels distribution, and biofuels end use) has been modeled as a standalone module but is linked to the others to receive and provide feedback. The feedstock production module simulates the production of biomass as well as other crops (corn, wheat, soybeans, cotton, and other grains) through farmer decision logic, land allocation dynamics, new agricultural practices, markets, and prices. The feedstock logistics system models the harvesting, collection, storage, preprocessing, and transportation of biomass feedstocks from the field (or forest) to the biorefinery. The conversion module has three conversion technologies [corn dry mill, biochemical (dilute acid enzymatic hydrolysis), and thermochemical (indirect gasification and mixed alcohol synthesis)] at four scales (pilot, demonstration, pioneer, and full-scale commercial). The ethanol produced during conversion is then distributed throughout the region(s). The model is solved numerically at a sub-monthly level and reports output for the timeframe of 2005 to 2050. Although the description herein implies a linear flow of information between the modules, in reality the modules receive and react to information in a complex, non-linear fashion that depends on, among other things, industrial learning, project economics, installed infrastructure, consumer choices, and investment dynamics. The model is geographically stratified, using the 10 U.S. Department of Agriculture (USDA) farm production regions as a basis, which facilitates analysis of regional differences in key variables. The BSM is particularly facile at addressing the following types of inquiries:

- Which sources of feedstock might plausibly contribute substantially to production in different regions of the United States?
- Under what combination of policies does the biofuels industry observe gradual, sustained growth?
- What gasoline price scenarios have the potential to increase biofuels adoption?

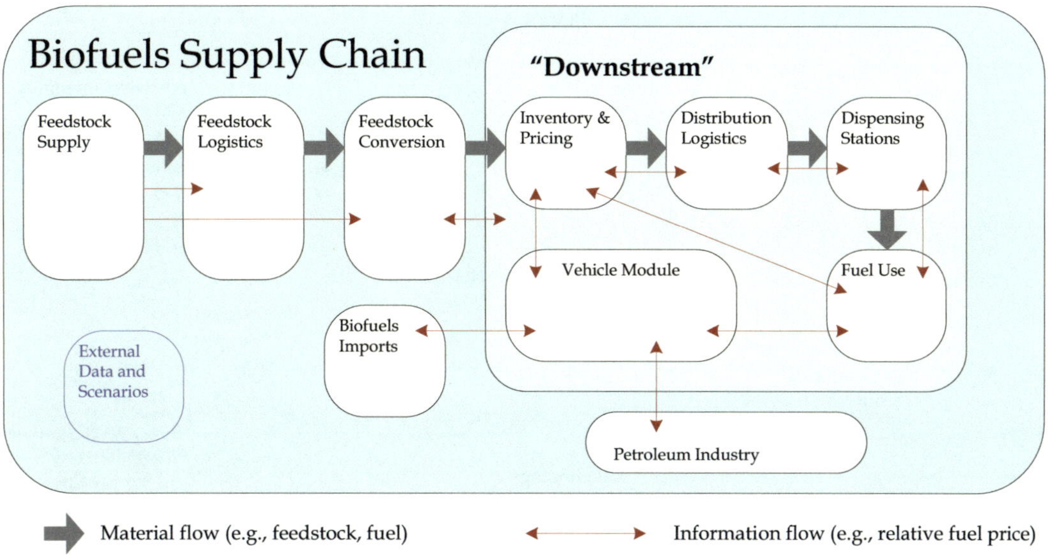

Fig. 2. Architecture of the BSM

3. System dynamics structure

Transitioning from the United States' current petroleum-based transportation fuel economy to one that incorporates significant amounts of alternative and renewable transportation fuels is characterized and addressed in the BSM as a "system of systems" problem. System dynamics, as a modeling discipline, focuses on the relationships and feedback among parts of a system and helps identify possible unintended consequences of certain inputs along with synergistic effects, bottlenecks, and leverage points for intervention; it is an established methodology for analyzing the behavior of complex, real-world feedback systems over time. Figure 3 shows a causal loop diagram, which is a visual way to explain key connections in a dynamic system for a simplified conception of the cellulosic ethanol supply chain. It also shows the direction of the main feedbacks in the system. Its broad, high-level approach captures the entire supply chain.

Within each module, the BSM contains several key decision-making variable interactions with complicated, yet understandable, logic. The major dynamic components that make up the model are described in this section.

3.1 Feedstock production dynamics

In the BSM, the production of both commodity and energy crops is governed by the dynamics of farmers' decision making, land allocations, crop markets, and farmers' transition to new agricultural practices (i.e., switching from growing traditional crops to

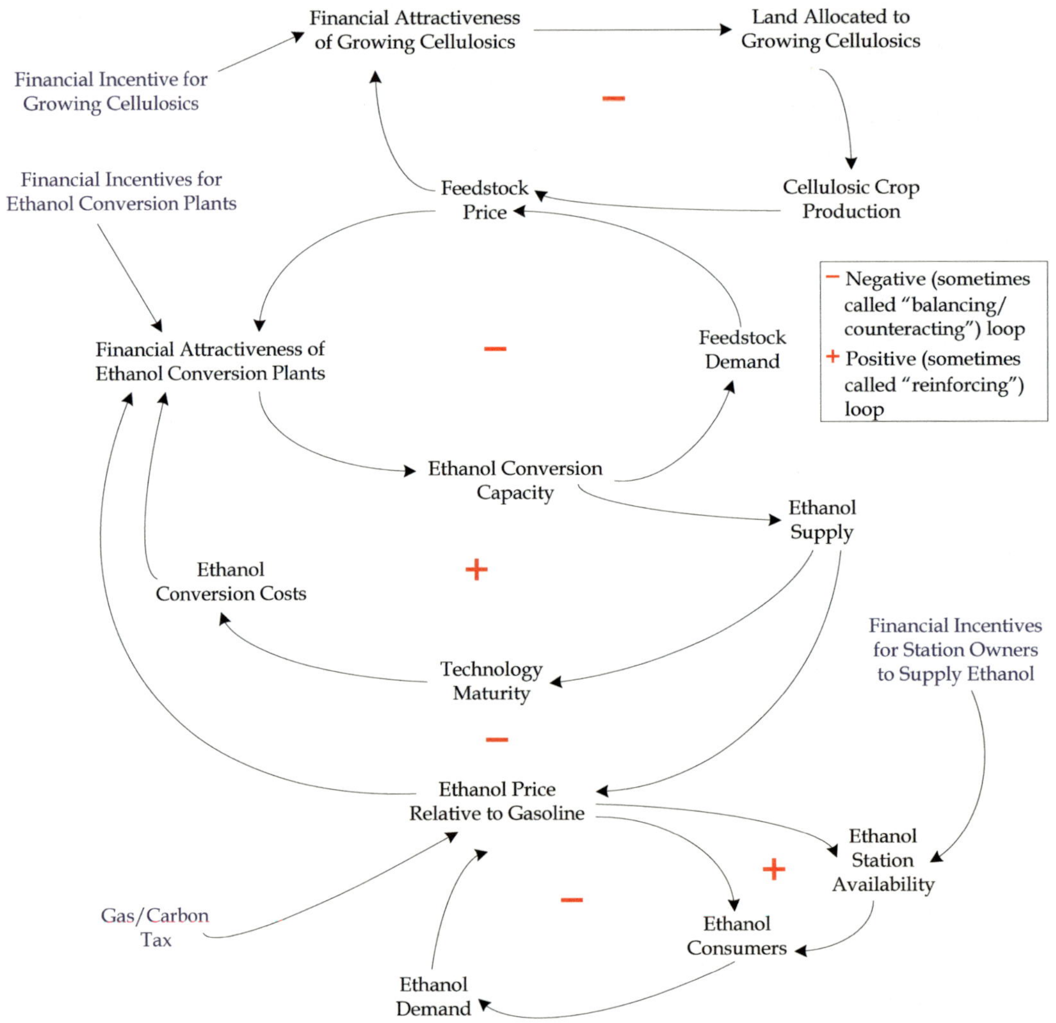

Fig. 3. Causal loop diagram representing key variable interactions in the BSM

either harvesting residues or explicitly growing energy crops). At a high level, there is a balancing loop (also known as a negative feedback loop) that controls feedstock production; this loop is shown in the upper right hand portion of the causal loop diagram in Figure 3. In this balancing loop, feedstock prices (as received by the farmers) directly affect the attractiveness of growing cellulosic crops; the higher the feedstock price being paid, the more attractive it is for farmers to reallocate land from commodity or hay crops to growing energy crops or harvesting agricultural residues. When land is allocated to growing energy crops, at the expense of producing commodity crops, the supply of the former is reduced, which can, in some situations, result in higher regional prices for the commodity crops. Conversely, as more farmers switch to new practices and allocate land to producing cellulosic crops, the availability of cellulosic material increases, which will cause the price paid for cellulosic feedstocks to be reduced, thereby diminishing the attractiveness of producing cellulosic crops; thus, the loop is balanced. Implicit in the very simple causal loop diagram depicted in Figure 3 and described above are numerous complex feedback loops, both balancing and reinforcing (also known as positive feedback loop), that interact across all modules of the BSM. Both types of feedbacks play important roles: balancing loops often encourage stability (in feedstock prices, for instance), and reinforcing loops encourage development (in growth of overall production capacity). The descriptions in Section 3.2–Section 3.5 provide more detail on the dynamics that underpin farmer decision making, land allocation, crop markets, and transitions to new practices. Key insights that pertain to these specific areas as well as to feedstock production, in general, will also be highlighted and discussed.

3.2 Farmers' decision making

Farmers make decisions each season that affect the supply side of agricultural markets. For example, each year individual farmers have to choose what crop(s) to plant and what portion(s) of their land they will utilize for these crops. At the farm level, these decisions are based on a myriad of factors including local climate, available equipment, land base, capital, past experience, tolerance to risk, and market cues such as future pricing and anticipated payments (e.g., farm subsidies). In the BSM, farmer decisions are nested in the Feedstock Supply Module (FSM). Farmers' decisions that are explicitly captured in the BSM include the type of crop grown during a particular year (commodity crop, no crop, hay crop, or bioenergy crop) and amount (area) of land dedicated to the production of the crop(s). Decisions on whether to cultivate bioenergy crops or to collect crop residues are based on endogenous net revenue calculations that are applied to a nested logit-based land-allocation model (Figure 4). Within the FSM, potential net payments to growers are calculated for each of the 10 USDA farm production regions. Net per-acre grower payments reflect the profitability of land, including subsidies, across its various uses less the costs of production, harvesting, storage, and transportation. Subsidies contained in the FSM include the Biomass Crop Assistance Program, which is administered by the U.S. Department of Agriculture and which provides a per-ton payment to farmers producing energy crops, an establishment payment for the establishment of woody and herbaceous crops, and a per-acre annual payment for those in designated project areas. The BSM contains different subsidy inputs based on current government programs and policies that could be potentially implemented. They can be altered depending on the user-defined scenario and are updated as regulations change. Two separate grower payments are considered for commodity crop (corn, wheat, soybeans, other grains, and cotton) production: (1) grower payments for production of primary and secondary crops and (2) grower payments for production of a primary crop with a crop residue potentially available for collection and processing for ethanol conversion. Production costs taken from the Policy Analysis System (POLYSYS) model (The University of Tennessee n.d.) for commodity crops, primary and secondary crops, and crop residues include collection and plant nutrient replacement (e.g., fertilization). POLYSYS is an agricultural economics simulation model that computes the volume of agricultural commodity production for a given farm gate price for each of the USDA farm production regions. The gross value of the primary crop is the crop price multiplied by the yield (tons per acre) plus any government subsidy; the gross value of the secondary crop is specified as a fraction of that for the primary crop. The net grower payment is calculated as the difference between the gross value and the production costs. Similarly, the value of residue from annual crops is the residue price minus the production costs.

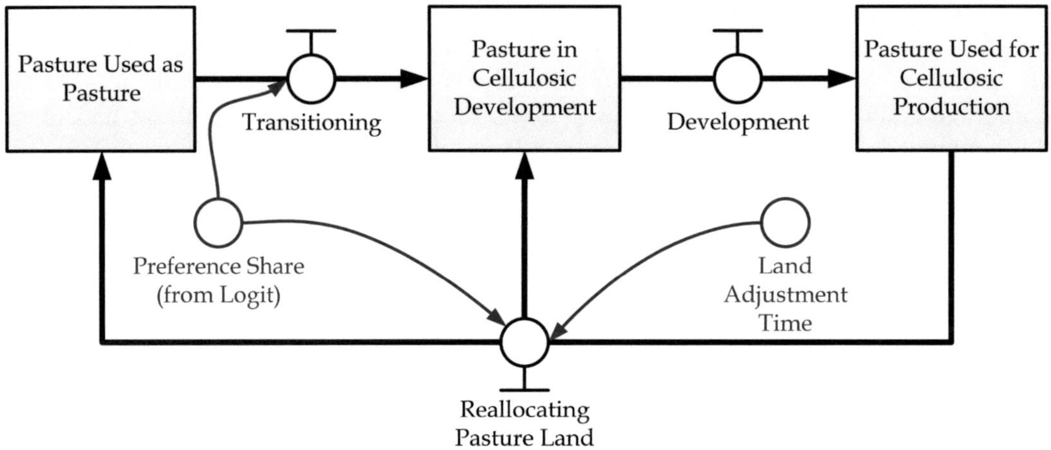

Fig. 4. Land allocation structure for pasture land allocation decisions

3.3 Land allocation

Land allocation in the BSM is driven by the farmers' decisions and is nested in the FSM, where land is allocated by a modified nested logit model that tracks correlations among decisions to allocate land to commodity crops (with or without the collection of crop residues), hay, and perennial energy crops. The logit model accounts for economic contributions (e.g., expected net revenue per acre) and non-economic contributions to the utility of land-allocation choices. For each production region, the model treats separately the distribution of land among pastureland and active cropland; within each of these categories, the "desired" allocation of land among specific crops was calculated from the nested logit, which was calibrated by comparison to long-term agricultural forecasts annually published by the USDA and where the nesting involves broad crop categories at the higher level and individual crops at the lower level. By determining distribution of land, we accounted for the fraction of land associated with growers who have adopted the new practice of producing cellulosic feedstocks (crop residue and/or perennial energy crops). Over time, the land allocations gradually adjust themselves toward the distribution indicated by the farmers' planting decisions. This logic works to reflect results of micro-level decision making by farmers and accounts for land area. The resolution with which land is tracked in the FSM is based on the USDA agricultural production regions and accounts for regional differences in production costs, yields, and potential feedstock supply. Available cropland is divided into three categories: active cropland, pastureland, and Conservation Reserve Program (CRP) (U.S. Department of Agriculture, 2011) land (shown in Table 1). Active land can be used to produce annual crops, perennial energy crops (herbaceous and woody), and hay. Five major types of cellulosic feedstocks are modeled: herbaceous energy crops, woody energy crops, crop residues, forest residues, and urban residues. Allocation of land within the FSM is based on net revenue calculations for the different crops, which are input to the logit function. Expected crop yield, price (i.e., grower payment), and production costs are all considered and integral to the net revenue calculations. The supply of commodity crops (wheat, corn, soybeans, cotton, and small grains) is similarly computed from the land base. The actual production from each source is calculated dynamically by regional price signals and competition among land uses. The model respects the fact that land-use change does not occur suddenly and that perennial energy crops pass through a development period where yields are lower than their mature production value. For each region, the desired separation of land among CRP, pastureland, and active cropland uses is determined after the distribution of land among specific crops is calculated. Determining the distribution of land accounts for the fraction of land that is associated with new practices (producing cellulosic feedstocks). Over time, the land allocations are adjusted toward the distribution "desired" by the farmers via a diffusion model with a single rate constant. This logic works to reflect results of micro-level decision making by farmers and accounts for the potential alternative uses of land area.

All Cropland	**Active Cropland**	Available for Traditional and Celluosic Crops	Annuals	Corn	*With residue collection*
					With secondary crop
				Soy	
				Wheat	*With residue collection*
					With secondary crop
				Other Grains	*With residue collection*
					With secondary crop
				Cotton	*With residue collection*
					With secondary crop
			Perennial Energy Crops	*Herbaceous*	*Immature*
					Mature
				Woody	*Immature*
					Mature
		Hay			
Pasture		Growing as Pasture		Used as Forage	
				Harvested for Cellulose	
		Planted with Energy Crops		*Immature*	
				Mature	
CRP		*Harvested for Cellulose*			
		Unharvested			
Excluded from FSM		High value cash crops, etc.			

Table 1. Land categories and corresponding production combinations that are tracked in the FSM

3.4 Crop markets

The market for commodity crops, hay, and energy crops is captured in the FSM; it provides a physical basis for generating a supply of biomass feedstocks for cellulosic ethanol production, while representing the economics of and physical constraints within the U.S. agricultural system. The FSM includes a market mechanism that provides feedback in the FSM, connecting production and demand for agricultural products. The basic feedbacks that drive changes in price for annuals are (1) the total inflow (roll-up of regional production and imports) relative to total consumption (domestic, export, shrinkage) and (2) the stock-to-use ratios relative to long-term ratios (allowing "target" stock-to-use ratios to float over time). For annuals, the regional production is rolled into aggregate inventory, and a single aggregate price index is generated for each annual; multipliers (derived from regional price variation in the incoming data set) are then used to provide regional price variations. The structures of the cellulosic feedstock and hay markets are similar to the annuals markets, but these are region-specific rather than national. Because transporting bulky feedstock over long inter-regional distances is costly, it was necessary to model separate markets for feedstock in each USDA region. The production levels of annuals and their prices were calibrated to USDA baseline projections. A simple diffusion structure within the module captures the adoption of new practices (crop residues and dedicated perennial energy crops) (Figure 5). The crop market is captured on a regional basis, based on the feedstock production capacities of the 10 USDA farm production regions and regional ethanol demands. In the BSM, the regional feedstock demands emerge endogenously from the simulation via the feedbacks between supply, demand, and logistics and associated capacity constraints. For perennial energy crops, the production costs vary annually over the life cycle of the project, which is generally 10 years, but can vary regionally. The supply/demand structure provides key feedbacks to the crop markets, connecting production and demand for agricultural products. The basic feedbacks that drive changes in price for commodity crops are (1) the total inflow (aggregate of regional production and imports) relative to total consumption (domestic, export, and shrinkage) and (2) the stock-to-use ratios relative to long-term ratios (allowing target stock-to-use ratios to float over time). For commodity crops, the regional production is represented as a single price index and is generated for each crop; multipliers are then used to provide regional price variation. The structures of the cellulosic feedstock and hay markets are similar to the commodities markets, but they are regional as opposed to national. The production volume of commodity crops and their prices are calibrated to USDA annual baseline projections (Interagency Agricultural Projections Committee, 2007). Within the crop market, cellulosic crops compete equally with commodity crops.

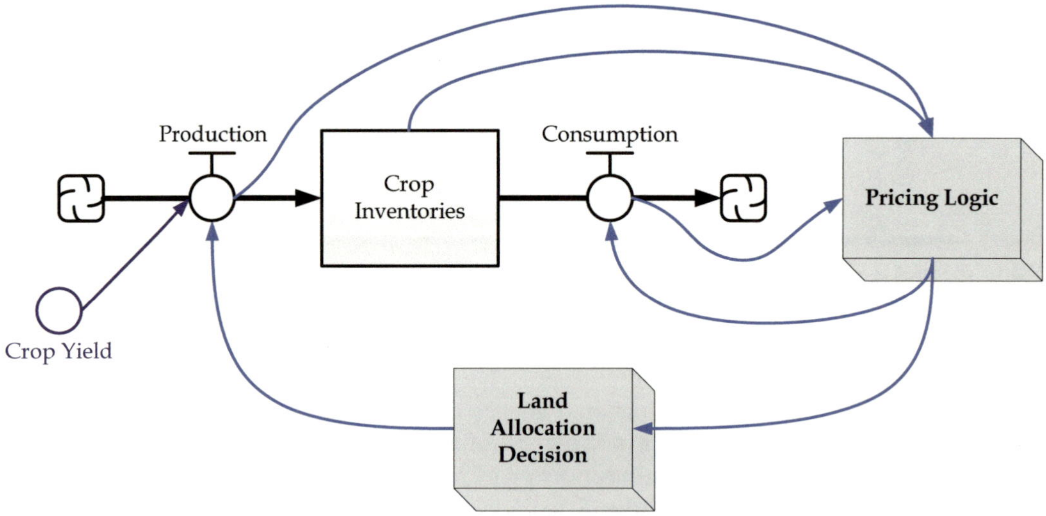

Fig. 5. Stock and flow structure showing the crop market, as modeled in the BSM

3.5 Transition to new practices

In the BSM old and new practices are defined as producing traditional commodity crops and producing biomass for conversion to fuel, respectively. Figure 6 shows the stock and flow structure that governs the transition to and from new practices. The fraction of old and new practice producers, within each region, is calculated endogenously by considering a combination of factors including expected net revenue and proximity to a biorefinery; only land that is within a biorefinery's collection radius (discussed in section 3.6) can shift to "new practices." As with other aspects of the FSM, much of the driving force behind transitioning to new practices is economic incentives in the form of grower payments received by the producers. The key drivers impacting the movement of producers to new practices are profitability per acre (for crop residues or cellulosic energy crops) as well as proximity to a biorefinery. New practices include both growing energy crops (herbaceous and woody) and collecting agricultural residues. The BSM accounts for potential presence of extremely risk-averse producers by tracking a subset of producers that will not shift to producing cellulosic material under any circumstances. The influences integral to the diffusion and uptake of new practices are represented by a simple modeling structure (Figure 6) that categorizes producers as employing "old" or "new" practice. The rate at which producers move from old to new practice is constrained by a scenario-dependent, exogenously-specified function of expected revenue. In the model, only the new practice farmers consider planting perennial energy crops, collecting crop residues, or harvesting cellulose from pasture or CRP land. The way in which the transition is modeled in the BSM allows for the analyses to explore questions around producers' conservatism. The amount of land in new practices is a key factor in cellulosic feedstock price stability in some regions; see Figure 7 for an example of the relationship between new practices and cellulosic ethanol production.

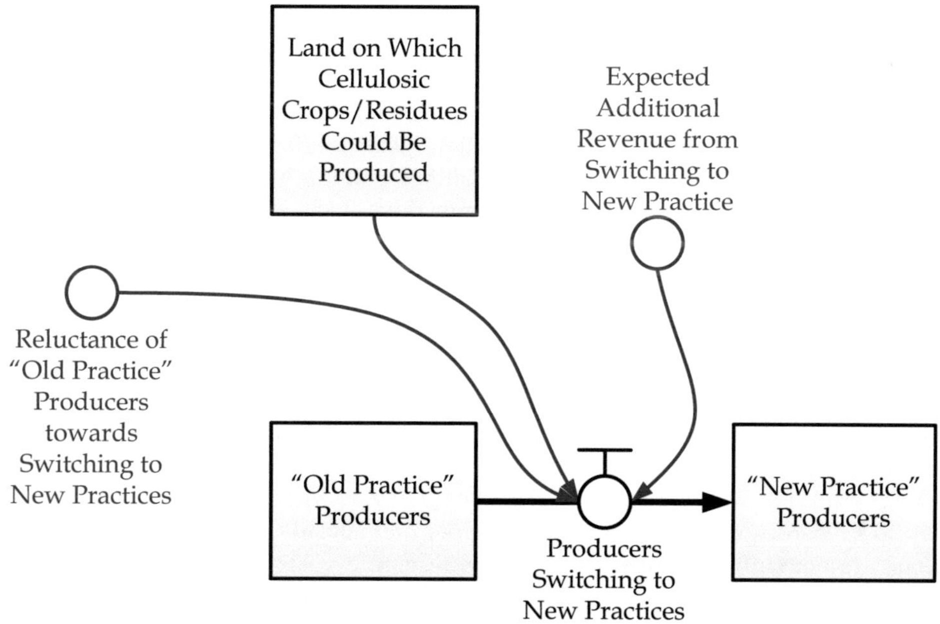

Fig. 6. Stock and flow structure showing the transition to new practices

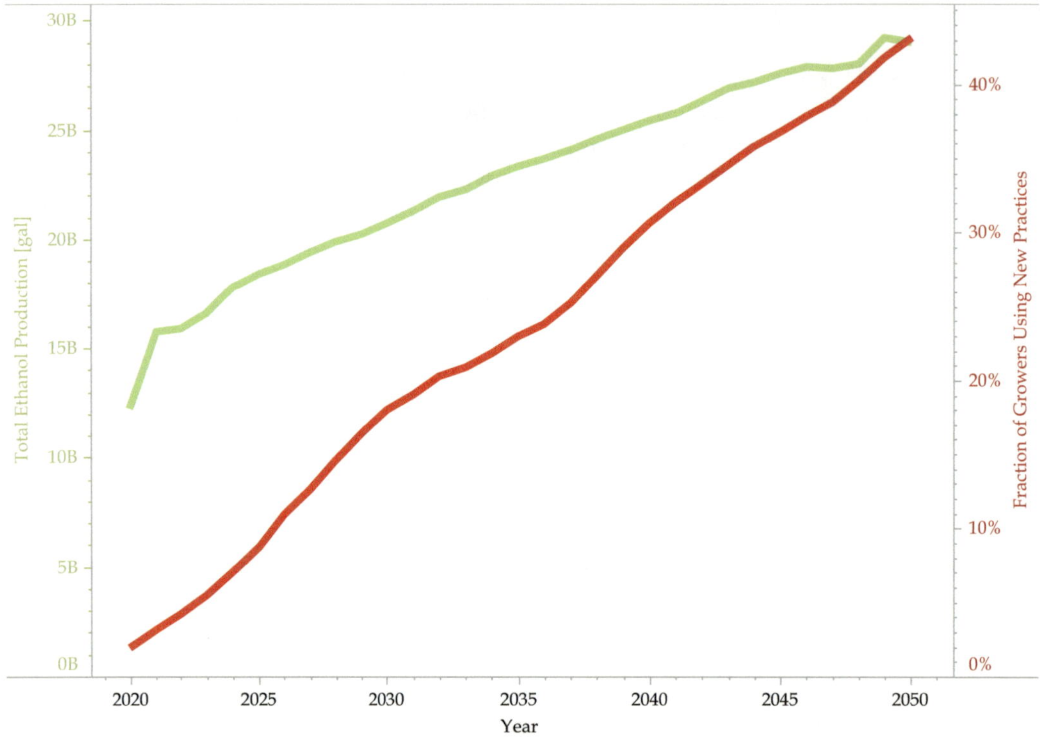

Fig. 7. Relationship between adoption of new practices and cumulative ethanol production; figure is based on results from the reference case (discussed in Section 4)

Multiple scenarios have been run using the BSM to provide insights into the biomass to biofuels supply chain; some important feedstock market insights include:

- **Feedstock Price Floor/Subsidy:** Subsides and policies focused on the feedstock production side of the ethanol supply chain alone do little towards pushing the system to meet the annual volumetric production goals outlined in the Renewable Fuel Standard II (RFS), which mandates 36 billion gallons by 2022 (U.S. Environmental Protection Agency, 2010). Establishing a price floor on cellulosic crops, where producers are guaranteed a minimum price, has little effect on transitioning to new practices and stimulating cellulosic feedstock production during the early years of industry development. However, a price floor does have a small but positive system-wide effect in the mid- and long-term if it is sustained. The greatest benefit is realized at a price floor above $70 per dry ton. At the $100-level, crop market instability is observed because the investment climate (e.g., willingness to invest in new biorefineries) causes the demand for cellulosic biomass to oscillate; market signals cause producers to over- and under-produce biomass. This oscillation is accentuated by the fact that perennial energy crops take between three to eight years to mature.
- **Initial Feedstock Prices:** As with a price floor and subsidy, high initial feedstock prices alone are not sufficient for meeting RFS volumetric goals. That said, in terms of total cellulosic feedstock production, an initial feedstock price $80 to $90 per ton does promote the production of cellulosic feedstocks throughout most of the USDA farm production regions. Lower initial feedstock prices (less than $80 per ton) cause growers

to not allocate land to producing cellulosic feedstocks, and the industry does not take off. At the other end of the spectrum, feedstock prices greater than $90 per dry ton cause downstream bottlenecks that feed back to the farmers causing them to not produce cellulosic feedstocks. At a regional level, cellulosic feedstock prices show considerable fluctuation when the cellulosic feedstock market is beginning to develop (because the market is not large enough to support stable prices over time); prices typically stabilize as regional demand becomes substantial and its trend of increase becomes more gradual.

- **Feedstock Supply:** In general, feedstock production resources are available to contribute significantly towards producing renewable fuels (both cellulosic and starch-based) as long as the payments received are high enough to alter the farmers' decisions with regard to land allocation. Under high-demand scenarios, feedstock production is nearly stretched to its maximum. The dominant sources of feedstock are herbaceous energy crops and forest residues. Neither crop residues nor urban residues contribute significantly to the overall feedstock supply. In general, competition between energy crops and commodity crops does not substantially increase annual crop prices unless feedstock demand is high.

3.6 Collection radius

Whether a feedstock producer can participate in the market for cellulosic materials depends greatly on the location of conversion facilities. Transport distances for cellulosic feedstock are estimated regionally from combining (1) the endogenously computed weighted average feedstock yields for cellulosic energy crops and agricultural residues with (2) biorefinery size, (3) an assessment of the fraction of arable land available for cellulosic harvesting, and (4) geometric factors accounting for the layout of the road network. The final three drivers of transportation distance are specified exogenously. Collection radii and transportation distances are typically observed in most scenarios at around 30–50 miles and often decrease as industry maturity and harvest yields increase, although the opposite behavior might be seen in some alternative feedstock logistics scenarios.

3.7 Experience at different stages

The BSM explicitly represents learning-by-doing in the refining of cellulosic feedstocks into ethanol via a modeling technique known as "cascading learning curves." This technique is an elaboration of common power-law representation of industrial learning that is typically expressed in the form $y = ax^b$, where y is the cumulative average cost per unit, x is the cumulative number of units produced, a is the cost of the first unit, and b is a constant characterizing the cost reduction that occurs with increasing experience. Learning occurs separately for each biofuels pathway (starch-ethanol, biochemically converted cellulosic ethanol, and thermochemically converted cellulosic ethanol), and the BSM actually tracks four scales of operation and maturity: (1) pilot, (2) demonstration, (3) pioneer commercial, and (4) full-scale commercial. Experience accumulates at each of these four scales, and each scale has a unique techno-economic characterization. In the cases of pilot and demonstration scale refineries, maturity is measured as the cumulative number of years of operation of plants at that scale; for pioneer and full-scale commercial plants, maturity is measured as cumulative industry output. Pilot- and demonstration-scale plants are specified exogenously to the BSM as scenario inputs, while pioneer and full-scale commercial plants are generated endogenously, although additional plants in these stages can also be added exogenously. The maturity, M, is related to a set of techno-economic multipliers, m, by the following equations:

$$M = \begin{cases} 1-(1-M_0)\left(\frac{E^*}{C}\right)^{\left(\frac{1-R}{\ln 2}\right)} & \text{for} \quad C \geq E^* \\ M_0 & \text{otherwise} \end{cases} \tag{1}$$

$$E^* = \max\{E, C_0\} \tag{2}$$

$$m = m_{early}(1-M) + m_{mature}M \tag{3}$$

Where M_0 indicates initial maturity, E is the minimum experience for learning, E* is the effective minimum experience for learning, C is the cumulative experience, C_0 is the initial cumulative experience, R is the progress ratio, m_{early} is the "early" multiplier and m_{mature} is the "mature multiplier".

Maturity is the ratio of current experience compared to the experience of the "n^{th} plant," or the infinitely mature biorefinery at that scale. With each doubling of experience, the gap remaining between current maturity and full maturity is decreased by a percentage derived from the progress ratio[1]. The progress ratio defines how much of the maturity gap remains after each doubling, meaning that maturity increases more slowly at higher progress ratio values. The multipliers are used to adjust the key techno-economic characteristics for the scale/stage as that stage matures. These learning curves "cascade" in the sense that the early multiplier for each stage equals the actual multiplier achieved at the previous scale; essentially, each subsequent stage builds upon the techno-economic learning that resulted from the previous stage. The key techno-economic biorefinery attributes affected by maturity at each stage are: (1) the conversion process yield, (2) the probability for technical yield, (3) input capacity, (4) capital cost for a new refinery, (5) technical risk in financial calculations, and (6) the portion of debt that can be financed through a loan to build a new plant. Within the model, the current attribute values are captured in the "state of the industry" variable and are the result of the maturity level across all stages; input data are taken from a variety of industry assessments, design reports, and research results. Figure 8 illustrates the interconnections between industrial development, learning, and investment in the BSM.

The structure of the BSM does not presuppose any particular evolution of the industry in terms of how pilot, demonstration, pioneer, and full-scale commercial operations are staged or scheduled. It is possible to "ride" the learning curve at any stage (scale of operation) of development if the introduction of new plants starting operation is carefully timed; investment at subsequent stages can compensate or substitute for a lack of investment in

[1]Technically, the progress ratio is defined as the ratio of the gap between the state of the industry after a doubling of experience (typically cumulative years in operation or production) and the current state of the industry.

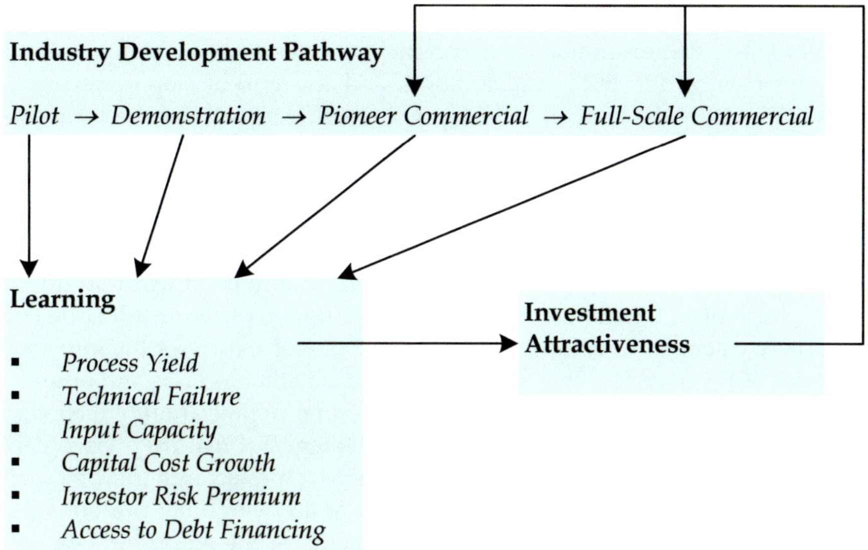

Fig. 8. Interaction in the BSM among learning, development, and investment

prior stages since an investment in a prior stage can affect the "starting point" for learning in subsequent stages. It is possible to construct development paths that are optimal in terms of cost or time by tracking learning curves and shifting investment to subsequent stages as the learning asymptote in a prior stage is approached. Figure 9 illustrates how process yield (quantity of ethanol produced per biomass input) and capital cost growth (ratio of the actual

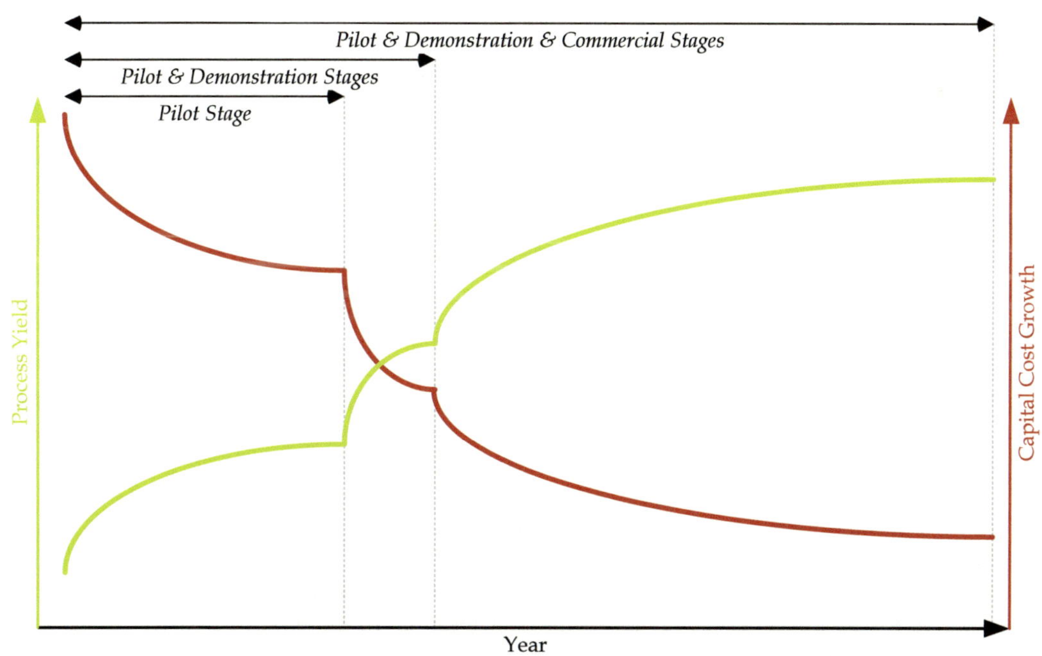

Fig. 9. Process yield and capital cost growth with different scales of operation

capital cost for a new plant to that cost when the industry is fully mature) rise or drop, respectively, as pilot, demonstration, and commercial experience accumulates. Analysis of realistic scenarios using the BSM has demonstrated the critical importance of industrial learning in the growth of the biofuels industry; policies that incentivize the early construction of even small numbers of pilot, demonstration, and pioneer plants typically have long-term positive impacts on ethanol production that are out of proportion with the cost of the policies.

3.8 Investment in biorefineries

The BSM uses a standard set of financial computations that mimic those that might be used by a potential investor to initiate the construction of a new pioneer or full-scale commercial biorefinery. These calculations compute the expected rate of return for the investment using major categories of revenue and expenses, assuming that ethanol price and other factors are constant over the plant lifetime. We assume straight-line depreciation, which significantly reduces detail complexity, constant tax and interest rates, and maturity-based capital costs and access to credit. The algorithm divides the biorefinery project into multiple periods (see Figure 10) whose present value is summed to arrive at an overall net present value for the prospective project. Figure 11 sketches the key elements of the algorithm. As the state-of-the-industry attributes improve through maturation, the estimated net present value increases and investment in new refineries becomes more attractive.

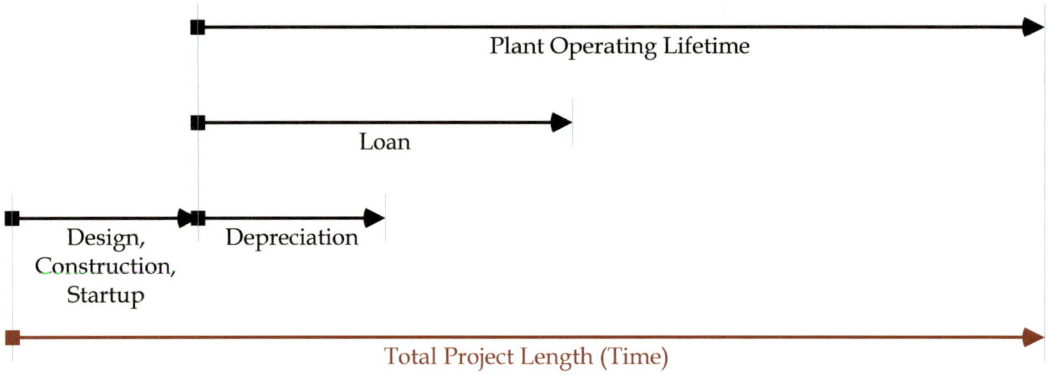

Fig. 10. Phases in the plant life cycle represented in the BSM

The conversion module in the BSM is responsive to a number of potential policies that improve the financial prospects of new biorefineries:

- **Product Subsidy:** Pays a fixed pre-tax amount to the ethanol producer at the plant gate for each gallon of cellulosic ethanol produced, improves the revenue stream in financial calculations, and enables regulators to indirectly manage the selling price of ethanol.
- **Feedstock Subsidy:** Pays a fixed amount to the non-corn feedstock suppliers (farmers) to lower the price paid by the cellulosic ethanol producer, which affects the expenses stream in the financial calculations and is not a direct subsidy of the ethanol production industry.
- **Capital Cost Reduction Subsidy:** Pays a percentage of the initial "cash" payment that is needed to start construction of a cellulosic ethanol pioneer-scale plant. The subsidy improves the construction cost of the pioneer-scale plants only and is a direct payment to the cellulosic ethanol producer.

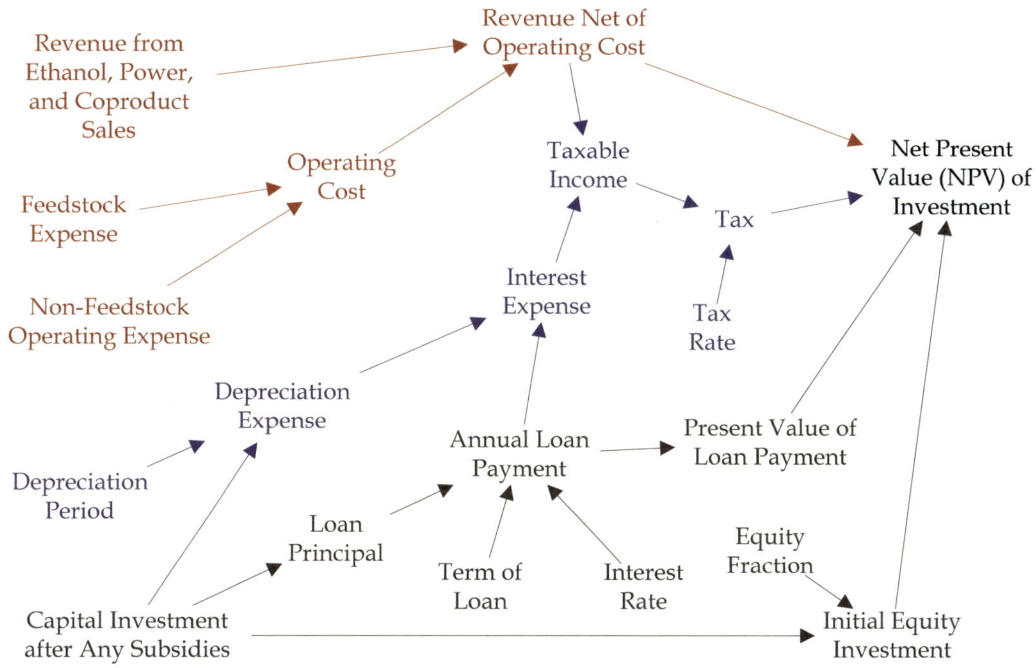

Fig. 11. Simplified representation of project economics computations in the BSM

- **Loan Guarantee:** Covers a fraction of loan given to a cellulosic ethanol producer to construct a pioneer-scale plant. The guarantee improves the ability of cellulosic ethanol producers to obtain financing for pioneer-scale facilities from banks and does not necessarily equate to a cost for the government if the ethanol plant is successful.

Typically one finds in scenario analysis that capital cost reduction and loan guarantee policies are somewhat substitutable and equally effective at addressing the cost barrier of the large amount of capital needed to build a biorefinery. Also, feedstock and product subsidies are mostly interchangeable and redundant. Rapid industry growth can be fostered by an early, but perhaps brief, implementation of policies, such as capital cost reduction and loan guarantees, followed by longer-term volumetric subsidies on feedstock or ethanol production that are gradually phased out.

3.9 Choice of plant type and location

Even if building a plant of a particular type is economical due to its net present value being favorable, it does not follow that such a plant would necessarily be built. Other biorefinery plant types might be more attractive or the general plant construction capacity (not just for biorefineries but for chemical plants and other large industrial facilities) in the nation might be constrained. In order to translate economic viability of the potential "next" plant into aggregate growth of the cellulosic ethanol industry by conversion option and region and to constrain the growth of industry in a "natural" way as overall industry runs into constraints imposed by capacity to produce new plants, we model the allocation of constrained production capacity among potential uses via a logit function that was calibrated to the historical experience of the starch-ethanol industry. This calculation determines the characteristics involved in the decision of whether to build a conversion facility including

(1) the several biomass-to-ethanol conversion pathways, (2) the pioneer and full commercial scales, (3) the 10 geographic regions represented in the BSM, and (4) other uses of plant-construction capacity. Key drivers of allocation include the net present value, the overall capacity to produce new plants (shared across technologies and regions), the maximum economically sustainable number of plants in the region, the regional feedstock availability, the potential market demand for products, and the extensiveness of the downstream logistics infrastructure. Once the decision to construct a particular type of plant in a particular region is made, additional model structure transforms this continuous signal into discrete plant additions. The BSM then tracks the progression of biorefineries from initiation, design and construction, start-up, and production while concurrently tracking the process yield. The logic assumes that as the industry develops, the yield for all of the plants improves in synchrony. Technical failures occur during the start-up phase of the plant and are captured within the model but do not feed back into any decision making in the industry. Once a biorefinery is built, it is not taken off-line, but it is assumed that the investor will continue to upgrade and maintain the biorefinery (see Figure 12). Figure 13 illustrates how maturity improves plant profitability, which in turn advances maturity.

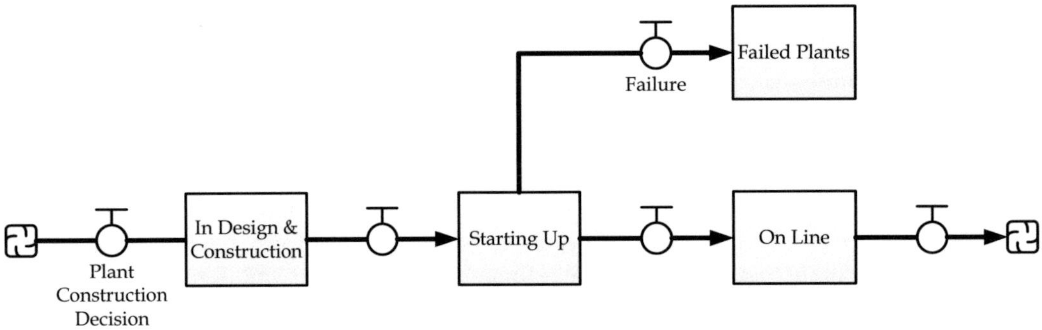

Fig. 12. Stock and flow diagram of biorefinery life cycle in the BSM

3.10 Building ethanol capability at terminals

From the conversion facilities, ethanol travels to terminal facilities. Regrettably, ethanol cannot be transported in the same pipelines as gasoline due to their differing chemical properties. Therefore, ethanol is generally transported via truck or rail. Once it arrives at the terminal, it is blended with gasoline and transferred to its final distribution location (see Figure 14). Not all terminals are suited to store or blend ethanol, so the BSM includes logic for terminals that do not yet have ethanol infrastructure and a means by which they can acquire that structure.

Contained in the Distribution Logistics Module, the ethanol terminal logic aims to provide a simple, high-level, physical, defensible representation of an evolving distribution network for biofuels, while respecting both economic and public policy considerations (see Figure 15). It takes regional ethanol production from the Feedstock Conversion Module and and determines whether existing terminal capacity is sufficient or if a gap exists between desired and existing capacity. Given current model input settings, the dynamic interaction shows potential evolution from the current "as is" world. Some of the key variables from Figure 15 include:

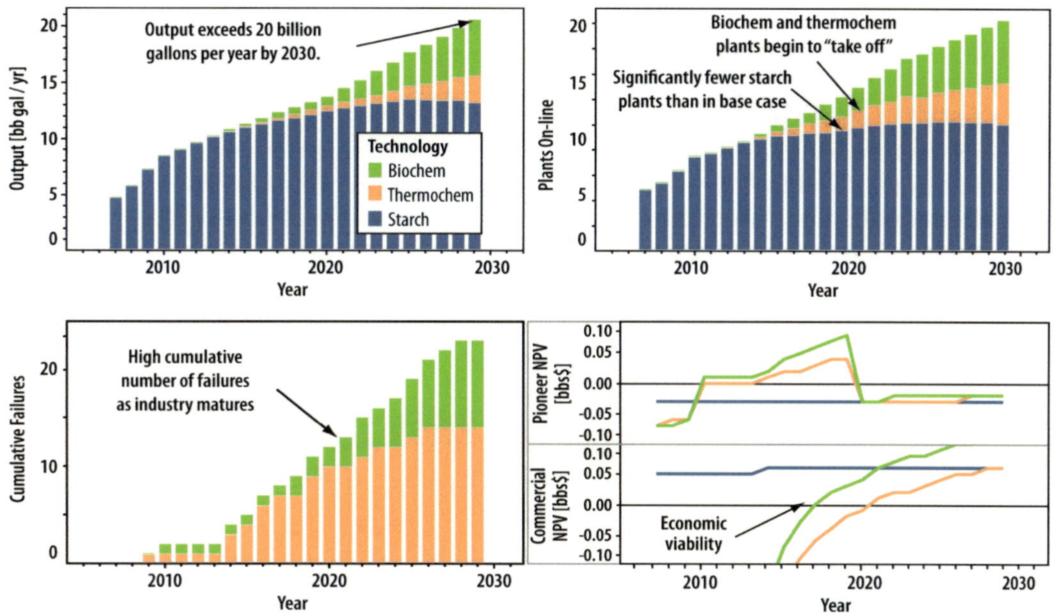

Fig. 13. Plant maturity logic in the BSM

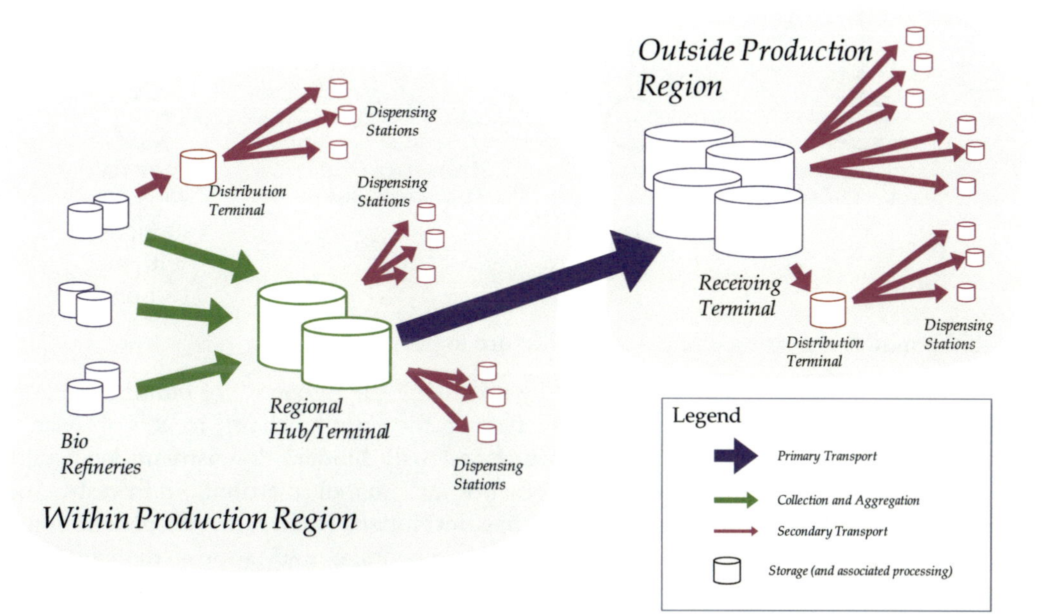

Fig. 14. Ethanol distribution system

- **Infrastructure Gap by Region:** For regions where there is ethanol production, shows the potential number of terminals that could have infrastructure, given current production capacity and potential gasoline consumption relative to the number of terminals with infrastructure.

- **Infrastructure Acquisition Rate:** Represents the effective rate at which the infrastructure gap is eliminated per year, given an increasing rate of adoption among terminals without ethanol infrastructure and external information on how quickly terminals are being upgraded.
- **Terminal Infrastructure Acquisition:** Shows the acquisition of ethanol infrastructure by terminals within a region, given the infrastructure gap and the rate at which infrastructure can be acquired.
- **Regional Ethanol Production without Terminal:** Addresses the situation when there is more ethanol capacity in a certain region than ethanol-capable terminals can store and transports that capacity to a different region.

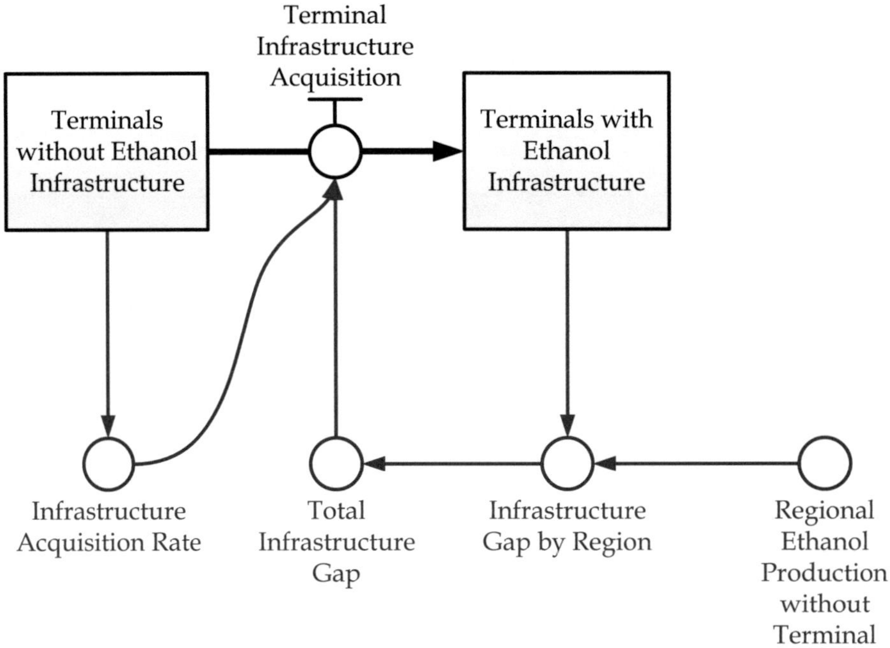

Fig. 15. Terminals acquiring ethanol infrastructure logic

Some powerful insights can be gained just by reviewing the dynamics of building ethanol-capable terminals in conjunction with the rest of the supply chain. In the absence of subsidies, the lack of distribution infrastructure seriously hinders downstream availability and adoption of high-blend fuel. The presence of ethanol distribution infrastructure supports consumption growth once demand has developed. Finally, overall penetration of high-blend fuel is constrained by infrastructure coverage and supply/demand/price considerations. In this idealized system, there is insufficient production capacity to cover potential demand so prices increase to stave off demand.

3.11 Gasoline stations having, considering, or not having high-blend fuel capabilities

In order for ethanol to be blended at the terminals, there must be sufficient distribution storage to accept the end product and dispensers to get the product to the consumers. Although many states do require a boutique blend of ethanol with gasoline due to emissions regulations, this alone will not cause the ethanol industry to expand; it will require a proliferation of fuels with a higher blend of ethanol to occur. But many gasoline station owners are hesitant to invest in high-blend tankage or repurpose existing tankage since they generally operate in a low-profit-margin environment.

There are three categories of gas station owners: those who have high-blend tankage, those who are considering it, and those who do not have it and are not considering it. Whether station owners actually invest depends upon the value proposition, which includes the current status of the regional distribution network, the potential demand for the high-blend fuel, and the opportunity cost of not investing. Figure 16 shows how this decision making is handled in the Dispensing Stations Module of the BSM.

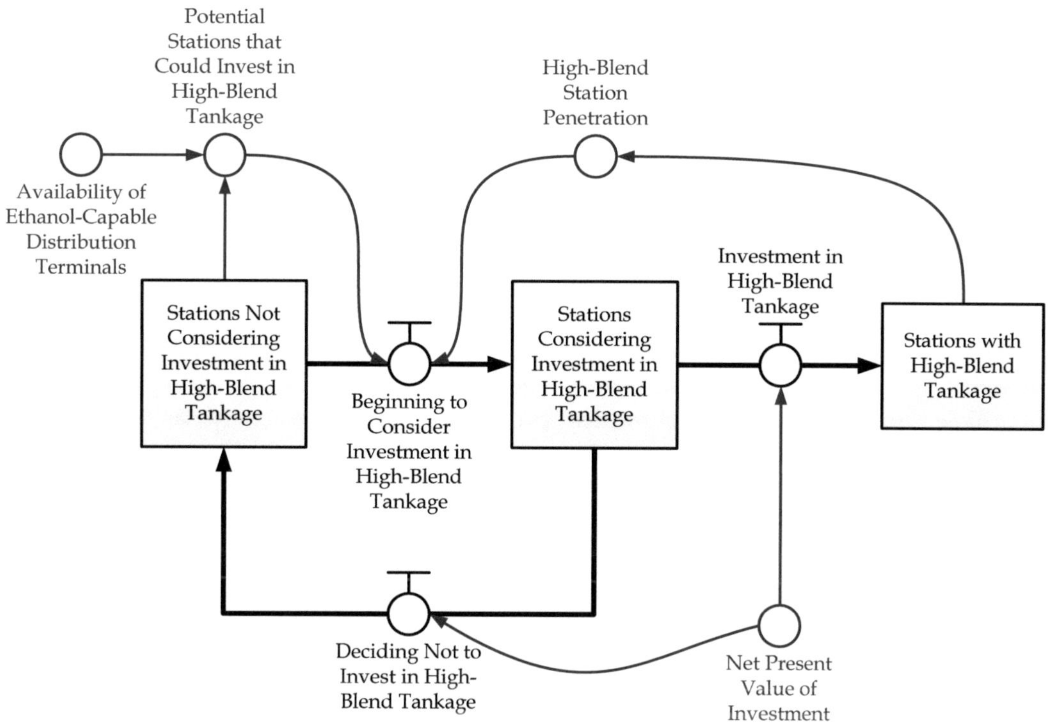

Fig. 16. BSM logic of investment in high-blend tankage

A similar logic is used for the repurposing of existing tankage. An important variable in Figure 16 that could be overlooked is "Net Present Value of Investment." The NPV calculation looks at the station owners' expected financials. The computation takes into account the expected incremental sales of conventional gasoline and high-blend fuel, sales of other items from the convenience store, expected taxes, financing from investing in high-blend tankage, and incentives for the investment.

There are many valuable takeaways that can be gained by evaluating how this decision logic fits into the system of the biofuels industry. Due to the small operating margins of refueling stations, comprehensive subsidies are essential in fostering the installation of high-blend refueling capacity. Even with significant external intervention, adjustment can take multiple years. The presence of ethanol-dispensing tankage supports consumption growth in the early years as long as the point-of-use prices remain low. In contrast, the lack of available tankage, dispensing equipment, and refueling stations hampers high-blend fuel adoption. Finally, the system appears to delicately-situated (especially in the case of unbranded independent gas stations because of relatively low margins and small sales volumes); small changes in incentives or input assumptions can lead to big changes in output behavior.

3.12 Ethanol and gasoline price—consumer decision making

Even if dispensing stations provide high-blend fuel as an option, there is no guarantee that flex-fuel vehicle owners will buy it. The choice to fill up using high-blend fuel depends mainly on price and availability. If there are no high-blend pumps in close proximity when the vehicle owner needs to fill his tank, he is unlikely to drive out of his way to find a facility that offers high-blend fuel. In addition, flex-fuel vehicle owners are unlikely to pay a premium on high-blend fuel. The relative price could confuse consumers, though, due to the different volumetric energy content of high-blend fuel in comparison with conventional gasoline.

In the BSM, the fuel choice logic is contained in the Fuel Use Module. Figure 17 gives a visual representation of how consumers choose what fuel to use. There are three types of fuel consumers: regular high-blend users, occasional high-blend users, and non-high-blend users. These stocks are all represented as a percent of total fuel consumers. One of the main constraints is a variable "Stations that Offer High-Blend Fuel," which excludes all consumers who do not have a station offering high-blend fuel in their areas. Since high-blend fuel's relative price compared with gasoline plays a major role in consumers' fuel selection, it is included in the decisions to either move to being a regular high-blend user from being an occasional user or to drop back to being an occasional user from being a regular user. Whether fuel purchasers are occasional or regular users is determined by the price advantage of ethanol over gasoline. The larger the price gap between high-blend fuel and gasoline, the more quickly the occasional users will become regular users (or the smaller the gap, the more quickly the regular users will drop back to occasional users).

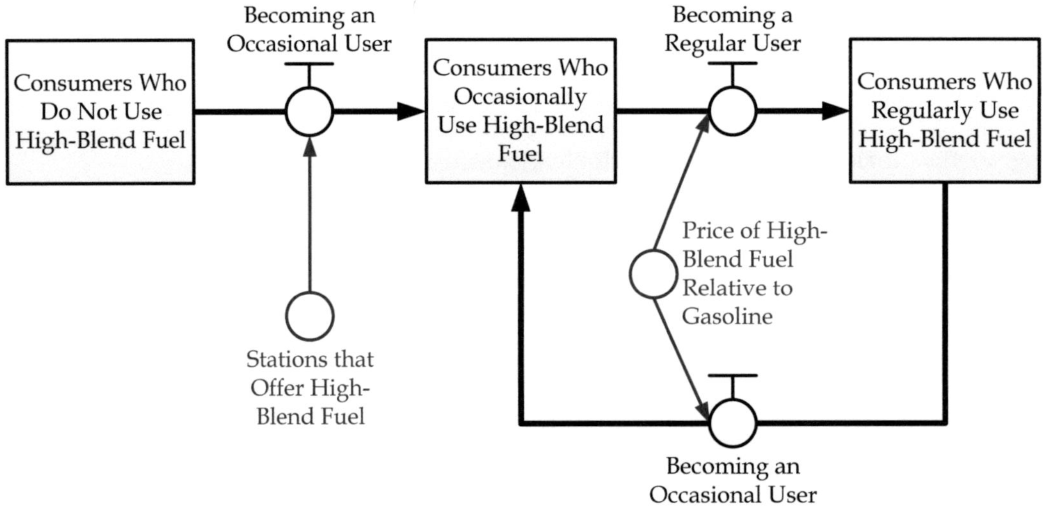

Fig. 17. High-blend fuel choice logic

As a base scenario, the model uses gasoline price projections from the 2010 Annual Energy Outlook (Energy Information Administration 2010), but it also provides the option of choosing from a variety of other gasoline price scenarios, some with price shocks, or for entering user-provided pricing data. The high-blend fuel price is estimated using a blend of two pricing strategies used by fuel station owners, as revealed through surveying owners in Minnesota (Anderson, 2009). The first strategy says that the price of E85 is a multiple of retail gasoline price. The second strategy determines E85 price by adding some constant markup factor from ethanol rack price. Therefore, the price of E85 can be expressed as a weighted average of the two strategies, shown as a mathematical relationship below, where P_{E85} is the retail price of E85, P_{gas} is the retail price of gasoline, P_{etoh} is the rack price of ethanol, f is the weight given to strategy 1 (a fraction between 0 and 1), b is the coefficient for strategy 1 (a discount on retail gasoline price), c is the coefficient for strategy 2 (a constant markup on ethanol rack price), and ε is the error term of the regression.

$$P_{E85} = f\,(b * P_{gas}) + (1\text{-}f)\,(c * P_{etoh}) + \varepsilon \qquad (4)$$

Policymakers can gain a different perspective on high-blend fuel consumption from the following system insights gained from evaluating different scenarios in the BSM. Even under the best conditions, adjustment can take multiple years, and 100% penetration is never reached. Both accessibility and price differential are needed to drive market penetration. Accessibility gives people the option to choose high-blend fuel, whereas price differential transforms occasional users into regular users and boosts high-blend share for each sub-group. The market for high-blend fuel does not persist in cases where high-blend price advantage or parity is lost; consumption quickly reverts mostly to gasoline when the price gap with high-blend fuel closes. In addition, select BSM runs show that short-term (less than one year) gasoline price shocks do not cause a large shift of users from gasoline to high-blend fuel, but long-term or repeating shocks do cause more consumers to switch to using high-blend fuel.

3.13 Maturation of fleet

In addition to the consumer's decision on which fuel to use, the number of flex fuel vehicles in commission has a large impact on the demand for biofuels. In order to address this element, the vehicle module aims to provide a physical basis for generating potential demand for transportation fuel arising from automobiles and light trucks. It has a simple structure to track influx, vintaging, and retirements from stock of vehicles. It does not integrate consumer vehicle purchasing decisions; rather, it provides an accounting framework to inspect potential policy pressures on vehicle stocks and related maximum potential ethanol consumption. Its purpose is to produce policy scenarios that can then be used in the downstream BSM. The maximum ethanol utilization potential under each scenario over time becomes an input to the integrated downstream model. The evolution of the system is driven by scenarios around volume, mix, and efficiency of new vehicles over time. The model contains vehicle tracking by fuel type (gasoline, diesel, flex fuel vehicle, gasoline hybrid electric vehicle, gasoline plug-in hybrid electric vehicle, diesel hybrid electric vehicle, and other) and by efficiency class (more/less efficient automobile and more/less efficient truck).

Figure 18 shows a simplification of the logic contained in the model. Although it only displays three vehicle cohorts (A, B, and C), there are 20 time periods represented in the actual model. After each year, a vehicle can either be removed from service or continue on to the subsequent year. In addition, the average vehicle efficiency can be altered on an annual basis depending on the characteristics of the cars and trucks that are retired. Whether a vehicle is taken out of service is dependent upon the vehicle survival rates, which are computed from historical data for the technology and efficiency classes. The baseline vehicle fleet is taken from the U.S. Energy Information Administration's *Annual Energy Outlook* (Energy Information Administration, 2010), but the parameters can be altered to meet the needs of any given scenario.

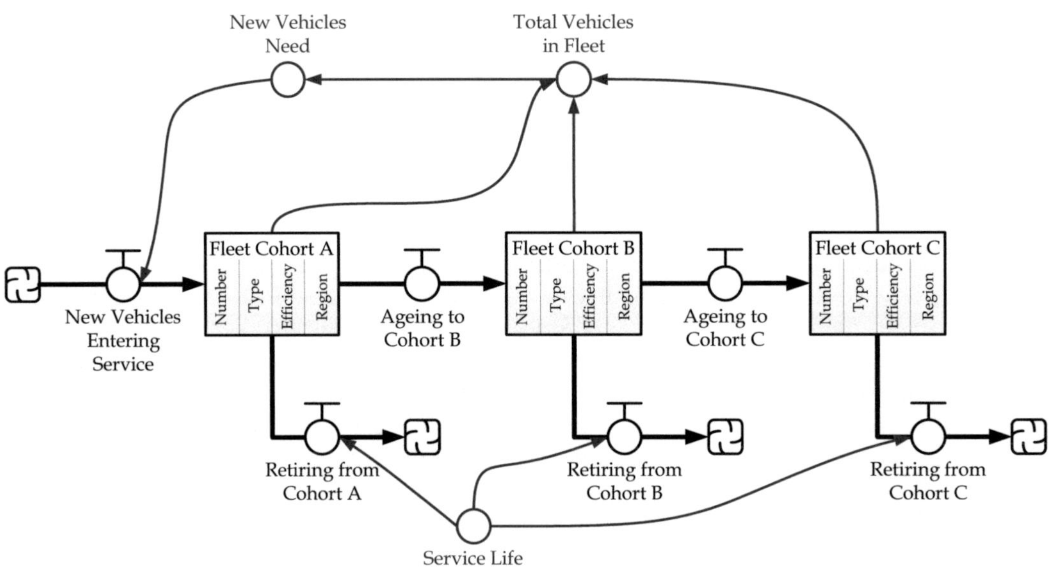

Fig. 18. Vehicle Module logic

Even though the Vehicle Module does not currently include a robust consumer choice component, insights can still be developed. In order for high-blend fuel to have a high amount of penetration, flex fuel vehicle adoption must be substantially higher than Annual Energy Outlook forecasts. For example, a policy that could increase the amount of flex fuel vehicles on the road is the Car Allowance Rebate System (also known as "cash for clunkers"): through this program, older vehicles are replaced by newer ones that have a high probability of having flex fuel capability. See recent analyses by Vimmerstedt et al. (2011) for further discussion on insights gained from the downstream portion of the BSM.

4. Analysis approach

Specific policy-relevant scenarios or past scenarios can be used to drive the BSM simulations, though the BSM is not limited to scenario analysis. Under a specified scenario, the BSM can be used to track the hypothetical development of the biofuels industry given the deployment of new technologies within various elements of the supply chain and the reaction of the investment community to those technologies and given the competing oil market, vehicle demand for biofuels, and various government policies over an extended timeframe. Note, however, that high-level models such as the BSM are not typically used to generate precise estimates but rather to (1) analyze and evaluate alternate policies, (2) generate highly effective scenarios, (3) identify high-impact levers and bottlenecks, and (4) focus discussion among policymakers, analysts, and stakeholders. When the model output includes unexpected system behaviors, modeling assumptions—particularly the behavioral aspects of decision making and the adequacy of the representation of feedback—need careful reexamination to distinguish potential insights from model limitations. The model itself often indicates what assumptions need the most scrutiny; hence, it helps define the research and learning agenda.

Although the BSM inputs can be altered to include any combination of policies, establishing a "reference policy case" to which subsequent scenarios are compared can be useful for determining what policies will have the greatest potential for producing substantial industry growth. The BSM reference policy case includes moderate incentives for ethanol production and a 50 cent per gallon gasoline tax (which could be interpreted as a "carbon" or GHG tax of approximately 51 dollars per ton of carbon dioxide). Policies are phased out in a staged manner, with the policies involving grants for capital equipment or loan guarantees ending earlier and the policies involving volumetric subsidies phasing out anywhere from 2020 to 2050. Each of the policies included in the reference case is based on historical precedence or future plausibility. Sensitivity, bottleneck, tipping-point, and other analyses are typically carried out with respect to a baseline scenario of existing policies and the reference policy case.

5. Scenario exploration and insight

Current experience with the biofuels industry and analysis of the supply chain components suggest that the cellulosic biofuels industry is unlikely to develop without substantial help from external sources. The BSM provides a valuable platform through which the possible effects of policy can be explored. Figure 19 shows many of the insights that have been

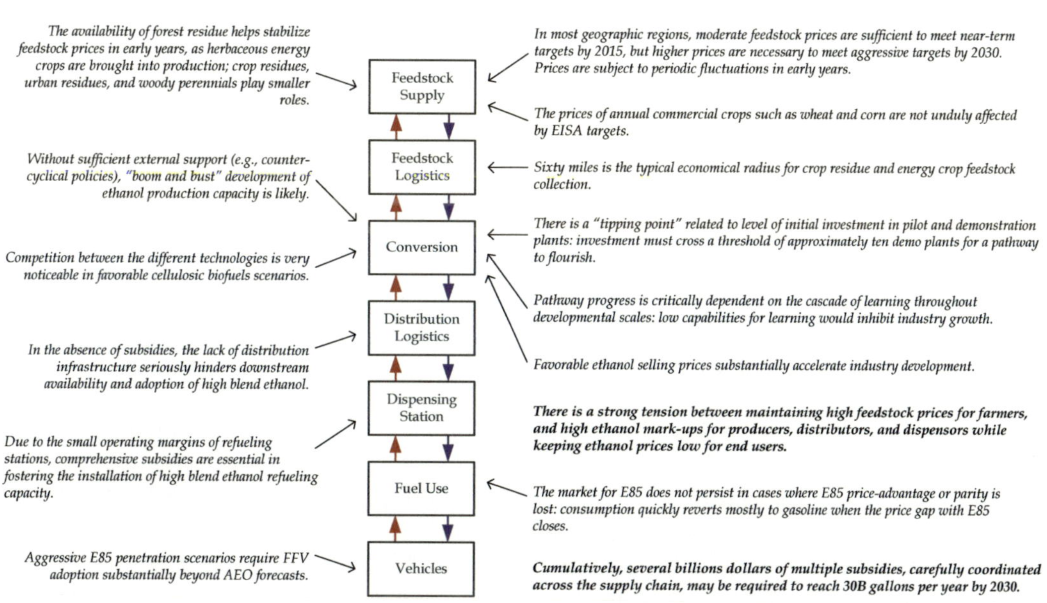

Fig. 19. Insights gained throughout the biofuels supply chain

gained from countless analyses and studies performed using the model. Each component of the supply chain has its own unique needs in terms of policy implementation, bottlenecks, and favorable points of intervention. Many of the insights were already discussed in the sections above dealing with dynamic components of the BSM.

Given the insights discovered in different areas of the supply chain, a picture begins to emerge dealing with a portfolio of policies that could lead to cellulosic ethanol industry takeoff. Based on a scenario without policy initiatives, the model results indicate a complete failure of the cellulosic ethanol industry to take off (left side of Figure 20) given the currently observed investments in demonstration and pioneer conversion plants. Furthermore, our analysis shows that overly aggressive or poorly targeted subsidies cause industry instability and only lead to a paltry increase in cellulosic biofuels production.

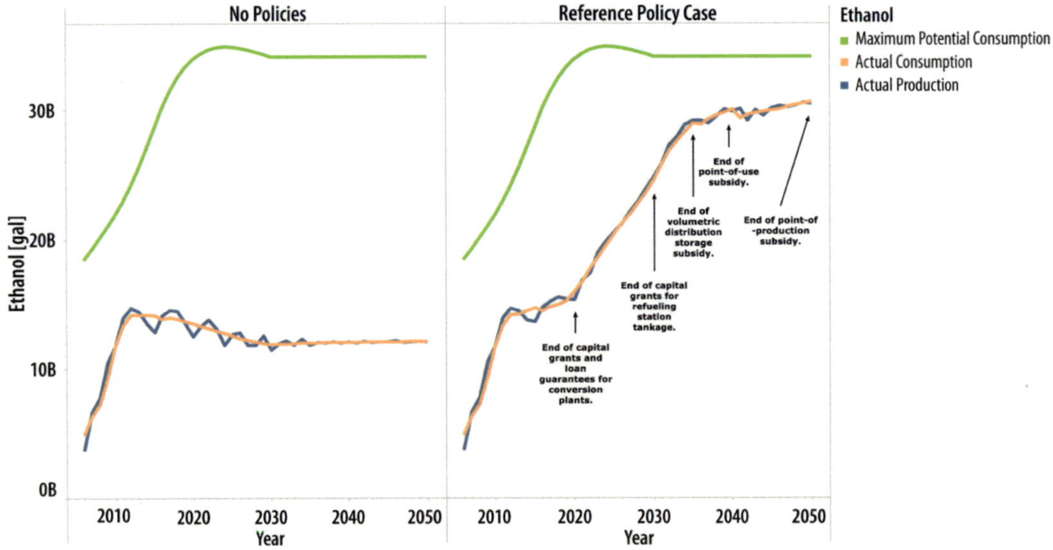

Fig. 20. Set of "reference policy" inputs produce a moderate, sustained industry takeoff ("B" = billion)

In general, policies are sequenced such that the more expensive or least cost-effective ones are phased out earlier; policies are phased out as early as possible so as to still achieve substantial industry growth. This policy configuration and these results are not intended to be prescriptive or to represent the minimal effective policy; instead, they simply form a basis against which the impact of potentially sensitive data inputs can be studied and against which an "on the margin" policy analysis can be conducted. Nevertheless, there is general historical precedence or future plausibility for each of the policies included in the reference and for the rough order of magnitude of the policies.

Figure 20 illustrates the resulting ethanol consumption and production in the BSM with and without the reference policy. Without the policies in place, the cellulosic ethanol industry does not take off; on the contrary, ethanol consumption actually falls with time due to the increased efficiency of the vehicle fleet. [Note that vehicle fleet and gasoline price forecasts from the Energy Information Administration's *Annual Energy Outlook* (Energy Information Administration, 2010) reference case are being used.] The leveling off of ethanol production around 2012 results from the saturation of ethanol demand for the blending of E10 and the

EISA-imposed 15B gallon limit on the production of starch-based ethanol. For scenarios with a higher level of ethanol production, resource constraints, such as availability of flex fuel vehicles and biomass supply, play a role in the leveling out of production in later years. This result is not fundamental to the model but rather an artifact of the given model setting of a certain scenario. Figure 21 provides a rough, preliminary estimate of the magnitude of the costs associated with the policies in the reference case. It is important to note that the revenue from the gasoline tax far exceeds the cost of these policies.

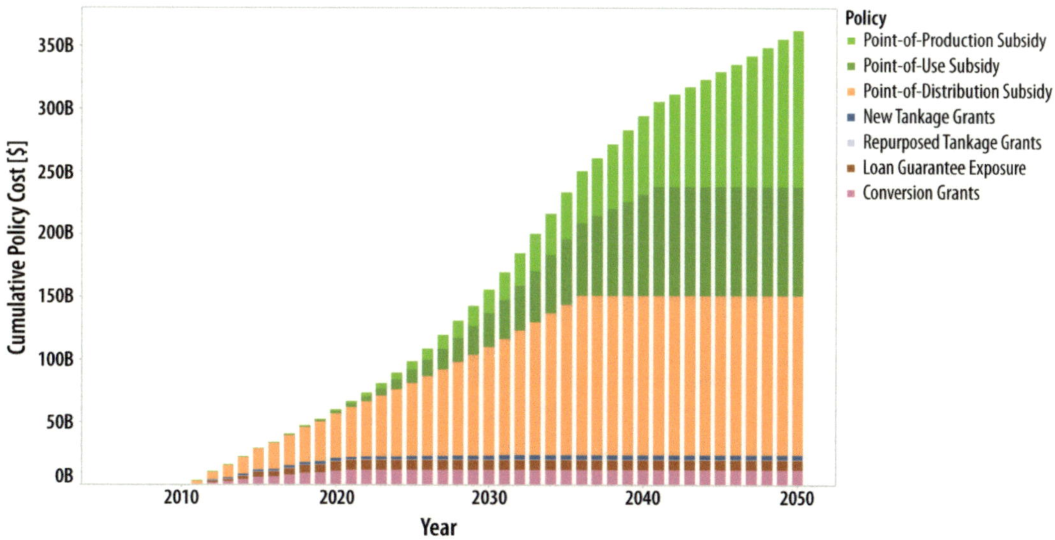

Fig. 21. Preliminary, rough estimate of the costs associated with the reference policy ("B" = billion)

In constructing the reference policy, we examined the effects of eliminating various grants or subsidies. Figure 22 shows the effect of removing each separately, suggesting that the availability of grants for new high-blend tankage at refueling stations is an essential component of this policy case. Other policies, particularly the gas/carbon tax are highly influential. In contrast to where single policies are "turned off," nearly every case where two policies are turned off results in ethanol consumption that is substantially lower than in the reference case.

Looking at the reference case and other runs of the BSM, insights can be gained in terms of what needs to be in place in order for the cellulosic ethanol industry to take off. In particular, the combination of initiatives is likely to require:

- Mechanisms to ensure a favorable-to-high-blend price spread as perceived by end users
- A high level of external investment in dispensing station infrastructure (tankage and related equipment)
- Aggressive initial external investment in pilot, demonstration, and pioneer-scale conversion facilities
- High rates of industry learning.

We have performed multiple analyses in the BSM dealing with different mechanisms for pricing high-blend fuel at the pump along with ethanol and gasoline price coupling. In general, high-blend fuel appears to be highly coupled with gasoline price. When short-term

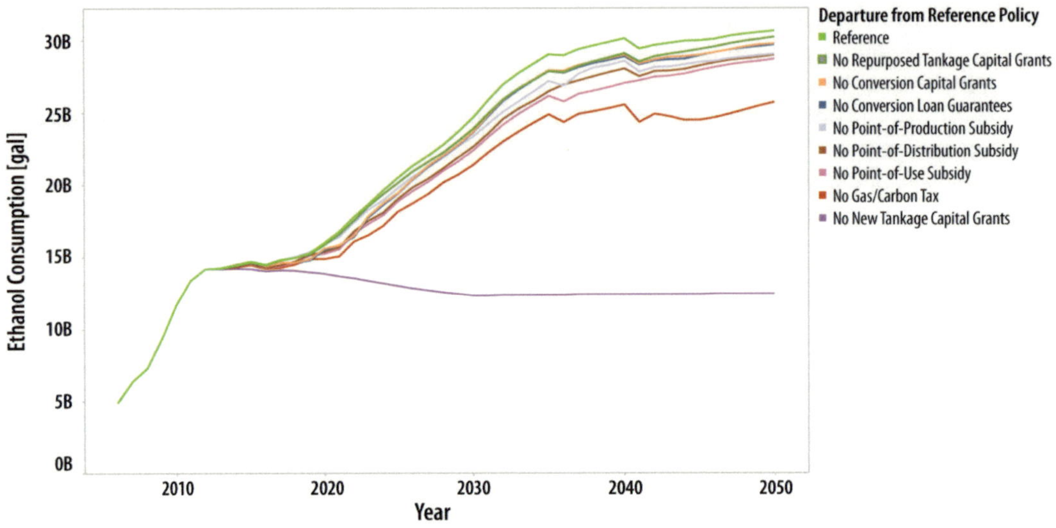

Fig. 22. Variations of the reference policy case, where one of the policies is "turned off" ("B" = billion)

gasoline price shocks are applied to the model, there is very little reaction by consumers to substitute high-blend fuel. Long-term, sustained gasoline price shocks do show some fuel switching but increases are less than 10% of the total ethanol consumption. As was discussed in Section 3.12 in recent research (Anderson, 2009), investigators found that there are two main methods by which gasoline station owners price high-blend fuel: as a discount on the gasoline price or as a mark-up on the rack ethanol price. The two-strategy pricing scheme was incorporated in the BSM and comparative tests were performed with and without the relationship in place. The link suggests tradeoffs. Price coupling implies a higher degree of profitability per gallon from high-blend fuel, which implies greater incentive for stations to invest in high-blend tankage and equipment. On the other hand, price coupling implies a smaller spread between gasoline and high-blend fuel, which suggests lower market penetration than in the previous formulation. Further exploration of price coupling effects between high-blend fuel and gasoline is warranted.

Bottlenecks in downstream distribution and dispensing infrastructure may significantly impede the growth of the cellulosic biofuels industry. Managing the biomass-to-biofuels supply chain involves a carefully-orchestrated arrangement of flow-through capacities at various stages in the chain. Bottlenecks result when those capacities are out of synch. For example, bottlenecks in downstream distribution and dispensing infrastructure may significantly impede the growth of the cellulosic biofuels industry. In the reference policy case, the critical importance of the capital grant subsidy for new high-blend tankage indicates the potential for bottlenecks in the downstream portion of the supply chain. To test this hypothesis, the fraction of the distribution system (terminals and transport) with ethanol infrastructure was set arbitrarily to 100% and the fraction of refueling stations with ethanol-capable tankage to 100%. This mimics the situation where ethanol would be a fungible or infrastructure-compatible fuel. Figure 23 indicates that removing these downstream constraints results in substantially more ethanol consumption, either with or without supportive polices. In general, a lack of appropriate downstream policies could limit the effectiveness of otherwise successful upstream ones, and vice versa. In addition, a carbon policy could potentially aid in elevating demand for high-blend ethanol. Although carbon policy has been discussed extensively in U.S. national policy considerations, the BSM does not currently implement it explicitly. Rather, carbon taxes and carbon caps can be simulated through a gasoline tax. It would be relatively simple to add a feedback controller, but the policy would have to be more well-defined. Currently it is not clear where the policy would affect the industry across the supply chain.

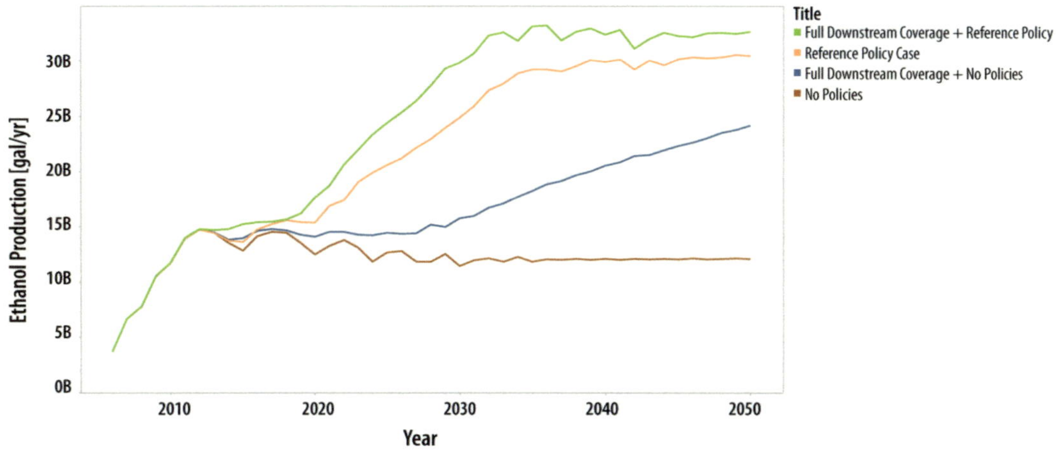

Fig. 23. Effect of removing downstream infrastructure constraints ("B" = billion)

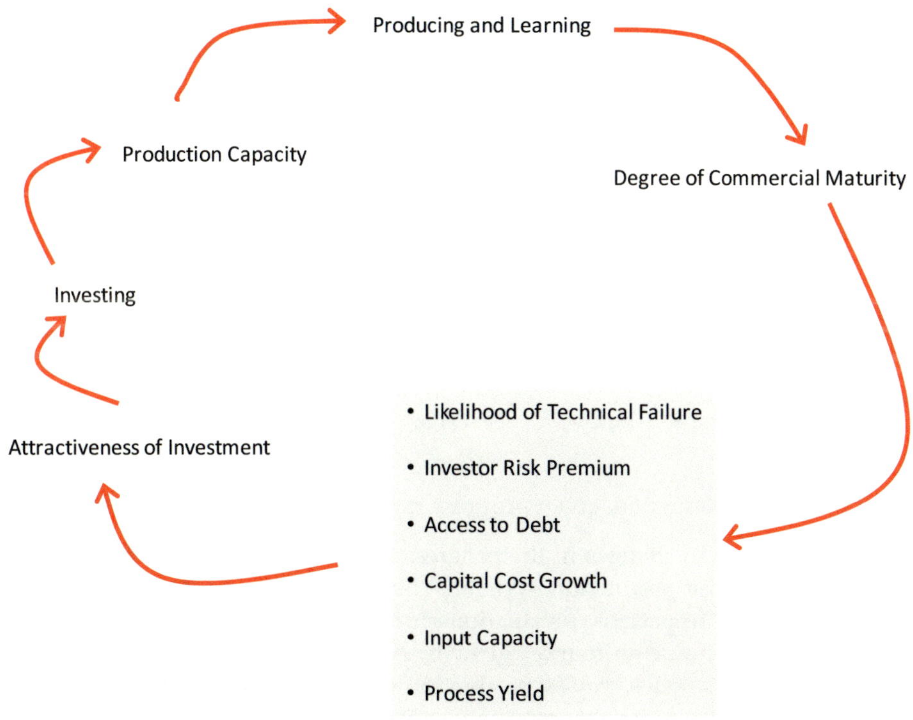

Fig. 24. Key feedback driving investment attractiveness

Although policies are important for industry takeoff, aggressive investment in conversion facilities and accelerated industry learning are also essential. In general, rapid learning in conversion technologies/plants "tips" the ethanol market. A focal point in the Conversion Module portion of the BSM is the concept of "cascading learning curves," which was already discussed in Section 3.7. The logic depicted in Figure 24 enables the model to capture the evolution of a conversion platform from an arbitrary initial state to "n^{th} plant" maturity. As shown in Figure 24, embedded within the learning curve logic is a rich set of positive/reinforcing feedbacks, all of which can drive a conversion technology toward a high degree of investment attractiveness and low production costs.

One critical component of the learning curve dynamics in the model is the concept of a progress ratio. Essentially, a progress ratio translates the activity basis of producing into learning and thereby into increases in commercial maturity. Figure 25 shows model simulations of total cellulosic ethanol production for five values of the progress ratio ranging from 75% to 95%. Smaller values for the progress ratio mean that cost falls faster with increases in cumulative production. Progress ratios in the range of 75% to 90% show similar levels of ethanol production, while a value of 95% causes production to stagnate. The response of production to different progress ratios is highly nonlinear. Moving from 95% to 90% causes a big response, while additional 5% percent decreases cause much smaller responses. Note that it is conceivable that certain types of government policies could alter industry progress ratios towards lower, more favorable values.

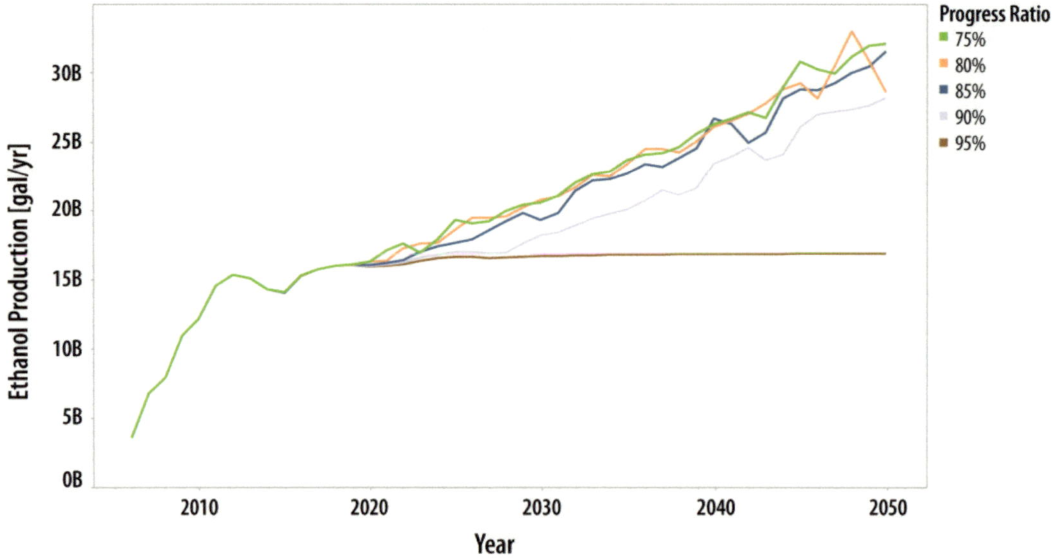

Fig. 25. Total ethanol production, given progress ratios of 75% to 95% ("B" = billion)

The nonlinear response to changes in the progress ratio in these simulations is the result of layered constraints in the production system. When the progress ratio is 95%, cumulative experience has a small impact on production cost and the high cost of cellulosic ethanol remains the main constraint on market growth. A progress ratio of 90% generates much lower production costs, which causes production to nearly double. Progress ratios below about 80% cause the system to encounter new constraints that limit market growth; ethanol production is no longer the binding constraint at lower progress ratio levels. One of the newly emerging constraints is the number of filling stations that have high-blend ethanol pumps. A second constraint is the supply and price of feedstock. The large increase in ethanol production causes feedstock prices to rise, which reduces the profitability of investments in new plants.

Figure 26 shows an ethanol price index over the period of interest. With faster rates of learning (as indicated by lower progress ratios), the system settles into a much lower price regime. With very slow rates of learning, prices remain very high (double the initial value) over the course of the simulation. The results for intermediate learning rates hint at increased price volatility. The oscillations observed at the start of the simulations are a result of the lag in the production response to demand signals in the early years of the cellulosic ethanol industry; the prices are not a forecast but rather an indication of dynamic interactions in the model.

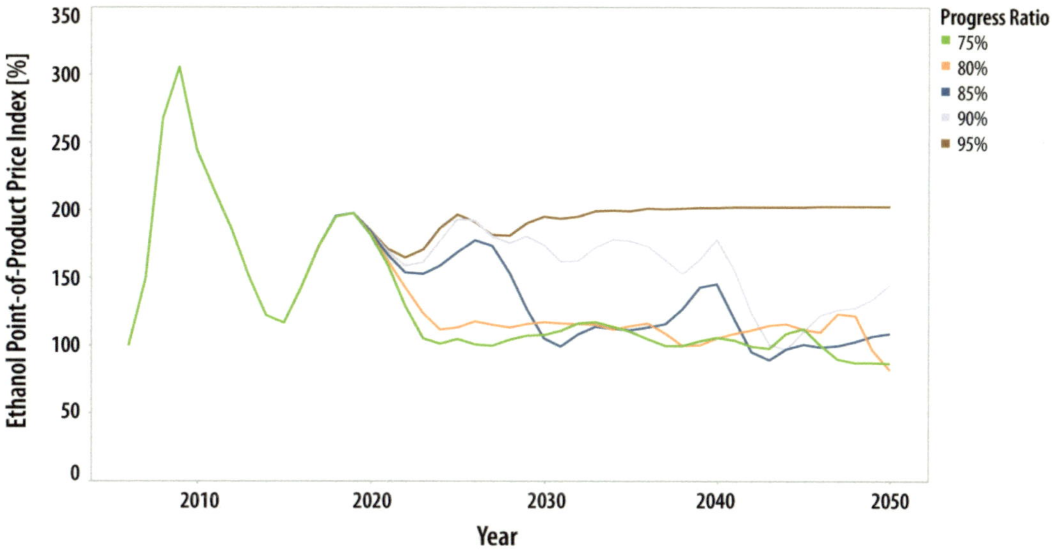

Fig. 26. Ethanol point-of-production price indices, given progress ratios of 75% to 95%

We have already discussed how certain policies must be in place in order to have significant cellulosic ethanol industry growth. There is also a dynamic element that policies alone are less effective than policies that are implemented in coordination with one another. The sum of benefits from each policy implemented in solitude is much less than the benefits attained when the policies are combined. In Figure 27, a policy focused on growers is combined with one targeting conversion facilities. When they are implemented in isolation, the policies are concretely less effective than when they are placed into service together.

This result highlights a significant advantage of a systems-focused simulation; rather than displaying just the summation of separate, static, disconnected analyses, systems modeling shows how the different sections of the supply chain work together to enhance policy outcomes. As installed capacity grows, it is easier to self-sustain growth in the industry. Because of multiple feedbacks, policy initiatives can potentially create interdependent benefits (see Figure 28).

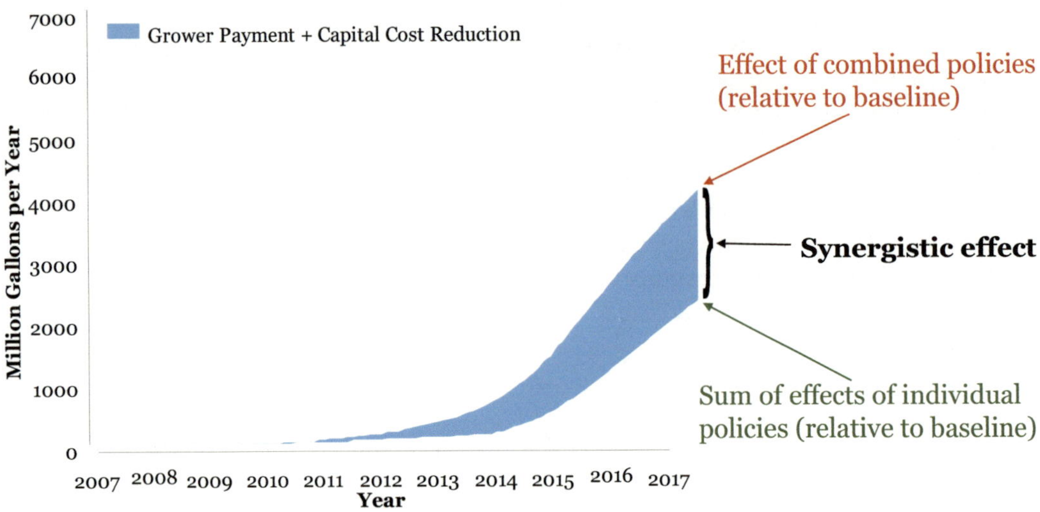

Fig. 27. Synergistic effect of coordinating policy implementation

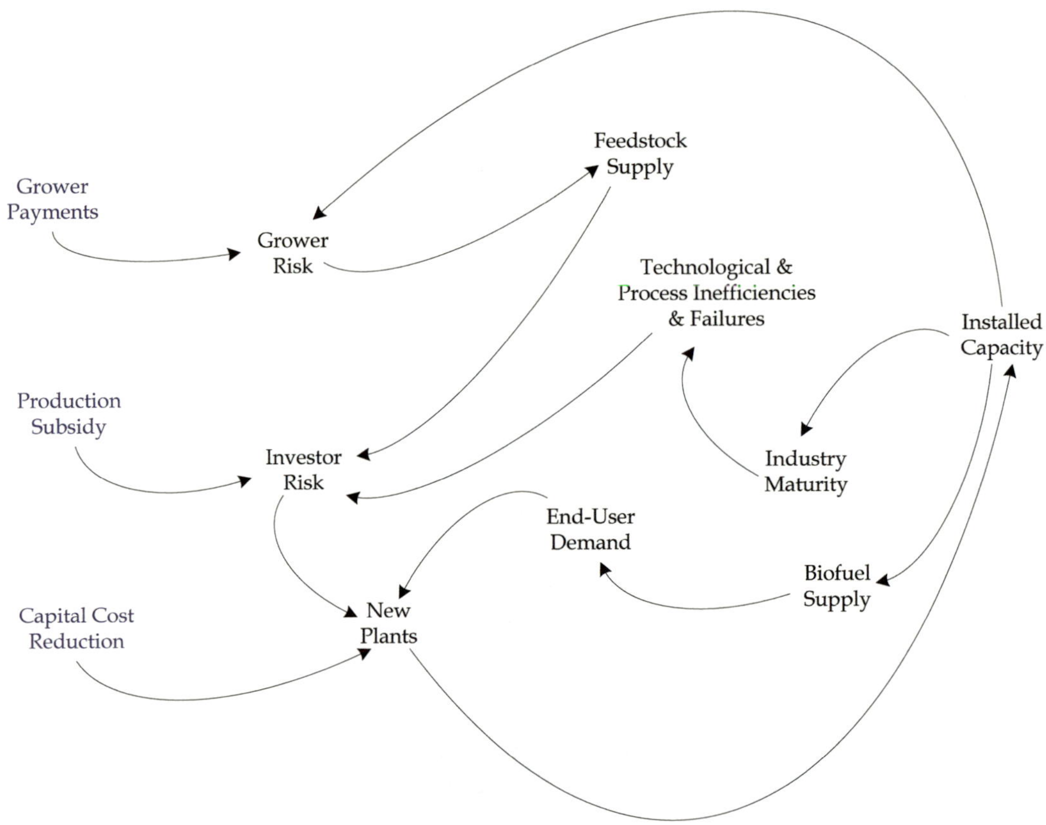

Fig. 28. Dynamic effects of synergistic policy interaction

6. Conclusion

The BSM is a powerful tool for gaining insights on the whole biomass-to-biofuels supply chain. It has unique mathematical formulations that comprise its variables with dynamic interconnections among them. A look at some of the specific influence diagrams representing complicated logic in the BSM shows how the interactions are modeled and gives a glimpse at how the BSM arrives at results after iteration. Each component of the supply chain is modeled so that it can either be run as a standalone model or with connection to the other components. Powerful insights have been gained from both the standalone modules and the entire system model. Overall, in the BSM simulations the cellulosic ethanol industry tends not to rapidly thrive without significant outside actions in early years of its evolution. An initial focus for jumpstarting the industry typically has strongest effects in the BSM in areas where effects of intervention have been identified to be multiplicative. Due to industrial learning dynamics, support for the construction of cellulosic ethanol conversion facilities in the near future encourages the industry to flourish. By accelerating the pace of development, industrial learning can grow substantially. In addition, without the alleviation of bottlenecks of high-blend fuel distribution and high-blend fuel pumps, the increased amount of ethanol produced may not have a viable market to serve. Future work includes additional analyses using the BSM and expanding the model to include infrastructure-compatible ("fungible") fuels such as biomass-based gasoline, diesel, and aviation fuel.

7. Acknowledgment

The authors would like to mention the people without whom the Biomass Scenario Model would not exist and the included analyses would not have been performed. Robert Wallace (Booz Allen Hamilton) and John Sheehan (The University of Minnesota) were instrumental in getting the initial model off the ground. Steve Peterson (Peterson Group) has been with the effort since its inception and has continued to be the main contributor to the modeling effort. Zia Haq and Sheila Moynihan of the U.S. Department of Energy have been tireless advocates for continuing analyses to shed light on the biofuels system. In addition, the other BSM team members have been invaluable: Corey Peck and Mark Paich of Lexidyne LLC and Laura Vimmerstedt, David Hsu, Dana Stright, Andrew Argo, and Amy Schwab, all of NREL.

8. References

Anderson, S. T. 2009. *The Demand for Ethanol as a Gasoline Substitute*. East Lansing, MI: Michigan State University, August.

Bush, B., M. Duffy, D. Sandor, and S. Peterson. 2008. *Using System Dynamics to Model the Transition to Biofuels in the United States: Preprint*. Preprint. Golden, CO: National Renewable Energy Laboratory.

Energy Information Administration. 2010. Annual Energy Outlook 2010. *Forecasts & Analyses*. May. http://www.eia.doe.gov/oiaf/archive/aeo10/index.html.

Interagency Agricultural Projections Committee. 2007. *USDA Agricultural Projections to 2016*. U.S. Department of Agriculture.

isee systems. 2010. STELLA: Systems Thinking for Education and Research Software. http://www.iseesystems.com/softwares/Education/StellaSoftware.aspx.

Office of the Biomass Program, and Energy Efficiency and Renewable Energy. 2008. *Multi-Year Program Plan March 2008*. Multi-Year Program Plan. U.S. Department of Energy, March.

The University of Tennessee. n.d. Agricultural Policy Analysis Center Research Tools - POLYSYS. *Agricultural Policy Analysis Center - The University of Tennessee.* http://www.agpolicy.org/polysys.html.

U.S. Department of Agriculture. 2011. Conservation Reserve Program. *Conservation Programs*. February. http://www.fsa.usda.gov/FSA/webapp?area=home&subject=copr&topic=crp.

U.S. Environmental Protection Agency. 2010. Renewable Fuels: Regulations & Standards. *Fuels and Fuel Additives*. December. http://www.epa.gov/otaq/fuels/renewablefuels/regulations.htm.

United States Government. 2007. *Energy Independence and Security Act of 2007.*

Vimmerstedt, L, B Bush, and S Peterson. 2011. Ethanol Distribution, Dispensing, and Use: Analysis of a Portion of the Biomass-to-Biofuels Supply Chain Using System Dynamics. Manuscript submitted for publication.

19

Distributed, Integrated Production of Second and Third Generation Biofuels

William Gibbons[1] and Stephen Hughes[2]
[1]*Biology & Microbiology Department, South Dakota State University, Brookings, SD*
[2]*National Center for Agricultural Utilization Research, USDA, Peoria, IL*
United States of America

1. Introduction

Materials that are burned directly for energy, such as firewood, wood chips, pellets, animal waste, forest and crop residues are considered primary biofuels. First generation biofuels also include bioethanol produced by fermentation of starch (from wheat, barley, corn, or potato) or sugars (from sugarcane or sugar beet), and biodiesel produced by transesterification of oil crops (including rapeseed, soybeans, sunflower, palm, coconut) and animal fats. Second generation biofuels include bioethanol and biodiesel produced from the residual, non-food parts of crops, and from other forms of lignocellulosic biomass such as wood, grasses, and municipal solid wastes (Inderwildi & King, 2009). Third generation biofuels include algae-derived fuels such as biodiesel from microalgae oil, bioethanol from microalgae and seaweeds, and hydrogen from green microalgae and microbes (Aylott, 2010; Dragone, et al., 2010). "Drop in" fuels like "green gasoline," "green diesel," and "green aviation fuel" produced from biomass are considered fourth generation biofuels (Kalita, 2008). Efforts are also underway to genetically engineer organisms to secrete these fourth generation hydrocarbon fuels.

Today, corn is the major source of first-generation bioethanol, with over 12 billion gallons of fuel ethanol produced in 2010 from approximately 4.6 billion bushels of corn (Anon, 2011b) in 190 operating facilities in 26 states. Most are located in the Midwest, near the site of feedstock production (Figure 1), however some are co-located with dairies or beef cattle feeding operations or dairies outside the Corn Belt. The typical size of corn ethanol plants is 50-100 million gallons per year. Table 1 provides a summary of industry growth in the US over the past decade. Ethanol is also the most important biofuel worldwide in terms of volume and market value (Licht, 2006).

In 2007, Congress passed the Renewable Fuels Standard 1 (RFS1), which mandated renewable fuel use of 7.5 billion gallons by 2012. Congress subsequently passed the Energy Independence and Security Act of 2007 (EISA), which made significant changes in the structure and magnitude of the renewable fuel program. The EISA (also called the RFS2) specified use of a total of 15.2 billion gallons/year of renewable fuel by 2012 and 36 billion gallons/year by 2022.[1] Also mandated were maximal amounts of corn-based ethanol,

---

[1] Federal Register, March 26, 2010, pp 14670-14904, Final Rule 40 CFR 80; 75 FR 14670; http://federalregister.gov/a/2010-3851.

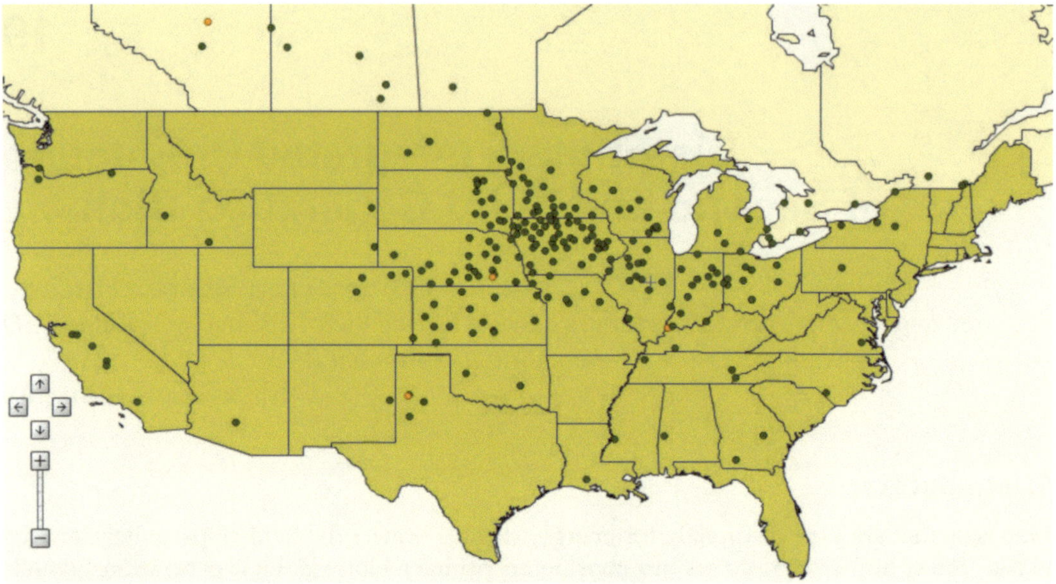

Fig. 1. United States and Canada Fuel Ethanol Plant Map[2]

Year	Total Ethanol Plants	Ethanol Production Capacity (BGY)	Plants Under Construction or Expanding	Capacity Under Construction or Expanding (MGY)	States with Ethanol Plants
1999	50	1.70	5	77	17
2000	54	1.75	6	92	17
2001	56	1.92	5	65	18
2002	61	2.35	13	391	19
2003	68	2.71	11	483	20
2004	72	3.10	15	598	19
2005	81	3.64	16	754	18
2006	95	4.34	31	1,778	20
2007	110	5.49	76	5,636	21
2008	139	7.89	61	5,536	21
2009	170[3]	12.48[4]	24	2,066	26
2010	187[3]	13.03[4]	15	1,432	26

Table 1. United States Ethanol Production Capacity[5]
BGY = Billion gallons per year; MGY = Million gallons per year.

[2]Anon, 2010b

[3]Operating plants

[4]Includes idled capacity

[5]Anon, 2011a; Federal Register, March 26, 2010, page 14674, Final Rule 40 CFR 80; 75 FR 14670; http://federalregister.gov/a/2010-3851

cellulosic biofuel, biomass-based diesel, advanced biofuels, and total renewable fuel that must be used in transportation fuel yearly from 2010 to 2022. Table 2 provides the annual renewable fuel volume requirements of RFS2. Other factors driving the demand for biofuels are greenhouse gas standards and regulations. According to EPA's regulatory action, lifecycle greenhouse gas emissions of cellulosic biofuels must be at least 60% less than the baseline lifecycle greenhouse gas emissions (Anon, 2010a). Tax subsidies, government financial support and fuel policies will also significantly affect economic competitiveness, consumption and production.

Production of 2nd generation biofuels has been a researched for 30 years, with the key impediment being the recalcitrance of the biomass itself (Sjostrom, 1993). Lignocellulose a structural material which nature has designed to resist breakdown. Scientists and engineers have focused on developing pretreatment processes to open the structure of biomass for enzymatic attack (Taherzadeh & Karimi, 2008), while molecular biologists have sought to increase the productivity and efficiency of cellulase enzymes that hydrolyze the fibers into fermentable sugars (Banerjee et al., 2010a; Banerjee et al., 2010b; Jorgensen et al., 2007; Taherzadeh & Karimi, 2007b). Microbes to efficiently convert the mixture of 5 and 6 carbon sugars to ethanol have been yet another technical hurdle (Ballesteros et al., 2004; Olofsson et al., 2008; Saha, 2003). Many of these systems are now being evaluated in pilot plants (Taherzadeh & Karimi, 2007a; Taherzadeh & Karimi, 2007b). Due to the infrastructure challenges associated with ethanol as a fuel, research efforts in the early 2000s expanded to production of third and fourth generation biofuels that are considered infrastructure compatible. Green gasoline, diesel, and jet fuels can be made from biomass, through either biochemical (enzymatic/fermentation) or thermochemical (gasification or pyrolysis) platforms.

Replacing petroleum-derived gasoline, diesel, and jet fuels with renewable fuels will have a wide range of environmental, societal, and economic impacts. The significance and timing of these impacts will be affected by the pace at which biofuels gain market share. This, in turn, will be affected by market forces (crude oil price and availability, feedstock prices), technology development, political conditions, and regulatory factors (Licht, 2006; Regalbuto, 2009) These impacts will be affected by: 1) the type of fuel produced and its use, 2) the types and locations of the feedstocks, 3) the locations, methods, and scale of conversion systems, 4) the yields of products and co-products from a given feedstock, and 5) the challenges associated with use of these feedstocks (Hahn-Hagerdal, et al., 2006).

To achieve the goal of secure and sustainable bioenergy and biofuel production by the middle of the 21st century, we must build upon on existing infrastructure, but also must make substantial improvements in infrastructure and process technology applied. Much of the challenge will revolve around feedstock supply and logistics, (Sims, 2003) and therefore we envision a decentralized, community-based system with integrated crossover to produce a range of third generation drop-in biofuels (long chain alkenes, alkanes, and alcohols) and bio-derived chemicals. These systems could be developed on green-field sites or built onto first/second generation biofuel (ethanol) facilities. Their smaller size and distributed locations will minimize transport of bulky biomass sources, and will spread the economic impact over a broader landscape.

Advancements are needed in sustainable feedstock production to ensure that plants are bred for rapid growth, minimal inputs, high yield, and desirable composition (Dufey, 2006) Perennial plants able to grow with minimal inputs (fertilizer, pesticides, irrigation water) on lower quality land would be most desirable in terms of sustainability (Ragauskas et al.,

Year	Cellulosic biofuel requirement	Biomass-based diesel requirement	Advanced biofuel requirement	Total renewable fuel requirement
2009	NA	0.5	0.6	11.1
2010	0.1	0.65	0.95	12.95
2011	0.25	0.80	1.35	13.95
2012	0.5	1.0	2.0	15.2
2013	1.0	a	2.75	16.55
2014	1.75	a	3.75	18.15
2015	3.0	a	5.5	20.5
2016	4.25	a	7.25	22.25
2017	5.5	a	9.0	24.0
2018	7.0	a	11.0	26.0
2019	8.5	a	13.0	28.0
2020	10.5	a	15.0	30.0
2021	13.5	a	18.0	33.0
2022	16.0	a	21.0	36.0
2023+	b	b	b	b

Table 2. Renewable Fuel Volume Requirements for RFS2 (billion gallons)
[a] To be determined by EPA through a future rulemaking, but no less than 1.0 billion gallons
[b] To be determined by EPA through a future rulemaking

2006). Crop residues and cover crops could provide dual-use of land, although there are some concerns regarding acceptable sustainability (Lal, 2005; Tilman et al., 2009) An emerging class of feedstocks includes single-celled phototrophs (eg. algae or cyanobacteria) that can fix CO_2 directly into oil (Aylott, 2010). These organisms are also being engineered to produce drop-in fuels or chemicals directly, hence providing a self-contained production/conversion system (Rosenberg, et al., 2008).

Improvements to conversion systems are needed to increase yield and minimize costs, while sizing systems for distributed processing. Efforts to improve the biochemical platform are focused on pretreatment strategies and engineering microbes and enzymes to deconstruct carbohydrate polymers and produce long chain hydrocarbons or alcohols. Thermochemical efforts are being directed towards integrated thermo-catalytic processes that can readily switch between a multitude of feedstocks. Conversion systems of the future must optimize value of products produced, minimize energy and water use, be scaleable to distributed processing networks (to minimize feedstock logistics challenges), and produce minimal wastes.

Can advanced biofuels of the future play a significant role in a world reconfigured to meet energy-related challenges? Many researchers now point to increasing evidence that commercial biofuel production can be reconciled with feeding humanity and preserving the environment, provided that we invest the time and effort needed to make the improvements necessary to achieve this goal (Lynd & de Brito Cruz, 2010). The biofuel production concept described in this chapter has the potential to meet this objective, by combining proven technologies with promising innovations that are currently under development.

2. Feedstock supply

Corn-based ethanol has been periodically criticized for causing high food prices, in spite of the fact that ethanol production uses only the carbohydrate portion of the kernel. The protein, fat, and minerals in corn are concentrated in the distillers' dried grains with solubles (DDGS), and fermentation yeast further improve the quality of this feed for use in livestock rations. Nevertheless, in the late 2000s, ethanol production was blamed for the dramatic increase in corn prices, even though reduced worldwide production and increased grain imports by China were key factors. Ethanol was also blamed for rapidly increasing food costs, even though subsequent analyses pinpointed higher petroleum prices (i.e., transportation costs) and increased profit margins of food manufacturers as the true underlying factors (Henderson, 2008; Perrin, 2008).

The transition from corn to biomass-based feedstocks for bioenergy production should greatly reduce the controversy over food vs. fuel, although some biomass feedstocks (or the landscapes on which they are produced) are currently used for livestock feed (Sims, 2003). For example grasses and some crop residues (e.g., corn stover) are used for cattle feed. Agricultural landscapes that are less productive, subject to erosion, or in more marginal climates are often used to produce livestock forage, and these same landscapes are under consideration for biofuel feedstock production. Therefore, biofuel producers may compete with livestock producers for herbaceous feedstocks (grasses and crop residues).

A further issue with crop residues is their value in providing fertility, tilth, and carbon sequestration when left on the soil. While a certain amount of crop residues (e.g., corn stover, wheat straw) can be sustainably removed, there has been uncertainty over the amounts needed to maintain soil productivity (Lal, 2005; Tilman, et al., 2009). Recently, soil scientists have increased their recommendations for crop residue amounts that should be left on the soil. Current guidance is that 2,500 pounds of organic matter per acre/yr (Gustafson, 2011). For high residue crops like corn and wheat, this would allow removal of substantial amounts of residue (2-3 tons/acre/yr). However, these high residue crops are frequently grown in rotations with low residue crops such as soybeans. Therefore to maintain a sustainable soil profile, some of the high residue crop cover must also be left to average out the low residue crop. For example, in a corn/soybean rotation it has been calculated that less than one ton/acre of residue could be removed from a 200 bushel/acre corn crop. However, major crop genetics companies predict that they will double average yields of corn by 2030, and this will also increase corn stover levels.

Perennial grasses have been proposed as a more sustainable supply for biofuel production, and plants such as switchgrass, big bluestem, prairie cordgrass, and miscanthus yield 5-10+ tons/acre (Ragauskas, et al., 2006; Schmer, et al., 2008). Far less inputs (tillage, fertilization, water, and herbicides/pesticides) are required for production of these feedstocks, compared to that required for corn (ethanol) or soybeans (biodiesel) (Tilman, et al., 2009; Yuan, et al., 2008; Karp, 2008). Many of these plants can be grown on landscapes subject to erosion, periodic flooding, or other factors that would prevent their use for crop production (Schmer, et al., 2008). Yields would likely be lower than could be achieved on highly productive farm ground, but this would avoid the food vs fuel issue. In many locations, perennial grasses are the natural climax community, and would thus provide significant ecosystem benefits in terms of water and soil quality, and wildlife habitat. Even in traditional agricultural regions, perennial grasses could provide an important role as buffer strips to protect water bodies.

Forest and wood processing wastes, and municipal solid waste components are also significant biomass resources that could be used for biofuels production (Isa, et al., 2004; Parikka, 2004). These do not compete for arable lands, nor are they subject to the food vs fuel controversy. Collection and transportation issues have already been resolved, as these feedstocks are byproducts or even negative value wastes of existing processes. Production of fast growing trees in agricultural areas has been suggested, (Schlamadinger & Marland, 1996) but if these landscapes were traditionally used for crop or forage production, then an argument could be made that fuel production would impact food production.

Microalgae offer great potential as a sustainable feedstock for the production of 3rd generation biofuels, as well as 4th generation fuels (Aylott, 2010; Dragone, et al., 2010; Rosenberg, et al., 2008). Recent studies have shown that microalgal biomass is one of the most promising sources of renewable biodiesel that is capable of meeting the global demand for transport fuels (Dragone, et al, 2010). Biodiesel production by microalgae will not compromise production of food, fodder and other products derived from crops. Microalgal biomass contains three main components: proteins, carbohydrates, and lipids (oil). However, several important scientific and technical barriers remain to be overcome before the large-scale production of microalgae-derived biofuels can become a commercial reality. Technological developments, including advances in photobioreactor design, microalgal biomass harvesting, drying, and processing are important areas that may lead to enhanced cost-effectiveness and therefore, effective commercial implementation of biofuel production from microalgae (Greenwell, et al, 2010; Pienkos & Darzins, 2009).

3. Logistics

A critical component of successful commercialization of bioenergy is a secure and reliable supply system for biomass-based feedstocks. In fact this may be the most significant constraint to 2nd/4th generation biofuels. Ample feedstock should be available to biorefineries at the appropriate time and at competitive prices, while assuring reasonable, steady profits to the biomass suppliers. Developing a consistent, economically viable feedstock supply system requires addressing and optimizing diverse harvesting, storage, preprocessing, and transportation scenarios (Kumar & Sokhansanj, 2007).

Unlike corn and other grains, most biomass resources lack a well developed transportation infrastructure. Perhaps the closest would be the logistical infrastructure for wood products industry. Some infrastructure is also present in localized markets for forage crops used for livestock feed. Issues that are problematic for biomass transport include its low bulk density, poor flowability, and susceptibility to physical degradation during storage (especially if moisture is present) (Rentizelas, et al, 2009; Richard, 2010). Therefore, highly efficient harvest, densification, and storage systems for biomass are critical for an efficient biomass-to-biorefinery supply chain that will minimize transportation costs and energy consumption, while maintaining a high quality feedstock for processing (Eksioglu, et al. 2009).

Each of the different types of proposed feedstocks has a unique set of challenges or advantages regarding harvest method and timing, densification, transportation, and storage (Sims, 2003). In the near term (2012~2018), crops residues such as corn stover and forest residues (wood chips) are expected to be the main resource for cellulosic fuels (Perlack, et al., 2005). Several companies (i.e., POET, ICM, Inc) are developing processes which could be "bolted-on" to existing corn biorefineries, to take advantage of synergisms (nutrient and thermal sharing). These cellulosic processes are being scaled so that readily available corn stover within a reasonable distance of the plant can be efficiently supplied by conventional systems (baling and truck transport). Challenges still remain, however, regarding collection and storage of corn stover bales (Sokhansanj, et al., 2002). Weather conditions during corn harvest are often challenging, and adding a second harvest pass for stover may be unachievable. Thus, farm equipment manufacturers are testing single pass systems that would collect both grain and stover (Shinners & Binversie, 2003) Questions also remain concerning where and how stover bales will be stored. Efforts to re-tool wood pulp operations for biofuel production have the advantage of existing infrastructure for wood harvest and transport, along with the obvious advantage that wood is more storable than herbaceous biomass (Perlack, et al, 2005).

As biomass utilization expands there will be growing pressure to maximize the efficiency at which these raw materials are harvested. Technically, the term woody biomass includes all the trees and woody plants in the forest, woodlands, or rangelands (Parikka, 2004). This biomass includes limbs, tops, needles, leaves, and other woody parts. In practice, woody biomass usually refers to material that has historically had a low value or no economic value and cannot be sold as timber or pulpwood. As markets change over time and from region to region, different kinds of materials may be considered woody biomass. The maintenance of site productivity is perhaps the key non-water quality issue when anticipating the expansion of the use of woody biomass (Groom, et al, 2008. If it is proven that the harvesting of woody biomass actually depletes the nutrients in certain soils, fertilization may become a standard management tool. Soil compaction and excessive rutting can also impact site productivity. Timing harvest operations to avoid wet soil conditions or minimizing equipment travel patterns can prevent such impacts.[6]

Ideally, biomass development will occur in a manner that maximizes efficiencies in energy production and minimizes energy consumption associated with transportation, storage, and raw material processing, while maintaining biodiversity and improving the environment (Perlack, et al., 2005; Rentizelas, et al., 2009). One of the central concerns in woody biomass removal is the reduction of the quantity of dead wood left on site (Kaltschmitt, et al., 1997; Perlack, et al., 2005). Dead wood plays an important role in the ecosystem, from wildlife habitat and nutrient cycling to carbon storage. Woody material on the ground decreases water run-off and erosion. If woody biomass harvesting becomes an issue, specific recommendations can be made to leave a certain amount/number of the desired material on-site. As biomass markets expand, more emphasis and attention may be placed on watershed management. In general, water quality and riparian concerns should not change with the addition of woody biomass removal to a harvest plan. Streams and wetlands should be protected by existing Best Management Practices (BMPs) for Forestry,[7] using the Clean Water Act as a fundamental base (Lynch & Corbett, 2007). The opportunity for forest-derived biomass to be part of the carbon solution is an important consideration in the planning and development of biomass projects, but if management practices are too costly, they are unlikely to be implemented on private lands (Groom, et al., 2008; Kaltschmitt, et al., 1997).

---

[6]http://www.forestry.alabama.gov/PDFs/WoodyBiomassHarvestingIssues.pdf

[7]http://www.fs.fed.us/im/directives/dughtml/fsm2000.html

The Biomass Crop Assistance Program (BCAP) is a program that responds to the added cost of transporting biomass to a certified facility. BCAP is part of the Farm Bill and Recovery Act and is administered by the USDA Farm Service Agency.[8]

In Phase 1, which is active, it provides financial assistance to producers that deliver eligible biomass material to designated biomass conversion facilities for use as heat, power, and/or bio-based products. Initial assistance is for the collection, harvest, storage, and transportation costs associated with the delivery of eligible materials through a direct matching of dollar for dollar of dry ton delivered to qualified facilities, up to $45 maximum over the next two years. The details of Phase 2 have not yet been made public.

Auburn University has received a grant worth up to $4.9 million from the US Department of Energy to design and demonstrate a high productivity system to harvest, process, and transport woody biomass from southern pine plantations.[9] Specific project objectives are to develop design improvements in tree length harvesting machines for energy plantations; configure and assemble a high-productivity, lowest-cost harvesting and transportation system for biomass and demonstrate at full industrial scale; and document performance of the systems.[10]

Over the longer term (2015+), dedicated, perennial bioenergy crops such as switchgrass, prairie cordgrass, and fast growing woody species are expected to become the preferred feedstocks, because of their higher yield and lower input requirements (Perlack, et al, 2005). However, it is also expected that these crops will be grown on marginal soils, leaving the most productive soils for food/feed crop production. Hence, dedicated energy crop production will be distributed across a broader landscape, and this will limit the total tonnage of biomass available in a given region (Kumar & Sokhansanj, 2007). Thus feedstock logistics is the main driver for smaller processing plants or a distributed network of pre-processing-densification plants to support a large centralized processing operation. This latter approach has been promoted by Carolan et al., (2007) in their regional biomass processing center (RBPC) model. Such smaller preprocessing facilities would have a 20-25 mile collection radius, instead of the 75-100+ mile radius common to large corn ethanol plants. These RBPC facilities would collect biomass directly from the farm, pretreat and densify the biomass into more storable and flowable billets, and then provide these to a centralized biofuel production facility. The ammonia fiber expansion (AFEX) treated biomass is also more digestible in livestock feeds, so the RBPC network could support cattle feeding operations and/or biorefineries (Carolan et al., 2007).

Harvest timing and storage are additional concerns for perennial grasses (Marten & Hovin, 1980). In the fall these plants translocate nutrients back down into the root system, to support growth the following year. Hence to maintain plant vigor over the long term, harvest should be delayed until late fall. This will also minimize moisture content in the biomass, enhancing storability. Processing operations will also prefer biomass with reduced mineral content, as this will limit accumulation of these minerals in process fluids and ash (which would need to be disposed of). Unfortunately, fall harvest will eliminate one of the key ecosystem benefits of native grasses, that being winter cover for wildlife (Roth, et al., 2005). Many of these grasses have been used as the backbone vegetation for the USDA

[8]www.fsa.usda.gov/FSA/webapp?area=home&subject=ener&topic=bcap

[9]http://nrmdi.wordpress.com/2009/09/04/auburn-universitys-center-for-bioenergy-and-bioproducts-awarded-4-9-million-grant-for-bioenergy-research

[10]www.supertrak.com/video/BIOBALER.wmv

Conservation Reserve Program (CRP) which has resulted in a significant boost to wildlife populations over the past 25 years. Early spring harvest of native grasses would be the logical solution, but wet weather conditions typical in spring may make this option very difficult.

Another logistical concern is in transport of biorefinery products to the consumer. Corn based ethanol has faced a significant hurdle, as it is not considered to be an infrastructure compatible fuel. The EPA has limited the concentration of ethanol that can be used in vehicles, originally to 10% and more recently to 15%. Ethanol also cannot be transported in pipelines used for petroleum products, due to its propensity to absorb water. Biomass based ethanol (2nd generation biofuel) will face the same issues. Third and fourth generation biofuels are infrastructure compatible, and should therefore move into the fuel distribution network seamlessly. However, these fuels will be produced in more rural areas, and therefore transport to metropolitan areas will be required.

Use of microalgal biomass has unique issues. Given the relatively low biomass concentration obtainable in microalgal cultivation systems due to the limit of light penetration (typically in the range of 1-5 g/L) and the small size of microalgal cells (typically in the range of 2-20 μm in diameter), costs and energy consumption for biomass harvesting are a significant concern that needs to be addressed (Greenwell, et al., 2010; Pienkos & Darzins, 2009). In this sense, harvesting of microalgal cultures has been considered a major bottleneck towards the industrial-scale processing of microalgae for biofuel production. The cost of biomass recovery from the broth can constitute 20–30% of the total cost of producing the biomass (Borowitzka, 1992). Microalgal biomass harvesting can be achieved in several physical, chemical or biological ways: flocculation, centrifugation, filtration, ultrafiltration, air-flotation, or autoflotation. Generally, microalgae harvesting is a two stage process. First, bulk harvesting is used to separate biomass from the bulk suspension. The concentration factors for this
operation are generally 100–800 times to reach 2–7 % total solid matter. This will depend on the initial biomass concentration and technologies employed, including flocculation, flotation or gravity sedimentation. Second, thickening is used to concentrate the slurry through techniques such as centrifugation, filtration and ultrasonic aggregation. This latter step is generally a more energy than bulk harvesting. Several essential issues must be addressed in photobioreactor (PBR) design, including effective and efficient provision of light; supply of CO_2 while minimizing desorption; efficient mixing and circulation of the culture; and the material used in the construction of the PBR. Light as the energy source for photoautotrophic life is the principal limiting factor in photobiotechnology. In addition, the supply of CO_2 to microalgal mass culture systems is one of the principal difficulties that must be solved. CO_2 must not reach the upper concentration that produces inhibition and, on the other hand, must never fall below the minimum concentration.

4. Conversion and recovery

A wide diversity of biochemical and thermochemical processes are under development for production of 2nd through 4th generation biofuels. Biochemical processes use physical or chemical pretreatments (Alvira, et al., 2010; Chandra, et al., 2007; Dongahai, et al., 2006; Hendriks & Zeeman, 2009; Mosier, et al., 2005; Taherzadeh & Karimi, 2008; Wyman, et al., 2005; Yang & Wyman, 2008), followed by enzymatic hydrolysis of polymers to simple sugars (Dale, et al., 1996; Vlasenko et al., 1997), with subsequent fermentation to fuel products (Ballesteros, et al., 2004; Saha & Cotta, 2007). Thermochemical processes use high temperatures and pressures to degrade biomass into simple compounds that are then reconstructed into hydrocarbon polymers using chemical catalysis. Some newer approaches link biochemical and thermochemical steps, to take advantage of synergisms.

Lynd et al. (2005), in their strategic analysis of biorefineries list the advantages of integrated multi-product biorefineries. First, integrated biorefineries enable maximizing the value generated from heterogeneous feedstocks, making use of component fractions. Second, revenues from high-value coproducts reduce the selling price of the primary product. Third, the economies of scale provided by an integrated biorefinery lowers the processing costs of low-volume, high-value coproducts, because common process elements are involved in producing fermentable carbohydrates, regardless of whether one or more products are produced, and coproduction can provide process integration benefits (e.g. meeting process energy requirements with electricity and steam cogenerated from process residues).

We propose that such an integrated biochemical/thermochemical process would be best suited for distributed, smaller scale production of biofuels (Fig 2). This design would facilitate use of multiple feedstocks which are likely to be the norm in smaller operations. We assume these feedstocks could include crop residues, forestry residues, components of municipal solid waste, or dedicated biomass sources such as fast growing trees or native grasses. The process design would also provide the flexibility to produce a broad range of products, which could be adjusted to meet market demands and generate the highest level of income.

The initial unit operations in this process would be similar to that in current dry mill corn ethanol plants (i.e., size reduction, fractionation, and hydrolysis). The primary difference will be in the size of these facilities. Instead of the 100-120 million gallon/yr (MMGY) corn ethanol plants, 2nd-4th generation biofuel facilities will be in the 10-40 MMGY range. The upper size limit will likely be determined by the amount of biomass that can be economically delivered to the plant gate (Carolan, et al., 2007; Eksioglu, et al., 2009). This will be affected by biomass yields per acre, competing uses/markets, price, and feedstock logistic issues.

Feedstocks would first be subject to particle size reduction, followed by a continuous solvent-based fractionation process to disrupt the protective matrix of lignin that surrounds the cellulose and hemicellulose fibers, and separate these three streams. NREL developed the clean fractionation process in the 1990s, and several companies and university research teams are working to develop alternative, lower cost solvents to make this process economical (Emmel, et al., 2003; Kim & Lee, 2006; Lee, et al., 2009; Li, et al., 2010; Moxley, et al., 2008; Pan, et al., 2006; Zhang, et al., 2007). We anticipate this could be commercialized by the late 2010s. Upstream fractionation of biomass would maximize downstream reactor productivity, since lignin would not dilute the sugar titer in cellulose and hemicellulose hydrolysates (Zhang, 2010b). Removing lignin will also prevent lignin from binding to and inactivating hydrolytic enzymes (Berlin, et al., 2006; Mes-Hartree & Saddler, 1983; Palmqvist & Hahn-Hagerdal, 2000), and will reduce production of chemicals inhibitory to yeast metabolism (Pfeifer, et al., 1984). Plus it will help ensure that each component is used for its highest value.

Lignin is an abundant, renewable, and amorphous natural polymer consisting of phenylpropane units (Boudet, et al., 2003). The units, primarily syringol, guaiacol, and p-hydroxyphenol, are linked together by ether and carbon–carbon inter-unit bonds to form a very complex three-dimensional polymer matrix. The lignin fraction contains a multitude of

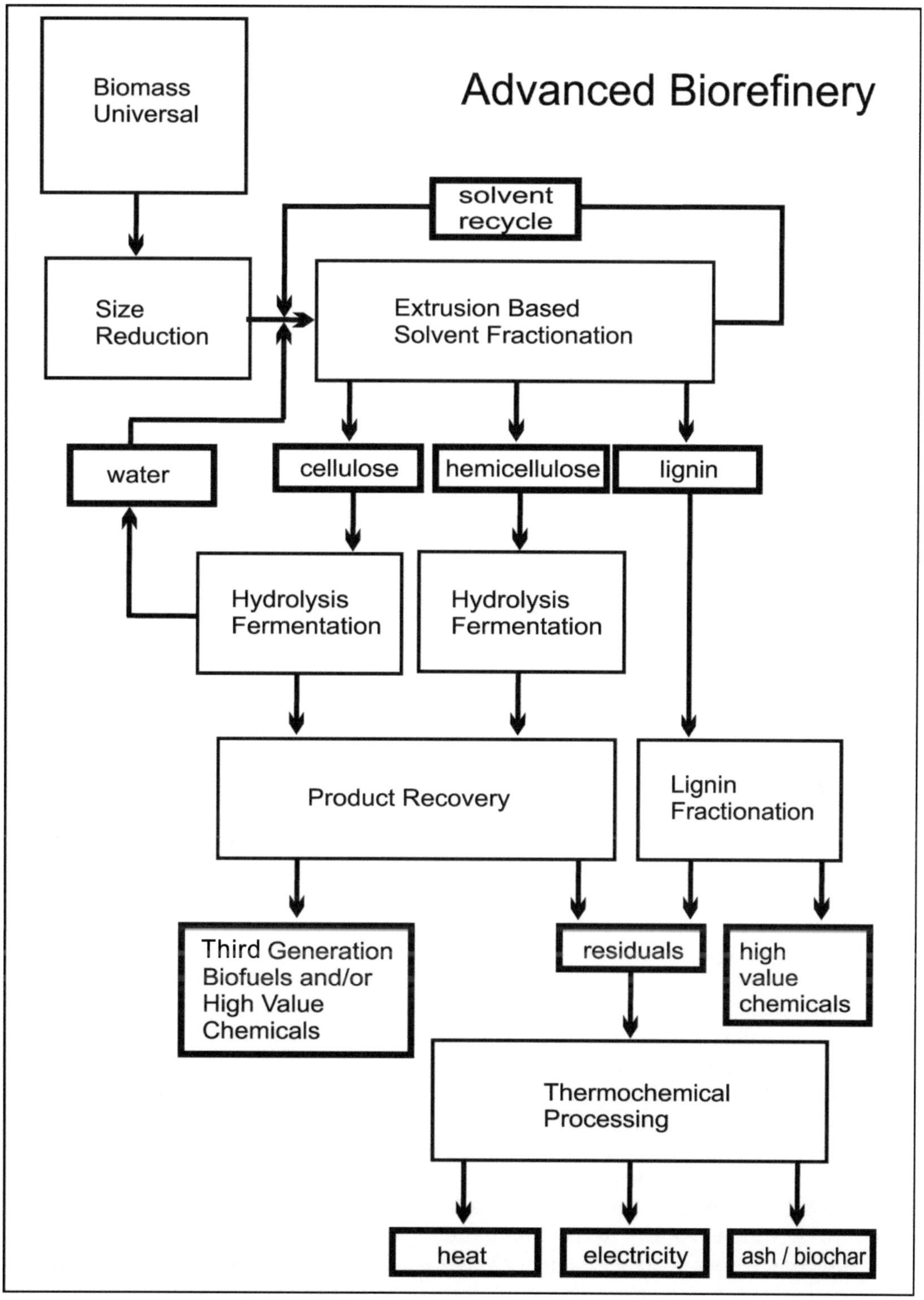

Fig. 2. Integrated advanced biorefinery platform

important and high value chemicals, (Taherzadeh & Karimi, 2008; Thring, et al., 2000) and researchers are developing separation methods (Pan et al., 2005). If these techniques become available at an appropriate scale, the lignin fraction would first be processed to recover these products (Katzen, et al., 1995; Sun & Cheng, 2002). Residual lignin would be thermochemically processed (combustion, pyrolysis, or gasification) to generate combined heat and power for the operation (as opposed to natural gas or other fossil fuels typically used in 1st generation biofuels). This will improve the carbon balance of 2nd-4th generation biofuels. Additional coproducts could be ash and/or biochar, which would be returned to the land to provide nutrients and carbon sequestration (Lehmann, et al., 2006). Alternatively, biochar could be converted into higher value products such as activated carbon.

The cellulose and hemicellulose fractions would then be subjected to chemical (Lavarack, et al., 2002; Torget, et al., 1991) or enzymatic (Banerjee, et al., 2010a; Banerjee, et al., 2010b; Banerjee, et al., 2010c; Fan, et al., 1980; Grethelin, 1985; Ragauskas, et al., 2006; Sun & Cheng, 2002), hydrolysis to convert the fiber polymers into fermentable sugars (Jorgensen et al., 2007). Having separate hydrolysates of glucose and xylose will provide the advantage of allowing for separate fermentations by organisms specifically adapted for these substrates. The glucose solution could be fermented by traditional yeast into ethanol (2nd generation biofuel) or by bacteria such as *Clostridium* into butanol (4th generation biofuel) (Qureshi & Ezeji, 2008). Researchers and companies such as Solazyme are developing other microbes (native or genetically modified) to ferment glucose into a range of infrastructure compatible, energy dense 4th generation biofuels such as longer chain alcohols, alkanes, and alkenes. The xylose fractions present a greater challenge, as fewer microbes have the needed metabolic machinery (Saha & Cotta, 2007). However, certain yeast strains (e.g., Pichia) are able to ferment xylose to ethanol, and research groups have engineered *Saccharomyces* for xylose fermentation. Other teams are using bacteria to convert xylose to ethanol (Ahring, et al, 1996). Perhaps the greater value for xylose would be fermentation by bacteria such as *Clostridium* into butanol or long chain length hydrocarbons.

One of the challenges with biomass hydrolysate fermentation is the low bulk density of biomass, which results in a high viscosity when solid loadings exceed 15-20% dry matter. This limits product concentrations that can be achieved in traditional submerged bioreactors, which in turn results in higher costs and energy consumption for downstream product recovery. It has been calculated that fermentation broth must contain at least 4% (w/w) ethanol for distillation to be economically viable (Larsen et al., 2008; Zhang, et al., 2010a). For many types of biomass this will mean a solid loading rate of at least 20% (w/w). At these high solid levels the high viscosity prevents traditional submerged bioreactors from achieving sufficient mixing and mass transfer, while resulting in localized solids build-up or caking. Low water activity and high concentrations of sugars, end products, and/or inhibitory chemicals can also inhibit enzymes and fermentation organisms.

Using a biomass fractionation pretreatment upstream will minimize viscosity problems and maximize solids loading during SSF, because the hydrolysate streams will consist primarily of cellulose and hemicellulose, respectively. For example, Zhang et al (2010c) conducted SSF of a pretreated corncob fraction at 19% solids and achieved 69 g/L ethanol. Fed-batch feeding of solids is another method to increase net solids loading (Hodge, et al., 2009; Hoyer, et al., 2010; Varga, et al., 2004). Intermittent feeding of a fractionated cellulose or hemicellulose stream can achieve higher overall solids loadings, while still maintaining low viscosity, because the substrate is continuously degraded to soluble sugars. This also prevents the buildup of sugars which can otherwise inhibit some fermentation organisms

(Hodge, et al., 2009; Olofsson, et al., 2008. Fed-batch feeding also effectively reduces enzyme dosage. Zhang et al., (2010c) conducted fed batch SSF, achieving a final solid loading of 25% and ethanol titer of 84.7 g/L.

To further increase product concentration, various types of high solids bioreactors have been proposed (Varga et al., 2004; Zhang, et al., 2010a). These have included gravitational tumbling in roller bottle reactors (Roche, et al., 2009a; Roche, et al., 2009b), horizontal paddle type bioreactors (Jorgensen, et al., 2007; Roche et al., 2009a;), scraped surface bioreactors (Dasari, et al., 2009), and stirred helical bioreactors (Jorgensen et al., 2007; Zhang et al. 2010a). Jorgensen et al. (2007) achieved 35% solid loading and 62 g/L ethanol in a bioreactor with a horizontal rotating shaft with paddles, and this design provides sufficient mixing at very low rotation rates, meaning less power consumption (Zhang et al., 2010a). Scraping blades can be used to prevent "dead zones," and keep the reactor surface clean to maximize heat/cooling transfer (Dasari, et al., 2009). Zhang et al (2010a) used a helical stirring system to 64.7 g/L ethanol from 30% solid loading of steam exploded corn stover.

Many of these sugar fermentation processes will produce CO_2 as a byproduct. In the case of ethanol fermentation, one third of the carbohydrate carbon is released as CO_2. The integrated process (Fig 2) could be adapted to include photobioreactors in which engineered microalgae or cyanobacteria would convert this CO_2 into 3rd generation fuels or solvents (Greenwell, et al., 2010; Pienkos & Darzins, 2009). In addition, these microalgae could be used to sequester CO_2 from flue gases emitted from fossil fuel-fired power plants and other sources, thereby reducing emissions of a major greenhouse gas (1 kg of dry algal biomass utilize about 1.83 kg of CO_2). The utilization of microalgae for biofuels production also offers the advantages that they: 1) synthesize and accumulate large quantities of neutral lipids (20–50 % dry weight of biomass) and grow at high rates; 2) are capable of year-round production, therefore, oil yield per area could greatly exceed the yield of oilseed crops; 3) need less water than terrestrial crops thus reducing the load on freshwater sources; 4) do not require herbicide or pesticide application; 5) bioremediate wastewater by removal of nitrogen from a variety of sources (e.g. agricultural run-off, concentrated animal feed operations, and industrial and municipal wastewaters); and 6) can be cultivated in saline/brackish water/coastal seawater on non-arable land, and do not compete for resources with conventional agriculture. Production would occur in recirculating photobioreactors, which would be located in adjacent greenhouses. Low grade heat from the biorefinery operations would maintain appropriate temperatures inside the greenhouse during cold weather.

Recovery of ethanol or butanol from fermented solutions would likely occur via the time-tested process of distillation. However, advancements in membrane technology may eventually allow for lower energy processes such as pervaporation (Lipnizki et al., 1999). Longer chain alcohols and hydrocarbons can be recovered from fermentation solutions by phase separation, if sufficiently high concentrations are produced. Non-fermentable solids in the residual fermentation broth would be separated by centrifugation or filtration and used to generate process steam and electricity via thermochemical processes, while process water would be recycled to the greatest extent possible.

5. Technology development and deployment

Projected costs for biomass ethanol (>$2-3/gal) have been substantially greater than actual costs ($1-1.50/gal) for corn-based ethanol (Eisentraut, et al., 2010; Sanderson & Ad, 2008).

For this reason the US DOE initiated the "Demonstration of Integrated Biorefinery Operations" program in 2006 to help cost-share construction costs for pilot or demonstration scale facilities to convert biomass to ethanol. Three rounds of grants have since been awarded, and the initial awardees are expected to begin production in 2011-2012. Six proposals were funded in 2007 at a federal investment of $385 million. Biochemical processes included POET, which is constructing a 25 MMGY corn cob/stover conversion facility adjacent to one of its 100 MMGY corn ethanol plants in Emmetburg, IA. Abengoa (Kansas) is constructing an 11.4 MMGY facility to use corn stover, wheat straw, and other feedstocks, while Iogen is establishing a similar facility in Idaho. In 2008, DOE announced that 9 second round awardees would receive a total of $240 million. Biochemical processes included Verenium's 1.5 MMGY facility in Louisiana, and Zeachem's 0.25 MMGY facility in Oregon. In 2009, DOE announced funding of $482.7 million for 18 pilot and demonstration scale facilities. Awardees planning biochemical processes included Amyris Biotechnologies (California), ADM (Illinois), ICM, Inc (Missouri) and Logos Technologies (California). These facilities are expected to begin operation in 2013-2014.

Production of 3rd and 4th generation biofuels from biomass is further behind, with butanol being the leading candidate. A British Petroleum and DuPont joint venture (Butamax) has refurbished an ethanol facility in Great Britain to make butanol from sugar using engineered yeast. Gevo has acquired the Agri-Energy corn ethanol plant in Luverne, MN and is pursuing a similar strategy. Companies such as Chevron and Weyerhaeuser are also exploring butanol production. Various research teams are working to develop improved microbes for butanol production, with pilot scale facilities already in operation, or scheduled for startup in 2011-2012. Besides butanol, other companies are exploring options to produce drop-in replacements for currently used fuels. For example, Flambau River Biofuels (WI) received funding in the second round of DOE's biorefinery program to construct a 6 MMGY wood-to-diesel facility that should be operational in 2012-2013. The company LS9 is engineering bacteria to produce other 3rd generation biofuels and plans to test pilot scale systems in the next 12-24 months.

Initial deployment of 2nd generation biofuel processes using corn stover is already occurring at certain corn-based ethanol facilities. Co-locating allows for several synergistic opportunities (Khanna, 2008). For example, lignin from the biomass plant can provide all its thermal energy, as well as meeting part of the needs for an adjacent corn-based plant. Excess nutrients from the corn ethanol plant can help enhance fermentation of biomass-derived sugars. Furthermore, downstream unit operations for ethanol recovery, purification, storage and shipping can be shared by the two processes. Similarly, some woody biomass conversion processes are being co-located at pulp mills to take advantage of delivery and pretreatment infrastructure.

6. Job opportunities in the bioeconomy

Production of 2nd – 4th generation biofuels is a sequential outgrowth of 1st generation biofuel production, and will have a substantial impact on jobs and market opportunities (Carr, et al., 2010). In areas such as biomass production and logistics, many additional jobs in traditional areas will be created to supply biorefineries with needed feedstock (DOE estimates 200,000 jobs nationwide). Similarly, the biorefineries themselves will create a significant number of additional jobs. Industries providing enzymes, microbes and other supplies for these facilities will create new and expanded job opportunities as well. These biorefineries will also increase job opportunities in traditional mechanical and plastic industries for manufacturing required equipment, instruments, and materials. These positions will require novel training and education program for workers currently employed in the 1st generation biorefinery industry and future or potential workers in labor pool.

A significant aspect of 2nd and 4th generation biofuel production will be the infrastructure to harvest, collect, storage, and transport of biomass (Table 3). The DOE has projected that at least 200,000 new jobs could be created for biomass logistics. This would be in addition to jobs that could be re-directed from providing woody wastes or forage materials to less competitive markets. Although there are only a few small demonstration-scale cellulosic ethanol plants in the U.S. presently, cellulosic ethanol and drop-in biofuel facilities will increase in number as costs decrease. Therefore, to supply feedstock to a 2nd and 4th generation biofuel facilities, jobs will be needed for feedstock production, harvesting, gathering, storing, transportation by road or rail, and quality monitoring. To achieve sustainable biomass production, agricultural workers, scientists, biochemists and engineers, who are in charge of operating or monitoring those biomass production and logistics processes will be needed. They should have education backgrounds in biochemistry, plant science, biological science, biochemical engineering or/and analytical sciences

Production stages	Instruments/Materials	Job positions	Jobs per plant
Biomass harvesting, densification, and storage	Agricultural machines such as harvesters, planters, irrigation systems (pump and piping), trucks, and hay balers, hydraulic compressor or extruder	Mechanic industries for producing needed equipment. Customized harvesters and planters may be needed. Mobile grinding, compression, or extrusion equipment for biomass densification.	15
Biomass transport	Trucking	Truck drivers, forklift operators	5-10
Biomass quality control, impact of removal	Testing instruments for quality and composition	Agricultural or Chemical Engineers	2

Table 3. Natural resource harvesting jobs for 2nd/4th generation biofuels produced via biochemical route

In the mid-2000s, there were dozens of design/engineering firms, each with hundreds of employees, that built out the corn ethanol industry. It is anticipated that these same firms, as well as those involved in chemical and other fuel related processes, will be the key suppliers of this technology to the 2nd - 4th generation biofuel industry. These firms will have the additional advantage that their intimate knowledge of ethanol production facilities will allow them to design-in adjacent 2nd - 4th generation biofuel systems to take the greatest advantage of potential synergies. These firms would also have the advantage of being able to provide on-going maintenance and repair services as they do now for 1st generation biofuel plants. Positions critical to these firms included: engineering and construction managers, mechanical engineers, electrical engineers, chemical engineers, civil and environmental engineers and technicians, computer control programmers and operators, tool and die makers, metal and plastic fabricators, forming/extruding/drawing machine operators, boilermakers, pipefitters, construction equipment operators and laborers, quality control inspectors and others. It is reasonable to expect that expansion of the 2nd - 4th generation biofuel industry will employ a similar number of people. The jobs impact for site preparation and construction of a 10-30 MMGY facility would be 300-500 positions.

As noted previously, the size of the biofuel facilities will be limited by the amount of feedstock that can be economically delivered to the plant gate. The consensus at this point is that 10-30 MMGY facilities will be the norm, and thus we based our labor projections on this size plant. Because these systems will generally mirror the conversion process used in corn ethanol, it is reasonable to use labor requirements of corn ethanol to estimate biomass fuel labor needs. On this basis a 10-30 MMGY 2nd/3rd generation biofuel facility would have 40-50 employees.

7. Technology development and incorporation constraints

The primary constraints to 2nd/4th generation biofuel production will be feedstock and water availability, and transportation infrastructure. In most cases, facility siting and size will be determined by the amount of feedstock available within an economical transportation radius (Rentizelas, et al., 2009; Richard, 2010). There is a diverse pool of un- or under-utilized biomass resources that could be directed to production of 2nd/3rd generation biofuels with minimal impact on other industries or interest groups (Perlack, et al., 2005). For example, forestry and wood wastes may not compete directly with food crops for land, but production of fast growing trees for biofuels could compete with production of timber for lumber markets. The DOE has estimated that over 1 billion tons of biomass would be available each year (Perlack, et al., 2005). Further expansion of 2nd/3rd generation biofuels may create some level of competition for feedstocks and/or land use. For example, use of crop residues (stover, straw) could compete with uses for livestock bedding or feed. Similarly, converting grasses to biofuels could compete with livestock feed use, or expanded grass production could displace crop production in some locations (Kumar & Sokhansan, 2007). Use of margin land or wildlife habitat for biomass production could significantly impact wildlife (Roth, et al, 2005). To minimize this, biomass sources and harvesting strategies that minimize negative impacts on wildlife systems will need to be employed. Diverse plantings of biomass crops would also be preferred over monocultures. Competition for these land and biomass resources will drive up prices for all users.

Production of feedstocks for 2nd and 4th generation biofuels will have fewer environmental constraints than corn production, however there are still potential issues (Eisentraut, 2010; Kaltschmitt, et al., 1997). Water is often a limiting resource, and biomass plants should be selected for drought and salt tolerance. In addition, the impacts of planting and harvesting practices on water resources and water quality need be evaluated. Biomass feedstocks are generally more nutrient efficient compared to corn, and do not require the same degree of annual fertilization. However, as with all plants, biomass feedstocks will respond favorably to fertilization. To minimize fertilizer use, several approaches are under development: 1) specific strains that require less minerals and nutrients, 2) recycling nutrient-rich coproducts and/or process water after biofuel production, to recycle most nutrients back to field, 3) integrating biomass production with animal or municipal waste water treatment systems to

provide nutrients, 4) developing specific harvest strategies to allow the plants to re-mobilize nutrients back into the roots before harvest of the above-ground parts of plants.
As with 1st generation biofuels, energy and chemical use are significant issues for 2nd - 4th generation biofuel processes. There are several key issues to address: 1) energy efficient densification technology and systems, 2) energy efficient pretreatment processes, which can produce fermentable sugars from lignocellulosic biomass with lower environmental footprint, 3) novel microbial strains with high product tolerance and expression levels, and 4) energy-saving product separation and recovery systems, which depend on development of new separation methods and materials. An additional concern is presenting the escape of genetically modified organisms that could be used in processing. Technologies such as high efficient air filtration systems, in-line chemical sterilization, steam in-place and sterile filtration will be necessary to maintain containment.

8. Summary

The DOE has estimated that 1.3 billion tons of un- or under-utilized biomass is available annually for use in producing biofuels (Perlack, et al., 2005). One resource is municipal solid waste, which is cheap, abundant, and available where the fuels would be used. Corn stover, cereal straws and other agricultural residues are being evaluated widely in the Midwest, for potential co-processing at corn ethanol plants. Elsewhere in the U.S., forest thinnings, pulp and paper mill waste, and yard waste are significant resources that could be converted into biofuels. Moreover, significant work is being conducted to develop dedicated energy crops such as switchgrass, cane, sorghum, poplar, and miscanthus. In addition to their high yield, some of these energy crops are perennial plants that can be grown on less productive sites, so as not to directly compete with food production. These feedstocks will be critical to achieving the renewable fuel production requirements set forth in RFS2.
Industry and the DOE have invested millions of dollars in designing and constructing the first cellulosic bioethanol demonstration and commercial scale plants. Most of these operations are being co-located with either wood pulping or corn-based ethanol plants, to take advantage of feedstock availability and process syngergies. These 2nd generation biofuel facilities are expected to begin operation in 2011-2013, and will serve as proving-grounds to help improve productivity and reduce costs. If successful, these facilities will encourage additional investment in lignocellulosic ethanol production. Based on current U.S. energy policy, it would appear that ethanol, derived from corn and lignocellulose, will be the main renewable liquid transportation biofuel through 2020.
Cellulosic ethanol will contribute significantly to RFS2 and the broader U.S. goals of creating a sustainable energy supply, reducing greenhouse gas emissions, assuring energy security, and promoting rural economic development. However, the infrastructure for distributing and using ethanol is limited, and expansion of the biofuel market share above 20% will likely involve production of advanced, drop-in biofuels (i.e., liquid hydrocarbons). Therefore the focus of many in industry and government has shifted to production of these infrastructure compatible biofuels that can also be produced from the same feedstocks being investigated for cellulosic ethanol production. Systems for producing these advanced biofuels will also take advantage of the feedstock logistic solutions that will be resolved by 2nd generation biofuels (Lynd & de Brito Cruz, 2010).
Algal oil can be produced by metabolizing biomass sugars or by fixing CO_2 via photosynthesis. These lipids (oils) can then be converted into 3rd generation biofuels such as

biodiesel or JP8. Alternatively, algae and cyanobacteria can be engineered to directly produce fuel compounds, instead of oil. Hydrocarbons and long-chain alcohols (4th generation biofuels) can be made from biomass sugars through microbial fermentation or liquid-phase catalysis, or directly from biomass via catalytic pyrolysis or gasification and Fisher Tropsch reactions. These biofuel replacements for gasoline, diesel, and jet fuel will give higher mileage than ethanol and biodiesel, and will work in existing engines and fuel distribution networks.

Logistical challenges of transporting, storing, and maintaining acceptable quality biomass will restrict the size of future biorefineries. These biorefineries are also likely to be most economical and energy efficient if they are able to produce a multitude of high value fuels and chemicals. Therefore we anticipate that a distributed, integrated platform technology for community-based production of advanced biofuels will prevail. This platform can be used in any location, due to its self-contained and autonomous design. The primary inputs will be biomass, CO_2, sunlight for photosynthesis, and solar or wind power to provide electricity. The integrated, community-based design would produce energy in an environmentally and socially sustainable manner.

9. References

Anon. (2010a). Draft Regulatory Impact Analysis: Changes to Renewable Fuel Standard Program; Assessment and Standards Division Office of Transportation and Air Quality U.S. Environmental Protection Agency.

Anon. (2010b). http://www.ethanolproducer.com/plantmap.

Anon. (2011a). http://www.ethanolrfa.org/pages/statistics (accessed February 2011).

Anon. (2011b). US corn surplus cut on higher ethanol output; smaller world crop is seen. http://www.bloomberg.com/news/2011-02-09/u-s-corn-surplus-cut-on-higher-ethanol-output-smaller-world-crop-is-seen.html.

Ahring, B.K., Jensen, K., Nielsen, P., Bjerre, A.B., and Schmidt, A.S. (1996). Pretreatment of wheat straw and conversion of xylose and xylan to ethanol by termophilic anaerobic bacteria. *Bioresource Technol.* 58, pp. 107-113.

Alvira, P., Tomas-Pejo, E., Ballestreros, M., and Negro, M.J. (2010). Pretreatment technologies for an efficient bioethanol production process based on enzymatic hydrolysis: A review. *Biores. Technol.* 101, pp. 4851-4861.

Aylott, M. (2010). Forget palm oil and soya, microalgae is the next big biofuel source. *The Ecologist.* (Sept 24). http://www.theecologist.org/blogs_and_comments/commentators/other_comments/609556/forget_palm_oil_and_soya_microalgae_is_the_next_big_biofuel_source.html. Retrieved 2011-02-22

Ballesteros, M., Oliva, J.M., Negro, M.J., Manzanares, P., and Ballesteros, I. (2004). Ethanol from lignocellulosic materials by a simultaneous saccharification and fermentation process (SSF) with *Kluyveromyces marxianus* CECT 10875. *Process Biochem.* 39, pp. 1843-1848.

Banerjee, G., Car, S., Scott-Craig, J.S., Borrusch, M.S., Aslam, N., and Walton, J.D. (2010a). Synthetic enzyme mixtures for biomass deconstruction: Production and optimization of a core set. *Biotechnol. Bioeng.* 106, pp. 707-720.

Banerjee, G., Car, S., Scott-Craig, J.S., Borrusch, M.S., Aslam, N., and Walton, J.D. (2010b). Synthetic multi-component enzyme mixtures for deconstruction of lignocellulosic biomass. *Biores. Technol.* 101, pp. 9097-9105.

Banerjee, G., Car, S., Scott-Craig, J.S. Borrusch, M.S., and Walton, J.D. (2010c). Rapid optimization of enzyme mixtures for deconstruction of diverse pretreatment/biomass feedstock combinations. *Biotechnol. Biofuels.* 3, pp. 22.

Berlin, A., Balakshin, M., Gilkes, N., Kadla, J., Maximenko, V., Kubo, S., and Saddler, J. (2006). Inhibition of cellulase, xylanase and beta-glucosidase activities by softwood lignin preparations. *J. Biotechnol.* 125, pp. 198-209.

Borowitzka, M.A. (1992). Algal biotechnology products and processes - matching science and economics. *J. Appl. Phycology.* 4, pp. 267-279.

Boudet, A.M., Kajita, S., Grima-Pettenati, J., and Goffner, D. (2003). Lignins and lignocellulosics: a better control of synthesis for new and improved uses. *Trends Pl. Sci.* 8, pp. 576-581.

Carolan, J.E., Joshi, S.V., and Dale, B.E. (2007). Technical and Financial Feasibility Analysis of Distributed Bioprocessing Using Regional Biomass Pre-Processing Centers. *J. Ag. Food Ind. Org.* 5, DOI: 10.2202/1542-0485.1203.

Carr, M., Davies, S., and Locke, B. (2010). Job creation and market opportunities for biobased chemicals and products. *Industrial Biotechnology.* 6, pp. 74-77. doi:10.1089/ind.2010.0003

Chandra, R., Bura, R., Mabee, W., Berlin, A., Pan, X., and Saddler, J. (2007). Substrate pretreatment: The key to effective enzymatic hydrolysis of lignocellulosics? *Adv. Biochem. Eng. Biotechnol.* 108, pp. 67-93.

Dale, B.E., Leong, C.K., Pham, T.K., Esquivel, V.M., Rios, I., and Latimer, V.M. (1996). Hydrolysis of lignocellulosics at low enzymes level: application of the afex process. *Bioresource Technol.* 56, pp. 111-116.

Dasari, R.K., Dunaway, K., and Berson, R.E. (2009). A scraped surface bioreactor for enzymatic saccharification of pretreated corn stover slurries. *Energy Fuels.* 23, pp. 492-497.

Dongahai, S., Junshe, S., Ping, L., and Yanping, L. (2006). Effects of different pretreatment modes on the enzymatic digestibility of corn leaf and corn stalk. *Chinese J. Chem. Eng.* 14, pp. 796-801.

Dragone, G., Fernandes, B., Vicente, A.A., and Teixeira, J.A. (2010). Third generation biofuels from microalgae. *Appl. Microbiology.* 2, pp. 1355-1366.

Dufey, A. (2006). Biofuels production, trade and sustainable development: emerging issues. International Institute for Environment and Development. London.

Eisentraut, A. (2010). Sustainable production of second-generation biofuels. Potential and perspectives in major economies and developing countries. IEA Energy papers. DOI:10.1787/20792581.

Eksioglu, S.D., Acharya, A., Leightley, L.E., and Arora, S. (2009). Analyzing the design and management of biomass-to-biorefinery supply chain. *Computers & Ind. Eng.* 57, pp. 1343-1352.

Emmel, A., Mathias, A.L., Wypysh, F., and Ramos, L.P. (2003). Fractionation of *Eucalyptus grandis* chips by dilute acid-catalysed steam explosion. *Bioresource Technol.* 86, pp. 105-115.

Fan, L.T., Lee, Y., and Beardmore, D.H. (1980). Mechanism of the enzymatic hydrolysis of cellulose: Effects of major structural features of cellulose on enzymatic hydrolysis. *Biotechnol. Bioeng*. 22, pp. 177-199.

Greenwell, H.C., Laurens, L.M.L., Shields, R.J., Lovitt, R.W., and Flynn, K.J. (2010). Placing microalgae on the biofuels priority list: a review of the technological challenges. *J. Royal Soc. Interface*. 7, pp. 703-726.

Grethelin, H.E. (1985). The effect of pore size distribution on the rate of enzymatic hydrolysis of cellulosic substrates. *Biotechnol*. 3, pp. 155-160.

Groom, M.J., Gray, E.M., and Townsend, P.A. (2008). Biofuels and Biodiversity: Principles for Creating Better Policies for Biofuel Production. *Conserv. Biol*. 100, pp. 1-8.

Gustafson, C. (2011). New energy economics: how much stover can be removed sustainably? NDSU Agriculture Communication. March 1, Fargo, ND.

Hahn-Hagerdal, B., Galbe, M., Gorwa-Grauslund, M.F., Liden, G., and Zacchi, G. (2006). Bio-ethanol – the fuel of tomorrow from the residues of today. *Trends Biotechnol*. 24, pp. 549-556.

Henderson, J. (2008). What is driving food price inflation? The Main Street Economist, Regional and Rural Analysis. The Federal Reserve Bank of Kansas City. 3, pp. 1-5.

Hendriks, A., and Zeeman, G. (2009). Pretreatments to enhance the digestibility of lignocellulosic biomass. *Biores. Technol*. 100, pp. 10-18.

Hodge, D.B., Karim, M.N., Shell, D.J., and McMillan, J.D. (2009). Model-based fed-batch for high-solids enzymatic cellulose hydrolysis. *Appl. Biochem. Biotechnol*. 152, pp. 88-107.

Hoyer, K., Galbe, M., and Zacchi, G. (2010). Effects of enzyme feeding strategy on ethanol yield in fed-batch simultaneous saccharification and fermentation of spruce at high dry matter. *Biotechnol. Biofuels*. 3, pp. 14.

Inderwildi, O.R., and King, D.A. (2009). Quo Vadis Biofuels. *Energy & Env. Sci*. 2, pp. 343.

Isa, B., Post, J., and Furedy, C. (2004). Solid waste management and recycling: actors, partnerships and policies in Hyderabad, India and Nairobi, Kenya. Kluwer Academic Publishers: Dordrecht, London, UK.

Jorgensen, H., Kristensen, J.B., and Felby, C. (2007). Enzymatic conversion of lignocellulose into fermentable sugars: challenges and opportunities. *Biofuels Bioprod. Biorefining*. 1, pp. 119-134.

Jorgensen, H., Vibe-Pedersen, J., Larsen, J., and Felby, C. (2007). Liquefaction of lignocellulose at high-solids concentrations. *Biotechnol. Bioeng*. 96, pp. 862-870.

Kalita, D. (2008). Hydrocarbon plant – New source of energy for future. *Renewable and Sustainable Energy Rev*. 12, pp. 455-471.

Kaltschmitt, M., Reinhardt, G.A., and Stelzer, T. (1997). Life cycle analysis of biofuels under different environmental aspects. *Biomass and Bioenergy*. 12, pp. 121-134.

Karp, A. (2008). Bioenergy from plants and the sustainable yield challenge. *New Phytologist*. 179, pp. 15-32.

Katzen, R., Madson, P.W., and Monceaux, D.A. (1995). Use of cellulosic feedstocks for alcohol production. In: The Alcohols Textbook. T.P. Lyons, J.E. Murtagh, and D.R. Kelsall (Eds). Northingham University Press. pp. 37-46.

Khanna, M. (2008). Cellulosic Biofuels: Are they economically viable and environmentally sustainable? *Choices*. 23, pp. 16-21.

Kim, T.H., and Lee, Y.Y. (2006). Fractionation of corn stover by hot-water and aqueous ammonia treatment. *Bioresource Technol.* 97, pp. 224-232.

Kumar, A., and Sokhansanj, S. (2007). Switchgrass (*Panicum vigratum*, L) delivery to a biorefinery using integrated biomass supply analysis and logistics (IBSAL) model. *Biores. Technol.* 98, pp. 1033-1044.

Lal, R. (2005). World crop residues production and implications of its use as a biofuel. *Env. Int.* 31, pp. 575-584.

Larsen, J., Petersen, M.O., Thirup, L., Li, H.W., and Iversen, F.K. (2008). The IBUS process - Lignocellulosic bioethanol close to a commercial reality. *Chem. Eng. Technol.* 31, pp. 765-772.

Lavarack, B.P., Giffin, G.J., and Rodman, D. (2002). The acid hydrolysis of sugarcane bagasse hemicellulose to produce xylose, arabinose, glucose and other products. *Biomass Bioeng.* 23, pp. 367-380.

Lee, S.H., Doherty, T.V., Linhardt, R.J., and Dordick, J.S. (2009). Ionic liquid-mediated selective extraction of lignin from wood leading to enhance enzymatic cellulose hydrolysis. *Biotechnol. Bioeng.* 102, pp. 1368-1376.

Lehmann, J., Gaunt, J., and Rodon, M. (2006). Bio-char sequestration in terrestrial ecosystems - a review. *Mitigation and Adaptation Strategies for Global Change.* 11, pp. 395-419.

Li, B., Asikkala, J., Filpponen, I., and Argyropoulos, D.S. (2010). Factors affecting wood dissolution and regeneration of ionic liquids. *Ind. Eng. Chem. Res.* 49, pp. 2477-2484.

Licht, F.O. (2006). World ethanol markets: the outlook to 2015. Agra Europe Special Report. Tunbridge Wells.

Lipnizki, F., Field, R.W., and Ten, P.K. (1999). Pervaporation-based hybrid process: a review of process design, applications and economics. *J. Membrane Sci.* 153, pp. 183-210.

Lynd, L.R., and de Brito Cruz, C.H. (2010). Make way for ethanol. *Science.* 330, pp. 1176.

Lynd, L.R., Johnson, D., and Landucci, R. (2005). Analysis of biorefineries, Subcontract Report NREL/SR-510-35578, October.

Lynch, J.A., and Corbett, E.S. (2007). Evaluation of Best Management Practices for controlling nonpoint pollution from silvicultural operations. *J. Am. Water Resources Assoc.* Published online 8 June 2007. DOI:10.1111/j.1752-1688.1990.tb01349.x

Marten, G.C., and Hovin, A.W. (1980). Harvest schedule, persistence, yield, and quality interactions among four perennial grasses. *Agron. J.* 72, pp. 378-387.

Mes-Hartree, M., and Saddler, J.M. (1983). The nature of inhibitory materials present in pretreated lignocellulosic substrates which inhibit the enzymic hydrolysis of cellulose. *Biotechnol. Lett.* 5, pp. 531-536.

Mosier, N., Wyman, C., Dale, B., Elander, R., Lee, Y.Y., Holtzapple, M., and Ladisch, M. (2005). Features of promising technologies for pretreatment of lignocellulosic biomass. *Bioresource Technol.* 96, pp. 673-686.

Moxley, G., Zhu, Z.G., and Zhang, Y.H.P. (2008). Efficient sugar release by the cellulose solvent-based lignocellulose fractionation technology and enzymatic cellulose hydrolysis. *J. Agric. Food. Chem.* 56, pp. 7885-7890.

Olofsson, K., Rudolf, A., and Liden, G. (2008). Designing simultaneous saccharification and fermentation for improved xylose conversion by a recombinant strain of *Saccharomyces cerevisiae*. *J. Biotechnol.* 134, pp. 112-120.

Palmqvist, E., and Hahn-Hagerdal, B. (2000). Fermentation of lignocellulosic hydrolysates. II: Inhibitors and mechanisms of inhibition. *Bioresource Technol.* 74, pp. 25-33.

Pan, X., Arato, C., Gilkes, N., Gregg, D., Mabee, W., Pye, K., Xiao, Z., Zhang, X., and Saddler, J. (2005). Biorefining of softwoods using ethanol organosolv pulping: Preliminary evaluation of process streams for manufacture of fuel-grade ethanol and co-coproduct. *Biotechnol. Bioeng.* 90, pp. 473-481.

Pan, X., Gilkes, N., Kadla, J., Pye, K., Saka, S., Gregg, D., Ehara, K., Xie, D., Lam, D., and Saddler, J. (2006). Bioconversion of hybrid poplar to ethanol and co-products using an organosolv fractionation process: optimization of process yields. *Biotechnol. Bioeng.* 94, pp. 851-861.

Parikka, M. (2004). Global biomass fuel resources. *Biomass Bioenergy.* 27, pp. 613-620.

Perlack, R.D., Wright, L.L., Turhollow, A.F., Graham, R.L., Stokes, B.J., and Erbach, D.C. (2005). Biomass as feedstock for a bioenergy and bioproducts industry: The technical feasibility of a billion-ton annual supply. Oak Ridge National Laboratory. Accession Number ADA436753.

Perrin, R.K. (2008). Ethanol and food prices - preliminary assessment. Faculty Publications: Agricultural Economics, Univ. Neb. Lincoln, pp 1-9.

Pfeifer, P.A., Bonn, G., and Bobleter, O. (1984). Influence of biomass degradation products on the fermentation of glucose to ethanol by *Saccharomyces carlsbergensis* W34. *Biotechol. Lett.* 6, pp. 541-546.

Pienkos, P.T., and Darzins, A. (2009). The promise and challenges of microalgal-derived biofuels. *Biofuels, Bioproducts and Biorefining.* 3, pp. 419-489.

Qureshi, N., and Ezeji, T.C. (2008). Butanol, 'a superior biofuel' production from agricultural residues renewable biomass: recent progress in technology. *Biofuels Bioprod. Biorefining.* 2, pp. 319-330.

Ragauskas, A.J., Williams, C.K., Davison, B.H., Britovsek, G., Cairney, J., Eckert, C.A., Frederick Jr., W.J., Hallett, J.P., Leak, D.J., Liota, C.L., Mielenz, J.R., Murphy, R., Templer R., and Tschaplinski. T. (2006). The path forward for biofuels and biomaterials. *Science.* 311, pp. 484-489.

Regalbuto, J.R. (2009). Cellulosic Biofuels - Got gasoline? *Science.* 325, pp. 822-824.

Rentizelas, A.A., Tolis, A.J., and Tatsiopoulos, I.P. (2009). Logistics issues of biomass: The storage problem and the multi-biomass supply chain. *Renewable and Sustainable Energy Rev.* 13, pp. 887-894.

Richard, T.L. (2010). Challenges in scaling up biofuels infrastructure. *Science.* 329, pp. 793-796.

Roche, C.M., Dibble, C.J., Knutsen, J.S., Stickel, J.J., and Liberatore, M.W. (2009a). Particle concentration and yield stress of biomass slurries during enzymatic hydrolysis at high-solids loadings. *Biotechnol. Bioeng.* 104, pp. 290-300.

Roche, C.M., Dibble, C.J., and Stickel, J.J. (2009b). Laboratory-scale method for enzymatic saccharification of lignocellulosic biomass at high-solids loadings. *Biotechnol. Biofuels.* 2, pp. 28.

Rosenberg, J.N., Oyler, G.A., Wilkinson, L., and Betenbaugh, M.J. (2008). A green light for engineered algae: redirecting metabolism to fuel a biotechnology revolution. *Cur. Opin. Biotechnol.* 19, pp. 430-436.

Roth, A.M., Sample, D.W., Ribic, C.A., Paine, L., Undersander, D.J., and Bartelt, G.A. (2005). Grassland bird response to harvesting switchgrass as a biomass energy crop. *Biomass and Bioenergy*. 28, pp. 490-498.

Saha, B.C. (2003). Hemicellulose bioconversion. *Ind. Microbiol. Biotechnol*. 30, pp. 279-291.

Saha, B.C., and Cotta, M.A. (2007). Enzymatic saccharification and fermentation of alkaline peroxide pretreated rice hulls to ethanol. *Enzyme Micro. Tech*. 41, pp. 528-532.

Sanderson, M.A., and Ad, P.R. (2008). Perennial forages as second generation bioenergy crops. *Int. J. Mol. Sci*. 9, pp. 768-788.

Schlamadinger, B., and Marland, G. (1996). Full fuel cycle carbon balances of bioenergy and forestry options. *Energy Conv. Management*. 37, pp. 813-818.

Schmer, M.R., Vogel, K.P., Mitchell, R.B., and Perrin, R.K. (2008). Net energy of cellulosic ethanol from switchgrass. *PNAS*. 105, pp. 464-469.

Shinners, K.J., and Binversie, B.N. (2003). Whole-plant corn harvesting for biomass: comparison of single-pass and multiple-pass harvest systems. 2003 ASAE Annual Meeting, Paper 036089.

Sims, R. (2003). Biomass and resources bioenergy options for a cleaner environment in developed and developing countries. Elsevier Science. London, UK.

Sjostrom, E. (1993). Wood chemistry: fundamentals and applications. Academic Press. San Diego, USA.

Sokhansanj, S., Turhollow, A., Cushman, J., and Cundiff, J. (2002). Engineering aspects of collecting corn stover for bioenergy. *Biomass Bioenergy*. 23, pp. 347-355.

Sun, Y., and Cheng. J. (2002). Hydrolysis of lignocellulosic materials for ethanol production: A review. *Bioresource Technol*. 83, pp. 1-11.

Taherzadeh, M.J., and Karimi, K. (2007a). Acid-based hydrolysis processes for ethanol from lignocellulosic materials: A review. *BioResources*. 2, pp. 472-499.

Taherzadeh, M.J., and Karimi, K. (2007b). Enzymatic-based hydrolysis proocesses for ethanol from lignocellulosic materials: A review. *BioResources*. 2, pp. 707-738.

Taherzadeh, M.J., and Karimi, K. (2008). Pretreatment of lignocellulosic wastes to improve ethanol and biogas production: A review. *Int. J. Mol. Sci*. 9, pp. 1621-1651.

Thring, R.W., Katikaneni, S.P.R., and Bakhsi, N.N. (2000). The production of gasoline range hydrocarbons from Alcell lignin using HZSM-5 catalyst. *Fuel Proc. Technol*. 62, pp. 17-30.

Tilman, D., Socolow, R., Foley, J.A., Hill, J., Larson, E., Lynd, L., Pacala, S., Reilly, J., Searchinger, T., Somerville, C., and Williams, R. (2009). Beneficial biofuels - The food, energy, and environment trilemma. *Science*. 325, pp. 270-271.

Torget, R., Himmel, M.E., and Grohmann, K. (1991). Dilute sulfuric acid pretreatment of hardwood bark. Bioresource *Technol*. 35, pp. 239-246.

Varga, E., Klinke, H.B., Reczey, K., and Thomsen, A.B. (2004). High solid simultaneous saccharification and fermentation of wet oxidized corn stover to ethanol. *Biotechnol. Bioeng*. 88, pp. 567-574.

Vlasenko, E.Y., Ding, H., Labavitch, J.M., and Shoemaker, S.P. (1997). Enzymatic hydrolysis of pretreated rice straw. *Bioresource Technol*. 59, pp. 109-119.

Wyman, C.E., Dale, B.E., Elander, R.T., Holtzapple, M., Ladisch, M.R., and Lee, Y.Y. (2005). Coordinated development of leading biomass treatment technologies. *Biores. Technol*. 96, pp. 1959-1966.

Yang, B., and Wyman, C.E. (2008). Pretreatment: the key to unlocking low-cost cellulosic ethanol. *Biofuels Bioprod. Biorefining*. 2, pp. 26-40.

Yuan, J.S., Tiller, K.H., Al-Amad, H., Stewart, N.R., and Stewart, Jr., C. N. (2008). Plants to power: bioenergy to fuel the future. *Trends Pl. Sci.* 13, pp. 421-429.

Zhang, J., Chu, D.Q., Huang, J., Yu, Z.C., Dai, C.G., and Bao, J. (2010a). Simultaneous saccharification and ethanol fermentation at high corn stover solids loading in a helical stirring bioreactor. *Biotechnol. Bioeng*. 105, pp. 718-728.

Zhang, Y.H.P., Ding, S.Y., Mielenz, J.R., Cui, J.B., Elander, R.T., Laser, M., Himmel, M.E., McMillan, J.R., and Lynd, L.R. (2007). Fractionating recalcitrant lignocellulose at modest reaction conditions. *Biotechnol.Bioeng*. 97, pp. 214-223.

Zhang, M.J., Su, R.X., Qi, W., and He, Z.M. (2010b). Enhanced enzymatic hydrolysis of lignocellulose by optimizing enzyme complexes. *Appl. Biochem. Biotechnol.* 160, pp. 1407-1414.

Zhang, M.J., Wang, F., Su, R.X., Qi, W., and He, Z.M. (2010c). Ethanol production from high dry matter corncob using fed-batch simultaneous saccharification and fermentation after combined pretreatment. *Biores. Technol.* 101, pp. 4959-4964.

20

Limitations and Challenges for Wheat-Based Bioethanol Production

Jessica Saunders[1], Marta Izydorczyk[2] and David B. Levin[1]
[1]*Department of Biosystems Engineering, University of Manitoba,*
[2]*Grain Research Laboratory - Canadian Grain Commission,*
Canada

1. Introduction

Bioethanol is currently the most widely used liquid biofuel in the world. Global ethanol production was ~19 billion L in 2000, and production has almost quadrupled over the past decade, with estimates for 2012 approaching 76 billion L [MRA, 2008]. In 2000, the total bioethanol produced in the U.S. represented 1.27% of the total gasoline pool by volume, and is expected to reach 7.5% of the gasoline pool by 2012 [EIA, 2007]. At present, bioethanol is produced exclusively via 1st generation technologies, utilizing sugar and starch-rich feedstocks, as no commercial size 2nd generation cellulosic ethanol facilities are presently in operation.

In countries like Canada, where wheat is locally available and abundant, the development of high yielding wheat-based bioenergy crops could contribute to reducing dependence on petroleum for transportation fuels and reduce green house gas (GHG) emissions. In order for wheat-based bioethanol to contribute maximally to the displacement of petroleum-based fuels, wheat varieties possessing characteristics optimized for end-use as bioethanol feedstocks are required. Wheat with characteristics tailored to the needs of the bioethanol industry would consist of high starch, low protein varieties, with the physicochemical parameters of the starch lending themselves to high conversion efficiencies under industrial conditions. However, no evaluative criteria of wheat starch quality, as it relates to maximized ethanol yield, is available. In this chapter, we review the rational for grain-based bioethanol production, the starch characteristics that would be optimal for a bioethanol feedstock, and propose selection criteria for development of dedicated, wheat-based, bioethanol feedstocks.

2. Background - the rational for 1st generation bioethanol

The largest ethanol producing industries, representing over 90% of the global 65.7 billion L produced in 2008 [CRFA, 2008], are located in Brazil and the United States of America (USA or US). Total production capacity in the U.S.A. is expected to reach about 90% of the 2015 goal of 56.2 billion L of corn ethanol production set in the 2007 Energy Independence and Security Act [Pryor, 2009]. In 2008, the USA produced 34 billion L (Table 1) of ethanol for use as fuel oxygenate. Most commercially made ethanol now comes from corn (~97.5% in

USA) [Zhao *et al.*, 2009] and the most commonly available blended gasoline contains 10% (E10) corn ethanol [Pryor, 2009; Tao & Aden, 2009]. Brazil produced approximately 24.5 billion L in 2008 but mandates 20-25% (E20-E25) blend ratios [Hahn, 2008]. According to Hahn [2008], the European Union (EU) has also set targets of 5.75% for blended gas, as has Argentina (at least 5%). China and India are also following this trend with nationa-wide fuel ethanol programs [Bai *et al.*, 2008], as is Canada, which will need to produce 2 billion L of ethanol by 2010 to meet the 5% federal renewable fuel standard recently mandated [CRFA, 2009].

Ethanol has been mandated for incorporation into many countries transportation economies largely in the hopes of addressing rising concern over anthropogenic greenhouse gas emissions, of which 80% is claimed accountable to fossil fuel combustion [Quadrelli & Peterson, 2007]. The role of ethanol as a blended gasoline product, in ratios of 5 – 20% (v/v) ethanol, referred to as gasohol, can be used without major modifications to pre-existing automobile engines and burns cleaner, due to the higher octane rating, reducing harmful emissions [Agarwal, 2007]. Most researchers agree that a net decrease in GHG emissions of 13 - 18% is likely to be observed with the incorporation of ethanol as E10-20 into the fuel supply [Farrel, 2006; Kim & Dale, 2004, Dale, 2008]. Fuel security, volatility in oil pricing and the development of local, rural economies, have all been cited as additional impetus for many nation's inclusion of biofuels into the energy policy discourse.

Country	Billions of Litres
USA	34.1
Brazil	24.5
European Union	2.8
China	1.9
Canada	0.9

Table 1. Global ethanol production. Adapted from RFA [2008].

Most governments have helped their biofuel industries flourish with subsidies, suggesting these technologies are, at present, economically unviable. For example, subsidies per liter of ethanol total more than $6 billion (USD) per year for US corn ethanol [Pimentel et al., 2009; Koplow, 2006]. In Canada, between 2006 and 2008, total support to biofuels was between $860 million and $1.02 billion (CDN), averaging $300 million (CDN) per year [Laan *et al.*, 2009]. However, at current energy prices, some agricultural feedstock have indeed already become competitive sources of energy, at least under certain production environments [Schmidhuber, 2006]. Sugarcane ethanol in Brazil is reported to offer higher energy return and greenhouse gas reductions per litre of ethanol than US-made corn ethanol [Rajagopal *et al.*, 2007], and is suggested to be cost competitive with petroleum at US$25 per barrel (bbl) [Schmidhuber, 2006]. Regional variability in agricultural conditions, however, dictates the fuel crops that can feasibly be produced in an area. Sugarcane, for example, does not grow outside of tropical or sub-tropical climates. Starch-based bioethanol production, utilizing wheat and corn as feedstock, dominates the North American biofuels market, as these ubiquitous cereal grains are well suited to that environment.

Neither biofuels, nor any other petroleum alternative, are able to compete with fossil fuels if the price of oil is ≤ US$20/bbl, as it has been for most of the past three decades [Dale, 2008]. Dale [2008] claims that at ≥ US$50/bbl, many alternatives make economic sense, including some biofuels and particularly cellulosic ethanol. Without subsidies, at corn prices of around $3.25 per bushel, ethanol as a high-octane fuel is competitive with oil at about US$60/bbl [Dale, 2008]. Schmidhuber [2006] puts the parity price closer to US$58/bbl for maize-based ethanol in the US. In the past 4 years (April 2006 - April 2010) the average per barrel oil cost is cited as US$72.55 [EIA, 2010], and in that time per barrel oil price has been less than US$60 for only 6.5 months, where it averaged US$54.84 [EIA, 2010]. In fact, the growth in ethanol production, according to an Iowa State University study, has caused retail gasoline prices to be $0.29 to $0.40 per gallon lower than would otherwise have been the case [Du & Hayes, 2008]. Wheat-based ethanol production, which represents a large fraction of production in the EU and Canada, has less favorable economics than corn-based production. In the EU, grain-based ethanol production cost was reported in 2006/2007 as ~ $0.578/L [Tao & Aden, 2009], compared to ~ $0.396/L for corn purchased at $3.35 per bushel [Tao & Aden, 2009]. Although wheat represents a large fraction of production in the EU and Canada, only 1.6% of the total wheat harvest in Europe and 2.9% of the total wheat harvest in Canada was used for bioethanol production in 2007 [Harlander, 2008; Husky Energy, 2009].

Fossil fuels at present provide 85% of the commercial energy consumed worldwide [Lackner & Sachs, 2005] and 40% of the total energy consumption in the world is in the form of liquid fuels [Tan *et al.*, 2008]. Global coal, oil, and natural gas reserves have been estimated to last for 218 years (coal), 41 years (oil), and 63 years (natural gas), under a business as usual scenario [Agarwal, 2007]. The world's supply of low cost, "sweet" crude oil, however, is dwindling, with 'peak oil' having occurred or likely to occur before 2010 [Lackner & Sachs, 2005]. As the population swells towards 9 billion, it is estimated that oil demand will double in the rapidly developing economies of China and India, resulting in an estimated 52% increase in global oil demand by 2025 [IEA, 2005]. Projections for the 30-year period from 1990-2020 indicate that vehicle travel, and consequently fossil-fuel demand, will almost triple [Agarwal, 2007].

The development of an energy supply that is local, renewable and sustainable is highly desirable for nations with growing transportation fuel demands and who already import large fractions of their supply. For example, of the roughly 20 million barrels of crude oil the USA consumes daily [EIA, 2009a], almost 60% is imported [EIA, 2009b], making the US highly vulnerable to oil market fluctuations. China's oil consumption has seen a purported 7.5% annual growth over the last several years, 7 times faster than the US [Luft, 2004]. In 2008, China consumed an estimated 7.8 million bbl/d and imported approximately 3.9 million bbl/d, roughly 50% of demand [EIA, 2010]. If the burgeoning economies of India and China place the predicted stress on oil availability and market price, this "would be the single most important aid and rationale for biofuels as a commercial reality" [Mousdale, 2008].

Despite the limitations of starch-based biofuels (see Section 4.0, below), the ease of ethanol's adaptation to present oil infrastructure suggests that as gasoline prices increase and emission regulations become more stringent, ethanol is likely to assume a role of escalating significance in a market that no longer has access to cheap and abundant petroleum products. First generation bioethanol technologies offer an imperfect solution to the world's long-term energy needs, whose utility must be viewed in context to locations that can feasibly support the diversion of food to energy crops. Adopting present processing technologies to utilize a feedstock, however, without the necessity of heavy cultivation and diversion of agricultural lands and foodstuffs, could represent a long-term solution to bioenergy generation and sustainable supply. Farrell [2006] claimed "large-scale use of ethanol for fuel will almost certainly require cellulosic (2nd generation) technology."

Transportation biofuels such as cellulosic ethanol, if produced from low-cost biomass grown on agriculturally marginal land or from waste biomass, could provide much greater supplies and environmental benefits than food-based biofuels [Hill *et al.*, 2006]. The US government, under the Energy Independence and Security Act of 2007, has mandated 16 billion gallons (~60 billion L) of "cellulosic ethanol" be included into the renewable fuel supply [Tao & Aden, 2009] by 2022. The cellulosic ethanol industry, however, remains nascent and has failed to emerge from the current demonstration phase to produce commercial-scale quantities. The US Environmental Protection Agency (EPA) recently acknowledged that, despite generous levels of taxpayer funding, cellulosic ethanol was not scaling up as quickly as had been hoped [Rapier, 2010] and previous targets outlined in 2007 have been significantly revised.

Starch-based ethanol production, although problematic, remains a mature technology capable of immediate contribution to the pressing global environmental and energy security needs. First-generation technologies are seen as an intermediate step to reduce GHG emissions and to diversify transport energy security [Antizar-Ladislao & Turrion-Gomez, 2008] whose production, however, is an undeniable aid in development of an ethanol infrastructure. Until cellulosic ethanol becomes a commercial reality, starch-based production will likely be of growing significance in many countries liquid fuel supply, especially for nations possessing large surplus quantities of grain, such as the EU, Canada and the USA.

3. Limitations to 1st generation bioethanol technologies

Arguments in favor of starch-based ethanol production are countered by myriad of concerns related to land-use patterns and diversion of food supply, which generates significant uncertainty in the long-term utility of starch-based ethanol production. According to Rajagopal *et al.* [2007], production of biofuels takes land away from its two other primary uses - food production and environmental preservation. Some even argue that biofuels will cause dramatic changes in land-use patterns which could offset any CO_2 savings derived from the utilization of biomass. Searchinger *et al.* [2008] contends that land-use changes will cause a net increase in GHG emissions with a doubling of GHG emissions over 30 years and increasing atmospheric CO_2 concentrations for 167 years.

A number of sweeping condemnations of corn-based bioethanol production in the USA have been published in the past several years. The criticisms are centered around resource consumption, such as water, and agricultural practices: it has been estimated that a 50 million gallon per year ethanol factory consumes 500 gallons of water per minute, and that intensive corn production uses more nitrogen fertilizer [NAS, 2004], has significant phosphorus requirements [USDA, 2007], and uses more insecticides [Mclaughlin & Walsh, 1998] and herbicides [Patzek, 2004] than any other crop grown. Compounding these environmental issues is the reality of ethanol's lower energy density. Ethanol delivers only two-thirds the energy that petroleum does and therefore more is required [Srinivasan, 2009].

Diverting food crops for use as substrate in biofuel production has caused unceasing controversy since the inception of the biofuels movement. In 2004, 13% of the US corn crop was diverted to produce ethanol [Patzek, 2004], in 2006 that number increased to 20%. In 2009/2010, 4.2 billion bushels of corn were used to produce ethanol, an increase from 3.6 billion bushels in 2008/2009 [USDA, 2009]. Overall, ethanol consumed 33% of the corn crop in 2009/2010, compared to 30% in 2008/2009. Some authorities have claimed that bioethanol production contributed to rising food prices, but these claims are controversial, and much uncertainty exists in the extent of the food price increases that may be attributed to bioethanol production [Sims, 2008]. For example, the use of corn for ethanol production was reported to have increased the prices of US beef, chicken, pork, eggs, breads, cereals, and milk by 10% to 20% [Brown, 2009]. However, in 2007 the UK, food prices increased even though no grain-based bioethanol was produced [Karl. 2010]

The switch to fuel crops, from other non-energy food crops, could cause additional food security issues. Projected corn ethanol production in 2016 would use 43% of the US corn land harvested for grain in 2004 [Searchinger *et al.*, 2009]. According to Searchinger *et al.* [2009], U.S. agricultural exports will decline sharply causing a myriad of problems for importing nations, who will be forced to become increasingly self-reliant, likely appropriating previously unused land for domestic agricultural production. In June 2007, due to concern over rising food prices, China's central government to ban the use of grain-based feedstocks for biofuel production and reoriented the country's bioenergy plans toward perennial crops grown on marginal land [Weyerhaeuser *et al.*, 2007]. In addition, Jacques Diouf, Director General of the UN Food and Agriculture Organization reported that using food grains to produce biofuels already is causing food shortages for the poor of the world [Diouf, 2007].

The most salient of arguments against 1st generation technologies are, however, (environmental and food diversion concerns aside), that grain-based bioethanol is "supply-limited" and cannot meet the expected US transportation fuel demand. Even if all current US soybean and corn production were dedicated to biofuels, only 12% of the gasoline demand and 6% of the diesel demand would be met [Srinivasan, 2009]. Globally seven crops (wheat, rice, corn, sorghum, sugarcane, cassava and sugar beet) account for 42% of cropland. If all land currently used to grow these crops were dedicated to biofuels, just over half of the global gasoline demand would be met.

4. The case for wheat-based bioethanol

As detailed above, feedstock sources for ethanol production vary around the world. Brazil, the largest ethanol producer, uses sugarcane as a primary feedstock source. Corn acts as the primary feedstock source in the United States. In Canada, corn is used as a feedstock for ethanol production in areas where grain corn can be produced. However, climatic conditions limit grain corn production in many areas. In western Canada, wheat is the most readily available, high quality feedstock. Therefore, the majority of proposals for expansion of ethanol production in western Canada include the construction of wheat-based ethanol plants.

Traditionally, Canadian wheat cultivars were developed to express high protein concentrations for functionality in the production of bread and pasta, extracting price premiums in the marketplace. High protein concentrations were accompanied, however, by low concentrations of starch, which make most small grain cereals less desirable for industrial ethanol production. Breeding efforts in the past 20 years have resulted in high-yielding wheat cultivars, such as those of the Canada Prairie Spring and the Canada Western Soft White Spring classes. Very high yielding triticale, hulless barley, and improved winter wheat cultivars have also been developed. All of these classes of small grains tend have protein concentrations between 10% and 14%, but their starch concentrations are mostly unknown [NRCan, 2003]. Starch concentrations are not generally determined, but

reported as "total carbohydrates", which are estimated as the difference between 100% and the sum of moisture, protein, crude fiber, fat and ash [Wolff, 1982]. In addition to starch and fermentable sugars, the "total carbohydrates" also contain water-soluble hemicellulose, β-glucans and pentosans, depending on the grain. Not all total carbohydrates are fermentable, and ethanol yields are overestimated when calculated from this parameter in cereal grains.
There is limited information regarding the potential ethanol yields of small grain cereals in western Canada. Wang *et al.* [1997] reported the spring triticale cultivar AC Copia yielded 362 to 367 L t^{-1} grain (14% moisture basis). Ethanol yields of 443 L t^{-1} of hulless barley (dry weight basis) have been reported [Thomas *et al.* 1995]. Thomas & Ingledew [1995] obtained 317 +/- 1.3 L t^{-1} on a dry weight basis from a hulled oat (cultivar unknown). Research in the USA reported that barley produced the greatest ethanol yield per hectare, slightly more than soft white spring wheat, while hard red and hard white spring classes produced the least [Lacerenza *et al.*, 2008].
More recently, however, McLeod *et al.* [2010] evaluated the potential of small grains in western Canada to supply feedstock to the ethanol industry. Thirty-one lines and cultivars of Canadian small grains were evaluated: eleven cultivars comprising five classes spring wheat, six cultivars of barley (feed, malting, and hulless varieties), eight cultivars of spring triticale, and six cultivars of oat were grown at seven locations in western Canada and evaluated as feedstock for ethanol production. Starch concentrations and, for certain grains, β-glucan and pentosans were determined and used to estimate ethanol yields in L t^{-1} and L ha^{-1}. On average, the ethanol yield in L t^{-1} was wheat > triticale > barley > oat.
Biologically, winter wheat has the highest yield potential of the *Triticum aestivum* wheat cultivars grown in western Canada. From 1993 to 2003 mean winter wheat yields were 107% of spring wheat yields in Alberta, 116% in Saskatchewan, and 127% in Manitoba [Canada Grains Council Statistical Handbook, 2003]. In the province of Manitoba, the mean yield of winter wheat was 140% of the Canada Western Red Spring wheat yield and 124% of the Canada Prairie Spring wheat yield from 1998 to 2004 (Yield Manitoba - Manitoba Crop Insurance Corporation). Therefore, commercially grown cultivars of winter wheat have a significant yield advantage over Canada Western Red Spring and Canada Prairie Spring wheat. The efficiency of ethanol production from grains, however, depends on process conditions, as well as the starch and fermentable sugar contents.

5. Cereal starch characteristics optimal for bioethanol feedstock

Starch is the predominant component of wheat grain, constituting 60-65% of the kernel. The functional properties of starch vary widely across botanical origin [Swinkels, 1985] and unique characteristics are desirable for different final product application [Franco *et al.*, 2002]. Lacerenza *et al.* [2008] recently pointed out that the traditional selection criteria for wheat-breeding, based on milling and baking quality, is not consistent with maximal ethanol yield per hectare. The lack of breeding programs for varieties designed specifically for ethanol production, suggests Swanston *et al.* [2007], is due to a lack of appropriate selection procedures due to limited understanding of the factors contributing to alcohol yield.
Recent research into optimized bioethanol production has focused on the development of new and improved cereal and maize hybrids with higher starch contents to increase ethanol yields [Wu *et al.*, 2006]. Grain preprocessing strategies are also being investigated and have the potential to 'increase throughput rate and capacity of ethanol plants' [Sosulski &

Sosulski, 1994; Kindred *et al.*, 2008] by decreasing non-starch carry-through. The outcome of present biofuel development is the inevitable processing of higher starch feed streams, achieved through adoption of new process technology or higher starch grains.

Starch content, although inarguably the most critical feature in determining ethanol conversion efficiency, is not necessarily the only influential parameter in understanding fermentation performance. Starch with high intrinsic resistance to enzymatic hydrolysis can yield a low sugar load to yeast, making it an erroneous selection as feedstock, especially when considering the industries move towards ever higher starch feed streams. Of particular interest to the bioethanol industry are the functional properties of starch that lend themselves to ease of amylolytic hydrolysis and high conversion efficiency of starch to fermentable sugars. The bioavailability of starch may differ among grain cultivars and may affect the conversion rate and final yield of ethanol [Moorthy, 2002]. A more thorough understanding of the influence starch structural and physicochemical properties have on the efficiency of gelatinization and liquefaction, the two most relevant industrial processes in the preparation of bioethanol feedstock, is therefore required.

5.1 The effect of physicochemical properties of wheat starch on enzymatic hydrolysis and fermentative alcohol yield

Presently no evaluative criteria of grain starch quality, as it relates to maximized ethanol yield, are available to bioethanol producers. The following five parameters have been shown to influence the functional properties of starch and are used as evaluative criteria in this study: amylose/amylopectin content [Zhao *et al.*, 2009; Wu *et al.*, 2006, Wu *et al.*, 2007; Lee *et al.*, 2001]; starch granule morphology [Liu *et al.*, 2007]; amylopectin fine structure [Zhang *et al.*, 2008a; Ao *et al.*, 2007; Sasaki *et al.*, 2002; Zhang *et al.*, 2008b]; thermal properties [Zhao *et al.*, 2009; Wu *et al.*, 2007]; pasting properties [Zhao *et al.*, 2009].

5.2 Amylose and amylopectin

Starch granules are composed of two types of alpha (α-)glucans, amylose (Figure 1) and amylopectin, which represent approximately 98-99% of the dry weight [Tester *et al.*, 2008]. The ratio of the two polysaccharides varies according to the botanical origin of the starch, but within wheat varieties maintains ranges of 25-28% amylose and 72-75% amylopectin [Hung *et al.*, 2006]. Starches with less than 5% amylose are found in "waxy" wheats, where as wheat varieties containing starches with greater than 35% amylose are considered "high-amylose" wheats [Wu *et al.*, 2006]. Amylose is an essentially linear molecule, consisting of α-(1,4)-linked D-glucopyranosol units with a degree of polymerization (DP) in the range of 50-6000 glucose residues. It is now well recognized that a fraction of the amylose molecules is slightly branched by α-(1,6)-linkages. In contrast, amylopectin is a very large, highly branched chain molecule with a DP ranging from 30,000 to 300,000 glucose units and consists of α-(1,6)-linked D-glucopyranosol units attached to glucose residues in the amylose chains [Zobel *et al.*, 1988]. Amylose and amylopectin strands are reported to have molecular weights in the range of 10^4-10^6 and 10^7-10^8 Daltons, respectively.

Wu *et al.* [2006] studied high-amylose starches and demonstrated that amylose content, more than protein or fiber content, had significant effect on ethanol fermentation efficiency. The study revealed that conversion efficiency decreased as amylose content increased. Starch in its native form is resistant to enzymatic digestion and must be gelatinized,

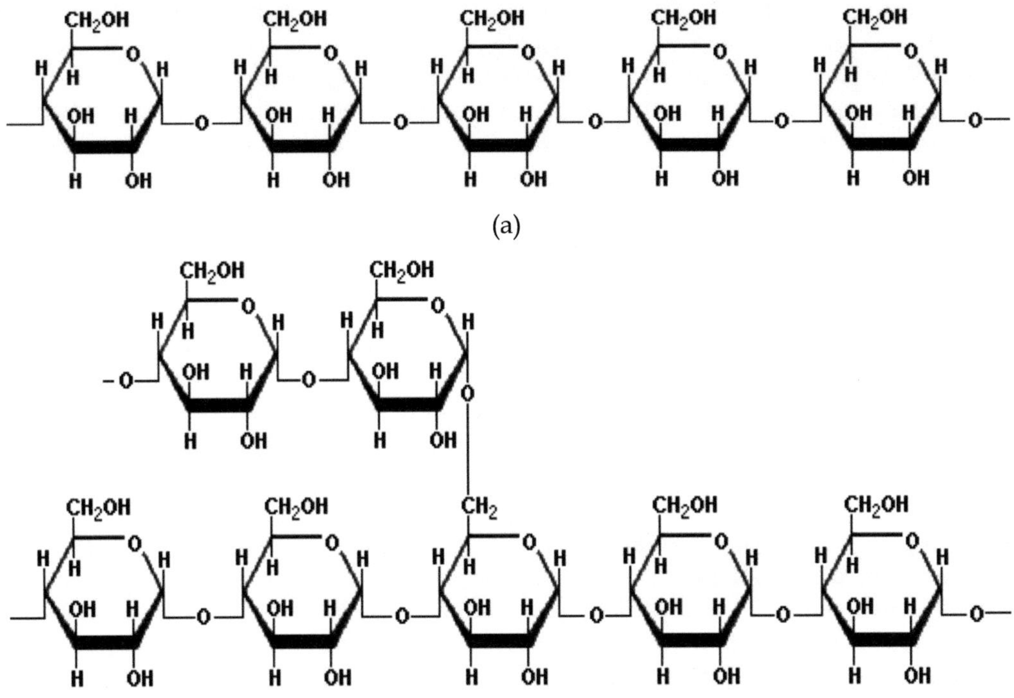

Fig. 1. Amylose molecules composed of glucose monomers connected as in **a)** via α-(1,4)-glycosidic bonds or as in **b)** with infrequent branch chains via α-(1,6)-glycosidic bonds. Adapted from Saunders [2010].

changing it into an amorphous mass, before it becomes susceptible to glucosidic enzymes. Because of the existence of starch granules with high gelatinizing temperatures, and the formation of amylose-lipid complexes (AML) and reassociation of amylose molecules during gelatinization and enzymatic hydrolysis, there is always some starch that escapes hydrolysis by amylolytic enzymes [Boltz & Thompson, 1999]. As much as 2% of the starch in industrial dextrose production remained undigested as insoluble particles in the hydrolysate [Hebeda & Leach, 1974]. Amylose is likely to form amylose-lipid complexes in the caryopsis or during mashing that are resistant to enzymatic hydrolysis [Wu *et al.*, 2007]. Wu *et al.* [2006] states that, in most instances, resistant starch content increases as the amylose content in starch increases. High amylose containing starches also inhibit fermentative ethanol production by yielding high viscosity mashes. Greater enzyme dosing, according to Wu *et al.* [2006], was not an effective strategy for increasing maltose and dextrin formation. It appears that high mash viscosity creates a bottleneck to enzymatic activity that can only be overcome with greater reaction time; greater reaction time, however, provides increased opportunity for deleterious retrogradation and amylose-lipid complexing, increasing the degree of insoluble sugars.

A reduction in amylose content has been positively correlated with enzymatic digestibility and represents the current accepted paradigm for the pattern of α-amylolysis on native starches. Fermentation studies consistently report higher ethanol yields on waxy (high amylopectin) substrates than non-waxy counterparts. A high percentage of amylopectin seems to be more susceptible to α-amylase during fermentation, with higher gas production of waxy wheat flour as compared with the non-waxy wheat flour [Lee *et al.*, 2001]. Wu *et al.* [2007] reported that waxy and hetero-waxy sorghum hybrids generally have higher conversion efficiencies than non-waxy hybrids. Zhao *et al.* [2009] studied the effects of waxy vs. non-waxy soft and hard wheats for fuel ethanol production and found results consistent with Lee *at al.* [2001] and Wu *et al.* [2007]. High ethanol conversion efficiencies of waxy-wheats were reported as compared to non-waxy wheats, (95.4 to 96.2% versus 92.6%, respectively) [Zhao *et al.*, 2009]. Wu *et al.* [2006] observed that conversion efficiencies increased as the amylose content decreased, especially when the amylose content was >35%. Employing waxy wheat as feedstock for fuel ethanol production has been recommended [Lacerenza *et al.*, 2008]. However, waxy-wheats have lower starch yields [Zhao *et al.* 2009] and no waxy wheat varieties have yet reached the commercial stage of development.

5.3 Granule size distribution

Present research suggests wheat starch has a trimodal distribution of granule sizes [41-43]. However, the existence of the smallest C-type granule remains somewhat putative with many authors reporting only the A and B-type granule populations [Ao & Jane, 2007; Morrison & Gadan, 1987; Peng *et al.*, 1999]. A-type granules make up the bulk of starch (~75% by weight), but are fewer in number than the smaller sized B- and C-type granules (~25% by weight) [Ao & Jane, 2007]. A-type granules have been reported as 10-35 µm in average spherical diameter and account for less than 10% of the granules by number, small B and C-type particles therefore constitute over 90% of the granules by number [Lindeboom *et al.*, 2004].

The large lenticular A-granules and the small, spherical B-granules have different physical, chemical and functional properties [Raeker *et al.*, 2007; Morrison & Gadan, 1987; Dronzek *et al.*, 1972; Kulp, 1973; Meredith, 1981; Soulaka & Morrison, 1985; Park *et al.*, 2004]. The two types of granules (A- and B-) differ in their ratio of amylopectin to amylose [Morrison & Gadan, 1987; Soulaka & Morrison, 1985; Tester & Morrison, 1990], and have differing ratios of amylose to bound lipids [Raeker *et al.*, 1998; Ao & Jane, 2007]. In cereal starches there are small quantities of naturally occurring lipids which are capable of forming complexes with amylose [Kwasniewska-Karolak *et al.*, 2008]. The presence of amylose-lipid complexes negatively influence production of glucose syrups because it reduces water binding and swelling of starch granules, thus impairing the access of amylolytic enzymes [Matser & Steeneken, 1998]. Liu *et al.* [2007] studied the *in vitro* digestibility of A- and B-type granules from soft and hard wheat flours and found higher resistant starch content in the A- type wheat granule as compared to the B-type granule. Several studies have reported higher amylose content in the A-type granule, explaining, in part, their increased resistance to enzymatic hydrolysis [Liu *et al.*, 2007; Peng *et al.*, 1999]. For example, absolute amylose content of wheat for the A-type granule was measured as 30.9% and that of the B-granules 25.5% [Peng *et al.*, 1999]. Liu *et al.* [2007] also reported apparent amylose content as 25.4-25.8% for A-type (soft and hard wheat, respectively) and 16.5-19.3% for B-type.

Based on the findings of Liu *et al.* [2007], it appears that B-type granules may contain less resistant starch, due to lower concentrations of amylose, and may therefore yield higher conversion efficiency of starch to fermentable sugars during industrial feedstock preparation. The ratio of A- to B-type granules appears to vary significantly across cultivar type. However, a comparison of reported values across cultivars is challenging given genotype, environment, and method of analysis [Stoddard, 1999], which all appear to be influential in assessing the proportions of starch type in the granule. Various studies have reported B-granule distribution in wheat starch as ranging from 25-40% [Dengate & Meredith, 1984], others report the range as 13-34% [Soulaka & Morrison, 1985]. Recent studies, however, have yielded more detailed information regarding variation between wheat cultivars. For example, hard red winter (HRW) wheat was found to have 48.4% by volume classified as B-type (<10 μm diameter) [Bechtel *et al.*, 1990]. In another study, B-granules occupied volumes in the range 28.5 - 49.1% (mean 39.9%) for HRW wheat, while hard red spring (HRS) wheat B-granules occupied volumes in the range 37.1 - 56.2% (mean 47.3%) [Park et al., 2009]. In a study by Raeker *et al.* [1998], thirty-four starch samples from soft wheat cultivars were investigated for differences in particle size distribution. It was found that contributions from the large granule population (>9.9 μm) to the total volume were 57.9 - 76.9%; B-type particles (<9.9 μm) therefore represent a range of 23.1 - 42.1% of total volume. However, Raeker *et al.* [1998] also reported a negative correlation between total starch and volume % of small granules (<10 μm) [Raeker *et al.*, 1998]. In general, according to these findings, as starch content increases, a trait highly amenable to bioethanol production, the volumetric quantity of A-type granules present in the endosperm appears to increase, as small B and C-type granules decrease, a trait not necessarily desirable for end-use as ethanol feedstock.

5.4 Amylopectin fine structure

The two key features of amylopectin fine structure are chain length distribution and branching pattern. According to Jenkins & Donald [1995], the currently accepted model for amylopectin structure involves short amylopectin chains forming double helices and associating into clusters. These clusters pack together to produce a structure of alternating crystalline (double helices) and amorphous lamellar composition (amylopectin branch points) [Jenkins & Donald,1995]. The branched chains of amylopectin, according to the cluster model of amylopectin proposed by Hizukuri [1986], can be fractionated into B3, B2, B1 and A chains that are described as follows: A-chains, DP 6-12; B1-chains, DP 13-24; B2-chains, DP 25-36; and B3-chains, DP >37. A and B1 chains dominate the distribution, forming double helices, with the longer B2 and B3 chains traversing two, three and four clusters (Figure 2) [Hizukuri,1986].

The dominance of certain fractions of side chain length dictates the type of crystallinity displayed during X-ray diffraction studies, referred to as A-type, B-type and C-type [Jenkins & Donald,1995]. Most cereal starches possess A-type crystallinity and have higher weights and number percents of short A chains [Chung *et al.*, 2008]. Amylose is thought to exist mainly in the non-crystalline state [Hizukuri *et al.*, 1996], but the exact location of amylose within the granule interior and the extent of its interaction with amylopectin is unclear [Gupta *et al.*, 2009]. It is likely that a large portion is found within the amorphous (lamellae), with only small amounts associated with the semi-crystalline (lamellae) [Jenkins & Donald, 1995].

During starch gelatinization, starch granular or supramolecular structure is disrupted, resulting in the pattern of enzymatic hydrolysis being predominantly related to the inherent molecular structure of amylopectin [Zhang *et al.*, 2008a]. The relationship between the molecular structure of starch (amylopectin fine structure) and its digestion rate after starch gelatinization is not well understood [Zhang *et al.*, 2008b]. Little variation in the branching pattern of amylopectin has been reported in normal wheat starches. Hanashiro *et al.* [1996] reported the following distribution of amylopectin side chains after debranching: DP 6-12, 27%; DP 13-24, 49%; DP 25-36, 14%; and DP >37, 10%. These findings were

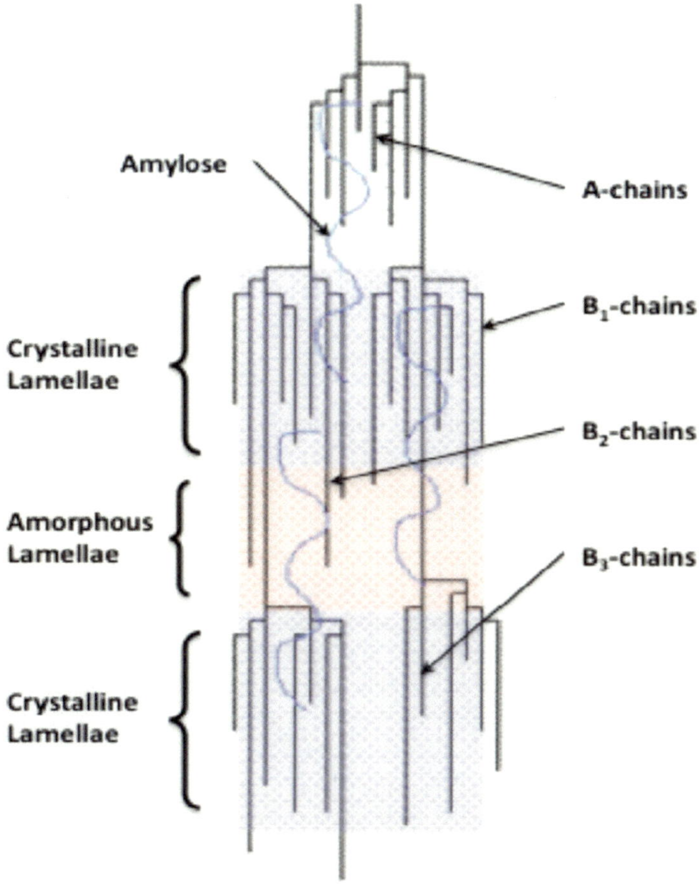

Fig. 2. Structure of amylopectin depicted as 'cluster model' proposed by Hizukuri (1986). A and B_1 chains dominate the distribution and occur as parallel strands, believed to wind into left-handed double helices. Larger B chains are thought to transverse two, three and four clusters. Short A-chains posses DP 6-12, B_1 chains range from DP 13-24, next B_2 chains occur at DP 25-36, and finally B_3 and longer chains at DP>36. Adapted from Hizukuri [1986].

supported in a study of 126 starch samples selected from the *Triticum-Aegilops* group, of which samples were derived from *Triticum aestivum*, or common bread wheat [Yasui *et al.*, 2005]. Additionally, in a study of 192 club and soft white winter wheat varieties, no measured difference in amylopectin side chain fractions was found between study samples [Lin & Czuchajowska, 1997]. These results were consistent with the findings reported by Hanashiro *et al.* [1996], and indicate that crystalline properties of starches considered should be indistinguishable.

Several studies have reported the effects of increasing branch density of amylopectin on decreased digestion rate of gelatinized starches through partial shortening of amylopectin exterior chains [Zhang *et al.*, 2008a; Zhang *et al.*, 2008b; Ao *et al.*, 2007]. Ao *et al.* [2007] reported that starch products exhibiting high branch densities, with shorter average chain lengths, showed reductions in rapidly digested starch of up to 30% and concomitant increases in slowly digested starch of up to 20%. Zhang *et al.* [2008b] found that amylopectin of maize starch with high proportions of short chain fraction (SF, DP < 13) to long chain fraction (LF, DP > 13) showed increased quantities of slowly digested starch. The inherent molecular structure of amylopectin with a higher density of branches and shorter chains (high SF/LF ratio) is not favorable for rapid enzyme digestion [Zhang *et al.*, 2008a]. In general, the findings of Ao *et al.* [2007] and Zhang *et al.* [2008a] suggest longer branch chain lengths of amylopectin may be ideal for a rapidly digestible starch. In fact, waxy starches (>95% amylopectin), indicated as having favorable ethanol conversion performance, tend to have lower proportions of DP 6-12 side chains and higher proportions of DP >35 compared with non-waxy starches [Sasaki *et al.*, 2002]. Future work is needed, however, to quantify the average amylopectin chain length optimal for rapid enzymatic digestion during liquefaction.

5.5 Thermal properties

Gelatinization precedes liquefaction in the fermentation process and describes the physical break down of granular starch into solubilized, amorphous polymers readily hydrolyzed by α-amylase and amyloglucosidase (AMG), the two enzymes responsible for the conversion of starch to sugar. This irreversible loss of native structure occurs when sufficient energy is applied to break intermolecular hydrogen bonds in the crystalline areas [Rooney & Pflugfelder, 1986]. Two endothermic peaks are seen when thermal properties are determined using differential scanning calorimetry (DSC) (Figure 3). The first peak represents the melting of amylopectin and the second peak corresponds to the melting of amylose-lipid complexes [Hung *et al.*, 2006]. Gelatinization temperatures and enthalpies associated with gelatinization endotherms vary between starches. In a study by Gupta *et al.* [2009] native wheat and corn starch, measured at 70% moisture content, were reported to have onset temperatures (To) of 60.19 °C and 70.12 °C, respectively, peak temperatures (Tp) of 64.06 °C and 73.85 °C, respectively, and conclusion temperatures (Tc) of 68.42 °C and 78.20 °C, respectively.

Starch transition temperatures and gelatinization enthalpies by DSC may be related to characteristics of the starch granule, such as the amount of double helical domains (amylopectin) and single helical structures (amylose-lipid complexes) that unravel and melt during heating of aqueous starch dispersion. Van Hung *et al.* [2007] reported that waxy wheat starch with a predominant amylopectin content requires higher energy for gelatinization caused by its higher crystallinity as compared to non-waxy and high-amylose wheat starches. The presence of amylose, conversely, lowers the melting (gelatinization) temperature by decreasing crystallinity [Gupta *et al.*, 2009].

Limited research has been performed to elucidate the relationship between gelatinization temperature and industrial fermentation efficiency. From an energy standpoint, low gelatinization temperature starch may be favorable as feedstock to produce fermentation-based products due to lower temperatures required to efficiently process the grain. Low gelatinization temperatures, however, are associated with high amylose content starch [Hung *et al.*, 2006; Noda *et al.*, 2002], previously demonstrated to be unfavorable to maximizing ethanol yield. Zhao *et al.* [2009] found waxy wheat starch to have complete disruption/dissolution of the granule at 70-80 °C, compared to non-waxy cultivars which showed evidence of intact granular structure under hot-stage microscopic visualization for temperatures as high as 90 °C. Wu *et al.* [2006] also reports that waxy starches easily gelatinize and have concomitantly high conversion efficiency. In regards to bioethanol production, the most salient thermal property is likely the point of complete

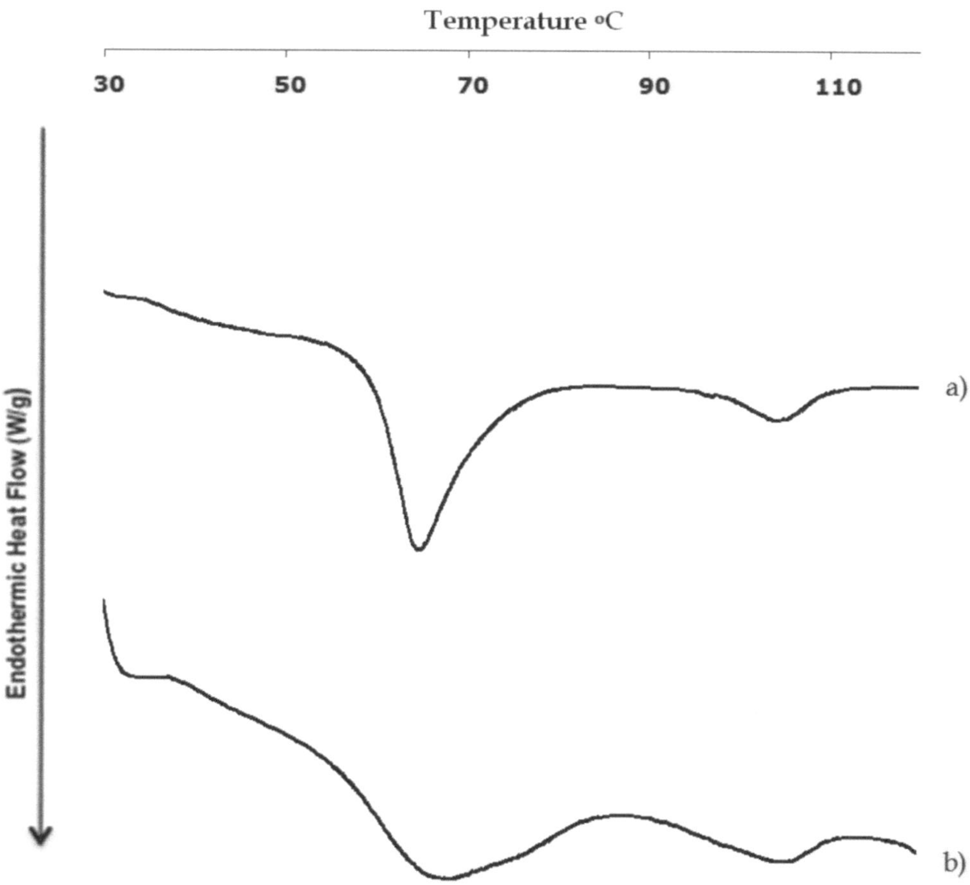

Fig. 3. Thermal properties of starch. Endotherm of **A**) wheat and **B**) maize starch, showing the first and second endothermic peaks, related to amylopectin gelatinization and amylose-lipid dissolution, respectively. Adapted from Saunders [2010].

disruption/dissolution, as pointed out by Zhao *et al.* [2009], and not necessarily traditional transition temperature ranges, as is generally reported. Amylopectin content appears to be the most influential feature dictating total granular disruption and is perhaps, in this regard, the best predictive metric for ethanol conversion.

The relationship between amylopectin fine structure and the thermal properties of starch has been well established [Franco *et al.*, 2002]. Starch that consists of amylopectin with a high proportion of long branch-chains purportedly displays higher gelatinization temperatures and enthalpy changes [Franco *et al.*, 2002; Jane *et al.*, 1999]. Several studies have reported the manipulation of branch chain length to modify thermal properties of starch. Amylopectin with increased quantities of longer branch chains produce more ordered double-helical crystallites, which require higher temperatures to uncoil and dissociate [Franco *et al.*, 2002; Song & Jane, 2000; Huang *et al.*, 2007]. Conversely, higher contents of extremely short chains within the amylopectin, DP 6 and 7, appear to lower To , Tp ,and ΔH (gelatinization enthalpy) [Noda *et al.*, 2002]. In regards to bioethanol production, driving down the gelatinization temperature by manipulating amylopectin branch chain length may lead to a substrate with lower energy requirements to achieve high conversion efficiency. However, the impact of amylopectin branch chain length on numerous other physicochemical properties [Franco *et al.*, 2002; Noda *et al.*, 2002], as well as enzymatic digestion rate [Zhang *et al.*, 2008; Ao *et al.*, 2007], suggest future work is needed to establish the amylopectin chain length distribution best suited for bio-ethanol end-use.

5.6 Pasting properties

Pasting viscosity profiles are analyzed using a Rapid Visco Analyzer (RVA). A typical profile is presented in Figure 4. The RVA curve describes pasting, a phenomenon following gelatinization, involving granular swelling, exudation of amylose and amylopectin, and total disruption of the starch granule. Pasting temperature is the point when the temperature rises above the gelatinization temperature, inducing starch granule swelling and resulting in increased viscosity. The peak viscosity indicates the maximum viscosity reached during the heating and holding cycle and is indicative of the water holding capacity of starch [Gupta *et al.*, 2009], and peak temperature occurs at peak viscosity. The breakdown viscosity is normally regarded as a measure of the disintegration of the starch granules as they are heated [Agu *et al.*, 2006] due to the rupture of granules and the release of soluble amylose. The degree of RVA breakdown is related to the solubility of the starch, and the more soluble the starch, the more it will thin on shearing [Hoseney, 1998]. As the mixture is cooled, re-association between starch molecules, especially amylose, results in the formation of a gel and the subsequent increase in viscosity. Total setback involves retrogradation, or re-ordering, of the starch molecule.

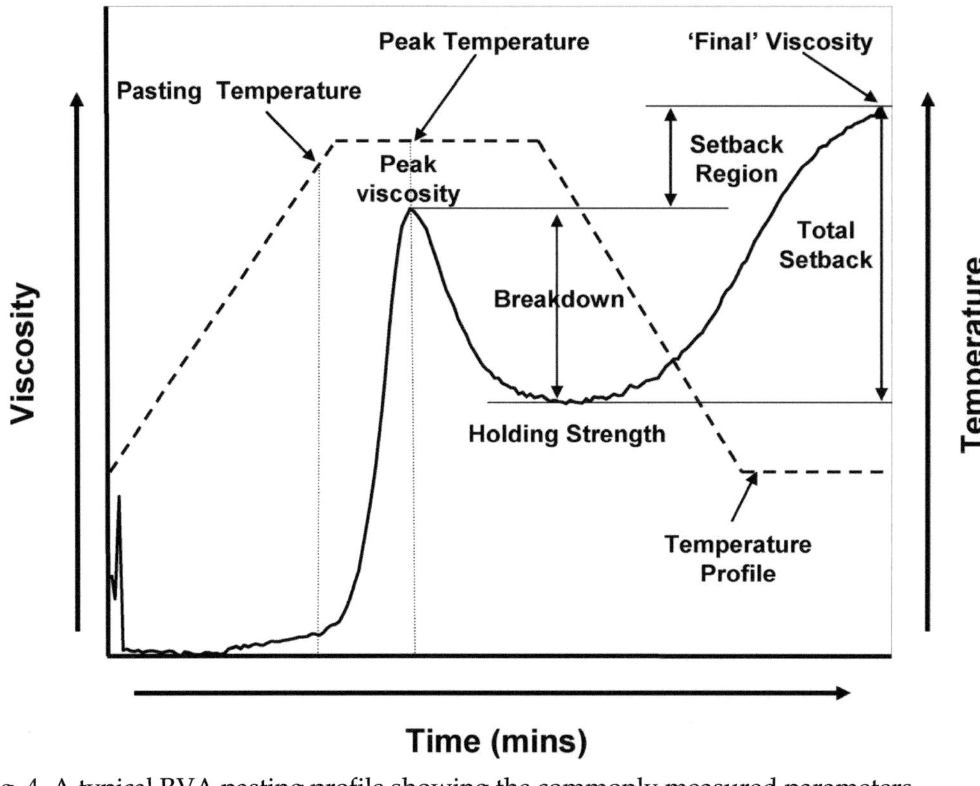

Fig. 4. A typical RVA pasting profile showing the commonly measured parameters. Adapted from Saunders [2010].

Pasting properties of starch are affected by amylose and lipid contents and by branch chain-length distribution of amylopectin [Gupta *et al.*, 2009]. Starches with larger amylose, lipid and phospholipid content have higher pasting temperatures, lower peak viscosity and shear-thinning (breakdown viscosity), and higher setback viscosity [Jane *et al.*, 1999; Zeng *et al.*, 1997]. Waxy wheat flour, conversely, has been shown to have significantly lower peak and pasting temperature, higher peak viscosity and lower setback viscosity than non-waxy or normal wheat flour [Gupta *et al.*, 2009; Zeng *et al.*, 1997; Abdel-Aal *et al.*, 2002]. Waxy starch swells rapidly and swollen granules degrade at lower temperature, indicating that waxy starch rapidly develops viscosity but cannot maintain the stability of paste viscosity [Gupta *et al.*, 2009]. Zhao *et al.* [2009] examined waxy starch granules using hot-stage microscopy and demonstrated that these starches rupture more extensively, even without mechanical shearing, and disperse more readily than non-waxy counterparts. Waxy wheat starch has been described as having lower final viscosity [Zeng *et al.*, 1997; Abdel-Aal *et al.*, 2002; Graybosch, 1998], an indication that it has less ability to retrograde and form strong gels.

Pasting properties, like thermal properties, are affected by the branch chain length distribution of amylopectin [Jane *et al.*, 1999]. According to Franco *et al.* [2002], amylopectin with longer branch chains display larger peak viscosity and lower pasting temperatures than shorter chain counterparts. Jane *et al.* [1999] also reported that long chains with DP > 50 accelerated retrogradation of amylopectin, whereas the short chains (DP 6-9) retarded it. It is plausible that very long chains of amylopectin mimic amylose to form helical complexes with lipids and intertwine with other branch chains to hold the integrity of starch granules during heating and shearing [Gupta *et al.*, 2009].

Wu *et al.* [2006, 2007] has described ideal feedstock for ethanol production as having rapid liquefaction characteristics and low viscosity during liquefaction. A high viscosity in the mash may impair the accessibility of starch to the enzyme and, thus, delay the liquefaction process [Wu *et al.*, 2007]. The ideal pasting properties, as depicted by RVA analysis, of a starch destined for use as bioethanol substrate include high solubility, demonstrated by a low viscosity after breakdown, and limited ability to retrograde upon cooling, demonstrated by a low final viscosity. Modified RVA analysis was performed by Zhao *et al.* [2009] to assess the viscosity of waxy versus non-waxy wheat during gelatinization and liquefaction. In this case, conventional RVA analysis was modified to include dosing with α-amylase, providing a metric for the balance between gelatinization and liquefaction. Zhao *et al.* [2009] states that "for waxy wheat cultivars, gelatinized starch granules were more susceptible to breakdown under liquefaction conditions; thus, starch molecules were more extensively exposed and more accessible to heat-stable α-amylase, so lower peak viscosities were obtained. Due to the low peak viscosity for waxy wheat during liquefaction, the dry-grind industry could thus increase the solids content in the mash, lower α-amylase dosages, or decrease energy requirements for stirring systems when waxy wheat is used as a feedstock."

The inherent pasting characteristics of starches, such as peak viscosity, are not distinctly observed in modern fuel ethanol production processes because of the addition of α-amylase to the cooking/liquefaction step, to aid pasting and reduce viscosity. Hence, the pasting properties of any particular starch may not have as much influence on cooking/liquefaction as one might anticipate, although accessibility of starch molecules to α-amylase would be a factor. The key is to have as much of the starch as possible (100%) converted to maltose and glucose in the saccharification step, which tends to be combined with fermentation in most modern processes.

6. Selection criteria for wheat as bioethanol feedstock

Ethanol yield, perhaps the most important fermentation performance criteria for the fuel ethanol industry, has been shown to be a starch related property of wheat [Zhao *et al.*, 2009; Lacerenza et al., 2008; Kindred *et al.*, 2008]. Obviously a cultivar with higher starch content in its grain is desirable because it will provide more ethanol per ton of grain and produce smaller amounts of DDGS, resulting in less residual material left over and a greater energy saving during DDGS drying [Zhao *et al.*, 2009]. Elite genotypes for ethanol production have been described as having rapid liquefaction characteristics, low viscosity during liquefaction, high fermentation speed and high fermentation efficiencies [Wu *et al.*, 2007].
Starch properties conferring high conversion efficiencies to fermentable sugars, based on the available literature, are indicated in Table 2. Of particular note is the relationship of amylopectin to the majority of indicated parameters, and the marked benefit to each, in regards to bioethanol application, of an increase in amylopectin and concomitant decrease in amylose content. The encompassing recommendation of the present review is the selection of wheat with the highest amylopectin content achievable, theoretically delivering starch optimized for both rapid and complete degradation by industrial enzymes.

Parameter	Ideal Condition for Bioethanol Production	Reference
Amylose/Amylose Content	> 75% amylopectin	Wu *et al.* [2006, 2007] Zhao *et al.* [2009]
Particle Size Distribution	High ratio of B-type granule	Liu *et al.* [2007]
Amylopectin Fine Structure	Increased long chain to short chain fraction	Ao *et al.* [2007], Zhang *et al.* [2008]
Thermal Properties	Total disruption/dissolution of starch granule	Wu *et al.* [2006], Zhao *et al.* [2009]
Pasting Properties	Low break-down, final & liquefaction viscosity	Zhao *et al.* [2009]

Table 2. Ideal Physicochemical Parameters of Starch Recommended for Use as Bioethanol Feedstock.

Identification of genetic factors within wheat cultivars contributing to these parameters is a topic that has received, to date, little attention. Wheat cultivars producing 'feed class' grain with high starch content, and thus relatively low protein content, have been highlighted as the preferred ideotype for ethanol production [Sosulski & Sosulski, 1994; Kindred *et al.*, 2008]. High starch, low protein content and high yield are reported as the most critical features of high ethanol producing wheat varieties. However, critical to starch conversion efficiency to fermentable sugars is the solubility of starch during gelatinization and the availability of solubilized material to liquefying enzymes. Therefore, a salient feature in the generation of varieties tailored to the needs of the bioethanol industry includes physicochemical parameters of starch lending themselves to high conversion efficiency under the conditions of liquefaction. High conversion efficiencies of starch to fermentable sugars will result in the greatest quantity of ethanol produced per unit of raw substrate when displayed in grains that exhibit both high yield and high starch content.

7. Future perspectives

Considering the probable economic parity of starch-based ethanol with petroleum, it is still unlikely that starch-based ethanol production has the potential to singularly address long-term, global transportation fuel demand. However, despite the great potential of cellulosic technologies to offset petroleum consumption in the future, cellulosic bioethanol production is not yet competitive with either sugar-cane or starch-based bioethanol production, and it is likely that starch-based bioethanol will continue to be a major source of fuel ethanol. As nations move toward increasing incorporation of bioethanol into their transportation fuel supplies, it is likely that starch-based technologies will play a growing role, at least in the near future, in fulfilling that demand. In countries where wheat is a major agricultural crop, wheat-based bioethanol would benefit from the development of high starch, low protein varieties of wheat with characteristics that are ideally suited for bioethanol production (high amylopectin and low protein content). This would ensure access to stable supply, increase ethanol yields, and thus increase the economic viability of wheat-based bioethanol production.

8. Executive summary

- Biofuels are of growing interest to many governments around the world as there is growing need to develop an energy supply that is local, renewable and independent of a financially volatile and potentially unreliable oil market.
- Cellulose-based (2nd generation) conversion technologies, although possessing tremendous potential to displace the demand for petroleum-derived transportation fuels, remains a nascent industry.
- Starch-based (1st generation) conversion technologies, although contentious in regards to diversion of food crops and land use patterns, are based on a mature industry capable of immediate contribution to pressing global, environmental and energy security needs.
- Recent developments in 1st generation technologies suggest that starch feedstocks are being processed increasingly with no clear understanding of the role starch structure plays on conversion efficiency to ethanol.
- Starch characteristics lending themselves to ease of amylolytic hydrolysis and high conversion efficiencies of starch to fermentable sugars would be desirable as bioethanol feedstock.
- Properties of starch that appear to influence susceptibility to amylolytic attack include: amylose to amylopectin ratio, particle size distribution, amylopectin fine structure, gelatinization and pasting properties.
- High amylopectin content appears to be the most meaningful predictive metric in assessing high conversion efficiency of starch to fermentable sugar.
- Starch with a high proportion of the B-type granule may contain starch that is less resistant to enzymatic hydrolysis.
- Amylopectin side-chain length fractions appear to vary little between wheat cultivars included in this review, making amylopectin fine structure an inappropriate metric for starch assessed as potential bioethanol feedstock.
- Gelatinization temperatures indicating a high quantity of amylopectin, i.e. exhibiting high pasting temperatures (Tp), are likely well suited to bioethanol application as they degrade more completely than high amylose counterparts.

- Pasting properties indicative of high amylopectin content suggest high solubility and low viscosity under the conditions of liquefaction.
- Cereal grain cultivars previously identified as preferred bioethanol feedstock include high starch and low protein content, but should also include, based on the findings of this review, high amylopectin content starch.

9. References

Abdel-Aal E, Hucl P, Chibbar R, Han H, Demeke T: Physicochemical and structural characteristics of flours and starches from waxy and nonwaxy wheats.*Cereal Chemistry* 79(3), 458-464 (2002).

Agarwal A: Biofuels (alcohols and biodiesel) applications as fuels for internal combustion engines.*Progress in Energy and Combustion Science* 33(3), 233-271 (2007).

Agu R, Bringhurst T, Brosnan J: Production of grain whisky and ethanol from wheat, maize and other cereals.*Journal of the Institute of Brewing* 112(4), 314-323 (2006).

Antizar-Ladislao B, Turrion-Gomez J: Second-generation biofuels and local bioenergy systems.*Biofuels Bioproducts & Biorefining-Biofpr* 2(5), 455-469 (2008).

Ao Z, Jane J: Characterization and modeling of the A- and B-granule starches of wheat, triticale, and barley.*Carbohydrate Polymers* 67(1), 46-55 (2007).

Ao Z, Simsek S, Zhang G, Venkatachalam M, Reuhs B, Hamaker B: Starch with a slow digestion property produced by altering its chain length, branch density, and crystalline structure.*Journal of Agricultural and Food Chemistry* 55(11), 4540-4547 (2007).

Bai F, Anderson W, Moo-Young M: Ethanol fermentation technologies from sugar and starch feedstocks.*Biotechnology Advances* 26(1), 89-105 (2008).

Bechtel D, Zayas I, Kaleikau L, Pomeranz Y: Size-distribution of wheat-starch granules during endopserm development.*Cereal Chemistry* 67(1), 59-63 (1990).

Boltz K, Thompson D: Initial heating temperature and native lipid affects ordering of amylose during cooling of high-amylose starches.*Cereal Chemistry* 76(2), 204-212 (1999).

Brown LR: Why ethanol production will drive food prices even higher in 2008. Earth Policy Institute. (2008). http://www.earth-policy.org/index.php?/plan_b_updates/2008/update69

Chung J, Han J, Yoo B, Seib P, Lim S: Effects of molecular size and chain profile of waxy cereal amylopectins on paste rheology during retrogradation.*Carbohydrate Polymers* 71(3), 365-371 (2008).

CRFA: Industry information. Canadian Renewable Fuels Association (2009). http://www.greenfuels.org/lists.php.

Dale B: Biofuels: Thinking clearly about the issues.*Journal of Agricultural and Food Chemistry* 56(11), 3885-3891 (2008).

Dengate H, Meredith P: Variation in size disbribution of starch granules from wheat-grain.*Journal of Cereal Science* 2(2), 83-90 (1984).

Diouf, J: Biofuels a disaster for world food. *In* EU Coherence: EU Coherence Policy for Development (2007). http://eucoherence.org/renderer.do/clearState/false/menuld/227351/return

Dronzek B, Bushuk W, Hwang P: Scanning electron-microscopy of starch from sprouted wheat. *Cereal Chemistry* 49(2), 232 (1972).

Du X, Hayes DJ: The impact of ethanol production on U.S. and regional gasoline prices and on the profitability of the U.S. Oil refinery industry. Working paper 08-WP 467. Center for Agricultural and Rural Development, Iowa State University (2008). http://www.card.iastate.edu/publications/synopsis.aspx?id=1076

EIA: Biofuels in the U.S. transportation sector. Official Energy Statistics From the U.S. Government. Energy Information Agency (2007). http://www.eia.doe.gov/oiaf/analysispaper/biomass.html

EIA: Petroleum consumption/sales. Official Energy Statistics From the U.S. Government. Energy Information Agency (2009). http://tonto.eia.doe.gov/dnav/pet/pet_cons_psup_dc_nus_mbblpd_a.htm

EIA: Imports/exports & movements. Official Energy Statistics From the U.S. Government. Energy Information Agency (2009). http://tonto.eia.doe.gov/dnav/pet/pet_move_wkly_dc_NUS-Z00_mbblpd_w.html

EIA: Country analysis briefs. Official Energy Statistics From the U.S. Government. Energy Information Agency (2010). http://www.eia.doe.gov/emeu/cabs/china/oil.html

Farrell A: Ethanol can contribute to energy and environmental goals. *Science* 312(5781), 1748-1748 (2006).

Franco C, Wong K, Yoo S, Jane J: Structural and functional characteristics of selected soft wheat starches.*Cereal Chemistry* 79(2), 243-248 (2002).

Graybosch R: Waxy wheats: Origin, properties, and prospects.*Trends in Food Science & Technology* 9(4), 135-142 (1998).

Gupta M, Bawa A, Semwal A: Morphological, thermal, pasting, and rheological properties of barley starch and their blends.*International Journal of Food Properties* 12(3), 587-604 (2009)

Hahn RW: Ethanol: Law, economics, and politics. Reg-Markets Center Working Paper No. 08-02 (2008). http://ssm.com/abstract=1082079.

Hanashiro I, Abe J, Hizukuri S: A periodic distribution of the chain length of amylopectin as revealed by high-performance anion-exchange chromatography.*Carbohydrate Research* 283, 151-159 (1996).

Harlander, K. 2008. Food vs. fuel - A turning point for bioethanol? *Acta Agronomica Hungarica* 56 (4), 429-433.

Hebeda R, Leach H: Nature of insoluble starch particles in liquefied corn-starch hydrolysates *Cereal Chemistry* 51(2), 272-281 (1974).

Hill J, Nelson E, Tilman D, Polasky S, Tiffany D: Environmental, economic, and energetic costs and benefits of biodiesel and ethanol biofuels.*Proceedings of the National Academy of Sciences of the United States of America* 103(30), 11206-11210 (2006).

Hizukuri S: Polymodal distribution of the chain lengths of amylopectins, and its significance.*Carbohydrate Research* 147(2), 342-347 (1986).

Hizukuri S, Abe J, Hanashiro I:Starch: Analytical aspects. In: *Carbohydrates in food*, Eliasson A (Ed.), Taylor & Francis Group, Boca Raton, FL, 305-470 (1996).

Hoseney RC: Starch. *In:* Principles of cereal science and technology. Hoseney RC (Ed.), American Association of Cereal Chemists, St. Paul, MN, Chapter 2 (1998).

Huang J, Schols H, Van Soest J, Jin Z, Sulmann E, Voragen A: Physicochemical properties and amylopectin chain profiles of cowpea, chickpea and yellow pea starches.*Food Chemistry* 101(4), 1338-1345 (2007).

Hung P, Maeda T, Morita N: Waxy and high-amylose wheat starches and flours - characteristics, functionality and application.*Trends in Food Science & Technology* 17(8), 448-456 (2006).

Husky Energy, 2009. Ethanol Fact Sheet. http://www.huskyenergy.com/operations/downstream/canada/facilities/lloydminster&minnedosaethanolplants.asp

IEA: Resources to reserves: Oil & gas technologies for the energy markets of the future. International Energy Agency (2005). http://www.iea.org/publications/free_new_desc.asp?pubs_ID=1568 832

Ingledew, W. M., Jones, A. M., Bhatty, R. S. and Rossnagel B. G. 1995. Fuel alcohol production from hull-less barley. *Cereal Chemistry* 72, 147-150.

Jane J, Chen Yy, Lee L *et al.*: Effects of amylopectin branch chain length and amylose content on the gelatinization and pasting properties of starch.*Cereal Chemistry* 76(5), 629-637 (1999).

Jenkins P, Donald A: The influence of amylose on starch granule structure. *International Journal of Biological Macromolecules* 17(6), 315-321 (1995).

Karl, 2010 Rising wheat prices raise fears over UK commitment to biofuels. September 13, 2010. http://www.makebiofuel.co.uk/news/the-soaring-price-of-wheat-has-raised-questions-about-the-uks-commitment-to-biofuels

Kim S, Dale B: Global potential bioethanol production from wasted crops and crop residues. *Biomass & Bioenergy* 26(4), 361-375 (2004).

Kindred D, Verhoeven T, Weightman R *et al.*: Effects of variety and fertiliser nitrogen on alcohol yield, grain yield, starch and protein content, and protein composition of winter wheat. *Journal of Cereal Science* 48(1), 46-57 (2008).

Koplow D: Biofuels – at what cost? Government support for ethanol and biodiesel in the United States. Global Subsidies Initiative (GSI) of the International Institute of Sustainable Development (IISD) (2006). http://www.globalsubsidies.org/img/pdf/biofuels_subsidies_us.pdf.

Kulp K: Characteristics of small-granule starch of flour and wheat. *Cereal Chemistry* 50(6), 666-679 (1973).

Kwasniewska-Karolak I, Nebesny E, Rosicka-Kaczmarek J: Characterization of amylose-lipid complexes derived from different wheat varieties and their susceptibility to enzymatic hydrolysis.*Food Science and Technology International* 14(1), 29-37 (2008).

Laan T, Litman TA, Steenblik R: Biofuels - at what cost? Government support for ethanol and biodiesel in Canada. Global Subsidies Initiative (GSI) of the International Institute of Sustainable Development (IISD) (2009). http://www.globalsubsidies.org/en/research/biofuelsubsidies-canada.

Lacerenza J, Martin J, Talbert L, Lanning S, Giroux M: Relationship of ethanol yield to agronomic and seed quality characteristics of small grains. *Cereal Chemistry* 85(3), 322-328 (2008).

Lackner K, Sachs J: A robust strategy for sustainable energy. *Brookings Papers on Economic Activity* 36(2), 215-284(2005).

Lee M, Swanson B, Baik B: Influence of amylose content on properties of wheat starch and bread making quality of starch and gluten blends. *Cereal Chemistry* 78(6), 701-706 (2001).

Lin P, Czuchajowska Z: Starch properties and stability of club and soft white winter wheats from the pacific northwest of the united states.*Cereal Chemistry* 74(5), 639-646 (1997).

Lindeboom N, Chang P, Tyler Rt Analytical, biochemical and physicochemical aspects of starch granule size, with emphasis on small granule starches: A review.*Starch-Starke* 56(3-4), 89-99 (2004).

Liu Q, Gu Z, Donner E, Tetlow I, Emes M: Investigation of digestibility in vitro and physicochemical properties of a- and b-type starch from soft and hard wheat flour. *Cereal Chemistry* 84(1), 15-21 (2007).

Luft G: Fueling the dragon: China's race into the oil market. Institute for the analysis of global security (2004).
http://www.iags.org/china.htm

Matser A, Steeneken P: Origins of the poor filtration characteristics of wheat starch hydrolysates.*Cereal Chemistry* 75(3), 289-293 (1998).

Mclaughlin S, Walsh M: Evaluating environmental consequences of producing herbaceous crops for bioenergy.*Biomass & Bioenergy* 14(4), 317-324 (1998).

McLeod, J. G., May, W. E., Salmon, D. F., Sosulski, K., Thomas, J. B., Brown, P. D. and Vera, C. L. 2010. Changes in ethanol production potential due to species, cultivar and location on the Canadian prairie. *Canadian Journal of Plant Science* 90, 163-171

Meredith P: Large and small starch granules in wheat - are they really different.*Starch-Starke* 33(2), 40-44 (1981).

Moorthy SN: Physicochemical and functional properties of tropical tuber starches: A review. *Starch-Starke* 54(12) , 559-592 (2002).

Morrison W, Gadan H: The amylose and lipid contents of starch granules in developing wheat endosperm *Journal of Cereal Science* 5(3), 263-275 (1987).

Mousdale DM: The economics of bioethanol. In: Biofuels: Biotechnology, Chemistry, and Sustainable Development, Mousdale DM (Ed.), Taylor & Francis Group, LLC, Glasgow, Scotland, Chapter 5 (2008).

MRA: World's ethanol production forecast 2008-2012. Market Research Analyst January 26 (2008).
http://www.marketresearchanalyst.com/2008/01/26/world-ethanol-production-forecast-2008-2012/

NAS: Frontiers in Agricultural Research: Food, Health, Environment, and Communities. Washington, DC, National Academy of Sciences November 5, 2004;
http://dels.nas.edu/rpt_briefs/frontiers_in_ag_final%20for%20print.pdf.

Noda T, Kimura T, Otani M *et al.*: Physicochemical properties of amylose-free starch from transgenic sweet potato.*Carbohydrate Polymers* 49(3), 253-260 (2002).

NRCan, 2003. The additional of ethanol from wheat to GHGenius. Natural Resources Canada. Office of Energy Efficiency. (S&T)^2 Consultants Inc.

Park S, Wilson J, Chung O, Seib P: Size distribution and properties of wheat starch granules in relation to crumb grain score of pup-loaf bread.*Cereal Chemistry* 81(6), 699-704 (2004).

Park S, Wilson J, Seabourn B: Starch granule size distribution of hard red winter and hard red spring wheat: Its effects on mixing and breadmaking quality.*Journal of Cereal Science* 49(1), 98-105 (2009).

Patzek T: Thermodynamics of the corn-ethanol biofuel cycle.*Critical Reviews in Plant Sciences* 23(6), 519-567 (2004).

Peng M, Gao M, Abdel-Aal E, Hucl P, Chibbar R: Separation and characterization of a- and b-type starch granules in wheat endosperm.*Cereal Chemistry* 76(3), 375-379 (1999).

Pimentel D, Marklein A, Toth M *et al.*: Food versus biofuels: Environmental and economic costs.*Human Ecology* 37(1), 1-12 (2009).

Prairie Registration Recommending Committee for Grain. 1996. Wheat, Rye & Triticale Subcommittee Report, Winnipeg, MB.

Pryor F: The economics of gasohol.*Contemporary Economic Policy* 27(4), 523-537 (2009).

Quadrelli R, Peterson S: The energy-climate challenge: Recent trends in co2 emissions from fuel combustion.*Energy Policy* 35(11), 5938-5952 (2007).

Raeker M, Gaines C, Finney P, Donelson T: Granule size distribution and chemical composition of starches from 12 soft wheat cultivars. *Cereal Chemistry* 75(5), 721-728 (1998).

RFA: FO Lichts ethanol production costs: A worldwide survey, a special study from FO Lichts and Agra CEAS consulting. Renewable Fuels Association (2008). http://www.agra-et.com/portal2/home.jsp?template=productpage&pubid=ag090

Rajagopal D, Sexton S, Roland-Holst D, Zilberman D: Challenge of biofuel: Filling the tank without emptying the stomach? *Environmental Research Letters* 2(4), (2007).

Rooney L, Pflugfelder R: Factors affecting starch digestibility with special emphasis on sorghum and corn *Journal of Animal Science* 63(5), 1607-1623 (1986).

Rapier R: Diminishing expectations: Broken promises in the development of cellulosic ethanol production. Global Subsidies Initiative (GSI) of the International Institue for Sustainable Development (IISD) (2010). http://www.globalsubsidies.org/en/subsidy watch/commentary/diminishing-expectations-broken-promises-development-cell

Sasaki T, Yasui T, Matsuki J, Satake T: Comparison of physical properties of wheat starch gels with different amylose content.*Cereal Chemistry* 79(6), 861-866 (2002).

Saunders J: Physicochemical properties of select western canadian wheat starches and their relationship to liquefaction and fermentative bio-ethanol performance. *M.Sc. Thesis, University of Manitoba, Winnipeg, Manitoba, Canada*, (2010).

Schmidhuber J: Impact of an increased biomass use on agricultural markets, prices and food security: A longer-term perspective.*Presented at: International symposium of Notre Europe* Paris, France, 27-29 November (2006).

Searchinger T, Heimlich R, Houghton R *et al.*: Use of us croplands for biofuels increases greenhouse gases through emissions from land-use change.*Science* 319(5867), 1238-1240 (2008).

Sims R: From 1st to 2nd generation biofuel technologies: An overview of current industry and RD&D activities. International Energy Agency (2008). http://www.iea.org/publications/free_new_desc.asp?pubs_id=2079

Song Y, Jane J: Characterization of barley starches of waxy, normal, and high amylose varieties.*Carbohydrate Polymers* 41(4), 365-377 (2000).

Sosulski K, Sosulski F: Wheat as a feedstock for fuel ethanol.*Applied Biochemistry and Biotechnology* 45(6), 169-180 (1994).

Soulaka A, Morrison W: The amylose and lipid contents, dimensions, and gelatinization characteristics of some wheat starches and their a-granul and b-granule fractions *Journal of the Science of Food and Agriculture* 36(8), 709-718 (1985).

Srinivasan S: The food v. fuel debate: A nuanced view of incentive structures.*Renewable Energy* 34(4), 950-954 (2009).

Stoddard F: Survey of starch particle-size distribution in wheat and related species.*Cereal Chemistry* 76(1), 145-149 (1999).

Swanston J, Smith P, Gillespie T, Brosnan J, Bringhurst T, Agu R: Associations between grain characteristics and alcohol yield among soft wheat varieties. *Journal of the Science of Food and Agriculture* 87(4), 676-683 (2007).

Swinkels J: Composition and properties of commercial native starches *Starch-Starke* 37(1), 1-5 (1985).

Tan K, Lee K, Mohamed A: Role of energy policy in renewable energy accomplishment: The case of second-generation bioethanol.*Energy Policy* 36(9), 3360-3365 (2008).

Tao L, Aden A: The economics of current and future biofuels. *In Vitro Cellular & Developmental Biology-Plant* 45(3), 199-217 (2009).

Tester R, Morrison W: Swelling and gelatinization of cereal starches. 1. Effects of amylopectin, amylose, and lipids *Cereal Chemistry* 67(6), 551-557 (1990).

Tester R, Karkalas J, Qi X: Starch structure and digestibility enzyme-substrate relationship.*World's Poultry Science Journal* 60(2), 186-195 (2004).

Thomas, K. C. and Ingledew, W. M. 1995. Production of fuel alcohol from oats by fermentation. *Journal of Industrial Microbiology* 15, 125-130.

Thomas, K. C., Dhas, A., Rossnagel, B. G. and Ingledew, W. M. 1995. Production of fuel ethanol from hull-less barley by very high gravity technology. *Cereal Chemistry* 72, 360-364.

USDA: Major land uses. Economic Research Services of the United States Department of Agriculture (2007). *http://www.ers.usda.gov/data/majorlanduses*

USDA: Agricultural research services. United States Department of Agriculture (2009). http://www.ars.usda.gov/main.html

Van Hung P, Maeda T, Morita N: Study on physicochemical characteristics of waxy and high-amylose wheat starches in comparison with normal wheat starch. *Starch-Starke* 59(3-4), 125-131 (2007).

Wang, S., Thomas, K. C., Ingledew, W. M., Sosulski, K. and Sosulski, F. W. 1997. Rye and triticale as feedstock for fuel ethanol production. *Cereal Chemistry* 74, 621-625.

Weyerhaeuser H, Tennigkeit T, Yufang S, Kahrl F: Biofuels in china: An analysis of the opportunities and challenges of *jatropha curcas* in Southwest China. Working Paper Number 53. World Agroforestry Centre (2007). www.worldagroforestrycentre.org

Wolff, I. A. (ed.). 1982. CRC handbook of processing and utilization in agriculture. CRC series in agriculture. Vol. 2. CRC Press, Boca Raton, FL. pp. 110-111.

Wu X, Zhao R, Wang D *et al.*: Effects of amylose, corn protein, and corn fiber contents on production of ethanol from starch-rich media. *Cereal Chemistry* 83(5), 569-575 (2006).

Wu X, Zhao R, Bean S *et al.*: Factors impacting ethanol production from grain sorghum in the dry-grind process. *Cereal Chemistry* 84(2), 130-136 (2007).

Yasui T, Sasaki T, Matsuki J: Variation in the chain-length distribution profiles of endosperm starch from triticum and aegilops. Species. *Starch-Starke* 57(11), 521-530 (2005).

Zhao R, Wu X, Seabourn B *et al.*: Comparison of waxy vs. Nonwaxy wheats in fuel ethanol fermentation.*Cereal Chemistry* 86(2), 145-156 (2009).

Zhang G, Sofyan M, Hamaker B: Slowly digestible state of starch: Mechanism of slow digestion property of gelatinized maize starch.*Journal of Agricultural and Food Chemistry* 56(12), 4695-4702 (2008a).

Zhang G, Ao Z, Hamaker B: Nutritional property of endosperm starches from maize mutants: A parabolic relationship between slowly digestible starch and amylopectin fine structure.*Journal of Agricultural and Food Chemistry* 56(12), 4686-4694 (2008b).

Zeng M, Morris C, Batey I, Wrigley C: Sources of variation for starch gelatinization, pasting, and gelation properties in wheat.*Cereal Chemistry* 74(1), 63-71 (1997).

Zobel H, Young S, Rocca L: Starch gelatinization - an x-ray diffraction study *Cereal Chemistry* 65(6), 443-446 (1988).